Equivalence, Invariants, and Symmetry

Equivalence, Invariants, and Symmetry

PETER J. OLVER

University of Minnesota

CAMBRIDGE
UNIVERSITY PRESS

CAMBRIDGE UNIVERSITY PRESS
Cambridge, New York, Melbourne, Madrid, Cape Town, Singapore, São Paulo, Delhi

Cambridge University Press
32 Avenue of the Americas, New York, NY 10013-2473, USA

www.cambridge.org
Information on this title: www.cambridge.org/9780521478113

First published 1995
Reprinted 1996
This digitally printed version 2008

A catalogue record for this publication is available from the British Library

ISBN 978-0-521-47811-3 hardback
ISBN 978-0-521-10104-2 paperback

To my family — Chehrzad, Parizad, Sheehan, and Noreen

Contents

Preface xi
Acknowledgments xv
Introduction 1

1. Geometric Foundations 7
Manifolds 7
Functions 10
Submanifolds 13
Vector Fields 17
Lie Brackets 21
The Differential 22
Differential Forms 23
Equivalence of Differential Forms 29

2. Lie Groups 32
Transformation Groups 35
Invariant Subsets and Equations 39
Canonical Forms 42
Invariant Functions 44
Lie Algebras 48
Structure Constants 51
The Exponential Map 52
Subgroups and Subalgebras 53
Infinitesimal Group Actions 55
Classification of Group Actions 58
Infinitesimal Invariance 62
Invariant Vector Fields 65
Lie Derivatives and Invariant Differential Forms 68
The Maurer–Cartan Forms 71

3. Representation Theory 75
Representations 75
Representations on Function Spaces 78
Multiplier Representations 81
Infinitesimal Multipliers 85

Relative Invariants 91
Classical Invariant Theory 95

4. Jets and Contact Transformations 105
Transformations and Functions 106
Invariant Functions 109
Jets and Prolongations 111
Total Derivatives 115
Prolongation of Vector Fields 117
Contact Forms 121
Contact Transformations 125
Infinitesimal Contact Transformations 129
Classification of Groups of Contact Transformations 134

5. Differential Invariants 136
Differential Invariants 136
Dimensional Considerations 139
Infinitesimal Methods 141
Stabilization and Effectiveness 143
Invariant Differential Operators 146
Invariant Differential Forms 153
Several Dependent Variables 157
Several Independent Variables 164

6. Symmetries of Differential Equations 175
Symmetry Groups and Differential Equations 175
Infinitesimal Methods 178
Integration of Ordinary Differential Equations 187
Characterization of Invariant Differential Equations 191
Lie Determinants 199
Symmetry Classification of Ordinary Differential Equations 202
A Proof of Finite Dimensionality 206
Linearization of Partial Differential Equations 209
Differential Operators 211
Applications to the Geometry of Curves 218

7. Symmetries of Variational Problems 221
The Calculus of Variations 222
Equivalence of Functionals 227
Invariance of the Euler–Lagrange Equations 230
Symmetries of Variational Problems 235

Invariant Variational Problems 238
Symmetry Classification of Variational Problems 240
First Integrals 242
The Cartan Form 244
Invariant Contact Forms and Evolution Equations 246

8. Equivalence of Coframes 252
Frames and Coframes 252
The Structure Functions 256
Derived Invariants 259
Classifying Functions 261
The Classifying Manifolds 266
Symmetries of a Coframe 274
Remarks and Extensions 276

9. Formulation of Equivalence Problems 280
Equivalence Problems Using Differential Forms 280
Coframes and Structure Groups 287
Normalization 291
Overdetermined Equivalence Problems 297

10. Cartan's Equivalence Method 304
The Structure Equations 304
Absorption and Normalization 307
Equivalence Problems for Differential Operators 310
Fiber-preserving Equivalence of Scalar Lagrangians 321
An Inductive Approach to Equivalence Problems 327
Lagrangian Equivalence under Point Transformations 328
Applications to Classical Invariant Theory 333
Second Order Variational Problems 337
Multi-dimensional Lagrangians 342

11. Involution 347
Cartan's Test 350
The Transitive Case 355
Divergence Equivalence of First Order Lagrangians 357
The Intrinsic Method 358
Contact Transformations 361
Darboux' Theorem 364
The Intransitive Case 366
Equivalence of Nonclosed Two-Forms 367

12. Prolongation of Equivalence Problems	372
The Determinate Case	373
Equivalence of Surfaces	377
Conformal Equivalence of Surfaces	385
Equivalence of Riemannian Manifolds	386
The Indeterminate Case	394
Second Order Ordinary Differential Equations	397
13. Differential Systems	409
Differential Systems and Ideals	409
Equivalence of Differential Systems	415
Vector Field Systems	416
14. Frobenius' Theorem	421
Vector Field Systems	421
Differential Systems	427
Characteristics and Normal Forms	428
The Technique of the Graph	431
Global Equivalence	440
15. The Cartan–Kähler Existence Theorem	447
The Cauchy–Kovalevskaya Existence Theorem	447
Necessary Conditions	449
Sufficient Conditions	455
Applications to Equivalence Problems	460
Involutivity and Transversality	465
Tables	472
References	477
Symbol Index	490
Author Index	499
Subject Index	504

Preface

The volume you are about to read will, I hope, prove to be a stimulating, unusual, and provocative blend of mathematical flavors. As its title indicates, the book revolves around three interconnected and particularly fertile themes, each arising in a wide variety of mathematical disciplines, and each having a wealth of significant and substantial applications. Equivalence deals with the determination of when two mathematical objects are the same under a change of variables. The symmetries of a given object can be interpreted as the group of self-equivalences. Conditions guaranteeing equivalence are most effectively expressed in terms of invariants, whose values are unaffected by the changes of variables. Issues of this generality naturally arise in all fields of mathematics, and, particularly in geometry, often lie at the heart of the subject. Although each of these concepts has a discrete counterpart, our primary focus will be on the continuous. The areas of immediate concern are analytical — differential equations, variational problems, vector fields, and differential forms — although algebraic objects, such as polynomials, matrices, and quadratic forms, also play an important role. This book will explore the available methods for systematically and algorithmically solving the problems of symmetry, equivalence, the classification of invariants, and the determination of canonical forms, thereby elucidating the many interconnections, some surprising, between the particular manifestations of these problems in seemingly unrelated situations.

The book naturally divides into four interconnected parts. The first, comprising Chapters 1–3, constitutes the algebro-geometric foundation of our subject. The first chapter provides a rapid survey of the basic facts from differential geometry, including manifolds, vector fields, and differential forms. Chapters 2 and 3 could, with some more fleshing out, form a basic course on Lie groups and representation theory. The primary omissions are the detailed structure theory of Lie groups and algebras, and the classification theory for irreducible representations, neither of which play a significant role in our applications. The second part, comprising Chapters 4–7, provides an in depth study of applications of symmetry methods to differential equations. We begin with a discussion of jet spaces, on which systems of differential equations are naturally realized as geometrical subsets, and conclude with a discussion of contact transformations. The next three chapters deal with the construction and classification of differential invariants, followed by applications to differential equations and variational problems. My earlier text, [**186**], can be used to supplement the material in this part with many additional applications. In the third part, Chapters 8–12, the focus shifts to equivalence problems, and the Cartan approach to their solution. The different possible branches that arise in Cartan's equivalence method are discussed in detail, and illustrated by a wide variety of applications. The final three chapters survey the required results from the theory of partial differential equations and differential systems. In particular, we describe the proofs of the fundamental existence theorems of Frobenius, and of Cartan and Kähler, that underlie all the methods.

I should mention that the arrangement of the chapters is slightly unnatural from a developmental standpoint. The most foundational material — the existence theorems for systems of partial differential equations — appears in the last three chapters. The geometry of coframes is used in part in the symmetry analysis of differential equations, particularly in the construction of differential invariants, which are used to classify equations admitting prescribed symmetry groups. However, I decided on a more pedagogical arrangement so as not to frighten off the prospective reader with overly technical topics at the outset. Thus, the book begins with the more readily digested methods from Lie group theory, then moves on to the more sophisticated approach of Cartan based on differential forms, and completes the repast with the hard analysis for dessert. I hope the more logically minded reader will not be too inconvenienced by this choice of presentation.

There is a wealth of related material in the mathematical literature, and one of the hardest tasks was choosing which results and applications to include, and which, for reasons of space, to omit. For example, at one time I envisioned a much more substantial section on classical invariant theory; however, this unfortunately had to be considerably shortened for the final version. (I hope, in the near future, to use this additional material as the basis for an introductory text on classical invariant theory and its applications.) Experts in any of the subjects touched on in the book will, no doubt, find many of their favorite theoretical developments or applications omitted or tersely commented upon. In many cases, this was made necessary by the space limitations and the need to choose results that were particularly representative, interesting, and/or interdisciplinary. I hope the omissions will not prevent anyone from enjoying what I did finally decide to include.

For clarity, I have adopted a fairly informal, discursive style to present the methods, with the hope that this will lead to new understanding of their utility and effectiveness. Too often in the recent literature, these powerful and constructive techniques have been obscured by elaborate, theoretical machinery that many have chosen to "rigorize" them with. In my opinion, in the practical realm, this only serves to obscure the fundamental issues, thereby interfering with a student's direct understanding. My own presentation relies ultimately on the original sources, particularly Lie and Cartan, which I wholeheartedly recommend to the genuine scholar. As a consequence, results are not always stated in complete generality, and occasionally some minor points of rigor are left for the reader to properly sort out. As long as one exercises the proper amount of caution, such technical details will rarely lead one astray in the practical applications.

Although a substantial fraction of the book covers results that are well known to followers of Lie and Cartan, the modern applicability of these methods is illustrated by a surprising variety of new applications. Much of Chapter 5 on the theory of differential invariants is new, and, apart from being surveyed in conference proceedings, has not appeared in the literature before. Some of the classification results for differential equations and variational problems based on symmetry are also new, and others have only appeared scattered in the literature. The applications of differential invariants to computer vision are the result of a recent collaboration with Allen Tannenbaum and Guillermo Sapiro. The integration of the Chazy equation using prolongation and symmetry appears in joint work with Peter Clarkson. Many of the applications of the Cartan equivalence method are based on results from my long-standing collaboration with Niky Kamran. The emphasis on the classifying manifolds in the solution to the equivalence problem for coframes is new, and offers a substantial advantage over the more traditional classifying function approach. The results on global equivalence and non-associative Lie groups are stated here for the first time. The applications of classical invariant theory appear in various contexts, and have been reformulated in a more consistent form; in particular, the connection with the Cartan solution to an equivalence problem from the calculus of variations is based on an earlier paper, [**183**].

The basic prerequisites for the book are multi-variable calculus — specifically the implicit and inverse function theorems and the divergence theorem — basic tensor and exterior algebra, and a smattering of group theory. Results from elementary linear algebra and complex analysis, and basic existence theorems for ordinary differential equations are used without comment. Many standard results are, for lack of space, left unproved, although ample references are supplied. It is hoped that the book will form the basis of an advanced graduate course in symmetry and equivalence problems. I have included a large number of exercises, which range in difficulty from the fairly obvious to quite substantial. References to the solutions of the more difficult exercises are provided.

It is my hope that this book will serve as a catalyst for the further development, both in theory and in applications, of this fascinating and fertile mathematic field. I am certain that there are many fundamental contributions yet to be made, and that the devoted student cannot help but play a role in its accelerating mathematical development, or in its ever-broadening range of applications.

Acknowledgments

The book has its origins in a series of ten London Mathematical Society Invited Lectures, delivered at the University of Bath, under the auspices of the London Mathematical Society, in April 1992. I should like to express my sincere appreciation to Geoffrey Burton and John Toland who played an indispensable role in the planning and organization. I should also like to thank the London Mathematical Society for their patience and understanding as the original lecture notes slowly metamorphosed into a full-fledged book, thereby delaying its appearance considerably beyond the original schedule.

Later versions of these notes were used as a basis for a course at the University of Maryland during spring semester 1993. I must thank the Maryland Mathematics Department, the participants in the course, and particularly John Maddocks, for the opportunity to further refine my ideas during the course.

I should like to express my gratitude to many people, whose ideas, insights, suggestions, comments, and corrections have helped mold the original rough lecture notes into what I hope is a coherent body of work. They include Alfred Aeppli, Sam Albert, Sergio Alvarez, Ian Anderson, Robert Bryant, Peter Clarkson, Jack Conn, Philip Doyle, Paul Edelman, Mikhail Foursov, Mark Fels, Robert Gardner, Artemio González–López, Raphael Heredero, Mark Hickman, Darryl Holm, Lucas Hsu, Willard Miller, Jr., Dmitry Ostrovsky, David Richter, Michael Singer, Vladimir Sokolov, and Yu Yuan.

I must particularly thank Niky Kamran, my dear friend and collaborator (not only mathematical, but also musical and spiritual). It was through his patient and lucid explanations that I first learned the Cartan equivalence method. Many of the applications of the Cartan equivalence method discussed in the book are based on joint work over the past seven years. For all this I owe him a profound debt of gratitude.

This book was typeset in plain TeX using a system of macros for automatic equation and reference numbering that I developed. I would like to thank Ian Anderson for sharing his own TeX macros, which formed the foundation of the OTeX system. The figures were drawn using the computer algebra system MATHEMATICA.

I should of course thank the very helpful and patient people at Cambridge University Press, particularly Alan Harvey and Sophia Prybylski, for their encouragement and support during the entire course of this project.

I must also offer thanks to the Bahá'í community of the Twin Cities for helping provide the spiritual foundations on which to ground my life during this difficult, yet unspeakably glorious, period in human history.

Finally, I reaffirm my love and gratitude to my wife, Cheri Shakiban, and my three children, Pari, Sheehan, and Noreen, for their patience, love, and understanding throughout the long and arduous process of writing the book. Without their unfailing support, I would never have completed this project.

Introduction

The most basic of the themes to be discussed is the concept of equivalence. The fundamental equivalence problem is to determine whether two geometric objects can be transformed into each other by a suitable change of variables. For example, one might be interested in whether two given differential equations are, in fact, the same equation rewritten in terms of different independent and dependent variables. As a particular instance, the linearization problem is to determine whether a differential equation can be mapped to a linear differential equation via a change of variables. In this manner, a solution to the general equivalence problem for differential equations will include, as a very particular case, the characterization of all linearizable differential equations. As a second example, in the calculus of variations, the basic equivalence problem is to determine when two variational problems can be identified under a change of variables. This problem forms a subcase of the differential equation equivalence problem, since the change of variables will also map their Euler-Lagrange equations to each other. In algebra, the problem of equivalence of (homogeneous) polynomials under linear transformations also falls under our broad purview. Geometric equivalence problems lie at the foundations of Riemannian geometry, where one wants to know whether two given Riemannian manifolds can be identified under a (local) diffeomorphism, and, similarly, in symplectic geometry, where one asks the same question for two symplectic manifolds.

The determination of the symmetry group of a geometric object can be regarded as a special case of the general equivalence problem. Indeed, provided it lies in the admissible class of changes of variables, a symmetry is merely a self-equivalence of the object. Thus, for instance, the solution to the equivalence problem for differential equations will include a determination of all symmetries of a given differential equation. Similarly, the isometries of a Riemannian manifold are merely the set of self-equivalences of the underlying metric. Besides addressing the general equivalence problem, we shall also develop methods to directly calculate the symmetry group of a given object. Two equivalent objects have isomorphic symmetry groups — indeed, conjugating any symmetry of the first object by the equivalence transformation produces a symmetry of the second. Thus, one means of recognizing equivalent objects is by inspecting their symmetry groups: If the two symmetry groups are not equivalent, e.g., they have different dimensions, or different structures, then the two objects cannot be equivalent. Of course, having isomorphic symmetry groups is no guarantee that the two objects are equivalent; nevertheless, in many highly symmetric cases, including linearization problems, the existence of a suitable symmetry group is both necessary and sufficient for the equivalence of the two objects.

The concept of an invariant proves to be of crucial importance in understanding an equivalence problem. By definition, an invariant is a quantity which is unaffected by the changes of variables. Consequently, two equivalent objects must necessarily have the same invariants. Conversely, in the regular case, if we know enough invariant functions, we can use them to completely characterize the equivalent objects, and thereby completely solve the equivalence problem. Thus, the construction of invariants and their characterization lies at the heart of most approaches to equivalence problems. In addition, every regular invariant system of equations (algebraic, differential, variational, etc.) can be characterized by functional relationships among the invariant functions, so the invariants form the fundamental building blocks which can be used to construct suitably symmetric objects, a process of immense utility in modern physics. Whereas invariant functions are the most important invariant quantities associated with the equivalence, many other invariant objects — vector fields, differential forms, differential operators, etc. — arise naturally and will be seen to play important roles.

Last, but not least, we shall consider the problem of determining canonical forms for the objects of interest. By definition, a "canonical form" is provided by a particularly simple equivalent object. Familiar examples include the Jordan canonical form for matrices, and the well-known canonical forms for quadratic and cubic polynomials. The determination of the canonical forms requires a solution to the equivalence problem; indeed, one way of characterizing two equivalent objects is that they have the same canonical form. The particular canonical form of a prescribed object is most readily characterized by its invariants; vice versa, the invariants themselves can often be found once a canonical form is known. For example, the eigenvalues and the Jordan block structure provide the proper invariants that will characterize those matrices equivalent to a given Jordan canonical form. In our discussion of equivalence problems, we shall, when feasible, attempt to describe a list of canonical forms.

It is these four issues — equivalence, symmetry, invariant, and canonical form — that form the core of the book. Our goal is to develop a variety of techniques, some very general, some quite specific, that will allow one to handle some or all of these problems in a wide variety of mathematical contexts. Particular attention will be paid to problems arising in geometry, differential equations, and the calculus of variations, although this will by no means restrict our discussion. These topics arise in many contexts, both in mathematics and in physical applications, and our survey will include problems from quantum mechanics, elasticity, computer vision, gas dynamics, relativity, optics, Hamiltonian mechanics, solitons, representation theory, and differential and projective geometry, to name a few. In a more theoretical direction, the theory of equivalence and invariants of ordinary polynomials, known as "classical invariant theory", has a long and distinguished history of its own, which, especially in view of recently discovered connections with the calculus of variations, is worth presenting in some detail.

In our study, the emphasis will be on the analytical, the continuous, and the local. Our geometric objects will be defined on smooth, finite-dimensional manifolds (most commonly Euclidean spaces), and depend smoothly or analytically on local coordinates. We shall exclusively deal with *local* equivalence problems, in which one asks for equivalence of the objects when restricted to sufficiently small open subsets. Moreover, our treatment will primarily deal with the regular case, away from singularities. This is not to say that the global and singular problems are less interesting (indeed, one can successfully argue that the opposite is true), but rather an indication of our necessarily restricted scope, imposed by space limitations. In consequence, we shall emphasize continuous symmetry groups, which, in the finite dimensional manifestation, appear as Lie groups of transformations, or, in the infinite-dimensional version, as "Lie pseudo-groups". Again, this is not because discrete groups are unimportant, but just to keep the length manageable. Even with our chosen restrictions, the proposed subject area is still vast, and we shall repeatedly be forced to choose particular subtopics.

In the history of our subject, two names tower over the rest — Sophus Lie and Élie Cartan. Its origins are to be found in Galois' simple, yet profound observation that the set of symmetries of a geometric object forms a group. Through a surprisingly tortuous route that began with a study of the integration of partial differential equations (see **[111]** for a fascinating account of the historical details), Lie was led to introduce the continuous groups that now bear his name, an accomplishment that has had remarkable consequences for mathematics and physics. In fact, Lie often viewed his mission as the extension of Galois theory to the study of differential equations. The last two decades have witnessed a long overdue and widespread popularization of Lie's constructive, infinitesimal symmetry methods, applied to an incredible variety of mathematical, physical, and other problems. Less well documented, though, is the role symmetry groups play in the classification of differential equations and variational problems. Of particular importance is Lie's classification of all possible Lie groups of transformations acting on one- and two-dimensional manifolds, discussed in Chapter 2; the full classification tables appear at the end of the book.

Although symmetry has clearly played the predominant role, equivalence problems have an equally long and distinguished mathematical history. For our purposes, a particularly important starting point is Riemann's seminal inaugural address that first formulated the fundamental concepts of n-dimensional geometry. Riemann's vision motivated the general isometric equivalence problem for Riemannian manifolds, which was successfully analyzed by Riemann himself, **[197]**, and, independently, by Christoffel, **[52]**. Remarkably, their solution led directly to the explicit formulas for the connection (Christoffel symbols) and the Riemann curvature tensor. Thus, in a very definite sense, modern differential geometry can trace its roots back to an equivalence problem. Various particular analytic equivalence problems were solved in the following fifty years, including examples appearing in the works of Lie and the monograph of Tresse, **[215]**, on the equivalence of second order scalar ordinary differential equations. Then, dramatically, in a profound and far-reaching study of infinite-dimensional pseudo-groups, Cartan, **[37]**, formulated and solved a completely general geometric equivalence problem — that of "coframes", which constitute spanning sets of differential one-forms. Remarkably, Cartan's coframe equivalence problem includes all other equivalence problems as particular cases. What makes Cartan's paper so astonishing is that, not only did he provide a complete and algorithmic solution to the coframe equivalence problem, understanding all its many twists and turns, but this solution necessitated a concurrent development of the newly emerging theory of differential forms and exterior differential systems. Cartan's equivalence method was subsequently applied by some of his disciples, but then lay dormant, a victim of its own computational complexity. However, because of its importance for solving equivalence problems in differential equations, the calculus of variations, control theory, and even classical invariant theory, Cartan's method has recently become the renewed focus of a flurry of research activity. One important goal of this book is to present an understandable introduction to the Cartan method, illustrated by numerous applications, with the hope of disseminating this powerful and useful theory yet wider in the mathematical community and beyond.

Equivalence problems can be handled by a variety of other approaches besides the methods of Lie and Cartan, some of which will be mentioned in the historical notes that appear throughout the text. Each method has its own advantages and disadvantages; in many cases, only by adroitly combining the different weapons in our arsenal can successful attacks on very complicated equivalence problems be mounted. For example, the Cartan method has the advantage of being able, in principle, to algorithmically solve completely general equivalence problems in a systematic fashion. The method only requires differentiation, whereas the Lie approach needs an integration of the determining equations for the symmetry groups. Moreover, in addition to providing a complete list of fundamental invariants, the Cartan approach pins down the dimension and structure of the relevant symmetry groups. Its disadvantages are that, in view of the complexity of the intervening computations, the full solution has been successfully implemented only for fairly simple equivalence problems. Moreover, the verification of the necessary and sufficient conditions for equivalence requires recognizing when two parametrized submanifolds are identical; in complicated situations, it is not so clear how to effect a solution to this problem. On the other hand, even when the equivalence computations are not carried through to completion, interesting and important invariant quantities can be determined, and applied in particular situations.

The Lie method for computing symmetries is more useful in complicated situations, leading to an effective means of computing symmetry groups of differential equations and variational problems. However, its applicability to the construction of invariants and the classification problems is much less straightforward. Lie's classification of transformation groups only applies in one and two dimensions, thereby severely limiting this approach. Direct methods for analyzing equivalence problems often have advantages over the more cumbersome Cartan approach when applied to particular situations, but suffer when attempting to analyze a general class of equivalence problems, due to their ad hoc nature. All of these methods can, to a greater or lesser extent, be automated by using a computer algebra system such as MATHEMATICA or MAPLE. However, they tend to tax even very large and fast machines. In the Lie case, the main difficulties lie in the size of the determining system of partial differential equations, which must be integrated in order to determine the explicit form of the symmetry group. In the Cartan method, which only requires differentiation, the difficulties are the appearance of rational algebraic functions, which are not handled particularly well by any of the systems, and intervening expression swell. Nonetheless, both methods can be successfully applied to particular problems with the help of computer algebra.

With this brief overview of our subject, we are now ready to plunge into the details. The first stage is to lay the foundations, introducing the basic language and concepts from differential geometry.

Chapter 1

Geometric Foundations

The study of symmetry groups and equivalence problems requires a variety of tools and techniques, many of which have their origins in geometry. Even our study of differential equations and variational problems will be fundamentally geometric in nature, in contrast to the analytical methods of importance in existence and uniqueness theory. We therefore begin our exposition with a brief review of the basic prerequisites from differential geometry which will be essential to the proper development of our subject. These include the definition and fundamental properties of manifolds and submanifolds, of vector fields and flows, and of differential forms. Even though most of our concerns will be local, nevertheless it will be extremely useful to adopt the coordinate-free language provided by the geometric framework. The advantage of this approach is that it frees one from excessive reliance on complicated local coordinate formulas. On the other hand, when explicit computations need to be done in coordinates, one has the added advantage of being able to choose a particular coordinate system adapted to the problem at hand.

Manifolds

A manifold is an object which, locally, just looks like an open subset of Euclidean space, but whose global topology can be quite different. Although most of our manifolds are realized as subsets of Euclidean space, the general definition is worth reviewing. Although almost all the important examples and applications deal with analytic manifolds, many of the constructions are valid for smooth, meaning infinitely differentiable (C^∞), manifolds, and it is this context that we take as our primary domain of exposition, restricting to the analytic category only when necessary.

Definition 1.1. An m-dimensional *manifold* M is a topological space which is covered by a collection of open subsets $W_\alpha \subset M$, called *coordinate charts*, and one-to-one maps $\chi_\alpha: W_\alpha \to V_\alpha$ onto connected open subsets $V_\alpha \subset \mathbb{R}^m$, which serve to define local coordinates on M. The manifold is *smooth* (respectively *analytic*) if the composite "overlap maps" $\chi_{\beta\alpha} = \chi_\beta \circ \chi_\alpha^{-1}$ are smooth (respectively analytic) where defined, i.e., from $\chi_\alpha[W_\alpha \cap W_\beta]$ to $\chi_\beta[W_\alpha \cap W_\beta]$.

We shall also have occasion to use complex (analytic) manifolds, which are defined in the same fashion, with $\mathbb{C}^m$ replacing $\mathbb{R}^m$; the overlap maps are assumed to be holomorphic. In order to keep the exposition straightforward, we will usually state our results for real manifolds — the reader can easily supply the corresponding complex version without major difficulty (beyond replacing $\mathbb{R}$ by $\mathbb{C}$).

The *topology* of a manifold is induced by that of $\mathbb{R}^m$. Thus, a subset $V \subset M$ is *open* if and only if its intersection $V \cap W_\alpha$ with every coordinate chart maps it to an open subset $\chi_\alpha[V \cap W_\alpha] \subset \mathbb{R}^m$, thereby making each χ_α a continuous invertible map. We will always assume that our manifolds are *separable*, meaning that there is a countable dense subset, and satisfy the Hausdorff topological separation axiom, meaning that any two distinct points $x \neq y$ in M can be separated by open subsets $x \in V$, $y \in W$, with empty intersection, $V \cap W = \varnothing$. A manifold is *connected* if it cannot be written as the disjoint union of two nonempty open subsets; many of our results will require connectivity of the manifolds in question. More generally, every manifold is the union of a countable number of disjoint connected components.

Besides the basic coordinate charts provided in the definition of a manifold, one can always adjoin many additional coordinate charts $\chi: M \to V \subset \mathbb{R}^m$ subject to the condition that, where defined, the corresponding overlap maps $\chi_\alpha \circ \chi^{-1}$ satisfy the same smoothness or analyticity requirements as M itself. For instance, composing a given coordinate map with any local diffeomorphism, meaning a smooth, one-to-one map defined on an open subset of $\mathbb{R}^m$, will give a new set of local coordinates. Often one expands the collection of coordinate charts to include all possible compatible charts, the resulting maximal collection defining an *atlas* on the manifold M. In practice, it is convenient to omit explicit reference to the coordinate maps χ_α and identify a point of M with its image in $\mathbb{R}^m$. Thus, the points in the coordinate chart W_α are identified with their local coordinate expressions $x = (x^1, \dots, x^m) \in V_\alpha$.

The changes of coordinates provided by the overlap maps are then given by local diffeomorphisms $y = \eta(x)$ defined on the overlap of the two coordinate charts.

Objects defined on manifolds must be defined intrinsically, independent of any choice of local coordinate. Consequently, manifolds provide us with the proper category in which most efficaciously to develop a coordinate-free approach to the study of their intrinsic geometry. Of course, the explicit formulae for the object may change when one goes from one set of coordinates to another. Thus, in one sense, any equivalence problem can be viewed as the problem of determining whether two different local coordinate expressions define the same intrinsic object on the manifold. In this language, the problem of determining canonical forms is that of finding local coordinates in which the object assumes a particularly simple form. Explicit examples of this general, underlying philosophy will appear throughout the book.

Example 1.2. The basic example of a manifold is, of course, $\mathbb{R}^m$ itself, or any open subset thereof, which is covered by a single coordinate chart. Another simple example is provided by the unit sphere $S^{m-1} = \{|x| = 1\} \subset \mathbb{R}^m$, which is an analytic manifold of dimension $m - 1$. It can be covered by two coordinate charts, obtained by omitting the north and south poles respectively; the local coordinates are provided by stereographic projection to $\mathbb{R}^{m-1}$. Alternatively, one can use local spherical coordinates on S^{m-1}, which are valid away from the poles.

Example 1.3. Another important example is provided by the projective space $\mathbb{RP}^{m-1}$ which is defined as the space of lines through the origin in $\mathbb{R}^m$. Two nonzero points $x, y \in \mathbb{R}^m$ determine the same point $p \in \mathbb{RP}^{m-1}$ if and only if they are scalar multiples of each other: $x = \lambda y$, $\lambda \neq 0$. (Alternatively, we can realize $\mathbb{RP}^{m-1}$ by identifying antipodal points on the sphere S^{m-1}.) Coordinate charts on $\mathbb{RP}^{m-1}$ are constructed by considering all lines with a given component, say x^i, nonzero; the coordinates are then provided by ratios $p^k = x^k/x^i$, $k \neq i$, which amounts to the choice of canonical representative of such a line given by normalizing its ith component to unity. In particular, the one-dimensional projective space $\mathbb{RP}^1$ can be identified with a circle, since each line through the origin in $\mathbb{R}^2$ is uniquely specified by the angle $0 \le \theta < \pi$ it makes with the horizontal. The coordinate chart consisting of nonvertical lines, i.e., with $y \neq 0$, has local coordinate $p = x/y$, and covers all but one point on the projective line; the horizontal x-axis is traditionally identified with the "point at infinity". In this way one regards $\mathbb{RP}^1 = \mathbb{R} \cup \{\infty\}$ as the completion of the real line $\mathbb{R}^1$. In a similar fashion $\mathbb{RP}^{m-1}$ is viewed as the completion of $\mathbb{R}^{m-1}$ obtained by adjoining all "directions at infinity". Similar considerations apply to the complex projective space $\mathbb{CP}^{m-1}$, which is defined as the space of all complex lines through the origin in $\mathbb{C}^m$. The complex projective line $\mathbb{CP}^1$ can be identified with the Riemann sphere S^2, which is obtained by adjoining a single point at infinity to the complex plane $\mathbb{C}$.

Exercise 1.4. Prove that if M and N are manifolds of respective dimensions m and n, then their Cartesian product $M \times N$ is an $(m+n)$-dimensional manifold.

Example 1.5. Let M be a manifold. A *vector bundle* over M is a manifold E whose local coordinate charts are of the form $W_\alpha \times \mathbb{R}^q$, where $W_\alpha \subset M$ is a local coordinate chart on M, so that the local coordinates have the form (x, u), where $x \in \mathbb{R}^m$ are referred to as the base coordinates and $u \in \mathbb{R}^q$ the fiber coordinates. The overlap functions are restricted to be linear in the fiber coordinates, $(y, v) = (\eta(x), \mu(x)u)$, where $\mu(x)$ is an invertible $q \times q$ matrix of functions defined on the overlap $W_\alpha \cap W_\beta$ of the two coordinate charts. Thus, locally a vector bundle looks like the Cartesian product $M \times \mathbb{R}^q$, although its global topology might be quite different. Every vector bundle comes with a natural projection $\pi: E \to M$ to its base manifold, defined by $\pi(x, u) = x$ in local coordinates. Two simple examples are provided by a cylinder $S^1 \times \mathbb{R}$, and a Möbius band, both of which are vector bundles with one-dimensional fiber (i.e., line bundles) over the circle S^1.

Although it is useful to have the full vocabulary of manifold theory at our disposal, many of our results will be local in nature. By the phrase "locally" we shall generally mean in a neighborhood U, usually a coordinate neighborhood, of a point $x_0 \in M$. Many of our maps $F: M \to N$ will only be defined locally, i.e., not on the entire manifold M but rather on an open subset $U \subset M$; nevertheless, it is convenient to retain the notation $F: M \to N$ even when the domain of F is a proper subset of M. In such cases, when we write $F(x)$ we are always implicitly assuming that x lies in the domain of F.

Functions

A map $F: M \to N$ between smooth manifolds is called smooth if it is smooth in local coordinates. In other words, given local coordinates

$x = (x^1, \ldots, x^m)$ on M, and $y = (y^1, \ldots, y^n)$ on N, the map has the form $y = F(x)$, or, more explicitly, $y^i = F^i(x^1, \ldots, x^m)$, $i = 1, \ldots, n$, where $F = (F^1, \ldots, F^n)$ is a C^∞ map from an open subset of $\mathbb{R}^m$ to $\mathbb{R}^n$. The definition readily extends to analytic maps between analytic manifolds.

Definition 1.6. The *rank* of a map $F: M \to N$ at a point $x \in M$ is defined to be the rank of the $n \times m$ Jacobian matrix $(\partial F^i / \partial x^j)$ of any local coordinate expression for F at the point x. The map F is called *regular* if its rank is constant.

Standard transformation properties of the Jacobian matrix imply that the definition of rank is independent of the choice of local coordinates. (Later we will provide a more intrinsic definition of the rank.) In particular, the set of points where the rank of F is maximal is an open submanifold of the manifold M (which is dense if F is analytic), and the restriction of F to this subset is regular.

The first of the equivalence problems which we encounter, then, is to determine whether two different maps $y = F(x)$ and $\bar{y} = \overline{F}(\bar{x})$ between manifolds of the same dimension are locally the same, meaning that they can be transformed into each other by appropriate changes of coordinates $\bar{x} = \chi(x)$, $\bar{y} = \psi(y)$. In the regular case, the Implicit Function Theorem solves the (local) equivalence problem, and, in fact, provides a canonical form for regular maps.

Theorem 1.7. *Let $F: M \to N$ be a regular map of rank r. Then there exist local coordinates $x = (x^1, \ldots, x^m)$ on M and $y = (y^1, \ldots, y^n)$ on N such that F takes the canonical form*

$$y = F(x) = (x^1, \ldots, x^r, 0, \ldots, 0). \tag{1.1}$$

Thus, all maps of constant rank are locally equivalent and can be linearized by the introduction of appropriate local coordinates. The places where the rank of a map decreases are *singularities*. The canonical forms at singularities are much more complicated — this is the subject of singularity theory (a.k.a. catastrophe theory), cf. [**14**], [**88**]. In order to keep the scope manageable, this book will be exclusively concerned with the regular cases — in this instance meaning regular maps — and will steer away from the more complicated (but perhaps even more interesting) investigation of singularities.

An important case occurs when M and N have the same dimension, and $F: M \to N$ is a regular map of rank $m = \dim M = \dim N$. The Inverse Function Theorem (which is a special case of Theorem 1.7) shows that F defines a local diffeomorphism between M and N, hence the inverse image $F^{-1}\{y\}$ of any point $y \in N$ is a discrete collection of points in M. If F is defined on all of $M =$ domain F, and, moreover, is onto $N =$ range F, then we shall call F a *covering map* and say that M *covers* N. For example, the map $F(t) = (\cos t, \sin t)$ provides a covering map from the real line $M = \mathbb{R}$ to the circle $N = S^1$. The reader should be warned that our definition of covering map is more general than the standard one, cf. [**221**; p. 98], in that N is not necessarily covered evenly by M — the cardinality of the inverse image $F^{-1}\{y\}$ of $y \in N$, can vary from point to point. For instance, according to our definition, the restriction of any covering map to any open subset $\widetilde{M} \subset M$ remains a covering map; thus the preceding example remains a covering map when restricted to any open interval of length at least 2π.

The notion of rank has a natural generalization to (infinite) families of smooth functions. First recall that the *differential* of a smooth function $f: M \to \mathbb{R}$ is given by the expression

$$df = \sum_{i=1}^{m} \frac{\partial f}{\partial x^i}\, dx^i. \tag{1.2}$$

Definition 1.8. Let $\mathcal{F}$ be a family of smooth real-valued functions $f: M \to \mathbb{R}$. The *rank* of $\mathcal{F}$ at a point $x \in M$ is the dimension of the space spanned by their differentials. The family is *regular* if its rank is constant on M.

Definition 1.9. A set $\{f_1, \ldots, f_k\}$ of smooth real-valued functions on a manifold M having a common domain of definition is called *functionally dependent* if, for each $x_0 \in M$, there is neighborhood U and a smooth function $H(z_1, \ldots, z_k)$, not identically zero on any subset of $\mathbb{R}^k$, such that $H(f_1(x), \ldots, f_k(x)) = 0$ for all $x \in U$. The functions are called *functionally independent* if they are not functionally dependent when restricted to any open subset of M.

For example $f_1(x, y) = x/y$ and $f_2(x, y) = xy/(x^2 + y^2)$ are functionally dependent on the upper half plane $\{y > 0\}$ since the second can be written as a function of the first: $f_2 = f_1/(1 + f_1^2)$. For a regular family of functions, the rank tells us how many functionally independent functions it contains.

Theorem 1.10. *If a family of functions $\mathcal{F}$ is regular of rank r, then, in a neighborhood of any point, there exist r functionally independent functions $f_1, \ldots, f_r \in \mathcal{F}$ with the property that any other function $f \in \mathcal{F}$ can be expressed as a function thereof: $f = H(f_1, \ldots, f_r)$.*

Proof: Given $x_0 \in M$, choose $f_1, \ldots, f_r \in \mathcal{F}$ such that their differentials $df_1, \ldots, df_r$ are linearly independent at x_0, and hence, by continuity, in a neighborhood of x_0. According to Theorem 1.7, we can locally choose coordinates (y, z) near x_0 such that $f_i(y, z) = y^i$, $i = 1, \ldots, r$. If $f(y, z)$ is any other function in $\mathcal{F}$, then, since the rank is r, its differential must be a linear combination of the differentials df_i, so that in the new coordinates $df = \sum_{i=1}^{r} h_i(y, z) dy^i$. We now invoke a simple lemma.

Lemma 1.11. *Let $U \subset \mathbb{R}^m$ be a convex open set. A function $f: U \to \mathbb{R}$ has differential $df = \sum_{i=1}^{r} h_i(x)\, dx^i$ given as a linear combination of the first r coordinate differentials if and only if $f = f(x^1, \ldots, x^r)$ is a function of the first r coordinates. In particular, $df = 0$ if and only if f is constant.*

Thus, by shrinking the neighborhood of x_0 if necessary so that it is convex in the (y, z)-coordinates, Lemma 1.11 implies that $f(y^1, \ldots, y^r)$ is a function of y alone. Reverting to the original x coordinates completes the proof of the Theorem. *Q.E.D.*

Consequently, if $f_1, \ldots, f_r$ is a set of functions whose $m \times r$ Jacobian matrix $(\partial f_i / \partial x^k)$ has maximal rank r at x_0, then, by continuity, they also have rank r in a neighborhood of x_0, and hence are functionally independent near x_0. Note also that Theorem 1.10 implies that, locally, there are at most m functionally independent functions on any m-dimensional manifold M.

Submanifolds

Naïvely, a submanifold of a manifold M is a subset which is a manifold in its own right. To be more precise, we need to parametrize the subset by a suitable map.

Definition 1.12. A smooth (analytic) n-dimensional *immersed submanifold* of a manifold M is a subset $N \subset M$ parametrized by a smooth (analytic), one-to-one map $F: \widetilde{N} \to N \subset M$, whose domain $\widetilde{N}$, the *parameter space*, is a smooth (analytic) n-dimensional manifold, and such that F is everywhere regular, of maximal rank n.

A submanifold is *regular* if, in addition to the regularity of the parametrizing map, at each point $x \in N$ there exist arbitrarily small neighborhoods $x \in U \subset M$ such that $F^{-1}[U \cap N]$ is a connected open subset of $\widetilde{N}$. The Implicit Function Theorem 1.7 provides an immediate canonical form for regular submanifolds.

Theorem 1.13. *An n-dimensional submanifold $N \subset M$ of an m-dimensional manifold M is regular if and only if for each $x_0 \in N$ there exist local coordinates $x = (x^1, \ldots, x^m)$ defined on a neighborhood U of x_0 such that $U \cap N = \{x \mid x^1 = \cdots = x^{m-n} = 0\}$.*

Thus, every regular n-dimensional submanifold of an m-dimensional manifold locally looks like an n-dimensional subspace of $\mathbb{R}^m$. We conclude that all regular n-dimensional submanifolds are locally equivalent. The topology on M induces the manifold topology on N, and so, in the regular case, we can effectively dispense with reference to the parameter space $\widetilde{N}$. Irregular submanifolds are trickier, since the same subset can be parametrized as a submanifold in several inequivalent ways — see Example 1.14 below.

It will occasionally be useful to enlarge our repertoire of submanifolds yet further to include submanifolds which have self-intersections. Consider an arbitrary regular map $F: \widetilde{N} \to N \subset M$ of maximal rank n, which is the dimension of the parameter space $\widetilde{N}$. We will call the image $N = F(\widetilde{N})$ a *submanifold with self-intersections* of M. According to Theorem 1.7, the image N of such a map F is, locally, a regular submanifold, but if F is not one-to-one, the submanifold N will intersect itself.

A *curve* in a manifold M is defined by a smooth map $\phi: I \to M$, where $I \subset \mathbb{R}$ is an open subinterval. The curve $C = \phi(I)$ defines a one-dimensional submanifold, with self-intersections, provided the regularity condition $\phi'(t) \neq 0$ holds. (Points at which the derivative of ϕ vanishes will, in general, be singularities of the curve.) If ϕ is one-to-one, then C is an ordinary one-dimensional submanifold, and is regular provided

$$\lim_{i \to \infty} \phi(t_i) = \phi(t) \qquad \text{if and only if} \qquad \lim_{i \to \infty} t_i = t.$$

Example 1.14. The curve $\phi_0(t) = (\sin t, 2\sin(2t))$ describes a figure eight in the plane, which is a one-dimensional manifold with self-intersections. The map $\phi_1(t) = (\sin(2\arctan t), 2\sin(4\arctan t))$

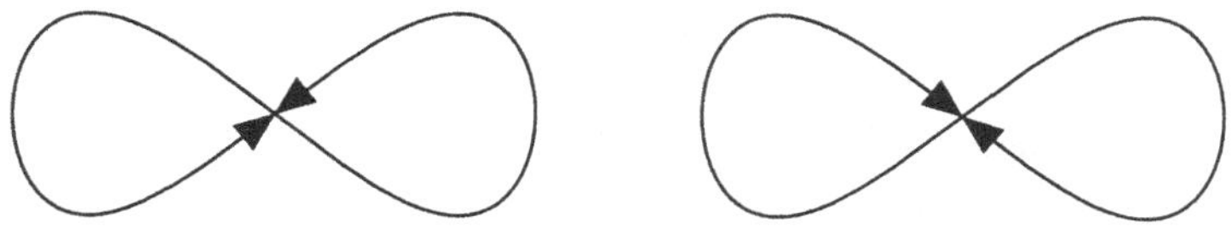

Figure 1. Inequivalent Submanifolds.

parametrizes the same figure eight, but is now one-to-one, and so describes an immersed submanifold, which, however, is not regular since $\lim_{t\to\infty} \phi(t) = 0 = \phi(0)$. Note further that the alternative one-to-one map $\phi_2(t) = (-\sin(2\arctan t), 2\sin(4\arctan t))$ defines an inequivalent means of parametrizing the figure eight. Thus, in the irregular case, a given subset can be parametrized as a submanifold in fundamentally different ways.

Exercise 1.15. Prove the map $\phi(t) = (e^{-t}\cos t, e^{-t}\sin t)$ describes a regular one-dimensional submanifold $N \subset \mathbb{R}^2$. Draw a picture of N.

Example 1.16. A *torus* is defined as the Cartesian product of circles. Consider the two-dimensional torus $T^2 = S^1 \times S^1$, with angular coordinates (θ, φ) with $0 \leq \theta, \varphi < 2\pi$. The curve $\phi(t) = (t, \kappa t) \bmod 2\pi$ is closed if κ/π is a rational number, and hence defines a regular submanifold of T^2, with parameter space S^1. On the other hand, if κ/π is irrational, then the curve forms a dense subset of T^2 and hence is not a regular submanifold.

Exercise 1.17. Let M and N be smooth manifolds of respective dimensions m and n. Prove that the *graph* $\Gamma = \{(x, F(x))\}$ of a smooth map $F: M \to N$ (which might only be defined on an open subset of M) forms a regular m-dimensional submanifold of the Cartesian product manifold $M \times N$.

Example 1.18. The preceding construction is generalized by the concept of a section of a vector bundle $\pi: E \to M$. A map $F: M \to E$ is called a *section* of E if $\pi \circ F = \mathbb{1}$ is the identity map on M, and, in the local coordinates $(x, u) \in W_\alpha \times \mathbb{R}^q$ of Example 1.5, it has the form $u = F(x)$ where $F: W_\alpha \to \mathbb{R}^q$ is a smooth function. The image $F(M)$ of a section is a smooth m-dimensional submanifold of E that intersects each fiber $E|_x = \pi^{-1}\{x\}$ in only one point. (*Exercise*: Is every submanifold having the latter property a section?)

Since manifolds are modeled on Euclidean space, it is not hard to see that every connected manifold is *pathwise connected* meaning that any two points can be joined by a smooth curve. A manifold is *simply connected* if every smooth curve can be smoothly contracted to a point. For example, the circle S^1 and the punctured plane $\mathbb{R}^2 \setminus \{0\}$ are not simply connected, whereas the sphere S^2 is simply connected.

The following definition will be of use in our analysis of the Cartan equivalence method.

Definition 1.19. Two n-dimensional submanifolds N and $\overline{N}$ of a manifold M are said to *overlap* if their intersection $N \cap \overline{N}$ is a nonempty n-dimensional submanifold of M. They are said to *strictly overlap* if their union $N \cup \overline{N}$ is also a connected n-dimensional submanifold (possibly with self-intersections) of M, containing N and $\overline{N}$ as open submanifolds.

Exercise 1.20. Prove that strictly overlapping submanifolds overlap. Prove that the converse to this statement is true for analytic submanifolds, but false for more general smooth manifolds.

An alternative to the parametric approach to submanifolds is to define them *implicitly* as a common level set of a collection of functions. In general, the *variety* $\mathcal{S}_\mathcal{F}$ determined by a family of real-valued functions $\mathcal{F}$ is defined to be the subset where they simultaneously vanish:

$$\mathcal{S}_\mathcal{F} = \{x \mid f(x) = 0 \text{ for all } f \in \mathcal{F}\}.$$

In particular, given $F: M \to \mathbb{R}^r$, the variety $\mathcal{S}_F = \{F(x) = 0\}$ is just the set of solutions to the simultaneous system of equations $F^1(x) = \cdots = F^r(x) = 0$ defined by the components of F. We will call the variety (or system of equations) *regular* if $\mathcal{S}_\mathcal{F}$ is not empty, and the rank of $\mathcal{F}$ is constant in a neighborhood of $\mathcal{S}_\mathcal{F}$; the latter condition clearly holds if $\mathcal{F}$ itself is a regular family. In particular, regularity holds if the variety is defined by the vanishing of a map $F: M \to \mathbb{R}^r$ which has maximal rank r at each point $x \in \mathcal{S}_F$, i.e., at each solution x to the system of equations $F(x) = 0$. The Implicit Function Theorem 1.10, coupled with Theorem 1.13, shows that a regular variety is a regular submanifold.

Theorem 1.21. *Suppose $\mathcal{F}$ is a family of functions defined on an m-dimensional manifold M. If the associated variety $S_{\mathcal{F}} \subset M$ is regular, then it defines a regular submanifold of dimension $m - r$.*

For example, the function $F(x, y, z) = x^2 + y^2 + z^2 - 1$ has rank one everywhere except at the origin, and hence its variety — the unit sphere — is a regular two-dimensional submanifold of $\mathbb{R}^3$. On the other hand, the function $F(x, y, z) = x^2 + y^2$ has rank zero on its variety — the z-axis — but rank one elsewhere; thus Theorem 1.21 does not apply (even though its variety is a submanifold, albeit of the "wrong" dimension). Finally, the function $F(x, y, z) = xyz$ is also not regular, and, in this case, its variety — the three coordinate planes — is not a submanifold.

Vector Fields

A *tangent vector* to a manifold M at a point $x \in M$ is geometrically defined by the tangent to a (smooth) curve passing through x. In local coordinates, the tangent vector $\mathbf{v}|_x$ to the parametrized curve $x = \phi(t)$ is determined by the derivative: $\mathbf{v}|_x = \phi'(t)$. The collection of all such tangent vectors forms the *tangent space* to M at x. Each tangent space $TM|_x$ is a vector space of the same dimension as M; they are "sewn" together in the obvious manner to form the *tangent bundle* $TM = \bigcup_{x \in M} TM|_x$ of the manifold, which is a $2m$-dimensional manifold, and forms a vector bundle over the m-dimensional manifold M.

Exercise 1.22. Prove that the tangent bundle $TS^1 \simeq S^1 \times \mathbb{R}$ to a circle is a cylinder. On the other hand, the tangent bundle to a sphere is not a trivial Cartesian product: $TS^2 \not\simeq S^2 \times \mathbb{R}^2$.

A *vector field* $\mathbf{v}$ is a smoothly (or analytically) varying assignment of tangent vectors $\mathbf{v}|_x \in TM|_x$, i.e., a section of the tangent bundle TM. We write the local coordinate formula for a vector field in the form

$$\mathbf{v} = \sum_{i=1}^{m} \xi^i(x) \frac{\partial}{\partial x^i}, \tag{1.3}$$

where the coefficients $\xi^i(x)$ are smooth (analytic) functions. The motivation for adopting the standard derivational notation in (1.3) will appear presently. In particular, the tangent vectors to the coordinate axes are denoted by $\partial/\partial x^i = \partial_{x^i}$, and form a basis for the tangent space $TM|_x$

at each point in the coordinate chart. If $y = \eta(x)$ is any change of coordinates, then the vector field (1.3) is, in the y coordinates, re-expressed using the basic change of variables formula

$$\mathbf{v} = \sum_{j=1}^{m} \left(\sum_{i=1}^{m} \xi^i(x) \frac{\partial \eta^j}{\partial x^i} \right) \frac{\partial}{\partial y^j}, \tag{1.4}$$

where the coefficients are evaluated at $x = \eta^{-1}(y)$. Equation (1.4) is a direct consequence of the chain rule applied to the original definition of tangent vectors to curves.

A parametrized curve $\phi: \mathbb{R} \to M$ is called an *integral curve* of the vector field $\mathbf{v}$ if its tangent vector coincides with the vector field $\mathbf{v}$ at each point; this requires that its parametrization $x = \phi(t)$ satisfy the first order system of ordinary differential equations

$$\frac{dx^i}{dt} = \xi^i(x), \qquad i = 1, \ldots, m. \tag{1.5}$$

Standard existence and uniqueness theorems for systems of ordinary differential equations imply that, through each point $x \in M$, there passes a unique, maximal integral curve. We shall employ the suggestive notation $\phi(t) = \exp(t\mathbf{v})x$ to denote the maximal integral curve passing through $x = \exp(0\mathbf{v})x$ at $t = 0$; the curve $\exp(t\mathbf{v})x$ may or may not be defined for all t.

Motivated by fluid mechanics, where $\mathbf{v}$ is the fluid velocity vector field, the family of (locally defined) maps $\exp(t\mathbf{v})$ is known as the *flow* generated by the vector field $\mathbf{v}$, which is classically referred to as the *infinitesimal generator* of the flow. The flow obeys standard exponential rules:

$$\begin{aligned} &\exp(t\mathbf{v})\exp(s\mathbf{v})x = \exp\big[(t+s)\mathbf{v}\big]x, && \exp(0\mathbf{v})x = x, \\ &\exp(t\mathbf{v})^{-1}x = \exp(-t\mathbf{v})x, && \frac{d}{dt}\exp(t\mathbf{v})x = \mathbf{v}\big|_{\exp(t\mathbf{v})x}, \end{aligned} \tag{1.6}$$

the first and third equations holding where defined. Conversely, given a flow obeying the first two equations in (1.6), we can reconstruct its generating vector field by differentiation:

$$\mathbf{v}|_x = \frac{d}{dt}\exp(t\mathbf{v})x\bigg|_{t=0}, \qquad x \in M. \tag{1.7}$$

In other words, identifying tangent vectors in Euclidean space with ordinary vectors, we have the local coordinate expansion

$$\exp(t\mathbf{v})x = x + t\mathbf{v}|_x + \mathrm{O}(t^2), \tag{1.8}$$

for the flow. Thus, the theory of first order systems of autonomous ordinary differential equations (1.5) is the same as the theory of flows of vector fields.

A point x where the vector field vanishes, $\mathbf{v}|_x = 0$, determines a *singularity* or *equilibrium point*. In this case, x is a fixed point under the induced flow: $\exp(t\mathbf{v})x = x$, for all t. Points at which $\mathbf{v}$ is not zero are called *regular*. The existence of flows implies that, away from singularities, all vector fields look essentially the same. Indeed, as with regular maps and submanifolds, there is a simple canonical form for a vector field in a neighborhood of any regular point.

Theorem 1.23. *Let* $\mathbf{v}$ *be a vector field defined on* M. *If* x_0 *is not a singularity of* $\mathbf{v}$, *so* $\mathbf{v}|_{x_0} \neq 0$, *then there exist local* rectifying coordinates $y = (y^1, \ldots, y^m)$ *near* x_0 *such that* $\mathbf{v} = \partial/\partial y^1$ *generates the translational flow* $\exp(t\mathbf{v})y = (y^1 + t, y^2, \ldots, y^m)$.

Theorem 1.23 provides a solution to the basic equivalence problem for systems of autonomous first order systems of ordinary differential equations. This result states that, away from equilibrium points, all such systems are locally equivalent, since they can all be mapped to the elementary system $dy^1/dt = 1$, $dy^2/dt = \cdots = dy^m/dt = 0$ by a suitable change of variables. Indeed, if $\xi^1(x_0) \neq 0$, then the rectifying y coordinates are defined so that $x_0 = 0$, and each $x = \exp(ty^1)(0, y^2, \ldots, y^m)$ lies on a unique integral curve emanating from the hyperplane $\{y^1 = 0\}$. However, Theorem 1.23 is of little practical use for actually solving the system of ordinary differential equations governing the flow, since finding the change of variables required to place the system in canonical form is essentially the same problem as solving it in the first place. As we shall see, the equivalence problem for higher order systems of ordinary differential equations is much more interesting.

Example 1.24. Consider the following three vector fields on $M = \mathbb{R}$. First, the vector field ∂_x generates the translation flow $\exp(t\partial_x)x = x + t$. The vector field $x\partial_x$ generates the scaling flow $\exp(tx\partial_x)x = e^t x$. In this case, away from the singularity at $x = 0$, the rectifying coordinate is given by $y = \log|x|$, in terms of which the vector field takes the form $\mathbf{v} = \partial_y$. Finally, the vector field $x^2\partial_x$ generates the "inversional" flow $\exp(tx^2\partial_x)x = x/(1-tx)$. Note that this vector field only generates a local flow on $\mathbb{R}$. In this case, the rectifying coordinate is $y = 1 - 1/x$.

Exercise 1.25. Prove that the three vector fields in Example 1.24 can be extended to the projective line $\mathbb{RP}^1$ and, in fact, define global flows there. (See [**172**], [**193**] for general results on globalizing flows and group actions.)

The problem of equivalence of vector fields at singularities is much more delicate. There is a large body of literature on the determination of normal forms for vector fields near equilibrium points; see, for instance, [**68**], [**105**]. The global equivalence problem is also of interest, but again more delicate since topological data come into play. For example, unless the Euler characteristic of M is trivial, every smooth vector field must have at least one singularity; see [**206**; Chapter 11].

Exercise 1.26. Let E be a vector bundle with local coordinates $(x,y) = (x^1,\ldots,x^m,y^1,\ldots,y^q)$. Show that a vector field on E depending linearly on the y coordinates,

$$\mathbf{v} = \sum_{i=1}^{m} \xi^i(x)\,\frac{\partial}{\partial x^i} + \sum_{j,k=1}^{q} \eta^k_j(x) y^j\,\frac{\partial}{\partial y^k}\,,$$

has a linear flow $\exp(t\mathbf{v})(x,y) = (\chi_t(x), \psi_t(x)y)$ in these variables.

Applying a vector field $\mathbf{v}$ to a function $f\colon M \to \mathbb{R}$ determines the infinitesimal change in f under the induced flow:

$$\mathbf{v}(f(x)) = \sum_{i=1}^{n} \xi^i(x)\,\frac{\partial f}{\partial x^i} = \frac{d}{dt}\,f(\exp(t\mathbf{v})x)\bigg|_{t=0}.$$

In this way, vector fields act as *derivations* on the smooth functions, meaning that they are linear and satisfy a Leibniz Rule:

$$\mathbf{v}(f+g) = \mathbf{v}(f) + \mathbf{v}(g), \qquad \mathbf{v}(fg) = f\mathbf{v}(g) + g\mathbf{v}(f). \qquad (1.9)$$

In particular, vector fields annihilate constant functions: $\mathbf{v}(c) = 0$. Indeed, an alternative definition of the tangent space $TM|_x$ is as the space of derivations on the smooth functions defined in a neighborhood of x, cf. [**206**; Chapter 3]. The action of the flow on a function can be reconstructed from its generating vector field using the *Lie series* expansion

$$f(\exp(t\mathbf{v})x) = f(x) + t\mathbf{v}(f(x)) + \tfrac{1}{2}t^2\mathbf{v}(\mathbf{v}(f(x))) + \cdots, \qquad (1.10)$$

which, in the analytic case, converges for t near 0.

Lie Brackets

The most important operation on vector fields is the Lie bracket or commutator.

Definition 1.27. Given vector fields $\mathbf{v}$ and $\mathbf{w}$ on a manifold M, their *Lie bracket* is the vector field $[\mathbf{v},\mathbf{w}]$ which satisfies $[\mathbf{v},\mathbf{w}]f = \mathbf{v}(\mathbf{w}(f)) - \mathbf{w}(\mathbf{v}(f))$ for any smooth function $f\colon M \to \mathbb{R}$.

The fact that the Lie bracket is a well-defined vector field rests on the readily verified fact that the commutator of two derivations is itself a derivation. In local coordinates, if

$$\mathbf{v} = \sum_{i=1}^{m} \xi^i(x)\,\frac{\partial}{\partial x^i}\,, \qquad \mathbf{w} = \sum_{i=1}^{m} \eta^i(x)\,\frac{\partial}{\partial x^i}\,,$$

then

$$[\mathbf{v},\mathbf{w}] = \sum_{i=1}^{m} \{\mathbf{v}(\eta^i) - \mathbf{w}(\xi^i)\}\,\frac{\partial}{\partial x^i} = \sum_{i=1}^{m}\sum_{j=1}^{m} \left\{\xi^j\,\frac{\partial \eta^i}{\partial x^j} - \eta^j\,\frac{\partial \xi^i}{\partial x^j}\right\}\frac{\partial}{\partial x^i}\,. \qquad (1.11)$$

The Lie bracket is anti-symmetric, $[\mathbf{v},\mathbf{w}] = -[\mathbf{w},\mathbf{v}]$, and bilinear; moreover it satisfies the important *Jacobi identity*

$$[\mathbf{u},[\mathbf{v},\mathbf{w}]] + [\mathbf{v},[\mathbf{w},\mathbf{u}]] + [\mathbf{w},[\mathbf{u},\mathbf{v}]] = 0, \qquad (1.12)$$

for any triple of vector fields $\mathbf{u},\mathbf{v},\mathbf{w}$.

The Lie bracket between two vector fields can be identified as the infinitesimal generator of the commutator of the two associated flows. This interpretation is based on the local coordinate expansion (see (1.8))

$$\exp(-t\mathbf{w})\exp(-t\mathbf{v})\exp(t\mathbf{w})\exp(t\mathbf{v})x = x + t^2[\mathbf{v},\mathbf{w}]\big|_x + \mathrm{O}(t^3), \qquad (1.13)$$

which follows directly from (1.10). For example, the Lie bracket of the scaling and translation vector fields on $\mathbb{R}$ is $[\partial_x, x\partial_x] = \partial_x$, reflecting the non-commutativity of the operations of translation and scaling. In particular, if the two flows commute, then the Lie bracket of their infinitesimal generators is necessarily zero. This statement admits an important converse, which is a consequence of the basic existence and uniqueness theorems for ordinary differential equations — see [**186**; Theorem 1.34].

Theorem 1.28. *Let* $\mathbf{v}$, $\mathbf{w}$ *be vector fields on a manifold* M. *Then* $[\mathbf{v}, \mathbf{w}] = 0$ *if and only if* $\exp(t\mathbf{v})\exp(s\mathbf{w})x = \exp(s\mathbf{w})\exp(t\mathbf{v})x$ *for all* $x \in M$ *and all* $t, s \in \mathbb{R}$ *where the equation is defined.*

The Differential

A smooth map $F: M \to N$ between manifolds will map smooth curves on M to smooth curves on N, and thus induce a map between their tangent vectors. The result is a linear map $dF: TM|_x \to TN|_{F(x)}$ between the tangent spaces of the two manifolds, called the *differential* of F. More specifically, if the parametrized curve $\phi(t)$ has tangent vector $\mathbf{v}|_x = \phi'(t)$ at $x = \phi(t)$, then the image curve $\psi(t) = F[\phi(t)]$ will have tangent vector $\mathbf{w}|_y = dF(\mathbf{v}|_x) = \psi'(t)$ at the image point $y = F(x)$. Alternatively, if we regard tangent vectors as derivations, then we can define the differential by the chain rule formula

$$dF\left(\mathbf{v}|_x\right)[h(y)] = \mathbf{v}[h \circ F(x)] \qquad \text{for any} \qquad h: N \to \mathbb{R}. \tag{1.14}$$

In terms of local coordinates,

$$dF\left(\mathbf{v}|_x\right) = dF\left(\sum_{i=1}^{m} \xi^i \frac{\partial}{\partial x^i}\right) = \sum_{j=1}^{n}\left(\sum_{i=1}^{m} \xi^i \frac{\partial F^j}{\partial x^i}\right)\frac{\partial}{\partial y^j}. \tag{1.15}$$

(Note that the change of variables formula (1.4) is a special case of (1.15).) Consequently, the differential dF defines a linear map from $TM|_x$ to $TN|_{F(x)}$, whose local coordinate matrix expression is just the $n \times m$ Jacobian matrix $\left(\partial F^j/\partial x^i\right)$ of F at x. In particular, the *rank* of the map F can now be defined intrinsically as the rank of the linear map determined by its differential dF. Note that the differential of the composition of two maps $F: M \to N$, and $H: N \to P$, is just the linear composition of the two differentials: $d(H \circ F) = dH \circ dF$; this fact is merely the coordinate-free version of the usual chain rule for Jacobian matrices.

An important remark is that, in general, unless F is one-to-one, its differential dF does *not* map vector fields to vector fields. Indeed if $\mathbf{v}$ is a vector field on M and x and $\widetilde{x}$ are two points in M with the same image $F(x) = F(\widetilde{x})$ in N, there is no reason why $dF(\mathbf{v}|_x)$ should necessarily agree with $dF(\mathbf{v}|_{\widetilde{x}})$. However, if $\mathbf{v}$ *is* mapped to a well-defined vector field $dF(\mathbf{v})$ on N, then the two flows match up, meaning

$$F[\exp(t\mathbf{v})x] = \exp(t\,dF(\mathbf{v}))\,F(x), \tag{1.16}$$

where defined. Moreover, the differential dF respects the Lie bracket operation:

$$dF\big([\mathbf{v},\mathbf{w}]\big) = [dF(\mathbf{v}), dF(\mathbf{w})], \tag{1.17}$$

whenever $dF(\mathbf{v})$ and $dF(\mathbf{w})$ are well-defined vector fields on N. This is a consequence of the commutator formulation; it can also be verified directly from (1.11), (1.15). A particular consequence is that the Lie bracket preserves tangent vector fields to submanifolds.

Proposition 1.29. *Suppose* $\mathbf{v}$ *and* $\mathbf{w}$ *are vector fields which are tangent to a submanifold* $N \subset M$. *Then their Lie bracket* $[\mathbf{v},\mathbf{w}]$ *is also tangent to* N.

Example 1.30. The vector fields $\mathbf{v} = x\partial_y - y\partial_x$, $\mathbf{w} = y\partial_z - z\partial_y$, generate the rotational flows around the z- and the x-axis respectively. They are both tangent to the unit sphere $S^2 \subset \mathbb{R}^3$, hence their Lie bracket $[\mathbf{v},\mathbf{w}] = x\partial_z - z\partial_x$, which generates the rotational flow around the y-axis, is also tangent to S^2.

Differential Forms

The dual objects to vector fields are differential forms. Given a point $x \in M$, a real-valued linear function $\omega\colon TM|_x \to \mathbb{R}$ on the tangent space is said to define a *one-form* at x. The evaluation of ω on a tangent vector $\mathbf{v}$ will be indicated by the bilinear pairing $\langle \omega\,;\mathbf{v}\rangle$. The space of one-forms is the dual vector space to the tangent space $TM|_x$, and is called the *cotangent space*, denoted $T^*M|_x$. The cotangent spaces are sewn together to form the *cotangent bundle* $T^*M = \bigcup_{x\in M} T^*M|_x$, which, like the tangent bundle, forms an m-dimensional vector bundle over the m-dimensional manifold M. A *differential one-form* or *Pfaffian form*, then, is just a (smooth or analytic) section of T^*M, i.e., a smoothly varying assignment of linear maps on the tangent spaces $TM|_x$. Given a smooth real-valued function $f: M \to \mathbb{R}$, its differential df, as given in (1.2), determines a one-form whose evaluation on any tangent vector (vector field) $\mathbf{v}$ is defined by $\langle df\,;\mathbf{v}\rangle = \mathbf{v}(f)$. In local coordinates $x = (x^1, \ldots, x^m)$, the differentials dx^i of the coordinate functions provide a basis of the cotangent space at each point of the coordinate chart, which forms the dual to the coordinate basis ∂_{x^j} of the tangent space; thus $\langle dx^i\,; \partial_{x^j}\rangle = \delta^i_j$, where δ^i_j denotes the *Kronecker delta*, which is 1 if $i = j$ and 0 otherwise. In terms of this basis, a general one-form takes the local coordinate form

$$\omega = \sum_{i=1}^{m} h_i(x)dx^i, \qquad \text{so that} \qquad \langle \omega\,; \mathbf{v}\rangle = \sum_{i=1}^{m} h_i(x)\xi^i(x) \tag{1.18}$$

defines its evaluation on the vector field (1.3).

Exercise 1.31. Let $\mathcal{S}_F = \{F_1(x) = \cdots = F_k(x) = 0\} \subset M$ be a regular variety. Prove that its tangent space is the common kernel of the differentials of its defining functions:

$$T\mathcal{S}_F|_x = \{\mathbf{v} \in TM|_x \mid \langle dF_\nu(x)\,; \mathbf{v}\rangle = 0, \nu = 1, \ldots, k\}.$$

Differential forms of higher degree are defined as alternating multi-linear maps on the tangent space. Thus a *differential k-form* Ω at a point $x \in M$ is a k-linear map

$$\Omega: \overbrace{TM|_x \times \cdots \times TM|_x}^{k \text{ times}} \longrightarrow \mathbb{R},$$

which is anti-symmetric in its arguments, meaning that

$$\langle \Omega\,; \mathbf{v}_{\pi 1}, \ldots, \mathbf{v}_{\pi k}\rangle = (\operatorname{sign}\pi)\,\langle \Omega\,; \mathbf{v}_1, \ldots, \mathbf{v}_k\rangle,$$

for any permutation π of the indices $\{1, \ldots, k\}$. We refer to k as the *degree* of the differential form Ω. A real-valued function is considered as a form of degree 0. The space of all k-forms at x is the k-fold *exterior power* of the cotangent space at x, denoted by $\bigwedge^k T^*M|_x$, and forms a vector space of dimension $\binom{m}{k}$. In particular, the only differential form whose degree is greater than the dimension of the underlying manifold is the trivial one $\Omega = 0$. These spaces are sewn together to form the kth exterior tangent bundle $\bigwedge^k T^* M$.

If $\omega^1, \ldots, \omega^k$ are one-forms at x, their *wedge product* defines a *decomposable* k-form $\omega^1 \wedge \cdots \wedge \omega^k$, which is defined by the determinantal formula

$$\langle \omega^1 \wedge \cdots \wedge \omega^k \, ; \mathbf{v}_1, \ldots, \mathbf{v}_k \rangle = \det\Big(\langle \omega^i \, ; \mathbf{v}_j \rangle \Big).$$

In particular, the one-forms $\omega^1, \ldots, \omega^k$ are linearly dependent if and only if their wedge product vanishes: $\omega^1 \wedge \cdots \wedge \omega^k = 0$. Not every k-form is decomposable, although they can all be written as linear combinations of decomposable forms. In local coordinates $x = (x^1, \ldots, x^m)$, the $\binom{m}{k}$ coordinate k-forms $dx^I = dx^{i_1} \wedge \cdots \wedge dx^{i_k}$ corresponding to all strictly increasing multi-indices $1 \leq i_1 < i_2 < \cdots < i_k \leq m$ form a basis for the exterior tangent space $\bigwedge^k T^* M|_x$. A general differential k-form, then, is a section of $\bigwedge^k T^* M$, and, in local coordinates, can be written as a sum

$$\Omega = \sum_I h_I(x) \, dx^I, \tag{1.19}$$

over the basis k-forms with smoothly varying coefficients $h_I(x)$. For example, every m-form on the m-dimensional manifold M is a multiple $h(x)\,dx$ of the standard volume form $dx = dx^1 \wedge \cdots \wedge dx^m$ on $\mathbb{R}^m$.

If $f_1, \ldots, f_k$ are smooth, real-valued functions on M, then the wedge product $df_1 \wedge \cdots \wedge df_k$ of their differentials forms a decomposable k-form on M, which vanishes at a point x if and only if the differentials are linearly dependent there. If we expand this form in terms of the coordinate differentials, as in (1.19), then the coefficient of the basis k-form dx^I is the Jacobian determinant

$$\frac{\partial(f_1, \ldots, f_k)}{\partial(x^{i_1}, \ldots, x^{i_k})} = \det\left(\frac{\partial F_j}{\partial x^{i_l}}\right).$$

Thus, Theorem 1.10 implies the following simple test for functional independence.†

† Actually, Theorem 1.10 only applies if the rank of $f_1, \ldots, f_k$ is constant. The proof of the second statement in Proposition 1.32 in the more general case when the rank is $< k$, but not necessarily constant, can be found in [**175**; Theorem 1.4.14].

Proposition 1.32. *If $f_1, \ldots, f_k$ satisfy $df_1 \wedge \cdots \wedge df_k \neq 0$, then they are functionally independent. On the other hand, if $df_1 \wedge \cdots \wedge df_k \equiv 0$ for all $x \in M$, then $f_1, \ldots, f_k$ are functionally dependent.*

If $\Omega = \omega^1 \wedge \cdots \wedge \omega^k$ and $\Theta = \theta^1 \wedge \cdots \wedge \theta^l$ are decomposable k- and l-forms, we define their *wedge product* to be the decomposable $(k+l)$-form $\Omega \wedge \Theta = \omega^1 \wedge \cdots \wedge \omega^k \wedge \theta^1 \wedge \cdots \wedge \theta^l$; this definition extends by linearity to arbitrary differential forms. The resulting wedge product between differential forms is bilinear and "super-symmetric": $\Omega \wedge \Theta = (-1)^{kl} \Theta \wedge \Omega$. In particular, the wedge product of a 0-form f, otherwise known as a smooth function, and a k-form Ω is the k-form $f\,\Omega$ obtained by multiplying Ω by f.

Remark: We shall *always* assume that our differential forms are of homogeneous degree. This means that, although we are allowed to take the wedge product of any pair of differential forms, we are only allowed to sum differential forms of the *same* degree.

Exercise 1.33. Let $\omega^1, \ldots, \omega^k$ be a set of linearly independent one-forms, so $\omega^1 \wedge \cdots \wedge \omega^k \neq 0$. Prove Cartan's Lemma, which states that the one-forms $\theta^1, \ldots, \theta^k$ satisfy $\sum_i \theta^i \wedge \omega^i = 0$ if and only if $\theta^i = \sum_j A^i_j \, \omega^j$ for some symmetric matrix of functions: $A^i_j = A^j_i$.

If $F\colon M \to N$ is a smooth map, then there is an induced map on differential forms, called the *pull-back* of F and denoted F^*, which maps a differential form on N *back* to a differential form on M. (*Warning*: The direction of the pull-back is reversed from that of F and its differential dF.) If $\theta \in T^*N|_y$ is a one-form at $y = F(x)$, then $\omega = F^*\theta \in T^*M|_x$ is a one-form on M defined so that $\langle \omega\,; \mathbf{v} \rangle = \langle F^*\theta\,; \mathbf{v} \rangle = \langle \theta\,; dF(\mathbf{v}) \rangle$ for any tangent vector $\mathbf{v} \in TM|_x$. In local coordinates, the pull-back of a one-form θ on N is the one-form

$$
\begin{aligned}
F^*\theta = F^*\left(\sum_{j=1}^n h_j(y)\, dy^j \right) &= \sum_{j=1}^n h_j(F(x))\, dF^j(x) \\
&= \sum_{i=1}^m \left(\sum_{j=1}^n h_j(F(x)) \frac{\partial F^j}{\partial x^i} \right) dx^i.
\end{aligned}
\tag{1.20}
$$

Thus, in local coordinates the pull-back is represented by the transpose of the Jacobian matrix of F. The pull-back extends to arbitrary differential forms by requiring that it commute with addition and the wedge product:

$$
\begin{aligned}
F^*(\Omega + \Theta) &= F^*(\Omega) + F^*(\Theta), \\
F^*(\Omega \wedge \Theta) &= F^*(\Omega) \wedge F^*(\Theta).
\end{aligned}
\tag{1.21}
$$

In particular, the pull-back of a smooth function $h: N \to \mathbb{R}$ (0-form) is given by composition, $F^*(h) = h \circ F$. In contrast to the behavior of vector fields under the differential, the pull-back of a differential k-form on N is *always* a well-defined smooth k-form on M. Note that the pull-back *reverses* the order of composition of maps: If $F: M \to N$ and $H: N \to P$, then $(H \circ F)^* = F^* \circ H^*$.

Exercise 1.34. Prove that the pull-back action of a smooth map $F: \mathbb{R}^m \to \mathbb{R}^m$ on the volume form $dx = dx^1 \wedge \cdots \wedge dx^m$ is given by $F^*(dx) = (\det J)\, dx$ where $J = \left(\partial F^i / \partial x^j\right)$ is the Jacobian matrix of F. Conclude that F is volume-preserving if and only if its Jacobian determinant has value ± 1.

The differential (1.2) of a smooth function has an important and natural extension to differential forms of higher degree. In general, the *differential* or *exterior derivative* of a k-form is the $(k+1)$-form defined in local coordinates by

$$
d\Omega = \sum_I dh_I(x) \wedge dx^I = \sum_{I,j} \frac{\partial h_I}{\partial x^j}\, dx^j \wedge dx^I, \tag{1.22}
$$

for Ω given by (1.19). For example, if $\omega = \sum h_i\, dx^i$ is a one-form, its differential is the two-form

$$
d\omega = \sum_{i=1}^m dh_i \wedge dx^i = \sum_{i<j} \left(\frac{\partial h_j}{\partial x^i} - \frac{\partial h_i}{\partial x^j} \right) dx^i \wedge dx^j. \tag{1.23}
$$

The differential satisfies the generalized ("super") derivational property

$$
d(\Omega \wedge \Theta) = d\Omega \wedge \Theta + (-1)^k\, \Omega \wedge d\Theta, \tag{1.24}
$$

whenever Ω is a k-form.

The fact that the differential is coordinate free is perhaps not so clear from the local coordinate definition (1.22). An alternative, intrinsic definition in the case of one-form is via the useful formula

$$
\langle d\omega\,; \mathbf{v}, \mathbf{w} \rangle = \mathbf{v}\langle \omega\,; \mathbf{w} \rangle - \mathbf{w}\langle \omega\,; \mathbf{v} \rangle - \langle \omega\,; [\mathbf{v}, \mathbf{w}] \rangle, \tag{1.25}
$$

valid for any pair of vector fields $\mathbf{v}, \mathbf{w}$. In (1.25), the first term on the right hand side is the result of applying the vector field $\mathbf{v}$ to the function defined by the evaluation of the one-form ω on the vector field $\mathbf{w}$, cf. (1.18). Formula (1.25) has the implication that, in a certain sense, the Lie bracket and the differential are dual operations — see Chapter 14 for details.

Exercise 1.35. Prove that formula (1.25) agrees with the local coordinate version (1.23). Find a generalization of (1.25) valid for arbitrary k-forms, cf. [**221**].

A *crucial* property of the differential is its invariance under smooth maps. This basic fact is of profound and fundamental importance in Cartan's approach to the solution of equivalence problems, to be discussed in the second half of the book. The proof is straightforward, using either local coordinates or the intrinsic approach.

Theorem 1.36. *Let $F: M \to N$ be a smooth map with pull-back F^*. If Ω is any differential form on N, then*

$$d[F^*\Omega] = F^*[d\Omega]. \tag{1.26}$$

Suppose $F: \widetilde{N} \to M$ parametrizes a submanifold $N \subset M$. We will call the pull-back $F^*\Omega$ of a differential form Ω on M the *restriction* of Ω to the submanifold N, and denote it by $\Omega \mid N$. Actually, we should call this the restriction of Ω to the parameter space $\widetilde{N}$, but the notation should not cause any confusion; besides, if N is a regular submanifold, we can, as remarked earlier, unambiguously identify N with $\widetilde{N}$.

Corollary 1.37. *Let $N \subset M$ be a submanifold. If Ω is any differential form which vanishes when restricted to N, so $\Omega \mid N = 0$, then so does its differential, $d(\Omega \mid N) = 0$.*

Finally, the anti-symmetry of the wedge product, coupled with the equality of mixed partial derivatives, implies that applying the differential twice in a row always produces zero; see, for instance, (1.23).

Theorem 1.38. *If Ω is any differential form, then*

$$d(d\Omega) = 0. \tag{1.27}$$

This final property provides the foundation of a remarkable, deep connection between the structure of the space of differential forms and the global topology of the underlying manifold. A differential form Ω is said to be *closed* if it has zero differential, $d\Omega = 0$. A k-form Ω is said to be *exact* if it is the differential of a $(k-1)$-form: $\Omega = d\Theta$. Theorem 1.38 implies that every exact form is closed, but the converse may not hold. Indeed, (1.27) implies that the differential defines a "complex", the *deRham complex* of the manifold M, whose cohomology, meaning, very roughly, the extent to which closed forms fail to be exact, depends, remarkably, only on the global topological character of M. For example, the Poincaré Lemma states that convex subdomains of Euclidean space have trivial cohomology.

Theorem 1.39. *Let $k > 0$. If $M \subset \mathbb{R}^m$ is a convex open subset, and Ω is any closed k-form defined on all of M, then Ω is exact, so that there exists a $(k-1)$ form Θ on M such that $\Omega = d\Theta$. A closed 0-form, i.e., a function satisfying $df = 0$, is constant.*

Thus, on a general manifold M, every closed form is locally, but perhaps not globally, exact. As our subsequent considerations are primarily local, we will not pursue this fascinating aspect of the theory of differential forms any further, but refer the interested reader to [**24**].

Example 1.40. Let $M = \mathbb{R}^2 \setminus \{0\}$. An easy calculation shows that the one-form $\omega = (x^2 + y^2)^{-1}[y\,dx - x\,dy]$ is closed, so $d\omega = 0$. However, there is *no* globally defined smooth function $f(x,y)$ satisfying $df = \omega$, so that ω is not exact. Indeed, locally $\omega = d\theta$ is the differential of the polar angle $\theta = \tan^{-1}(y/x)$, but, of course, θ is not a globally defined single-valued function on M. It is not hard to show that ω is essentially the unique closed one-form with this property; any closed one-form η can be written as $\eta = c\omega + df$ for some constant c and some smooth, globally defined function f. This result reflects the fact that M is a topological surface with a single hole.

Equivalence of Differential Forms

At first glance, because they are essentially "dual objects", one might expect vector fields and one-forms to behave very similarly under coordinate changes. However, we have already seen that they have very different transformational properties under smooth maps: differential forms pull back to differential forms, whereas vector fields are not necessarily mapped forward to other vector fields. Another surprise comes in the solution to the two equivalence problems. Away from singularities, there is just one canonical form for any vector field. However, the "dual problem" of finding (local) canonical forms for (nonsingular) differential one-forms is of a completely different character, as we now describe.

We begin by noting that any one-form ω on an m-dimensional manifold M defines, using the differential and wedge product, a sequence of auxiliary differential forms:

$$\omega_1 = \omega, \qquad \omega_2 = d\omega, \qquad \omega_3 = \omega \wedge d\omega,$$
$$\omega_4 = d\omega_3 = d\omega \wedge d\omega, \qquad \omega_5 = \omega \wedge d\omega \wedge d\omega, \qquad \text{etc.}$$

Note that each ω_k is a k-form, so $\omega_{m+1} = 0$ automatically. Clearly, in view of the invariance of wedge products and differentials under general coordinate changes, cf. (1.24), (1.26), if ω is mapped to $\overline{\omega}$ under a change of variables, so is ω_k mapped to $\overline{\omega}_k$. In particular, if $\omega_k = 0$, then $\overline{\omega}_k = 0$ also. Consequently, these auxiliary forms provide important restrictions that any equivalent pair of one-forms must satisfy.

Definition 1.41. The *rank* of a one-form ω at a point x is the integer $0 \leq r = r(x) \leq m$ such that $\omega_k|_x \neq 0$ for $k \leq r$, whereas $\omega_k|_x = 0$ for all $k > r$.

Note that the rank of a one-form is not always well defined, since if k is odd, it is possible that $\omega_k|_x = 0$ while $\omega_{k+1}|_x \neq 0$. However, if $\omega_k = 0$ in a neighborhood of x, then $\omega_{k+1} = d\omega_k = 0$ also. Thus points at which the rank is not defined form singularities of the one-form, and will not be discussed here. Clearly, by the preceding remarks, the rank of a one-form is invariant under general coordinate changes, and so forms of different ranks cannot be mapped to each other by a change of variables. A one-form is *regular* if its rank is constant. The following theorem is due to Darboux, [**58**], who extended the original results of Pfaff, [**194**]; it provides a complete list of canonical forms for regular one-forms.

Theorem 1.42. *Suppose ω is a one-form of constant rank r on an m-dimensional manifold M. Then there exist local coordinates $x = (x^1, \ldots, x^m)$ such that ω has the canonical form*

$$\omega = \begin{cases} x^1\, dx^2 + \cdots + x^{2s-1}\, dx^{2s}, & r = 2s, \\ x^1\, dx^2 + \cdots + x^{2s-1}\, dx^{2s} + dx^{2s+1}, & r = 2s+1. \end{cases} \tag{1.28}$$

depending on whether the rank is even or odd.

Thus, every regular one-form of the same rank is locally equivalent. See also [**207**; §III.6] for a direct inductive proof of Theorem 1.42. In Chapter 11 we will present an alternative proof, based on the Cartan equivalence method which, however, only applies to analytic one-forms.

Warning: The usual Darboux canonical form (1.28) has rank r provided $x^{2j-1} \neq 0$, $j = 1, \ldots, s$. Thus, to avoid singularities, we should appropriately restrict its domain of definition; alternatively, we can replace x^i by e^{x^i} for i odd.

Exercise 1.43. A function f is known as an *integrating factor* of a one-form ω if the product $f\omega$ is an exact one-form. Find (in the regular case) necessary and sufficient conditions that a given one-form admit an integrating factor.

As a corollary of Theorem 1.42, we obtain the well-known Hamiltonian version of Darboux' Theorem, which describes canonical forms for closed two-forms. First, define the *rank* of a two-form Ω at x to be the smallest integer $0 \leq r \leq \frac{1}{2}m$ such that

$$\overbrace{\Omega \wedge \cdots \wedge \Omega}^{r \text{ times}}\,|_x \neq 0, \qquad \text{whereas} \qquad \overbrace{\Omega \wedge \cdots \wedge \Omega}^{r+1 \text{ times}}\,|_x = 0. \tag{1.29}$$

Again, the rank is the only local invariant of a regular two-form.

Theorem 1.44. *If Ω is a closed two-form, $d\Omega = 0$, such that* $\operatorname{rank}\Omega = r$ *is constant, then there exist coordinates $x = (x^1, \ldots, x^m)$ such that*

$$\Omega = dx^1 \wedge dx^2 + \cdots + dx^{2r-1} \wedge dx^{2r}. \tag{1.30}$$

In particular, if M has even dimension, $m = 2n$, a *symplectic* two-form is defined to be a closed two-form of maximal rank n. Such forms play a fundamental role in Hamiltonian mechanics, [**12**], [**186**]. Theorem 1.44 implies that any symplectic form can, locally, be put into the canonical Darboux form $dp^1 \wedge dq^1 + \cdots + dp^n \wedge dq^n$, using suitable local coordinates $(p, q) = (p^1, \ldots, p^n, q^1, \ldots, q^n)$ on M. The canonical p's and q's are identified with the momentum and position variables of classical mechanics, [**227**], and lie at the foundation of the passage from classical to quantum mechanics, cf. [**224**]. The corresponding equivalence problem for nonclosed two-forms will be discussed in Chapter 11. For results on canonical forms at singularities of differential forms, the reader should consult [**166**], [**232**], [**233**].

Chapter 2

Lie Groups

The symmetry groups that arise most often in the applications to geometry and differential equations are Lie groups of transformations acting on a finite-dimensional manifold. Since Lie groups will be one of the cornerstones of our investigations, it is essential that we gain a basic familiarity with these fundamental mathematical objects. The present chapter is devoted to a survey of a number of fundamental facts concerning Lie groups and Lie algebras, and their actions on manifolds. More detailed presentations can be found in a variety of references, including [**186**], [**195**], [**221**].

Recall first that a *group* is a set G that has an associative (but not necessarily commutative) multiplication operation, denoted $g \cdot h$ for group elements $g, h \in G$. The group must also contain an (necessarily unique) identity element, denoted e, and each group element g has an inverse g^{-1} satisfying $g \cdot g^{-1} = g^{-1} \cdot g = e$.

The continuous nature of Lie groups is formalized by the requirement that, in addition to satisfying the basic group axioms, they are also endowed with the stucture of a smooth manifold.

Definition 2.1. A *Lie group* G is a smooth manifold which is also a group, such that the group multiplication $(g, h) \mapsto g \cdot h$ and inversion $g \mapsto g^{-1}$ define smooth maps.

Analytic Lie groups are defined by analytic manifolds, with analytic multiplication and inversion maps. Most of our examples are, in fact, analytic; indeed, any smooth Lie group can be endowed with an analytic structure, [**195**]. Often, an r-dimensional Lie group is referred to as an r parameter group, the "group parameters" referring to a choice of local coordinates on the group manifold.

Example 2.2. The simplest example of an r parameter Lie group is the abelian (meaning commutative) Lie group $\mathbb{R}^r$. The group operation is given by vector addition. The identity element is the zero vector, and the inverse of a vector x is the vector $-x$.

Example 2.3. The prototypical example of a real Lie group is the general linear group $\mathrm{GL}(n, \mathbb{R})$ consisting of all invertible $n \times n$ real matrices, with matrix multiplication defining the group multiplication, and matrix inversion defining the inverse. Equivalently, $\mathrm{GL}(n, \mathbb{R})$ can be regarded as the group of all invertible linear transformations on $\mathbb{R}^n$, where composition serves to define the group operation. Note that $\mathrm{GL}(n, \mathbb{R})$ is an n^2-dimensional manifold, simply because it is an open subset (namely, where the determinant is nonzero) of the space of all $n \times n$ matrices, which is itself isomorphic to $\mathbb{R}^{n^2}$. The group operations are clearly analytic. Similarly, the prototypical complex Lie group is the group $\mathrm{GL}(n, \mathbb{C})$ of invertible $n \times n$ complex matrices. We will often employ the notation $\mathrm{GL}(n)$ to mean either the real or complex general linear group — in such usage, the precise version will either be irrelevant or clear from the context.

A subset $H \subset G$ of a group is a *subgroup* provided the multiplication and inversion operations restrict to it. The subgroup H is called a *Lie subgroup* if it is also an (immersed) submanifold parametrized by a smooth group homomorphism $F: \widetilde{H} \to H \subset G$. The term "homomorphism" means that F respects the group operations: $F(g \cdot h) = F(g) \cdot F(h)$, $F(e) = e$, $F(g^{-1}) = F(g)^{-1}$, so the parameter space $\widetilde{H}$ is a Lie group isomorphic to H. Most (but not all!) Lie groups can be realized as Lie subgroups of the general linear group $\mathrm{GL}(n)$; these are the so-called "matrix Lie groups". The following result is useful for analyzing matrix (and other) subgroups; see [**221**; Theorem 3.42] for a proof.

Proposition 2.4. *If $H \subset G$ is a subgroup of a Lie group G, which is also a (topologically) closed subset, then H is a Lie subgroup of G.*

In particular, if G is a Lie group, then the connected component of G containing the identity element, denoted G^+, is itself a Lie group. For example, the real general linear group consists of two disconnected components, indexed by the sign of the determinant. The subgroup $\mathrm{GL}(n, \mathbb{R})^+ = \{A \in \mathrm{GL}(n, \mathbb{R}) \mid \det A > 0\}$, consisting of orientation-preserving linear transformations, is the connected component containing the identity matrix. On the other hand, the complex general linear group $\mathrm{GL}(n, \mathbb{C})$ is connected. The other connected components of a general Lie group are recovered by multiplying the connected component G^+ by a single group element lying therein. For example, every orientation-reversing matrix in $\mathrm{GL}(n, \mathbb{R})$ can be obtained by multiplying an orientation-preserving matrix by a fixed matrix with negative determinant, e.g., the diagonal matrix with entries $-1, +1, \ldots, +1$.

Example 2.5. The most important Lie groups are the three families of "classical groups". The *special linear* or *unimodular group* is $\mathrm{SL}(n) = \{A \in \mathrm{GL}(n) \mid \det A = 1\}$ consisting of all volume-preserving linear transformations. In other words, $\mathrm{SL}(n)$ is the group of linear symmetries of the standard volume form $dx = dx^1 \wedge \cdots \wedge dx^n$. Both the real and complex versions are connected, and have dimension $n^2 - 1$. The *orthogonal group* $\mathrm{O}(n) = \{A \in \mathrm{GL}(n) \mid A^T A = 1\!\!1\}$ is the group of norm-preserving linear transformations — rotations and reflections — and forms the group of linear symmetries of the Euclidean metric $ds^2 = (dx^1)^2 + \cdots + (dx^n)^2$ on $\mathbb{R}^n$. The component containing the identity forms the *special orthogonal group* $\mathrm{SO}(n) = \mathrm{O}(n) \cap \mathrm{SL}(n)$, consisting of just the rotations. As we shall see, $\mathrm{O}(n)$ and $\mathrm{SO}(n)$ have dimension $\frac{1}{2}n(n-1)$. The *symplectic group* is the $r(2r+1)$-dimensional Lie group

$$\mathrm{Sp}(2r) = \left\{ A \in \mathrm{GL}(2r, \mathbb{R}) \;\middle|\; A^T \mathrm{J} A = \mathrm{J} \right\}, \tag{2.1}$$

consisting of linear transformations which preserve the canonical nonsingular skew-symmetric matrix $\mathrm{J} = \begin{pmatrix} 0 & -1\!\!1 \\ 1\!\!1 & 0 \end{pmatrix}$. In particular, $\mathrm{Sp}(2) \simeq \mathrm{SL}(2)$. We remark that a linear transformation $L\colon x \mapsto Ax$ lies in $\mathrm{Sp}(2r)$ if and only if it is a canonical transformation, in the sense of Hamiltonian mechanics, hence $\mathrm{Sp}(2r)$ forms the group of linear symmetries of the canonical symplectic form $\Omega = dx^1 \wedge dx^2 + \cdots + dx^{2r-1} \wedge dx^{2r}$.

Exercise 2.6. Let G be a group. If $H \subset G$ is a subgroup, then the *quotient space* G/H is defined as the set of all left cosets $g \cdot H = \{g \cdot h \mid h \in H\}$, for $g \in G$. Show that if H is a *normal* subgroup, meaning that it equals its conjugate subgroups: $gHg^{-1} = H$ for all $g \in G$, then G/H can be given the structure of a group. If H is an s-dimensional closed subgroup of an r-dimensional Lie group G, then G/H can be endowed with the structure of a smooth manifold of dimension $r - s$, which is a Lie group provided H is a normal closed subgroup.

At the other extreme, a *discrete subgroup* $\Gamma \subset G$ of a Lie group is a subgroup whose intersection with some neighborhood $\{e\} \subset U \subset G$ of the identity element consists only of the identity: $\Gamma \cap U = \{e\}$. Examples include the integer lattices $\mathbb{Z}^r \subset \mathbb{R}^r$, and the group $\mathrm{SL}(n, \mathbb{Z})$ of integer matrices of determinant 1. Although discrete groups can be regarded as zero-dimensional Lie groups, they are totally disconnected, and so cannot be handled by any of the wonderful tools associated with connected Lie groups. The quotient group G/Γ by a discrete normal subgroup, then, is a Lie group which is locally isomorphic to G itself; see Theorem 2.56 below for more details.

Transformation Groups

In most instances, groups are not given to us in the abstract, but, rather, concretely as a family of transformations acting on a space. In the case of Lie groups, the most natural setting is as groups of transformations acting smoothly on a manifold.

Definition 2.7. A *transformation group* acting on a smooth manifold M is determined by a Lie group G and smooth map $\Phi\colon G \times M \to M$, denoted by $\Phi(g, x) = g \cdot x$, which satisfies

$$e \cdot x = x, \qquad g \cdot (h \cdot x) = (g \cdot h) \cdot x, \qquad \text{for all } x \in M,\ g \in G. \tag{2.2}$$

Condition (2.2) implies that the inverse group element g^{-1} determines the inverse to the transformation defined by the group element g, so that each group element g induces a diffeomorphism from M to itself. Definition 2.7 assumes that the group action is *global*, meaning that $g \cdot x$ is defined for every $g \in G$ and every $x \in M$. In applications, though, a group action may only be defined "locally", meaning that, for a given $x \in M$, the transformation $g \cdot x$ is only defined for group elements g sufficiently near the identity. Thus, for a *local transformation group*, the map Φ is defined on an open subset $\{e\} \times M \subset \mathcal{V} \subset G \times M$, and the conditions (2.2) are imposed wherever they make sense.

Example 2.8. An obvious example is provided by the usual linear action of the general linear group $\mathrm{GL}(n, \mathbb{R})$, acting by matrix multiplication on column vectors $x \in \mathbb{R}^n$. This action clearly induces a linear action of any subgroup of $\mathrm{GL}(n, \mathbb{R})$ on $\mathbb{R}^n$. (The classification and analysis of linear group actions is known as representation theory and will be discussed in more detail in the following chapter.) Since linear transformations map lines to lines, there is an induced action of $\mathrm{GL}(n,\mathbb{R})$, and its subgroups, on the projective space $\mathbb{RP}^{n-1}$. Of particular importance is the planar case, $n = 2$, so we discuss this in some detail. The linear action of $\mathrm{GL}(2,\mathbb{R})$ on $\mathbb{R}^2$ is

$$(x, y) \longmapsto (\alpha x + \beta y, \gamma x + \delta y), \qquad A = \begin{pmatrix} \alpha & \beta \\ \gamma & \delta \end{pmatrix} \in \mathrm{GL}(2). \tag{2.3}$$

Now, as in Example 1.3, we can identify the projective line $\mathbb{RP}^1$ with a circle S^1. If we use the projective coordinate $p = x/y$, the induced action is given by the *linear fractional* or *Möbius transformations*

$$p \longmapsto \frac{\alpha p + \beta}{\gamma p + \delta}, \qquad A = \begin{pmatrix} \alpha & \beta \\ \gamma & \delta \end{pmatrix} \in \mathrm{GL}(2). \tag{2.4}$$

In this coordinate chart, the x–axis in $\mathbb{R}^2$ is identified with the point $p = \infty$ in $\mathbb{RP}^1$, and the linear fractional transformations (2.4) have a well-defined extension to include the point at infinity. Alternatively, we can regard (2.4) as defining a local action of $\mathrm{GL}(2,\mathbb{R})$ on the real line $\mathbb{R}$, defined on the subset $\mathcal{V} = \{(A, p) \mid \gamma p + \delta \neq 0\} \subset \mathrm{GL}(2,\mathbb{R}) \times \mathbb{R}$.

Similarly, the complex general linear group $\mathrm{GL}(n,\mathbb{C})$ acts linearly on $\mathbb{C}^n$, and there is an induced action on the complex projective space $\mathbb{CP}^{n-1}$. In particular, the action (2.3) of $\mathrm{GL}(2,\mathbb{C})$ on $\mathbb{C}^2$ induces an action on complex projective space $\mathbb{CP}^1 \simeq S^2$, given by complex linear fractional transformations (2.4). As in the real case, it restricts to define a local action of $\mathrm{GL}(2)$ on the complex plane $\mathbb{C}$.

Example 2.9. An important, but almost tautological example, is provided by the action of a Lie group on itself by multiplication. In this case, the manifold M coincides with G itself, and the map $\Phi: G \times G \mapsto G$ is given by left multiplication: $\Phi(g, h) = g \cdot h$. Alternatively, we can let G act on itself by right multiplication via $\widetilde{\Phi}(g, h) = h \cdot g^{-1}$; in this case, the inverse ensures that the composition laws (2.2) remain valid.

Example 2.10. Let $\mathbf{v}$ be a vector field on a manifold M. The properties (1.6) imply that the flow $\exp(t\mathbf{v})$ defines a (local) action of the one-parameter group $\mathbb{R}$, parametrized by the "time" t, on the manifold M. For example, if $M = \mathbb{R}$, with coordinate x, then the vector fields ∂_x, $x\partial_x$ and $x^2\partial_x$ generate flows, given explicitly in Example 1.24, which form one-parameter subgroups of the projective group (2.4) (identifying p with x). In fact, these three particular subgroups — translations, scalings, and inversions — serve to generate the full projective group, a fact that will become clear from the Lie algebra methods discussed below.

Example 2.11. The (real) *affine group* $\mathrm{A}(n)$ is defined as the group of affine transformations $x \mapsto Ax + a$ in $\mathbb{R}^n$. Thus, the affine group is parametrized by a pair (A, a) consisting of an invertible matrix $A \in \mathrm{GL}(n)$ and a vector $a \in \mathbb{R}^n$. The affine group $\mathrm{A}(n)$ has dimension $n(n+1)$, being isomorphic, as a manifold, to the Cartesian product space $\mathrm{GL}(n) \times \mathbb{R}^n$. However, $\mathrm{A}(n)$ is *not* the Cartesian product of the groups $\mathrm{GL}(n)$ and $\mathbb{R}^n$ since the group multiplication law is given by $(A, a) \cdot (B, b) = (AB, a + Ab)$. The affine group can be realized as a subgroup of $\mathrm{GL}(n+1)$ by identifying the group element (A, a) with the $(n+1) \times (n+1)$ matrix $\left(\begin{smallmatrix} A & a \\ 0 & 1 \end{smallmatrix}\right)$.

The affine group provides an example of a general construction known as the "semi-direct product". In general, if G and H are, respectively, r- and s-dimensional Lie groups, their Cartesian product $G \times H$ is an $(r+s)$-dimensional Lie group with the group operation being defined in the obvious manner: $(g, h) \cdot (\widetilde{g}, \widetilde{h}) = (g \cdot \widetilde{g}, h \cdot \widetilde{h})$. If, in addition, G acts as a group of transformations on the Lie group H, satisfying $g \cdot (h \cdot \widetilde{h}) = (g \cdot h) \cdot (g \cdot \widetilde{h})$, then the *semi-direct product* $G \ltimes H$ is the $(r+s)$-dimensional Lie group which, as a manifold just looks like the Cartesian product $G \times H$, but whose multiplication is given by $(g, h) \cdot (\widetilde{g}, \widetilde{h}) = (g \cdot \widetilde{g}, h \cdot (g \cdot \widetilde{h}))$. In particular, the affine group is the semi-direct product of the general linear group $\mathrm{GL}(n)$ acting on the abelian group $\mathbb{R}^n$, written $\mathrm{A}(n) = \mathrm{GL}(n) \ltimes \mathbb{R}^n$. Another important example is provided by the *Euclidean group* $\mathrm{E}(n) = \mathrm{O}(n) \ltimes \mathbb{R}^n$, which is generated by the groups of orthogonal transformations and translations, and thus forms a subgroup of the full affine group. Its connected component is $\mathrm{SE}(n) = \mathrm{SO}(n) \ltimes \mathbb{R}^n$. The Euclidean group has as alternative characterization as the group of isometries, meaning norm-preserving transformations, of Euclidean space, and thus, according to Klein's characterization of geometry based on groups, lies at the foundation of Euclidean geometry. We also define the special affine subgroup $\mathrm{SA}(n) = \mathrm{SL}(n) \ltimes \mathbb{R}^n$, which consists of volume-preserving affine transformations, and forms the basis of affine geometry, cf. [**106**].

Given an action of the Lie group G on a manifold M, the *isotropy* subgroup of a point $x \in M$ is $G_x = \{g \mid g \cdot x = x\} \subset G$ consisting of all group elements g which fix x. Proposition 2.4 demonstrates that G_x is a Lie subgroup, and can be viewed as the symmetry subgroup for the point x. For example, the isotropy subgroup of the projective group (2.4) fixing the origin $p = 0$ is the group of invertible lower triangular matrices. In particular, $G_x = G$ if and only if x is a *fixed point* of the group action. If $g \cdot x = y$, then the isotropy group at y is conjugate to that at x, meaning $G_y = g \cdot G_x \cdot g^{-1}$. A transformation group acts *freely* if the isotropy subgroups are all trivial, $G_x = \{e\}$ for all $x \in M$, which means that, for $e \neq g \in G$, we have $g \cdot x \neq x$ for any $x \in M$. The action is *locally free* if this holds for all $g \neq e$ in a neighborhood of the identity; equivalently, the isotropy subgroups are discrete subgroups of G. A transformation group acts *effectively* if different group elements have different actions, so that $g \cdot x = h \cdot x$ for all $x \in M$ if and only if $g = h$; this is equivalent to the statement that the only group element acting as the identity transformation is the identity element of G. The effectiveness of a group action is measured by its *global isotropy subgroup* $G_M = \bigcap_{x \in M} G_x = \{g \mid g \cdot x = x \text{ for all } x \in M\}$, so that G acts effectively if and only if $G_M = \{e\}$. Clearly, a free group action is effective, although the converse is certainly not true. Slightly more generally, a Lie group G is said to act *locally effectively* if G_M is a discrete subgroup of G, which is equivalent to the existence of a neighborhood U of the identity e such that $G_M \cap U = \{e\}$.

Proposition 2.12. *Suppose G is a transformation group acting on a manifold M. Then the global isotropy subgroup G_M is a normal Lie subgroup of G. Moreover, there is a well-defined effective action of the quotient group G/G_M on M, which "coincides" with that of G in the sense that two group elements g and $\widetilde{g}$ have the same action on M, so $g \cdot x = \widetilde{g} \cdot x$ for all $x \in M$, if and only if they have the same image in $\widehat{G}$, so $\widetilde{g} = g \cdot h$ for some $h \in G_M$.*

Thus, if a transformation group G does not act effectively, we can, without any significant loss of information or generality, replace it by the quotient group G/G_M, which does act effectively, and in the same manner as G does. For a locally effective action, the quotient group G/G_M is a Lie group having the same dimension, and the same local structure, as G itself. A group acts *effectively freely* if and only if G/G_M acts freely; this is equivalent to the statement that every local isotropy subgroup equals the global isotropy subgroup: $G_x = G_M$, $x \in M$.

Example 2.13. The general linear group $\mathrm{GL}(n,\mathbb{R})$ acts effectively on $\mathbb{R}^n$. The isotropy group of a nonzero point $0 \neq x \in \mathbb{R}^n$ can be identified with the affine group $\mathrm{A}(n-1)$ of Example 2.11; the isotropy group of 0 is all of $\mathrm{GL}(n,\mathbb{R})$. The induced action on projective space $\mathbb{RP}^{n-1}$ is no longer effective since the diagonal matrices $\lambda \mathbb{1}$ act trivially on lines through the origin. The quotient subgroup $\mathrm{PSL}(n,\mathbb{R}) = \mathrm{GL}(n,\mathbb{R})/\{\lambda \mathbb{1}\}$ is called the *projective group*. Note that if n is odd, we can identify $\mathrm{PSL}(n,\mathbb{R}) \simeq \mathrm{SL}(n,\mathbb{R})$ with the special linear group. However, if n is even, $\mathrm{SL}(n,\mathbb{R})$ only acts locally effectively on $\mathbb{RP}^{n-1}$ since $-\mathbb{1} \in \mathrm{SL}(n,\mathbb{R})$, and hence $\mathrm{PSL}(n,\mathbb{R}) = \mathrm{SL}(n,\mathbb{R})/\{\pm \mathbb{1}\}$.

Exercise 2.14. Discuss the corresponding action of $\mathrm{GL}(n,\mathbb{C})$ on complex projective space $\mathbb{CP}^{n-1}$.

Example 2.15. Consider the action of $\mathrm{GL}(n,\mathbb{R})$ on the space of all real $n \times n$ matrices given by $X \mapsto AXA^T$ for $A \in \mathrm{GL}(n)$. The orthogonal Lie group is the isotropy group for the identity matrix $\mathbb{1}$. Similarly, if $n = 2r$ is even, the isotropy subgroup of the canonical nondegenerate skew-symmetric matrix J given in (2.1) is the symplectic group $\mathrm{Sp}(2r)$. The isotropy group of the diagonal matrix with p entries equaling $+1$ and $m-p$ entries equaling -1 is the pseudo-orthogonal group $\mathrm{O}(p, m-p)$; for example $\mathrm{O}(1,1)$ is the one-dimensional group of hyperbolic rotations $\begin{pmatrix} \cosh t & \sinh t \\ \sinh t & \cosh t \end{pmatrix}$.

Invariant Subsets and Equations

Let G be a (local) group of transformations acting on the manifold M. A subset $S \subset M$ is called *G–invariant* if it is unchanged by the group transformations, meaning $g \cdot x \in S$ whenever $g \in G$ and $x \in S$ (provided $g \cdot x$ is defined if the action is only local). The most important classes of invariant subsets are the varieties defined by the vanishing of one or more functions. A group G is called a *symmetry group* of a system of equations

$$F_1(x) = \cdots = F_k(x) = 0, \tag{2.5}$$

if and only if the variety $\mathcal{S}_F = \{x \mid F_1(x) = \cdots = F_k(x) = 0\}$ is a G-invariant subset of M. Thus, a symmetry group of a system of equations maps solutions to other solutions: If $x \in M$ satisfies (2.5) and $g \in G$ is any group element such that $g \cdot x$ is defined, then the transformed point $g \cdot x$ is also a solution to the system. Knowledge of a symmetry group of a system of equations allows us to construct new solutions from old ones, a fact that will be particularly useful when we apply these methods to systems of differential equations.

An *orbit* of a transformation group is a minimal (nonempty) invariant subset. For a global action, the orbit through a point $x \in M$ is just the set of all images of x under arbitrary group transformations: $\mathcal{O}_x = \{g \cdot x \mid g \in G\}$. More generally, for a local group action, the orbit is the set of all images of x under arbitrary finite sequences of group transformations: $\mathcal{O}_x = \{g_1 \cdot g_2 \cdots g_n \cdot x \mid g_i \in G, n \geq 0\}$. If G is connected, its orbits are connected. Clearly, a subset $S \subset M$ is G-invariant if and only if it is the union of orbits. The group action is called *transitive* if there is only one orbit, so (assuming the group acts globally), for every $x, y \in M$ there exists at least one $g \in G$ such that $g \cdot x = y$. At the other extreme, a fixed point is a zero-dimensional orbit; for connected group actions, the converse holds: Any zero-dimensional orbit is a fixed point.

Example 2.16. For the usual linear action of $\mathrm{GL}(n)$ on $\mathbb{R}^n$, there are two orbits: The origin $\{0\}$ and the remainder $\mathbb{R}^n \setminus \{0\}$. The same holds for $\mathrm{SL}(n)$ since we can still map any nonzero vector in $\mathbb{R}^n$ to any other nonzero vector by a matrix of determinant 1. The orthogonal group $\mathrm{O}(n)$ is a bit different: The orbits are spheres $\{|x| = \text{constant}\}$ (and the origin), and any other invariant subset is the union of spheres. The induced projective actions of each of these three groups on $\mathbb{RP}^{n-1}$ are all transitive.

A particularly important class of transitive group actions is provided by the *homogeneous spaces*, defined as the quotient space G/H of a Lie group G by a closed subgroup; see Exercise 2.6. The left multiplication of G induces a corresponding globally defined, transitive action of the group G on the homogeneous space G/H. For example, the left multiplication action of the three-dimensional rotation group $\mathrm{SO}(3)$ on itself induces the standard action on the two-dimensional sphere $S^2 = \mathrm{SO}(3)/\mathrm{SO}(2)$ realized as the quotient space by any of the (non-normal) two-dimensional rotation subgroups. In fact, every global transitive group action can be identified with a homogeneous space.

Theorem 2.17. *A Lie group G acts globally and transitively on a manifold M if and only if $M \simeq G/H$ is isomorphic to the homogeneous space obtained by quotienting G by the isotropy subgroup $H = G_x$ of any designated point $x \in M$.*

Remark: In [**172**], Mostow proves that every local, transitive group action on a manifold M of dimension $m \leq 4$ is locally isomorphic to a homogeneous space G/H, where $\dim G - \dim H = m$. However, it is not true that M itself is globally isomorphic to an open subset of a homogeneous space — see Example 14.29. Mostow shows that even the local result is not true if $m \geq 5$. See [**193**] for further results on the globalization of local group actions.

Example 2.18. Consider the action of $\mathrm{GL}(n)$ on $M = \mathbb{R}^n \setminus \{0\}$. The isotropy group of the point $e_1 = (1, 0, \ldots, 0)$ is the subgroup of matrices whose first column equals e_1, which can be identified with the affine group $\mathrm{A}(n-1)$, as in Example 2.11. Theorem 2.17 implies that we can identify $\mathbb{R}^n \setminus \{0\}$ with the homogeneous space $\mathrm{GL}(n)/\mathrm{A}(n-1)$.

Exercise 2.19. Determine the homogeneous space structures of the sphere S^{n-1}, and of the projective space $\mathbb{RP}^{n-1}$, induced by the transitive group actions of $\mathrm{GL}(n)$, $\mathrm{SL}(n)$, and $\mathrm{SO}(n)$.

In general, the orbits of a Lie group of transformations are all submanifolds of the manifold M. A group action is called *semi-regular* if all its orbits have the same dimension. The action is called *regular* if, in addition, each point $x \in M$ has arbitrarily small neighborhoods whose intersection with each orbit is a connected subset thereof. The condition that each orbit be a regular submanifold is necessary, but not sufficient, for the regularity of the group action.

Example 2.20. Let $T = S^1 \times S^1$ be the two-dimensional torus, with angular coordinates (θ, φ), $0 \leq \theta, \varphi < 2\pi$. Consider the one-parameter group action $(\theta, \varphi) \mapsto (\theta + t, \varphi + \kappa t) \bmod 2\pi$, $t \in \mathbb{R}$, where $0 \neq \kappa \in \mathbb{R}$. If κ/π is a rational number, then the orbits of this action are closed curves, diffeomorphic to the circle S^1, and the action is regular. On the other hand, if κ/π is irrational, then the orbits never close, and, in fact, each orbit is a dense subset of the torus. The action in the latter case is semi-regular, but not regular.

Example 2.21. Consider the one-parameter group action,

$$(r, \theta) \mapsto \left(\frac{re^t}{1 + r(e^t - 1)}, \theta + t \right),$$

defined in terms of polar coordinates, on the punctured plane $M = \mathbb{R}^2 \setminus \{0\}$. The orbits of this group action are all regular one-dimensional submanifolds of M — they consist of the unit circle $r = 1$ and two families of spirals. However, the group action is *not* regular — indeed, all the spiral orbits intersect any small neighborhood of each point on the unit circle in infinitely many disconnected components.

Proposition 2.22. *An r-dimensional Lie group G acts locally freely on a manifold M if and only if its orbits have the same dimension r as G itself. The group acts effectively freely if and only if its orbits have dimension $s = \dim G - \dim G_M$.*

The orbits of regular and semi-regular group actions have a particularly simple local canonical form, generalizing the rectifying coordinates for a one-parameter group of transformations in Theorem 1.23.

Theorem 2.23. *Let G be a Lie group acting regularly on a manifold M with s-dimensional orbits. Then, near every point of M, there exist* rectifying *local coordinates $(y, z) = (y^1, \ldots, y^s, z^1, \ldots, z^{m-s})$ having the property that any orbit intersects the coordinate chart in at most one slice $N_c = \{z^1 = c_1, \ldots, z^{m-s} = c_{m-s}\}$, for constants $c = (c_1, \ldots, c_{m-s})$. If the action is semi-regular, the same statement holds except that an orbit may intersect the chart in more than one such slice.*

This fundamental result is a direct consequence of Frobenius' Theorem, to be proved in Chapter 14. Theorem 2.23 demonstrates that the orbits form a *foliation* of the underlying manifold. In practice, the rectifying coordinates are most readily constructed using the infinitesimal methods to be discussed below.

Example 2.24. For the orthogonal group $O(3)$ acting on $\mathbb{R}^3 \setminus \{0\}$, the orbits are the two-dimensional spheres, and rectifying coordinates are given by spherical coordinates, with the angular coordinates θ, φ playing the role of y, and the radius r the role of z in Theorem 2.23.

Canonical Forms

Given a group of transformations acting on a manifold M, by a *canonical form* of an element $x \in M$ we just mean a distinguished, simple representative x_0 of the orbit containing x. Of course, there is no uniquely specified canonical form, and some choice, usually based on one's æsthetic judgment of "simplicity", must be exercised. For example, the well-known Jordan canonical form of a matrix corresponds to the conjugation action $X \mapsto AXA^{-1}$ of the complex general linear group

$\mathrm{GL}(n,\mathbb{C})$ on the space of $n \times n$ complex matrices X. However, other less commonly known canonical forms, including the older "rational canonical form", can also be advantageously utilized, [**216**]. If the canonical form is unique, then it provides a solution to the associated equivalence problem: Two objects are equivalent if and only if they have the same canonical form.

Example 2.25. As a simple example, the canonical form of a vector $x \in \mathbb{R}^n$ under the action of the rotation group $\mathrm{SO}(n)$ can be chosen to be a nonnegative multiple of the first unit basis vector, re_1, with $r = |x| \geq 0$. Clearly there is no particular reason to choose the first basis vector for the canonical form. As for the actions of $\mathrm{GL}(n)$ and $\mathrm{SL}(n)$, each vector has just two possible canonical forms — 0 or, say, e_1.

Example 2.26. According to Sylvester's Theorem of Inertia, the action $X \mapsto AXA^T$ of the general linear group $\mathrm{GL}(n,\mathbb{R})$ on the space of real symmetric $n \times n$ matrices has a discrete set of orbits, which are classified by the matrix' signature, meaning the number of positive, zero, and negative eigenvalues. Canonical forms are provided by the diagonal matrices with $0 \leq p \leq n$ entries of $+1$, followed by $0 \leq q \leq n-p$ entries of -1, followed by $n-p-q$ zeros on the diagonal. In particular, the orbit containing the identity matrix is the space of symmetric, positive definite matrices. If we restrict this action to the orthogonal subgroup $\mathrm{O}(n) \subset \mathrm{GL}(n)$, then we can still diagonalize any symmetric matrix X, leading to a canonical form $D = \mathrm{diag}(\lambda_1, \ldots \lambda_n)$. Note that this canonical form is not quite uniquely determined, since we can rearrange the order of the entries λ_i.

Example 2.27. The action $X \mapsto AXA^T$ of $\mathrm{GL}(n)$ on the space of *skew-symmetric* $n \times n$ matrices can be identified with its induced action on the space of two-forms $\bigwedge^2 \mathbb{R}^n$. The orbits are classified by the rank of the two-form, cf. (1.29), and any two-form of rank r has algebraic canonical form

$$\Omega = \sum_{j=1}^{r} e_j \wedge e_{j+r}, \tag{2.6}$$

in terms of a basis $\{e_1, \ldots, e_n\}$ of $\mathbb{R}^n$.

Exercise 2.28. Determine the isotropy group of the canonical two-form (2.6).

Invariant Functions

An invariant of a transformation group is defined as a real-valued function whose values are unaffected by the group transformations. The determination of a complete set of invariants of a given group action is a problem of supreme importance for the study of equivalence and canonical forms. In the regular case, the orbits, and hence the canonical forms, for a group action are completely characterized by its invariants. Consequently, a significant portion of the book will be devoted to characterizing, classifying, analyzing, and applying invariants of group actions. We begin with the general definition.

Definition 2.29. Let G be a transformation group acting on a manifold M. An *invariant* of G is a real-valued function $I: M \to \mathbb{R}$ which satisfies $I(g \cdot x) = I(x)$ for all transformations $g \in G$.

Proposition 2.30. *Let $I: M \to \mathbb{R}$. The following conditions are equivalent:*

(*i*) *I is a G-invariant function.*

(*ii*) *I is constant on the orbits of G.*

(*iii*) *The level sets $\{I(x) = c\}$ of I are G-invariant subsets of M.*

For example, in the case of the orthogonal group $\mathrm{O}(n)$ acting on $\mathbb{R}^n$, the orbits are spheres $r = |x| =$ constant, and hence any orthogonal invariant is a function of the radius: $I = F(r)$. If G acts transitively on the manifold M, then there are no nonconstant invariants. If G acts transitively on a dense subset $M_0 \subset M$, then the only *continuous* invariants are the constants. For instance, the only continuous global invariants of the irrational flow on the torus, cf. Example 2.20, are the constants, since every orbit is dense in this case. Similarly, the only continuous invariants of the standard action of $\mathrm{GL}(n, \mathbb{R})$ on $\mathbb{R}^n$ are the constant functions, since the group acts transitively on $\mathbb{R}^n \setminus \{0\}$.

Note that the canonical form x_0 of any element $x \in M$ must have the same invariants: $I(x_0) = I(x)$; this condition is also sufficient if there are enough invariants to distinguish the orbits, i.e., x and y lie in the same orbit if and only if $I(x) = I(y)$ for every invariant I, which, according to Theorem 2.34 below, is the case for regular group actions. However, singular orbits are often not completely distinguished by invariants alone, and more sophisticated algebraic features must be utilized.

Example 2.31. Consider the action $X \mapsto AXA^T$ of $\mathrm{GL}(n, \mathbb{R})$ on the space of symmetric matrices discussed in Example 2.26. Since the group orbits are discrete, the only invariant functions are the matrix signatures, which serve to characterize the canonical forms. Restricting to the orthogonal subgroup $O(n) \subset GL(n)$, we see that any symmetric function of the eigenvalues of the matrix provides an invariant; again these are sufficient to completely characterize its canonical diagonal form. (An alternative system of invariants is provided by the traces of the powers of the matrix: $\operatorname{tr} A^k$, $k = 1, \ldots, n$.) On the other hand, consider the conjugation action $X \mapsto AXA^{-1}$ of $GL(n, \mathbb{C})$ on the space of $n \times n$ matrices. Again, the symmetric functions of the eigenvalues provide invariants of the action, but, if the eigenvalues are repeated, these are not sufficient to distinguish the different canonical forms, and one must introduce additional discrete invariants to properly characterize the Jordan block structure.

A fundamental problem is the determination of *all* the invariants of a group of transformations. Note that if $I_1(x), \ldots, I_k(x)$ are invariants, and $H(y_1, \ldots, y_k)$ is any function, then $I(x) = H(I_1(x), \ldots, I_k(x))$ is also invariant. Therefore, we need only find a complete set of functionally independent invariants, cf. Definition 1.9, having the property that any other invariant can be written as a function of these fundamental invariants. In many cases, globally defined invariants are not so readily available or easy to find, and so we will often have to be content with the description of locally defined invariants.

Definition 2.32. Let G be a Lie group acting on a manifold M. A function $I: M \to \mathbb{R}$ defined on an open subset $U \subset M$ is called a *local invariant* of G if $I(g \cdot x) = I(x)$ for all $x \in U$ and all group transformations $g \in V_x$ in some neighborhood $V_x \subset G$ (possibly depending on x) of the identity element. If $I(g \cdot x) = I(x)$ for all $x \in U$ and all $g \in G$ such that $g \cdot x \in U$, then I is called a *global invariant* (even though it is only defined on an open subset of M).

Example 2.33. Consider the one-parameter group acting on the torus as discussed in Example 2.20. If κ is irrational, then, on the open subset $0 < \theta, \varphi < 2\pi$, the difference $\varphi - \kappa\theta$ is a local invariant of the group action that is clearly not globally invariant.

According to Theorem 2.23, the number of independent local invariants of a regular transformation group is completely determined by the orbit dimension. Indeed, in terms of the rectifying coordinates (y, z), the coordinate functions $z^1, \ldots, z^{m-s}$, and any function thereof, provide a complete set of local invariants for the group action. Thus, Theorem 2.23 implies the basic classification result for the invariants of regular group actions, which is destined to play a key role in many later developments.

Theorem 2.34. *Let G be a Lie group acting semi-regularly on the m-dimensional manifold M with s-dimensional orbits. At each $x \in M$, there exist $m-s$ functionally independent local invariants $I_1, \ldots, I_{m-s}$, defined on a neighborhood U of x, with the property that any other local invariant I defined on U can be written as a function of the fundamental invariants: $I = H(I_1, \ldots, I_{m-s})$. If G acts regularly, then we can choose the I_ν's to be global invariants on U. Moreover, in the regular case, two points $y, z \in U$ lie in the same orbit of G if and only if the invariants all have the same value, $I_\nu(y) = I_\nu(x)$, $\nu = 1, \ldots, m-s$.*

Theorem 2.34 provides a complete answer to the question of local invariants of group actions. Global considerations are more delicate. For example, consider the elementary one-parameter scaling group $(x, y) \mapsto (\lambda x, \lambda y)$, $\lambda \in \mathbb{R}^+$. Locally, away from the origin, the ratio x/y, or y/x, or any function thereof (e.g., $\theta = \tan^{-1}(y/x)$) provides the only independent invariant. However, if we include the origin, then there are no nonconstant continuous invariants. (A discontinuous invariant is provided by the function which is 1 at the origin and 0 elsewhere.) On the other hand, the scaling group $(x, y) \mapsto (\lambda x, \lambda^{-1} y)$ does have a global invariant: xy.

Example 2.35. Consider the projective action of $\mathrm{SL}(2, \mathbb{C})$ on the n-fold Cartesian product $\mathbb{CP}^1 \times \cdots \times \mathbb{CP}^1$

$$(p_1, \ldots, p_n) \mapsto \left(\frac{\alpha p_1 + \beta}{\gamma p_1 + \delta}, \ldots, \frac{\alpha p_n + \beta}{\gamma p_n + \delta} \right), \qquad \begin{pmatrix} \alpha & \beta \\ \gamma & \delta \end{pmatrix} \in \mathrm{SL}(2, \mathbb{C}).$$

Let us concentrate on the open subset $M = \{p_i \neq p_j, i \neq j\}$ consisting of distinct n-tuples of points. For $n \leq 3$, the action is transitive on M, and there are no nonconstant invariants. Indeed, we can map any three distinct points (p_1, p_2, p_3) on the Riemann sphere to any desired canonical form, e.g., $(0, 1, \infty)$, by a suitable linear fractional transformation. For $n = 4$, the *cross-ratio*

$$[\, p_1, p_2, p_3, p_4 \,] = \frac{(p_1 - p_2)(p_3 - p_4)}{(p_1 - p_3)(p_2 - p_4)}, \tag{2.7}$$

is invariant, as can be verified directly. (If one of the points is infinite, (2.7) is computed in a consistent way.) Now only the first three points can be fixed, so a canonical form could be $(0,1,\infty,z)$ where z can be any other point, whose value is fixed by the cross-ratio $1/(z-1)$. An alternative choice is $(s,-s,s^{-1},-s^{-1})$, which is unique if we restrict $s>1$; the cross-ratio is now $-4/(s-s^{-1})^2$. For $n>4$, it can be proved that every invariant can be written as a function of the cross-ratios of the points p_j, taken four at a time. According to Theorem 2.34, only $n-3$ of these cross-ratios are functionally independent, and we can clearly take $[\,p_1,p_2,p_3,p_k\,]$, $k=4,\ldots,n$, as our fundamental invariants.

Exercise 2.36. Determine how to express other cross-ratios, e.g., $[\,p_1,p_3,p_4,p_5\,]$, in terms of the fundamental cross-ratios.

Exercise 2.37. Discuss the action of $\mathrm{SL}(2,\mathbb{R})$ on $\mathbb{RP}^1\times\cdots\times\mathbb{RP}^1$, and on $\mathbb{CP}^1\times\cdots\times\mathbb{CP}^1$, viewed as a real manifold of dimension $2n$.

The cross-ratios are a special case of the general concept of a joint invariant. If G acts simultaneously on the manifolds $M_1,\ldots,M_n$, then by a *joint invariant* we mean an ordinary invariant $I(x_1,\ldots,x_n)$ of the Cartesian product group action of G on $M_1\times\cdots\times M_n$; in other words, $I(g\cdot x_1,\ldots,g\cdot x_n)=I(x_1,\ldots,x_n)$ for all $g\in G$. In most cases of interest, the manifolds $M_\kappa=M$ are identical, with identical actions of G.

The invariants of a regular group action can be used to completely characterize the invariant submanifolds. First, if G acts on M, and $I_1,\ldots,I_k$ are any invariants, then Proposition 2.30 implies that their common level set $\{I_1(x)=c_1,\ldots,I_k(x)=c_k\}$ is an invariant subset of M. Conversely, if the group action is regular, Theorem 2.23 implies that any invariant submanifold can be expressed (locally) as the vanishing set for some collection of invariant functions.

Theorem 2.38. *Let G act regularly on an m-dimensional manifold M. A regular n-dimensional submanifold $N\subset M$ is G-invariant if and only if at each point $x\in N$ there exists a neighborhood U and invariants $I_1,\ldots,I_{m-n}$ such that $N\cap U=\{I_1(x)=\cdots=I_{m-n}(x)=0\}$.*

Proof: Choose rectifying local coordinates (y,z) as in Theorem 2.23. Since any invariant submanifold must be a collection of orbits, which, in the given coordinate chart, are just the slices where the invariant coordinates z are constant, we can characterize its intersection with the coordinate chart by the vanishing of one or more functions depending on $z^1, \ldots, z^{m-s}$. But any function of the z's is a (local) invariant of the group action, so the result follows immediately. *Q.E.D.*

Lie Algebras

Besides invariant functions, there are many other important invariant objects associated with a transformation group, including vector fields, differential forms, differential operators, etc. The most important of these are the invariant vector fields, since they serve as the "infinitesimal generators" of the group action.

Definition 2.39. Let G be a group acting on the manifold M. A vector field $\mathbf{v}$ on M is called *G-invariant* if it is unchanged by the action of any group element, meaning that $dg(\mathbf{v}|_x) = \mathbf{v}|_{g\cdot x}$ for all $g \in G$, and all $x \in M$ such that $g \cdot x$ is defined.

The most important example is provided by the action of a Lie group G on itself by left or right multiplication, as described in Example 2.9. Here, the invariant vector fields determine the Lie algebra or "infinitesimal" Lie group, which plays an absolutely crucial role in both the general theory of Lie groups and its many applications. Given $g \in G$, we let $L_g: h \mapsto g \cdot h$ and $R_g: h \mapsto h \cdot g$ denote the associated left and right multiplication maps. A vector field $\mathbf{v}$ on G is called *left-invariant* if $dL_g(\mathbf{v}) = \mathbf{v}$, and *right-invariant* if $dR_g(\mathbf{v}) = \mathbf{v}$, for all $g \in G$.

Definition 2.40. The *left* (respectively *right*) *Lie algebra* of a Lie group G is the space of all left-invariant (respectively right-invariant) vector fields on G.

Thus, associated to any Lie group, there are two different Lie algebras, which we denote by $\mathfrak{g}_L$ and $\mathfrak{g}_R$ respectively. Traditionally, one refers to "the" Lie algebra associated with a Lie group, and denotes it by $\mathfrak{g}$, but this requires the adoption of either a left or a right convention. The more common is to use left-invariant vector fields for the Lie algebra, although some authors do prefer the right-invariant ones. In this book, both Lie algebras play a role, although the right-invariant one is by far the more useful of the two, due, perhaps surprisingly, to our convention that Lie groups act on the left on manifolds; the reason for this apparent switch will become clear below. Therefore, in the sequel, when we talk about the Lie algebra associated with a Lie group, we shall mean the right-invariant Lie algebra, and use $\mathfrak{g} = \mathfrak{g}_R$ to denote it.

Every right-invariant vector field $\mathbf{v}$ is uniquely determined by its value at the identity e, because $\mathbf{v}|_g = dR_g(\mathbf{v}|_e)$. Thus we can identify the right Lie algebra with the tangent space to G at the identity element, $\mathfrak{g}_R \simeq TG|_e$, so $\mathfrak{g}_R$ is a finite-dimensional vector space having the same dimension as G. A similar statement holds for left-invariant vector fields, providing a similar isomorphism $\mathfrak{g}_L \simeq TG|_e$. Thus, a given tangent vector $\mathbf{v}|_e \in TG|_e$ determines both a left- and a right-invariant vector field on the Lie group, which are (usually) different.

Each Lie algebra associated with a Lie group comes equipped with a natural bracket, induced by the Lie bracket of vector fields. This follows immediately from the invariance (1.17) of the Lie bracket under diffeomorphisms, which implies that if both $\mathbf{v}$ and $\mathbf{w}$ are right-invariant vector fields, so is their Lie bracket $[\mathbf{v}, \mathbf{w}]$. The basic properties of the Lie bracket translate into the defining properties of an (abstract) Lie algebra.

Definition 2.41. A *Lie algebra* $\mathfrak{g}$ is a vector space equipped with a bracket operation $[\,\cdot\,,\,\cdot\,] : \mathfrak{g} \times \mathfrak{g} \to \mathfrak{g}$ which is bilinear, anti-symmetric, $[\mathbf{v}, \mathbf{w}] = -[\mathbf{w}, \mathbf{v}]$, and satisfies the Jacobi identity

$$[\mathbf{u}, [\mathbf{v}, \mathbf{w}]] + [\mathbf{v}, [\mathbf{w}, \mathbf{u}]] + [\mathbf{w}, [\mathbf{u}, \mathbf{v}]] = 0. \tag{2.8}$$

Example 2.42. The Lie algebra $\mathfrak{gl}(n)$ of the general linear group $\mathrm{GL}(n)$ can be identified with the space of all $n \times n$ matrices. In terms of the coordinates provided by the matrix entries $X = (x_{ij}) \in \mathrm{GL}(n)$, the left-invariant vector field associated with a matrix $A = (a_{ij}) \in \mathfrak{gl}(n)$ has the explicit formula

$$\widehat{\mathbf{v}}_A = \sum_{i,j,k=1}^{n} x_{ik} a_{kj} \frac{\partial}{\partial x_{ij}}. \tag{2.9}$$

The Lie bracket of two such vector fields is $[\widehat{\mathbf{v}}_A, \widehat{\mathbf{v}}_B] = \widehat{\mathbf{v}}_C$, where $C = AB - BA$, so the left-invariant Lie bracket on $\mathrm{GL}(n)$ can be identified with the standard matrix commutator $[A, B] = AB - BA$. On the other hand, the right-invariant vector field associated with a matrix $A \in \mathfrak{gl}(n)$ is given by

$$\mathbf{v}_A = \sum_{i,j,k=1}^{n} a_{ik} x_{kj} \frac{\partial}{\partial x_{ij}}. \tag{2.10}$$

Now the Lie bracket is $[\mathbf{v}_A, \mathbf{v}_B] = \mathbf{v}_{\widehat{C}}$, where $\widehat{C} = -C = BA - AB$ is the *negative* of the matrix commutator. Thus, the matrix formula for the Lie algebra bracket on $\mathfrak{gl}(n)$ depends on whether we are dealing with its left-invariant or right-invariant version!

Exercise 2.43. Prove that the right and left Lie algebras for the abelian Lie group $G = \mathbb{R}^r$ are identical, each isomorphic to the abelian Lie algebra $\mathfrak{g} = \mathbb{R}^r$ with trivial Lie bracket: $[\mathbf{v}, \mathbf{w}] = 0$ for all $\mathbf{v}$, $\mathbf{w}$.

The right and left Lie algebras associated with a Lie group are, in fact, isomorphic as abstract Lie algebras. The inversion map $\iota(g) = g^{-1}$ provides the explicit isomorphism since it interchanges the roles of left and right. Thus, its differential $d\iota$ maps right-invariant vector fields to left-invariant ones, and vice versa, while preserving the Lie bracket.

Proposition 2.44. *If G is a Lie group, then the differential of the inversion map defines a Lie algebra isomorphism between its associated left and right Lie algebras $d\iota\colon \mathfrak{g}_L \simeq \mathfrak{g}_R$.*

As above, we identify an invariant vector field $\mathbf{v}_R$ or $\mathbf{v}_L$ with its value $\mathbf{v} = \mathbf{v}_R|_e = \mathbf{v}_L|_e$ at the identity. Since $d\iota|_e = -\mathbb{1}$, the differential $d\iota$ maps the right-invariant vector field $\mathbf{v}_R$ to *minus* its left-invariant counterpart $\mathbf{v}_L = -d\iota(\mathbf{v}_R)$. Consequently, the left and right Lie brackets on a Lie group always have *opposite* signs: $[\mathbf{v}, \mathbf{w}]_L = -[\mathbf{v}, \mathbf{w}]_R$. This explains the observation in Example 2.42.

The operations of right and left multiplication commute: $L_g \circ R_h = R_h \circ L_g$ for all $g, h \in G$. Therefore, according to Theorem 1.28, the corresponding infinitesimal generators commute: $[\mathbf{v}_L, \mathbf{w}_R] = 0$ for all $\mathbf{v}_L \in \mathfrak{g}_L$, $\mathbf{w}_R \in \mathfrak{g}_R$. In fact, it is not hard to see that the two Lie algebras are uniquely characterized in terms of each other in this manner. The following result is a special case of Theorem 2.81 below.

Proposition 2.45. *The right Lie algebra $\mathfrak{g}_R$ of a Lie group G can be characterized as the set of vector fields on G which commute with all left-invariant vector fields:*

$$\mathfrak{g}_R = \{\mathbf{w} \mid [\mathbf{v}_L, \mathbf{w}] = 0 \text{ for all } \mathbf{v}_L \in \mathfrak{g}_L\}.$$

The same result holds with left and right interchanged.

Example 2.46. Consider the two parameter group $\mathrm{A}(1)$ of affine transformations $x \longmapsto ax + b$ on the line $x \in \mathbb{R}$, as in Example 2.11.

The group multiplication law is given by $(a,b)\cdot(c,d) = (ac, ad+b)$, and the identity element is $e = (1,0)$. The right and left multiplication maps are therefore given by $R_{(a,b)}(c,d) = (c,d)\cdot(a,b) = (ac, bc+d)$, $L_{(a,b)}(c,d) = (a,b)\cdot(c,d) = (ac, ad+b)$. A basis for the right Lie algebra $\mathfrak{a}(1)_R$, corresponding to the coordinate basis $\partial_a|_e$, $\partial_b|_e$ of $T\mathrm{A}(1)|_e$, is provided by the pair of right-invariant vector fields

$$\mathbf{v}_1 = dR_{(a,b)}\big[\partial_a|_e\big] = a\partial_a + b\partial_b, \qquad \mathbf{v}_2 = dR_{(a,b)}\big[\partial_b|_e\big] = \partial_b. \tag{2.11}$$

These satisfy the commutation relation $[\mathbf{v}_1, \mathbf{v}_2] = -\mathbf{v}_2$. A similar basis for the left Lie algebra $\mathfrak{a}(1)_L$ is provided by the pair of left-invariant vector fields

$$\widehat{\mathbf{v}}_1 = dL_{(a,b)}\big[\partial_a|_e\big] = a\partial_a, \qquad \widehat{\mathbf{v}}_2 = dL_{(a,b)}\big[\partial_b|_e\big] = a\partial_b, \tag{2.12}$$

satisfying the negative commutation relation $[\widehat{\mathbf{v}}_1, \widehat{\mathbf{v}}_2] = \widehat{\mathbf{v}}_2$. Note that the vector fields (2.11) commute with those in (2.12): $[\mathbf{v}_i, \widehat{\mathbf{v}}_j] = 0$, confirming Proposition 2.45. Finally, the inversion map is given by $\iota(a,b) = (a,b)^{-1} = (1/a, -b/a)$, and, in accordance with the above remarks, its action on the two Lie algebras is $d\iota(\mathbf{v}_1) = -\widehat{\mathbf{v}}_1$, $d\iota(\mathbf{v}_2) = -\widehat{\mathbf{v}}_2$, as can be checked by an explicit computation.

Exercise 2.47. Define the semi-direct product of Lie algebras, and show that $\mathfrak{a}(1) = \mathbb{R} \ltimes \mathbb{R}$ can be identified as a semi-direct product of two one-dimensional Lie algebras.

Exercise 2.48. Prove that every two-dimensional Lie algebra is isomorphic to either the abelian algebra $\mathbb{R}^2$ or the affine algebra $\mathfrak{a}(1)$.

Structure Constants

Let $\mathbf{v}_1, \ldots, \mathbf{v}_r$ be a basis of a Lie algebra $\mathfrak{g}$. We define the associated *structure constants* C^k_{ij} by the bracket relations

$$[\mathbf{v}_i, \mathbf{v}_j] = \sum_{k=1}^{r} C^k_{ij}\mathbf{v}_k. \tag{2.13}$$

Anti-symmetry of the Lie bracket and the Jacobi identity imply the basic identities

$$C^k_{ij} = -C^k_{ij}, \qquad \sum_{l=1}^{r}\left\{C^l_{ij}C^m_{lk} + C^l_{ki}C^m_{lj} + C^l_{jk}C^m_{li}\right\} = 0, \tag{2.14}$$

which must be satisfied by the structure constants of any Lie algebra. Conversely, given $\frac{1}{2}r^2(r-1)$ constants satisfying the identities (2.14), we can reconstruct the Lie algebra $\mathfrak{g}$ by introducing a basis $\mathbf{v}_1, \ldots, \mathbf{v}_r$, and then imposing the bracket relations (2.13). In turn, as stated in Theorem 2.56 below, one can always construct an associated Lie group whose right (or left) Lie algebra coincides with $\mathfrak{g}$. Thus (from an admittedly reductionist standpoint) the theory of Lie groups can be essentially reduced to the study of its structure constants. This fact plays a key role in the detailed structure theory of Lie groups, cf. [**119**].

Exercise 2.49. Find a formula showing how the structure constants change under a change of basis of the Lie algebra $\mathfrak{g}$.

Exercise 2.50. Show that, relative to a fixed basis of the tangent space $TG|_e$, the structure constants for the left and right Lie algebras of a Lie group differ by an overall sign.

The Exponential Map

Given a right-invariant vector field $\mathbf{v} \in \mathfrak{g}_R$ on the Lie group G, we let $\exp(t\mathbf{v}): G \to G$ denote the associated flow. An easy continuation argument can be used to prove that this flow is globally defined for all $t \in \mathbb{R}$. Applying the flow to the identity element e serves to define the *one-parameter subgroup* $\exp(t\mathbf{v}) \equiv \exp(t\mathbf{v})e$; the vector field $\mathbf{v}$ is known as the *infinitesimal generator* of the subgroup. The notation is not ambiguous, since the flow through any $g \in G$ is the same as *left* multiplication by the elements of the subgroup, so $\exp(t\mathbf{v})g$ can be interpreted either as a flow or as a group multiplication. This fact is a consequence of the invariance of the flow, as in (1.16), under the right multiplication map R_h. Therefore, the right-invariant vector fields form the infinitesimal generators of the action of G on itself by *left* multiplication. Vice versa, the infinitesimal generators of the action of G on itself by right multiplication is the Lie algebra $\mathfrak{g}_L$ of left-invariant vector fields. This interchange of the role of infinitesimal generators and invariant vector fields is one of the interesting peculiarities of Lie group theory. Finally, note that although the left- and right-invariant vector fields associated with a given tangent vector $\mathbf{v} \in TG|_e$ are (usually) different, and have different flows, nevertheless the associated one-parameter groups coincide: $\exp(t\mathbf{v}_R) = \exp(t\mathbf{v}_L)$.

Example 2.51. Consider the general linear group $\mathrm{GL}(n)$ discussed in Example 2.42. The flow corresponding to the right-invariant vector field $\mathbf{v}_A$ given by (2.10) is given by left multiplication by the usual matrix exponential: $\exp(t\mathbf{v}_A)X = e^{tA}X$. Conversely, the flow corresponding to the left-invariant vector field $\widehat{\mathbf{v}}_A$ given by (2.9) is given by right multiplication $\exp(t\widehat{\mathbf{v}}_A)X = Xe^{tA}$. In either case, the one-parameter subgroup generated by the vector field associated with a matrix $A \in \mathfrak{gl}(n)$ is the matrix exponential $\exp(t\mathbf{v}_A) = \exp(t\widehat{\mathbf{v}}_A) = e^{tA}$.

From now on, in view of subsequent applications, we shall restrict our attention to the right-invariant vector fields, so that $\mathfrak{g}$ will always denote the right Lie algebra of the Lie group G. Evaluation of the flow $\exp(t\mathbf{v})$ at $t = 1$ for each $\mathbf{v} \in \mathfrak{g}$ serves to define the *exponential map* $\exp\colon \mathfrak{g} \to G$. Since $\exp(0) = e$, $d\exp(0) = \mathbb{1}$, the exponential map defines a local diffeomorphism in a neighborhood of $0 \in \mathfrak{g}$. Consequently, all Lie groups having the same Lie algebra look locally the same in a neighborhood of the identity; only their global topological properties are different. (Indeed, in Lie's day, one only considered "local Lie groups", the global version being a more recent introduction, cf. [**35**].) Globally, the exponential map is not necessarily one-to-one nor onto. However, if a Lie group is connected, it can be completely recovered by successive exponentiations.

Proposition 2.52. *Let G be a connected Lie group with Lie algebra $\mathfrak{g}$. Every group element can be written as a product of exponentials: $g = \exp(\mathbf{v}_1)\exp(\mathbf{v}_2)\cdots\exp(\mathbf{v}_k)$, for $\mathbf{v}_1, \ldots, \mathbf{v}_k \in \mathfrak{g}$.*

This result forms the basis of the infinitesimal or Lie algebraic approach to symmetry groups and invariants. It is proved by first noting that the set of all group elements $g \in G$ of the indicated form forms an open and closed subgroup of G, and then invoking connectivity.

Subgroups and Subalgebras

By definition, a *subalgebra* of a Lie algebra is a subspace $\mathfrak{h} \subset \mathfrak{g}$ which is invariant under the Lie bracket. Every one-dimensional subspace forms a subalgebra, and generates an associated one-parameter subgroup of the associated Lie group via exponentiation. More generally, each subalgebra $\mathfrak{h} \subset \mathfrak{g}$ generates a unique connected Lie subgroup $H \subset G$, satisfying $\mathfrak{h} \simeq TH|_e \subset TG|_e \simeq \mathfrak{g}$.

Theorem 2.53. *Let G be a Lie group with Lie algebra $\mathfrak{g}$. There is a one-to-one correspondence between connected s-dimensional Lie subgroups $H \subset G$ and s-dimensional Lie subalgebras $\mathfrak{h} \subset \mathfrak{g}$.*

Example 2.54. In particular, every matrix Lie group $G \subset \mathrm{GL}(n)$ will correspond to a subalgebra of $\mathfrak{g} \subset \mathfrak{gl}(n)$ the Lie algebra of $n \times n$ matrices, cf. Example 2.42. To determine the subalgebra, we need only find the tangent space to the subgroup at the identity matrix. For instance, the standard formula $\det \exp(tA) = \exp(t \operatorname{tr} A)$ implies that the Lie algebra $\mathfrak{sl}(n)$ of the unimodular subgroup $\mathrm{SL}(n)$ consists of all matrices with trace 0. The orthogonal groups $\mathrm{O}(n)$ and $\mathrm{SO}(n)$ have the same Lie algebra, $\mathfrak{so}(n)$, consisting of all skew-symmetric $n \times n$ matrices. This result proves our earlier claim that $\mathrm{SO}(n)$ and $\mathrm{O}(n)$ have dimension $\frac{1}{2}n(n-1)$. The reader should verify that the indicated subspaces are indeed subalgebras under the matrix commutator.

Exercise 2.55. Determine the Lie algebra $\mathfrak{sp}(2r)$ of the symplectic group $\mathrm{Sp}(2r)$.

Ado's Theorem, cf. [**124**], says that any finite-dimensional Lie algebra $\mathfrak{g}$ is isomorphic to a subalgebra of $\mathfrak{gl}(n)$ for some n. Consequently, $\mathfrak{g}$ is realized as the Lie algebra of the associated matrix Lie group $G \subset \mathrm{GL}(n)$. The full connection between Lie groups and Lie algebras is stated in the following fundamental theorem, whose proof can be found in [**195**].

Theorem 2.56. *Each finite-dimensional Lie algebra $\mathfrak{g}$ corresponds to a unique connected, simply connected Lie group $\widetilde{G}$. Moreover, any other connected Lie group G having the same Lie algebra $\mathfrak{g}$ is isomorphic to the quotient group of $\widetilde{G}$ by a discrete normal subgroup Γ, so that $G \simeq \widetilde{G}/\Gamma$.*

Note that the projection $\pi \colon \widetilde{G} \to G$ defined in Theorem 2.56 defines a (uniform) covering map, so that $\widetilde{G}$ is as a *covering group* for any other connected Lie group G having the same Lie algebra. Consequently, any two Lie groups having the same Lie algebra $\mathfrak{g}$ are not only locally isomorphic, but are in fact both covered by a common Lie group.

Exercise 2.57. The subgroup $C = \{g \in G \mid ghg^{-1} = h, \text{ for all } h \in G\}$ consisting of all group elements which commute with every element in a group G is known as the *center* of G. Prove that the Lie algebra associated with the center of a Lie group G is the abelian subalgebra $\{\mathbf{v} \in \mathfrak{g} \mid [\mathbf{v}, \mathbf{w}] = 0 \text{ for all } \mathbf{w} \in \mathfrak{g}\}$, known as the *center* of $\mathfrak{g}$.

Exercise 2.58. Let G be a Lie group. Prove that a vector field $\mathbf{v}$ on G lies in the intersection $\mathfrak{g}_L \cap \mathfrak{g}_R$ of the left and right Lie algebras, and so is both left- and right-invariant, if and only if it lies in the center of both $\mathfrak{g}_L$ and $\mathfrak{g}_R$.

Exercise 2.59. Let $H \subset G$ be a Lie subgroup with Lie subalgebra $\mathfrak{h} \subset \mathfrak{g}$. Prove that the *normalizer subgroup* $G_H = \{g \mid gHg^{-1} \subset H\}$ is a normal subgroup with Lie algebra $\mathfrak{g}_H = \{\mathbf{v} \in \mathfrak{g} \mid [\mathbf{v}, \mathbf{w}] \in \mathfrak{h}, \text{ for all } \mathbf{w} \in \mathfrak{h}\}$.

Exercise 2.60. A subspace $\mathfrak{i} \subset \mathfrak{g}$ of a Lie algebra $\mathfrak{g}$ is an *ideal* if $[\mathbf{v}, \mathfrak{i}] \subset \mathfrak{i}$ for every $\mathbf{v} \subset \mathfrak{g}$. Prove that an ideal is a subalgebra, but the converse is not necessarily true. Show that ideals in $\mathfrak{g}$ are in one-to-one correspondence with connected normal Lie subgroups of G.

A Lie algebra of dimension greater than 1 is called *simple* if it contains no proper nonzero ideals. Each of the classical Lie algebras $\mathfrak{sl}(n)$, $n \geq 2$, $\mathfrak{so}(n)$, $3 \leq n \neq 4$, and $\mathfrak{sp}(2r)$, $r \geq 2$, is known to be simple, although $\mathfrak{so}(4) \simeq \mathfrak{so}(3) \oplus \mathfrak{so}(3)$ is the direct sum of two simple algebras. In fact, besides the three infinite families of "classical" complex Lie groups, there exist just five additional "exceptional" simple Lie groups. See [**119**] for the precise classification theorem, due to Killing and Cartan.

Exercise 2.61. The *derived algebra* $\mathfrak{g}'$ of a Lie algebra $\mathfrak{g}$ is defined as the subalgebra spanned by all Lie brackets $[\mathbf{v}, \mathbf{w}]$ for all $\mathbf{v}, \mathbf{w} \in \mathfrak{g}$. Prove that $\mathfrak{g}'$ is an ideal in $\mathfrak{g}$, hence, if $\mathfrak{g}$ is simple, then $\mathfrak{g}' = \mathfrak{g}$. Prove that the derived algebra of $\mathfrak{gl}(n)$ is $\mathfrak{sl}(n)$. Show that the corresponding subgroup of the associated Lie group G is the *derived subgroup* G', generated by all commutators $ghg^{-1}h^{-1}$ for $g, h \in G$. (*A subtle point*: G' is *not* equal to the set of all commutators, as this may not even be a subgroup.)

Infinitesimal Group Actions

Just as a one-parameter group of transformations is generated as the flow of a vector field, so a general Lie group of transformations G acting on a manifold M will be generated by a set of vector fields on M, known as the *infinitesimal generators* of the group action. Each infinitesimal generator's flow coincides with the action of the corresponding one-parameter subgroup of G. Specifically, if $\mathbf{v} \in \mathfrak{g}$ generates the one-parameter subgroup $\{\exp(t\mathbf{v}) \mid t \in \mathbb{R}\} \subset G$, then we identify $\mathbf{v}$ with the infinitesimal generator $\widehat{\mathbf{v}}$ of the one-parameter group of transformations or flow $x \mapsto \exp(t\mathbf{v}) \cdot x$. According to (1.13) the infinitesimal generators of the group action are found by differentiation:

$$\widehat{\mathbf{v}}|_x = \frac{d}{dt} \exp(t\mathbf{v})x \Big|_{t=0}, \qquad x \in M, \quad \mathbf{v} \in \mathfrak{g}. \tag{2.15}$$

Consequently, $\widehat{\mathbf{v}}|_x = d\Phi_x(\mathbf{v}|_e)$, where $\Phi_x: G \to M$ is given by $\Phi_x(g) = g \cdot x$. Since $\Phi_x \circ R_h = \Phi_{h \cdot x}$, if $\mathbf{v} \in \mathfrak{g} = \mathfrak{g}_R$ is any *right-invariant* vector field on G, then $d\Phi_x(\mathbf{v}|_g) = \widehat{\mathbf{v}}|_{g \cdot x}$, where defined. The differential $d\Phi_x$ preserves the Lie bracket between vector fields; therefore the resulting vector fields form a finite-dimensional Lie algebra of vector fields on the manifold M, satisfying the same commutation relations as the right Lie algebra $\mathfrak{g}$ of G (and hence the negative of the commutation relations of the left Lie algebra — a fact that reflects our convention that group elements act on the left). The infinitesimal generators of the group action are not quite isomorphic to the Lie algebra $\mathfrak{g}$ since some of the nonzero Lie algebra elements may map to the trivial (zero) vector field on M. It is not hard to prove that this possibility is directly connected to the effectiveness of the group action, leading to an infinitesimal test for the local effectiveness of group actions.

Theorem 2.62. *Let G be a transformation group acting on a manifold M. The linear map ρ taking an element $\mathbf{v} \in \mathfrak{g} = \mathfrak{g}_R$ to the corresponding vector field $\widehat{\mathbf{v}} = \rho(\mathbf{v})$ on M defines a Lie algebra homomorphism: $\rho([\mathbf{v}, \mathbf{w}]) = [\rho(\mathbf{v}), \rho(\mathbf{w})]$. Moreover, the image $\widehat{\mathfrak{g}} = \rho(\mathfrak{g})$ forms a finite-dimensional Lie algebra of vector fields on M which is isomorphic to the Lie algebra of the effectively acting quotient group G/G_M, where G_M denotes the global isotropy subgroup of G. In particular, G acts locally effectively on M if and only if ρ is injective, i.e.,* $\ker \rho = \{0\}$.

Thus, if $\mathbf{v}_1, \ldots, \mathbf{v}_r$ forms a basis for $\mathfrak{g}$, the condition for local effectiveness is that the corresponding vector fields $\widehat{\mathbf{v}}_i = \rho(\mathbf{v}_i)$ be linearly independent over $\mathbb{R}$, i.e., $\sum_i c_i \widehat{\mathbf{v}}_i = 0$ on M for constant $c_i \in \mathbb{R}$ if and only if $c_1 = \cdots = c_r = 0$. Usually, we will not distinguish between an element $\mathbf{v} \in \mathfrak{g}$ in the Lie algebra and the associated infinitesimal generator $\widehat{\mathbf{v}} = \rho(\mathbf{v})$ of the group action of G, which we also denote as $\mathbf{v}$ from now on. When the action is locally effective, which, according to Proposition 2.12, can always be assumed, this identification does not result in any ambiguities.

Just as every Lie algebra generates a corresponding Lie group, given a finite-dimensional Lie algebra of vector fields on a manifold M, we can always reconstruct a (local) action of the corresponding Lie group via the exponentiation process. See [**195**; Theorem 98] for a proof.

Theorem 2.63. *Let $\mathfrak{g}$ be a finite-dimensional Lie algebra of vector fields on a manifold M. Let G denote a Lie group having Lie algebra $\mathfrak{g}$. Then there is a local action of G whose infinitesimal generators coincide with the given Lie algebra.*

Example 2.64. Consider the action of the group $\mathrm{SL}(2)$ acting by linear fractional transformations (2.4) on $\mathbb{RP}^1$. Its Lie algebra $\mathfrak{sl}(2)$ consists of all 2×2 matrices of trace 0, and hence is spanned by the three matrices

$$J^- = \begin{pmatrix} 0 & 1 \\ 0 & 0 \end{pmatrix}, \qquad J^0 = \begin{pmatrix} 1 & 0 \\ 0 & -1 \end{pmatrix}, \qquad J^+ = \begin{pmatrix} 0 & 0 \\ 1 & 0 \end{pmatrix}, \tag{2.16}$$

having the commutation relations

$$[J^-, J^0] = -2J^-, \qquad [J^+, J^0] = 2J^+, \qquad [J^-, J^+] = J^0.$$

The corresponding one-parameter subgroups and their infinitesimal generators are

Translations:	$p \longmapsto p + t$	$\mathbf{v}_- = \partial_p,$
Scalings:	$p \longmapsto e^{2t}p$	$\mathbf{v}_0 = 2p\partial_p,$
Inversions:	$p \longmapsto \dfrac{p}{tp+1}$	$\mathbf{v}_+ = -p^2\partial_p.$

The vector fields $\mathbf{v}_-$, $\mathbf{v}_0$, $\mathbf{v}_+$, obey the same commutation relations as the matrices J^-, J^0, J^+, except for an overall sign:

$$[\mathbf{v}_-, \mathbf{v}_0] = 2\mathbf{v}_-, \qquad [\mathbf{v}_+, \mathbf{v}_0] = -2\mathbf{v}_+, \qquad [\mathbf{v}_-, \mathbf{v}_+] = -\mathbf{v}_0,$$

since the right Lie algebra bracket is the negative of the matrix commutator, cf. Example 2.42. Although the three infinitesimal generators $\mathbf{v}_-$, $\mathbf{v}_0$, $\mathbf{v}_+$ are pointwise linearly dependent (since the underlying manifold is merely one-dimensional), there is no nontrivial *constant coefficient* linear combination $c_-\mathbf{v}_- + c_0\mathbf{v}_0 + c_+\mathbf{v}_+ = 0$ which vanishes identically. Therefore, Theorem 2.62 reconfirms the fact that $\mathrm{SL}(2)$ acts locally effectively on $\mathbb{RP}^1$. On the other hand, if we extend the linear fractional action to all of $\mathrm{GL}(2)$, the generator $\mathbb{1} \in \mathfrak{gl}(2)$ of the scaling subgroup maps to the trivial (zero) vector field, reflecting the ineffectiveness of this action.

The infinitesimal generators also determine the tangent space to, and hence the dimension of, the orbits of a group action.

Proposition 2.65. *Let G be a Lie group acting on a manifold M with Lie algebra $\mathfrak{g}$ of G. Then, for each $x \in M$, the tangent space to the orbit through x is the subspace spanned by the infinitesimal generators: $\mathfrak{g}|_x = \{\widehat{\mathbf{v}}|_x \mid \mathbf{v} \in \mathfrak{g}\} \subset TM|_x$. In particular, the dimension of the orbit equals the dimension of $\mathfrak{g}|_x$.*

Corollary 2.66. *A Lie group acts transitively on a connected manifold M if and only if $\mathfrak{g}|_x = TM|_x$ for all $x \in M$.*

There is an important connection between the dimension of the isotropy subgroup at a point and the dimension of the orbit through that point.

Proposition 2.67. *If G is an r-dimensional Lie group acting on M, then the isotropy subgroup G_x of any point $x \in M$ has dimension $r - s$, where s is the dimension of the orbit of G through x. In particular, G acts semi-regularly if and only if all its isotropy subgroups have the same dimension.*

Exercise 2.68. Prove that if $x \in M$, then the isotropy subgroup G_x has Lie algebra $\mathfrak{g}_x = \ker d\Phi_x \subset \mathfrak{g}$. Use this to prove Proposition 2.67.

Exercise 2.69. Prove that a group acts locally freely if and only if its infinitesimal generators are pointwise linearly independent: for each $x \in M$, $\sum_i c_i \widehat{\mathbf{v}}_i|_x = 0$ if and only if $c_1 = \cdots = c_r = 0$.

Classification of Group Actions

Let M be an m-dimensional manifold. The problem of classifying all possible Lie group actions on M is, according to Theorem 2.63, essentially the same as the problem of classifying finite-dimensional Lie algebras of vector fields on M. The classification results to be discussed here are local, up to a general change of variables. Thus, two Lie algebras of vector fields, defined on open subsets of $\mathbb{R}^m$, are considered to be equivalent if there is a local diffeomorphism on $\mathbb{R}^m$ taking one to the other. For instance, a one-dimensional Lie group is generated by a vector field, and hence, according to Theorem 1.23, away from singularities, all such group actions/Lie algebras are isomorphic to the group of translations in a single coordinate direction. The general classification problem was investigated in great detail by Lie, who, at one time, viewed it as the principal goal of his mathematical career. He succeeded in completely classifying all nonsingular (i.e., without fixed points) finite-dimensional Lie group actions in one and two dimensions, determining a complete list of (local) canonical forms for Lie algebras of vector fields on one- and two-dimensional manifolds, [**151**]. Further, in the third volume of his treatise on transformation groups, [**155**], Lie claimed to have completed the three-dimensional classification, and presents a large fraction thereof. Unfortunately, he states that, while he has completed the rest of the classification, the results are too long to be included in the book. As far as I can determine, despite the evident importance of this problem (and despite unsubstantiated rumors appearing sporadically in the literature), the complete classification in three dimensions was never published by Lie, and, to this day, remains unknown! See [**111**] for further fascinating historical details.

In the case of a one-dimensional manifold, either real or complex, there are just three different possible Lie group actions, and they are all equivalent to subgroups of the three-dimensional group of projective transformations.

Theorem 2.70. *Let G be a connected finite-dimensional Lie group acting locally effectively on a one-dimensional real or complex manifold, without fixed points. Then, locally, G is equivalent to one of the following group actions:*

	Action	Group	Generators
1. Translation	$x \longmapsto x+\beta$	$\mathbb{R}$	∂_x
2. Affine	$x \longmapsto \alpha x+\beta$	A(1)	$\partial_x, x\partial_x$
3. Projective	$x \longmapsto (\alpha x+\beta)/(\gamma x+\delta)$	SL(2)	$\partial_x, x\partial_x, x^2\partial_x$

Proof: We outline the proof in the complex case, the real case being similar but slightly more complicated. The assumption that G has no fixed points means that the vector fields in the Lie algebra $\mathfrak{g}$ do not all simultaneously vanish at a point. Fix a point x_0 and choose $\mathbf{v}_1 \in \mathfrak{g}$ such that $\mathbf{v}_1|_{x_0} \neq 0$. According to Theorem 1.23, we can introduce a local coordinate x near x_0 which straightens out $\mathbf{v}_1 = \partial_x$. Every other infinitesimal generator in $\mathfrak{g}$ has the form $\mathbf{v}_f = f(x)\partial_x$ for some function $f(x)$; let $\mathcal{F} = \{f(x)\}$ denote the finite-dimensional vector space spanned by these functions. Since the commutator $[\mathbf{v}_1, \mathbf{v}_f] = f'(x)\partial_x$

must also be an infinitesimal generator, we deduce that $f' \in \mathcal{F}$ whenever $f \in \mathcal{F}$. Smoothness implies that $\mathcal{F}$ is a finite-dimensional translationally invariant space of functions: If $f(x) \in \mathcal{F}$ and $c \in \mathbb{R}$, then $f(x+c) \in \mathcal{F}$. The following useful lemma now comes into play.

Lemma 2.71. *A finite-dimensional space $\mathcal{F}$ of smooth functions of a single variable x is translationally invariant if and only if it is the space of solutions to a linear, homogeneous, constant coefficient ordinary differential equation.*

Proof: Let $f_1(x), \ldots, f_n(x)$ be a basis for $\mathcal{F}$. Each derivative $f_i'(x)$ belongs to $\mathcal{F}$, and is therefore a constant coefficient linear combination of the basis elements: $f_i' = \sum_{k=1}^n a_i^k f_k$. Therefore, the column vector $f(x) = (f_1(x), \ldots, f_n(x))^T$ solves the first order linear system $f' = Af$, where $A = (a_k^i)$ is a constant matrix. Let $p(\lambda) = \det(\lambda \mathbb{1} - A)$ denote the characteristic polynomial of the matrix A. The Cayley–Hamilton Theorem demonstrates that each entry $f_i(x)$ satisfies the nth order linear homogeneous constant coefficient ordinary differential equation $p(D)y = 0$ obtained by replacing λ by the derivative operator $D = d/dx$. Moreover, the solution space to the nth order differential equation has dimension n, and so coincides with $\mathcal{F}$. *Q.E.D.*

Consequently, over $\mathbb{C}$, our Lie algebra has, as a basis, a collection of vector fields of the form $x^i e^{\lambda x} \partial_x$, where $\lambda \in S$, a finite set of complex numbers (which necessarily includes 0 since $\partial_x \in \mathfrak{g}$), and $0 \le i \le n_\lambda$. (The real case is similar, but also involves sines and cosines.) We compute the commutator of two such vector fields:

$$[x^i e^{\lambda x}\, \partial_x, x^j e^{\mu x}\, \partial_x] = \big[(\mu - \lambda)x^{i+j} + (j-i)x^{i+j-1}\big] e^{(\lambda+\mu)x}\, \partial_x. \quad (2.17)$$

The simplest case occurs when $S = \{0\}$, so that $\mathfrak{g}$ is spanned by $x^i \partial_x$ for $0 \le i \le n_0$. Taking $i = n_0$, $j = n_0 - 1$ in (2.17) implies $x^{2n_0 - 2}\partial_x \in \mathfrak{g}$, which implies $2n_0 - 2 \le n_0$, hence $n_0 \le 2$. The cases $n_0 = 0, 1, 2$ are precisely the three Lie algebras stated in the theorem. If $S = \{0, \lambda\}$ contains just one nonzero complex number, then, using (2.17), it is easy to see that $n_0 = n_\lambda = 0$; in this case $\mathfrak{g}$ is spanned by $\partial_x, e^{\lambda x}\partial_x$, and is mapped to the affine algebra by the change of variables $x \mapsto e^{\lambda x}$. The only other possible case is $S = \{0, \lambda, -\lambda\}$, with $n_0 = n_\lambda = n_{-\lambda} = 0$, which is mapped to the projective algebra by the same change of variables. *Q.E.D.*

Similar, but considerably more involved arguments, can be used to give a complete local classification of all nonsingular finite-dimensional Lie group actions on a two-dimensional manifold. The genericity hypothesis is that there are no fixed points of the group action (i.e., no zero-dimensional orbits), which is equivalent to the assumption that not all vector fields in the associated Lie algebra vanish at a point. The full list of possible local group actions appears in the Tables at the end of the book; these tables have been adapted from those appearing in Lie, **[152]**, **[151]**, in Mostow, **[172]**, and in **[95]**, to which we refer for proofs. We first distinguish between transitive and intransitive actions, the latter being given in Table 3. The transitive group actions are further divided into primitive actions and imprimitive actions. In general, a group action is called *imprimitive* if there is an invariant foliation of the manifold. For example, the group generated by $\{\partial_x, \partial_y, x\partial_x + \alpha y\partial_y\}$ admits two invariant foliations of $\mathbb{R}^2$ — by horizontal lines and by vertical lines. Invariance is the fact that each group transformation maps every horizontal (vertical) line to another horizontal (vertical) line. For the imprimitive group actions, the real and complex classifications coincide, and are displayed in Table 1. As for the primitive group actions, there are three different complex actions, given in Table 2, and eight distinct real actions, presented in Table 6; see **[94]** for details. In view of Theorem 2.17, each of the transitive group actions has a global model as a homogeneous space G/H, where H is a closed subgroup of codimension 2. Thus, the classification of transitive group actions is the same as the classification of two-dimensional homogeneous spaces; see **[140]**. Beyond this, the primitive group actions on manifolds of arbitrary dimension have been completely classified; see **[87]**, **[139]**.

Exercise 2.72. Suppose that G is a Lie group acting on M. Prove that the level sets of a function $F: M \to \mathbb{R}^k$ form an invariant foliation for G if and only if for each infinitesimal generator $\mathbf{v}$ of G there is a function $h_{\mathbf{v}}: \mathbb{R}^k \to \mathbb{R}^k$ such that $\mathbf{v}(F) = h_{\mathbf{v}}(F)$.

Exercise 2.73. Consider the Cartesian product action of SL(2) generated by the vector fields $\partial_x + \partial_u$, $x\partial_x + u\partial_u$, $x^2\partial_x + u^2\partial_u$. Determine which Lie algebra in the tables this is equivalent to, and find the explicit change of variables mapping one to the other. (See **[54]**.)

Infinitesimal Invariance

The fundamental feature of (connected) Lie groups is the ability to work infinitesimally, thereby effectively linearizing complicated invariance criteria. Indeed, the practical applications of Lie groups all ultimately rely on this basic method, and its importance cannot be overestimated. We begin by stating the infinitesimal criterion for the invariance of a real-valued function.

Theorem 2.74. *Let G be a connected group of transformations acting on a manifold M. A function $I: M \to \mathbb{R}$ is invariant under G if and only if*

$$\mathbf{v}[I] = 0, \tag{2.18}$$

for all $x \in M$ and every infinitesimal generator $\mathbf{v} \in \mathfrak{g}$ of G.

Proof: Let $\mathbf{v} \in \mathfrak{g}$ be fixed. We then differentiate the invariance condition $I[\exp(t\mathbf{v})x] = I(x)$ with respect to t and set $t = 0$ to deduce the infinitesimal condition (2.18). Conversely, if (2.18) holds, then $d(I[\exp(t\mathbf{v})x])/dt = 0$ for all t (where defined), and hence $I[\exp(t\mathbf{v})x]$ is constant under the one-parameter subgroup generated by $\mathbf{v}$. The invariance of I under arbitrary group elements then follows from Proposition 2.52. *Q.E.D.*

Thus, according to Theorem 2.74, the invariants $u = I(x)$ of a one-parameter group with infinitesimal generator $\mathbf{v} = \sum_i \xi^i(x)\partial_{x^i}$ satisfy the first order, linear, homogeneous partial differential equation

$$\sum_{i=1}^{m} \xi^i(x) \frac{\partial u}{\partial x^i} = 0. \tag{2.19}$$

The solutions of (2.19) are effectively found by the method of characteristics. We replace the partial differential equation by the characteristic system of ordinary differential equations

$$\frac{dx^1}{\xi^1(x)} = \frac{dx^2}{\xi^2(x)} = \cdots = \frac{dx^m}{\xi^m(x)}. \tag{2.20}$$

The general solution to (2.20) can be (locally) written in the form $I_1(x) = c_1, \ldots, I_{m-1}(x) = c_{m-1}$, where the c_i are constants of integration. It is not hard to prove that the resulting functions $I_1, \ldots, I_{m-1}$ form a complete set of functionally independent invariants of the one-parameter group generated by $\mathbf{v}$.

Example 2.75. Consider the (local) one-parameter group generated by the vector field

$$\mathbf{v} = -y\,\frac{\partial}{\partial x} + x\,\frac{\partial}{\partial y} + (1+z^2)\,\frac{\partial}{\partial z}\,.$$

The group transformations (or flow) are

$$(x,y,z) \longmapsto \left(x\cos t - y\sin t,\, x\sin t + y\cos t,\, \frac{\sin t + z\cos t}{\cos t - z\sin t} \right). \tag{2.21}$$

Note that if we fix a point (x,y,z), then (2.21) parametrizes the integral curve passing through the point. The characteristic system (2.20) for this vector field is

$$\frac{dx}{-y} = \frac{dy}{x} = \frac{dz}{1+z^2}\,.$$

The first equation reduces to a simple separable ordinary differential equation $dy/dx = -x/y$, with general solution $x^2 + y^2 = c_1$, for c_1 a constant of integration; therefore the cylindrical radius $r = \sqrt{x^2+y^2}$ provides one invariant. To solve the second characteristic equation, we replace x by $\sqrt{r^2-y^2}$, and treat r as constant. The solution is $\tan^{-1} z - \sin^{-1}(y/r) = \tan^{-1} z - \tan^{-1}(y/x) = c_2$, hence $\tan^{-1} z - \tan^{-1}(y/x)$ is a second invariant. A more convenient choice is provided by the tangent of this invariant. We deduce that, for $yz + x \neq 0$, the functions $r = \sqrt{x^2+y^2}$, $w = (xz-y)/(yz+x)$ form a complete system of functionally independent invariants, whose common level sets describe the integral curves.

Exercise 2.76. Let $\mathbf{v}$ be a nonvanishing vector field on a manifold M. Prove that the rectifying coordinates $y = \eta(x)$ of Theorem 1.23 satisfy the partial differential equations $\mathbf{v}(\eta^1) = 1$, $\mathbf{v}(\eta^i) = 0$, $i > 1$. Thus the coordinates $y^i = \eta^i(x)$, $i = 2, \ldots, m$ are the functionally independent invariants of the one-parameter group generated by $\mathbf{v}$.

For multi-parameter groups, the invariants are simultaneous solutions to an overdetermined system of linear, homogeneous, first order partial differential equations. One solution method is to look for invariants of each generator in turn and try to re-express subsequent generators in terms of the invariants. However, this can become quite complicated to implement in practice. We illustrate the technique with a simple example.

Example 2.77. Consider the action of the unimodular group $\mathrm{SL}(2,\mathbb{R})$ on $\mathbb{R}^3$ generated by the three vector fields

$$\mathbf{v}_- = 2y\partial_x + z\partial_y, \qquad \mathbf{v}_0 = -2x\partial_x + 2z\partial_z, \qquad \mathbf{v}_+ = x\partial_y + 2y\partial_z. \tag{2.22}$$

Away from the origin $x = y = z = 0$, the vector fields (2.22) span a two-dimensional space, and hence, according to Proposition 2.65, the orbits of the group action are (except for the origin) all two-dimensional. Therefore, we expect to find one independent invariant. Solving first the characteristic system for $\mathbf{v}_0$, which is $dx/(2x) = dy/0 = dz/(-2z)$, we derive invariants $w = xz$ and y. Thus, our desired invariant must have the form $I = F(w,y) = F(xz,y)$. Applying $\mathbf{v}_+$ to I, we find $2xyF_w + xF_y = 0$, hence the desired invariant is $I = y^2 - xz$. (Why don't we need to check the invariance of I under $\mathbf{v}_-$?)

Remark: This action can be identified with the action of $\mathrm{SL}(2,\mathbb{R})$ on the space of quadratic polynomials. In this interpretation, the invariant I coincides with the standard discriminant, and we have proved that every invariant is a function thereof. See Chapter 3 for details.

Infinitesimal techniques are also effective for the determination of invariant submanifolds and invariant systems of equations. The starting point is the following basic result on the invariance of submanifolds.

Theorem 2.78. *A closed submanifold $N \subset M$ is G-invariant if and only if the space $\mathfrak{g}$ of infinitesimal generators is tangent to N everywhere, i.e., $\mathfrak{g}|_x \subset TN|_x$ for every $x \in N$.*

The proof of Theorem 2.78 follows immediately from the uniqueness of the flow, which implies that if $\mathbf{v}$ is tangent to N, then $\exp(t\mathbf{v})N \subset N$. If the submanifold $N \subset M$ is not closed, then the infinitesimal criterion of Theorem 2.78 implies that N is merely *locally G-invariant*, meaning that for every $x \in N$ and every $g \in G_x$ in a neighborhood of the identity, possibly depending on x, we have $g \cdot x \in N$. For example, if G is the group of translations $(x,y) \mapsto (x+t,y)$, then any horizontal line segment, e.g., $\{(x,0) \mid 0 < x < 1\}$ is locally, but not globally, translationally invariant.

An important consequence of Theorem 2.78 is the following crucially important characterization of symmetry groups of regular systems of equations — see Theorem 1.21.

Theorem 2.79. *A connected Lie group G is a symmetry group of the regular system of equations $F_1(x) = \cdots = F_k(x) = 0$ if and only if*

$$\mathbf{v}[F_\nu(x)] = 0, \quad \nu = 1, \ldots, k, \quad \textit{whenever} \quad F_1(x) = \cdots = F_k(x) = 0, \tag{2.23}$$

for every infinitesimal generator $\mathbf{v} \in \mathfrak{g}$ of G.

Using the formula $\mathbf{v}(F) = \langle dF\,; \mathbf{v} \rangle$, we see that Theorem 2.79 follows directly from Theorem 2.78 and Exercise 1.31. The following example provides a simple illustration of how the infinitesimal criterion (2.23) is verified in practice. However, the real power of Theorem 2.79 must await our applications to symmetry groups of differential equations, which is the subject of Chapter 6.

Example 2.80. The equation $x^2 + y^2 = 1$ defines a circle, which is rotationally invariant. To check the infinitesimal condition, we apply the generator $\mathbf{v} = -y\partial_x + x\partial_y$ to the defining function $F(x,y) = x^2 + y^2 - 1$. We find $\mathbf{v}(F) = 0$ everywhere (since F is an invariant). Since dF is nonzero on the circle, the equation is regular, and hence its solution set is rotationally invariant. A more complicated example is provided by $H(x,y) = x^4 + x^2y^2 + y^2 - 1$. Now, $\mathbf{v}(H) = -2xy(x^2+1)^{-1}H$, hence $\mathbf{v}(HF) = 0$ whenever $H = 0$, hence the set of solutions to $H(x,y) = 0$ is rotationally invariant. (What is this set?) To see the importance of the regularity condition, consider the function $K(x,y) = y^2$. The solution set is the x-axis, which is clearly not rotationally invariant, even though $\mathbf{v}(K) = 2xy = 0$ vanishes when $y = 0$, and so the infinitesimal condition (2.23) is satisfied.

Invariant Vector Fields

Although the invariant functions and systems of equations for transformation groups form our primary focus of interest, we shall also find it worthwhile to consider the existence and construction of other types of invariant objects. When a Lie group acts on itself via right or left multiplication, the existence of invariant vector fields, i.e., the Lie algebra, was discussed above. For more general group actions, invariant vector fields $\mathbf{w}$ on the underlying manifold are characterized by the condition $dg(\mathbf{w}|_x) = \mathbf{w}|_{g\cdot x}$ for all group elements $g \in G$ and all x in the domain of g. The infinitesimal invariance criterion for such a vector field is handled by the Lie bracket.

Theorem 2.81. *Let G be a connected transformation group acting on M. A vector field $\mathbf{w}$ on M is G-invariant if and only if $[\mathbf{v}, \mathbf{w}] = 0$ for every infinitesimal generator $\mathbf{v} \in \mathfrak{g}$.*

Proof: Let $\mathbf{v}$ be any infinitesimal generator of G. We need to compare the value of $\mathbf{w}$ at a point $\exp(t\mathbf{v})x$ with its value at x. Since these two vectors lie in different tangent spaces, to effect the comparison we must map $\mathbf{w}|_{\exp(t\mathbf{v})x}$ back to x using the differential $d\exp(-t\mathbf{v})$. We expand the resulting expression in powers of t, using formula (1.10), and deduce that

$$d\exp(-t\mathbf{v})\left[\mathbf{w}|_{\exp(t\mathbf{v})x}\right] = \mathbf{w}|_x + t\,[\mathbf{v}, \mathbf{w}]|_x + \cdots. \tag{2.24}$$

As a consequence, the Lie bracket $[\mathbf{v}, \mathbf{w}]$ represents the infinitesimal change of the vector field $\mathbf{w}$ under the action of the flow induced by the vector field $\mathbf{v}$. (The reader might contemplate what the anti-symmetry of the Lie bracket implies in this context.) In particular, if $\mathbf{w}$ is invariant under the flow induced by $\mathbf{v}$ then the Lie bracket vanishes. The converse follows from the uniqueness theorem for ordinary differential equations. Finally, if the infinitesimal invariance condition holds for every infinitesimal generator, the vector field $\mathbf{w}$ is invariant under every one-parameter subgroup of G, so the theorem follows immediately from Proposition 2.52. *Q.E.D.*

Proposition 2.82. *If $\mathbf{w}$ is a G-invariant vector field, and I is any invariant function, then $I\mathbf{w}$ is also a G-invariant vector field. More generally, if $\mathbf{w}_1, \ldots, \mathbf{w}_k$ are pointwise linearly independent invariant vector fields, then $\mathbf{w} = \sum_{i=1}^k I_k \mathbf{w}_k$ is G-invariant if and only if the coefficients I_k are G-invariant functions.*

In contrast to the situation for invariant functions, or for invariant vector fields on Lie groups, the existence of invariant vector fields of general transformation groups is not so straightforward. For, example, consider the three transformation groups acting on $\mathbb{R}$ determined in Theorem 2.70. For the translation group, any constant multiple of the translation vector field ∂_x is an invariant vector field. The other two groups, however, do not possess any nonzero invariant vector fields. Indeed, the only case in which a definitive count of the number of invariant vector fields is available occurs when the group acts freely, in which case we are guaranteed a maximal collection of invariant vector fields.

Thus, the fact that the affine and projective groups do not act freely on $\mathbb{R}$ serves to explain the nonexistence of invariant vector fields.

Definition 2.83. A *frame* on an m-dimensional manifold M is an ordered set of vector fields $\mathcal{V} = \{\mathbf{v}_1, \ldots, \mathbf{v}_m\}$ having the property that they form a basis for the tangent space $TM|_x$ at each point $x \in M$.

For example, the coordinate vector fields $\partial/\partial x^i$, $i = 1, \ldots, m$, define a frame on the associated local coordinate chart. The geometry of frames will be discussed in more detail in Chapter 8.

Theorem 2.84. *Let G be an r-dimensional Lie group acting effectively freely on the m-dimensional manifold M. Then, locally, there exist m pointwise linearly independent G-invariant vector fields $\mathbf{w}_1, \ldots, \mathbf{w}_m$, forming a G-invariant frame on M.*

Theorem 2.84 is a particular case of the general Theorem 3.36 governing the existence of relative invariants for multiplier representations. An alternative approach is to use the rectifying coordinates from Frobenius' Theorem 2.23, and thereby identify each orbit with a neighborhood of the identity of the freely acting quotient group G/G_M. The invariant vector fields are then provided by the left-invariant vector fields on G/G_M, mapped to the orbits, together with the invariant vector fields $\partial/\partial z^i$ transverse to the orbits.

Example 2.85. Consider the action of the connected component of the Euclidean group SE(2) on $M = \mathbb{R}^3$, which is generated by the vector fields

$$\mathbf{v}_1 = \partial_x, \qquad \mathbf{v}_2 = \partial_y, \qquad \mathbf{v}_3 = -y\partial_x + x\partial_y + (1 + z^2)\partial_z. \tag{2.25}$$

Note that $\mathbf{v}_1$, $\mathbf{v}_2$, $\mathbf{v}_3$ are pointwise linearly independent, and so form a frame. Thus the action is locally transitive, and there are no nonconstant invariant functions. Since the orbit dimension is the same as that of the group, the group action is (locally) freely, hence Theorem 2.84 implies the existence of a complete system of three linearly independent invariant vector fields. There are two possible approaches to their construction. The direct method looks for vector fields $\mathbf{w} = \xi\partial_x + \eta\partial_y + \zeta\partial_z$ which commute with the three infinitesimal generators of the group action. This implies that the coefficients ξ, η, ζ are independent of x and y, and satisfy the coupled system of ordinary differential equations

$$(1 + z^2)\xi_z = -\eta, \qquad (1 + z^2)\eta_z = \xi, \qquad (1 + z^2)\zeta_z = 2z\zeta.$$

Therefore, the general invariant vector field is a constant coefficient linear combination of

$$\mathbf{w}_1 = \frac{\partial_x + z\partial_y}{\sqrt{1+z^2}}, \qquad \mathbf{w}_2 = \frac{z\partial_x - \partial_y}{\sqrt{1+z^2}}, \qquad \mathbf{w}_3 = (1+z^2)\partial_z. \tag{2.26}$$

An alternative approach is to use the transitivity of the group action to identify the underlying manifold $M = \mathbb{R}^3$ with a neighborhood of the identity of the group. According to Proposition 2.44, the group inversion ι maps the infinitesimal generators (2.25), which, as in Theorem 2.62, are identified with a basis of the right Lie algebra on G, to the corresponding left-invariant vector fields, which, by Proposition 2.45, form a basis for the invariant vector fields on M. In order to carry out this procedure, we parametrize SE(2) by $g = (a, b, \theta)$, with a, b representing the translation, and θ the rotational angle. The group inversion map is

$$\iota(a, b, \theta) = (a\cos\theta + b\sin\theta, -a\sin\theta + b\cos\theta, -\theta). \tag{2.27}$$

The map $g \mapsto g \cdot 0 = (a, b, \tan\theta) = (x, y, z)$ defines a local diffeomorphism that identifies M with the open subset $\{\theta \neq \pm\frac{1}{2}\pi\} \subset G$, with 0 corresponding to the identity element. Formula (2.27) implies that, under this identification,

$$\iota(x, y, z) = \left(\frac{x + yz}{\sqrt{1+z^2}}, \frac{y - xz}{\sqrt{1+z^2}}, -z \right).$$

A short computation proves that the differential $d\iota$ maps the infinitesimal generator $\mathbf{v}_i$ to the invariant vector field $\mathbf{w}_i = d\iota(\mathbf{v}_i)$, reconfirming the previous formulae (2.26).

Lie Derivatives and Invariant Differential Forms

A differential form Ω on M is called *G-invariant* if it is unchanged under the pull-back action of the group: $g^*(\Omega|_{g\cdot x}) = \Omega|_x$ for all $g \in G$, $x \in M$. In particular, an invariant 0-form is just an ordinary invariant function. Formulae (1.21) demonstrate that the sum and wedge product of invariant forms are also invariant; in particular, multiplying an invariant k-form by an invariant function produces another invariant k-form. Furthermore, the fundamental invariance property (1.26) of the exterior derivative proves that the differential $d\Omega$ of any invariant k-form is an invariant $(k+1)$-form. In particular, we find:

Proposition 2.86. *If I is an invariant function, its differential dI is an invariant one-form.*

There is, of course, an infinitesimal criterion for the invariance of a differential form under a connected group of transformations. This condition is formalized by the definition of the Lie derivative of a differential form with respect to a vector field $\mathbf{v}$, which indicates how the form varies infinitesimally under the associated flow $\exp(t\mathbf{v})$.

Definition 2.87. Let $\mathbf{v}$ be a vector field on the manifold M with flow $\exp(t\mathbf{v})$. The *Lie derivative* $\mathbf{v}(\Omega)$ of a differential form Ω with respect to $\mathbf{v}$ is defined as

$$\mathbf{v}(\Omega)|_x = \frac{d}{dt} \exp(t\mathbf{v})^* \left(\Omega|_{\exp(t\mathbf{v})x}\right)\bigg|_{t=0}. \tag{2.28}$$

Note that the pull-back $\exp(t\mathbf{v})^*$ moves the form at the point $\exp(t\mathbf{v})x$ back to the point x, enabling us to compute the derivative consistently; see the discussion following (2.24). Thus, we find the series expansion

$$\exp(t\mathbf{v})^* \left(\Omega|_{\exp(t\mathbf{v})x}\right) = \Omega|_x + t\,\mathbf{v}(\Omega)|_x + \cdots, \tag{2.29}$$

the higher order terms being provided by higher order Lie derivatives of Ω. In particular, the Lie derivative $\mathbf{v}(f)$ of a function $f\colon M \to \mathbb{R}$ (or 0-form) coincides with the action of the vector field on f, and (2.29) reduces to the earlier expansion (1.10).

Exercise 2.88. Define the Lie derivative of a vector field $\mathbf{w}$ with respect to the vector field $\mathbf{v}$. Prove that the Lie derivative coincides with the Lie bracket $[\mathbf{v}, \mathbf{w}]$.

The explicit local coordinate formula for the Lie derivative are most readily deduced from its basic linearity, derivation, and commutation properties:

$$\begin{aligned} \mathbf{v}(\Omega + \Theta) &= \mathbf{v}(\Omega) + \mathbf{v}(\Theta), \\ \mathbf{v}(\Omega \wedge \Theta) &= \mathbf{v}(\Omega) \wedge \Theta + \Omega \wedge \mathbf{v}(\Theta), \\ \mathbf{v}(d\Omega) &= d\mathbf{v}(\Omega). \end{aligned} \tag{2.30}$$

These all follow directly from the basic definition (2.28) using the properties of the pull-back map. In fact, the properties (2.30) along with the action on functions serve to uniquely define the Lie derivative operation. For example, the Lie derivative of a one-form (1.18) with respect to the vector field (1.3) is given by

$$\mathbf{v}(\omega) = \sum_{i=1}^{m} \Big[\mathbf{v}(h_i)\, dx^i + h_i\, d\xi^i \Big], \qquad \mathbf{v} = \sum_{i=1}^{m} \xi^i(x) \frac{\partial}{\partial x^i}, \quad \omega = \sum_{i=1}^{m} h_i(x) dx^i.$$

Exercise 2.89. Prove that the exterior derivative of a differential form can be written in local coordinates $(x^1, \ldots, x^m)$ as

$$d\Omega = \sum_{i=1}^{m} dx^i \wedge \frac{\partial \Omega}{\partial x^i}, \tag{2.31}$$

the latter term being the Lie derivative of Ω with respect to the coordinate frame vector fields $\partial/\partial x^i$. Can this formula be generalized to other frames?

Exercise 2.90. Given a k-form Ω, its *interior product* with a vector field $\mathbf{v}$ is the $(k-1)$-form $\mathbf{v} \mathbin{\lrcorner} \Omega$ defined so that

$$\langle \mathbf{v} \mathbin{\lrcorner} \Omega\, ; \mathbf{w}_1, \ldots, \mathbf{w}_{k-1} \rangle = \langle \Omega\, ; \mathbf{v}, \mathbf{w}_1, \ldots, \mathbf{w}_{k-1} \rangle, \tag{2.32}$$

for any set of $k-1$ vector fields $\mathbf{w}_1, \ldots, \mathbf{w}_{k-1}$. Using this, prove the following important formula relating the Lie derivative and the exterior derivative of a differential form:

$$\mathbf{v}(\Omega) = \mathbf{v} \mathbin{\lrcorner} (d\Omega) + d(\mathbf{v} \mathbin{\lrcorner} \Omega). \tag{2.33}$$

Theorem 2.91. *A differential form Ω is invariant under a connected Lie group of transformations G if and only if its Lie derivative with respect to every infinitesimal generator $\mathbf{v} \in \mathfrak{g}$ vanishes: $\mathbf{v}(\Omega) = 0$.*

The proof of this result is similar to that of Theorem 2.81, and is left to the reader.

Example 2.92. Let $dx = dx^1 \wedge \cdots \wedge dx^m$ denote the volume form on $\mathbb{R}^m$. The Lie derivative of dx with respect to a vector field $\mathbf{v} = \sum_{i=1}^{p} \xi^i(x) \partial_{x^i}$ is given as $\mathbf{v}(dx) = (\operatorname{div} \xi) dx$, where $\operatorname{div} \xi = \sum_i \partial \xi^i / \partial x^i$ is the divergence of the coefficients of $\mathbf{v}$. Therefore, a transformation group is volume-preserving on $\mathbb{R}^m$ if and only if each of its infinitesimal generators is divergence free: $\operatorname{div} \xi = 0$. (See also Exercise 1.34.)

The existence of invariant forms for general transformation groups parallels that of invariant vector fields. Only when the group acts freely are we guaranteed a (complete) collection of invariant one-forms.

Theorem 2.93. *Let G be an r-dimensional Lie group acting effectively freely on the m-dimensional manifold M. Then, locally, there exist m pointwise linearly independent G-invariant one-forms $\omega^1, \ldots, \omega^m$.*

At each point $x \in M$, the one-forms $\omega^1, \ldots, \omega^m$ form a basis for the cotangent space $T^*M|_x$, and so, in the language of Chapter 8, form a *G-invariant coframe*. Their wedge product $\omega^1 \wedge \cdots \wedge \omega^m \neq 0$ defines a nonvanishing G-invariant volume form on M. Theorem 2.93 follows immediately from Theorem 2.84. Indeed, the vector fields $\mathbf{v}_1, \ldots, \mathbf{v}_m$ form a G-invariant frame on M if and only if the dual one-forms $\omega^1, \ldots, \omega^m$, defined so that $\langle \omega^i ; \mathbf{v}_j \rangle = \delta^i_j$, form a G-invariant coframe.

Example 2.94. Consider the action of SE(2) on $M = \mathbb{R}^3$ described earlier in Example 2.85. The invariant one-forms are dual to the invariant vector fields (2.26), and hence are given by

$$\omega^1 = \frac{dx + z\,dy}{\sqrt{1+z^2}}, \qquad \omega^2 = \frac{dy - z\,dx}{\sqrt{1+z^2}}, \qquad \omega^3 = \frac{dz}{1+z^2}.$$

Alternatively, one could directly construct the invariant one-forms by analyzing the Lie derivative condition of Theorem 2.91.

The Maurer–Cartan Forms

Of particular importance in the theory of Lie groups are the invariant differential forms associated with the right (and left) actions of the Lie group on itself. By definition, a differential form Ω on a Lie group G is right-invariant if it is unaffected by right multiplication by group elements: $(R_g)^*\Omega = \Omega$ for all $g \in G$. In particular, the right-invariant one-forms on G are known as the (right-invariant) *Maurer–Cartan forms*. The space of Maurer–Cartan forms is naturally dual to the Lie algebra of G, and hence forms a vector space of the same dimension as the Lie group. If we choose a basis $\mathbf{v}_1, \ldots, \mathbf{v}_r$ of the Lie algebra $\mathfrak{g}$, then there is a dual basis (or coframe) $\alpha^1, \ldots, \alpha^r$, consisting of Maurer–Cartan forms, satisfying $\langle \alpha^i ; \mathbf{v}_j \rangle = \delta^i_j$. As a direct consequence of formula (1.25), and duality, the Maurer–Cartan forms are seen to satisfy the fundamental *structure equations*

$$d\alpha^k = -\frac{1}{2} \sum_{i,j=1}^{r} C_{ij}^k \, \alpha^i \wedge \alpha^j = - \sum_{\substack{i,j=1 \\ i<j}}^{r} C_{ij}^k \, \alpha^i \wedge \alpha^j, \qquad k = 1, \ldots, r. \tag{2.34}$$

The coefficients C_{ij}^k are the *same* as the structure constants corresponding to our choice of basis of the Lie algebra $\mathfrak{g}$.

If the group G is given as a parametrized matrix Lie group, then a basis for the space of Maurer–Cartan forms can be found among the entries of the matrix of one-forms

$$\gamma = dg \cdot g^{-1}, \qquad \text{or} \qquad \gamma_j^i = \sum_{k=1}^{r} dg_k^i (g^{-1})_j^k. \tag{2.35}$$

Each entry γ_j^i is clearly a right-invariant one-form since if h is any fixed group element, then

$$(R_h)^* \gamma = d(g \cdot h) \cdot (g \cdot h)^{-1} = dg \cdot g^{-1} = \gamma.$$

It is also not hard to see that the number of linearly independent entries of γ is the same as the dimension of the group, and hence the entries provide a complete basis for the space of Maurer–Cartan forms.

Of course, one can also define left-invariant Maurer–Cartan forms on a Lie group, dual to the Lie algebra of left-invariant vector fields. The construction goes through word for word. Note that the structure constants appearing in the left-invariant structure equations are those corresponding to the left-invariant Lie algebra, and so have the opposite sign to the right-invariant structure constants. Replacing formula (2.35) is the matrix $\widehat{\gamma} = g^{-1} \cdot dg$ whose entries provide a basis for the left-invariant Maurer–Cartan forms.

Example 2.95. As in Example 2.11, the two-dimensional affine group $\mathrm{A}(1)$ can be identified with the group of 2×2 matrices of the form $g = \left(\begin{smallmatrix} a & b \\ 0 & 1 \end{smallmatrix}\right)$. According to (2.35), the right-invariant Maurer–Cartan forms on $\mathrm{A}(1)$ are provided by the independent entries of the matrix

$$dg \cdot g^{-1} = \begin{pmatrix} da & db \\ 0 & 0 \end{pmatrix} \begin{pmatrix} a^{-1} & -ba^{-1} \\ 0 & 1 \end{pmatrix} = \begin{pmatrix} a^{-1}\, da & a^{-1}(a\, db - b\, da) \\ 0 & 0 \end{pmatrix}.$$

Therefore, the two Maurer–Cartan forms are

$$\alpha^1 = \frac{da}{a}, \qquad \alpha^2 = \frac{a\,db - b\,da}{a},$$

forming the dual basis to the Lie algebra basis $\{\mathbf{v}_1, \mathbf{v}_2\}$ of $\mathfrak{a}(1)$ found in (2.11) above. The Maurer–Cartan structure equations (2.34) for the group A(1) are then

$$d\alpha^1 = 0, \qquad d\alpha^2 = \frac{da \wedge db}{a} = \alpha^1 \wedge \alpha^2,$$

reconfirming the Lie algebra commutation relation $[\mathbf{v}_1, \mathbf{v}_2] = -\mathbf{v}_2$. Similarly, the left-invariant Maurer–Cartan forms are the matrix entries of

$$g^{-1} \cdot dg = \begin{pmatrix} a^{-1} & -ba^{-1} \\ 0 & 1 \end{pmatrix} \begin{pmatrix} da & db \\ 0 & 0 \end{pmatrix} = \begin{pmatrix} a^{-1}\,da & a^{-1}\,db \\ 0 & 0 \end{pmatrix}.$$

Therefore,

$$\widehat{\alpha}^1 = \frac{da}{a}, \qquad \widehat{\alpha}^2 = \frac{db}{a},$$

form the dual basis to the left Lie algebra basis in (2.12). Note that the left Maurer–Cartan structure equations have the opposite sign: $d\widehat{\alpha}^1 = 0$, $d\widehat{\alpha}^2 = -\widehat{\alpha}^1 \wedge \widehat{\alpha}^2$, in accordance with the effect of left and right invariance on the structure constants of the Lie algebra.

Exercise 2.96. Find the right and left Lie algebras and the corresponding Maurer–Cartan forms for the three-dimensional Heisenberg group U(3) consisting of all 3×3 upper triangular matrices of the form

$$g = \begin{pmatrix} 1 & x & y \\ 0 & 1 & z \\ 0 & 0 & 1 \end{pmatrix}.$$

Exercise 2.97. Let $\rho^i_j = -\rho^j_i$ be the standard basis for the Maurer–Cartan forms for the orthogonal group SO(n), given as the matrix entries of the skew-symmetric matrix of one-forms $\boldsymbol{\rho} = dR \cdot R^{-1}$, $R \in$ SO(n). Prove that the associated structure equations are $d\rho^i_j = \sum_k \rho^i_k \wedge \rho^k_j$, often written in matrix form as $d\boldsymbol{\rho} = \boldsymbol{\rho} \wedge \boldsymbol{\rho}$.

Exercise 2.98. Prove that a one-form α on G is a right-invariant Maurer–Cartan form if and only if it satisfies $\widehat{\mathbf{v}}(\alpha) = 0$ for all *left*-invariant vector fields $\widehat{\mathbf{v}} \in \mathfrak{g}_L$.

As we shall see, the right-invariant Maurer–Cartan forms on a Lie group play an absolutely crucial role in the solution to equivalence problems based on the Cartan method. Further details on their use and construction can be found later, starting in Chapter 8.

Chapter 3

Representation Theory

Linear actions of Lie groups play a particularly important role, both for theoretical developments, as well as in an astounding variety of applications to both physics and mathematics. In order to distinguish them from the more general nonlinear transformation group actions, linear actions are referred to as "representations" of the group. Representation theory is a vast field of mathematical research, and we shall only have time to survey the most elementary aspects, concentrating on those which are relevant to our subsequent development. More complete treatments can be found in numerous reference texts, including [**119**] and [**223**]. To give some flavor of the remarkable variety of applications of representation theory, we discuss in some detail the representations of importance in classical invariant theory and apply these to equivalence problems for differential operators and the geometry of curves. There is, of course, a huge variety of additional applications, including, for example, special function theory, [**220**], and quantum mechanics, [**160**], [**110**], that we, unfortunately, do not have time to go into here.

Representations

We begin with the basic definition of a representation of a group.

Definition 3.1. A *representation* of a group G is defined by a group homomorphism $\rho: G \to \mathrm{GL}(V)$ from G to the space of invertible linear transformations on a vector space V.

In other words, the representation must satisfy

$$\rho(g \cdot h) = \rho(g)\rho(h), \qquad \rho(e) = \mathbb{1}, \qquad \rho(g^{-1}) = \rho(g)^{-1}, \qquad g, h \in G.$$

The vector space on which the representation acts is not necessarily finite-dimensional, although, in view of our particular applications, we can safely ignore all of the technical and analytical complications inherent in the infinite-dimensional context.

Example 3.2. Consider the general linear group $\mathrm{GL}(n, \mathbb{R})$. The simplest possible representation is the trivial one-dimensional representation that assigns to each matrix $A \in \mathrm{GL}(n)$ the real number 1. Slightly more interesting is the one-dimensional determinantal representation $\rho(A) = \det A$, or, more generally, any power $\rho^i(A) = (\det A)^i$ of the determinant. An obvious n-dimensional representation is provided by the identity representation $\rho_e(A) = A$ acting on $V = \mathbb{R}^n$. A second n-dimensional representation is the so-called *contragredient* or inverse transpose representation $\rho_c(A) = A^{-T}$. Higher dimensional representations can be constructed by a variety of techniques. For example, the conjugation action $X \mapsto AXA^{-1}$ defines a representation of $\mathrm{GL}(n)$ on the space of $n \times n$ matrices, as does the action $X \mapsto AXA^T$. The latter representation restricts to the subspaces consisting of all symmetric, or all skew-symmetric matrices. The action of $\mathrm{GL}(n)$ on $\mathbb{R}^n$ induces representations on all of the tensor spaces over $\mathbb{R}^n$. For example, on the exterior powers $\bigwedge^k \mathbb{R}^n$, the induced representation is given by $v_1 \wedge \cdots \wedge v_k \mapsto (Av_1) \wedge \cdots \wedge (Av_k)$. Note that the determinantal representation can be identified with the exterior representation of $\mathrm{GL}(n)$ on $\bigwedge^n \mathbb{R}^n \simeq \mathbb{R}$, whereas the representation on $\bigwedge^2 \mathbb{R}^n$ coincides with the skew-symmetric matrix representation mentioned above. All these constructions have their complex counterparts, valid for $\mathrm{GL}(n, \mathbb{C})$.

Clearly, if $G \subset \mathrm{GL}(n)$ is any matrix Lie group, then each of the above representations of $\mathrm{GL}(n)$ restricts to define a representation of the subgroup G. However, different representations of $\mathrm{GL}(n)$ may give rise to the same representation of the subgroup G. For instance, the determinantal representation becomes trivial when restricted to the special linear group $\mathrm{SL}(n)$.

Different representations can also be combined using the basic tensorial operations. Given two representations ρ on V and σ on W respectively, one can form the direct sum $\rho \oplus \sigma$, which is representation on the sum $V \oplus W$ of the two representation spaces. Similarly, the tensor product $\rho \otimes \sigma$ defines a representation on the tensor product space $V \otimes W$. Moreover, one can compose any finite-dimensional representation $\rho: G \to \mathrm{GL}(n)$ with any of the representations of $\mathrm{GL}(n)$ to induce yet further representation of G, such as $\det \rho$, $\bigwedge^k \rho$, etc.

Exercise 3.3. Prove that the tensor product of the contragredient representation of GL(2) with the determinantal representation is the same as the identity representation. Is this result true for GL(n)?

Exercise 3.4. Let G be a Lie group and $\mathfrak{g}$ its Lie algebra. Prove that the differential of the conjugation action $h \mapsto ghg^{-1}$ of G on itself defines a representation of G on its Lie algebra $\mathfrak{g}$, called the *adjoint representation* of G. Prove that the adjoint representation of GL(n) coincides with the conjugation representation $X \mapsto AXA^{-1}$.

If $\rho: G \to \mathrm{GL}(V)$ is any representation, and $W \subset V$ is an invariant subspace, meaning that $\rho(g)W \subset W$ for all $g \in G$, then the representation ρ reduces to a subrepresentation of G on W. Trivial subrepresentations are when $W = \{0\}$ or $W = V$. A *reducible* representation is one that contains a nontrivial subrepresentation; an *irreducible* representation, then, is one that has no nontrivial invariant subspaces (subrepresentations). For example, the direct sum of two representations is reducible since each appears as a subrepresentation therein. In some cases — for instance, when the group is compact, or the general tensor representations of GL(n) — all finite-dimensional reducible representations can be decomposed into a direct sum of irreducible subrepresentations. However, this is not uniformly true; a simple counterexample is provided by the reducible two-dimensional representation $\rho(A) = \begin{pmatrix} 1 & \log|\det A| \\ 0 & 1 \end{pmatrix}$ of the general linear group GL(n). The classification of irreducible representations of Lie groups is a major topic in representation theory. However, to keep our exposition reasonably brief, we shall be content to concentrate on a few particular examples in this book.

Each representation of a connected Lie group G induces, and is generated by, a corresponding "infinitesimal" representation of its Lie algebra. Because of our convention that group elements in GL(V) act on the left on V, we shall consistently use the right Lie algebra $\mathfrak{g} = \mathfrak{g}_R$ on the group G; see our earlier discussion in Chapter 2. This requires us to adopt the bracket $[A, B] = BA - AB$, which is the negative of the usual matrix commutator, on the Lie algebra $\mathfrak{gl}(V)$ of linear operators on V — see Example 2.42.

Definition 3.5. A *representation* of a Lie algebra $\mathfrak{g}$ is defined by a Lie algebra homomorphism $\rho: \mathfrak{g} \to \mathfrak{gl}(V)$ to the space of linear maps on a vector space V.

In other words, a Lie algebra representation is a linear map which preserves the Lie bracket operation: $\rho([\mathbf{v},\mathbf{w}]) = \rho(\mathbf{w})\rho(\mathbf{v}) - \rho(\mathbf{v})\rho(\mathbf{w})$. Given a representation ρ of a Lie group G, the infinitesimal version

$$\rho(\mathbf{v})x = \frac{d}{dt}\,\rho(\exp(t\mathbf{v}))x \,\bigg|_{t=0}, \qquad x \in V, \quad \mathbf{v} \in \mathfrak{g},$$

defines the representation of the Lie algebra $\mathfrak{g}$. If the representation $\rho: G \longrightarrow \mathrm{GL}(n)$ acts on $\mathbb{R}^n$, with coordinates $x = (x^1, \ldots, x^n)$, its infinitesimal generators $\mathbf{v}_A = \sum_{i,j=1}^n a^i_j x^j \partial_{x^i}$ are linear vector fields on $\mathbb{R}^n$. The Lie algebra representation $\rho: \mathfrak{g} \longrightarrow \mathfrak{gl}(n)$ is obtained by identifying the vector field $\mathbf{v}_A$ with the matrix $A = (a^i_j) \in \mathfrak{gl}(n)$. As above, the Lie bracket corresponds to the negative of the usual matrix commutator.

Exercise 3.6. Define irreducibility for a Lie algebra representation. Prove that a representation of a connected Lie group G is irreducible if and only if the associated Lie algebra representation is irreducible.

Exercise 3.7. Prove that the infinitesimal adjoint representation of a Lie algebra $\mathfrak{g}$ on itself is given by the Lie bracket $\rho(\mathbf{v})\mathbf{w} = [\mathbf{v}, \mathbf{w}]$.

Representations on Function Spaces

One easy way to turn a nonlinear group action into a linear representation is to look at its induced action on the functions on the manifold. (Indeed, this is how symmetries manifest themselves in the passage from classical to quantum mechanics.) Let G be a Lie group acting on a manifold M. Then there is a naturally induced representation of G on the space $\mathcal{F} = \mathcal{F}(M)$ of real-valued functions $F: M \longrightarrow \mathbb{R}$, which maps the function F to the function $\overline{F} = g \cdot F$ defined by

$$\overline{F}(\bar{x}) = F(g^{-1} \cdot \bar{x}), \qquad \text{or, equivalently,} \qquad \overline{F}(g \cdot x) = F(x). \tag{3.1}$$

The introduction of the inverse g^{-1} in this formula ensures that the action of G on $\mathcal{F}$ is a group homomorphism: $g \cdot (h \cdot F) = (g \cdot h) \cdot F$ for all $g, h \in G$, and $F \in \mathcal{F}$. This construction clearly extends to vector-valued functions $F: M \longrightarrow \mathbb{R}^k$.

Exercise 3.8. Prove that if ρ is a representation on a space of functions $F(x)$ on a manifold M, then (up to sign) the Lie algebra representation can be identified with the action of the infinitesimal generators of G on functions, so $\rho(\mathbf{v})F = \mathbf{v}(F)$ for $\mathbf{v} \in \mathfrak{g}$.

The representation of G on the (infinite dimensional) function space $\mathcal{F}$ will usually decompose into a wide variety of important subrepresentations, including representations on spaces of polynomial functions, representations on spaces of smooth (C^∞) functions, analytic functions, normalizable (L^2) functions, etc.

Example 3.9. A particularly important example is the standard representation of the general linear group $\mathrm{GL}(2) = \mathrm{GL}(2,\mathbb{R})$ acting on $\mathbb{R}^2$, and its complex counterpart $\mathrm{GL}(2,\mathbb{C})$ acting on $\mathbb{C}^2$. The induced representation (3.1) on the space of real-valued functions $Q\colon \mathbb{R}^2 \to \mathbb{R}$ is

$$\overline{Q}(\bar{x},\bar{y}) = \overline{Q}(\alpha x + \beta y, \gamma x + \delta y) = Q(x,y), \quad \text{where} \quad \begin{pmatrix} \alpha & \beta \\ \gamma & \delta \end{pmatrix} \in \mathrm{GL}(2). \tag{3.2}$$

In particular, if Q is a homogeneous polynomial, so is $\overline{Q}$, so that the representation (3.2) includes the fundamental finite-dimensional subrepresentations $\rho_n = \rho_{n,0}$ of $\mathrm{GL}(2)$ on $\mathcal{P}^{(n)}$, the space of homogeneous polynomials of degree n. The analysis of these representations of $\mathrm{GL}(2)$ (and their generalizations to $\mathrm{GL}(n)$) forms the focus of classical invariant theory. In the classical literature, e.g., [**115**], [**99**], a homogeneous polynomial is called a "form". (Not to be confused with a differential form!) Thus a real-valued *binary form* of degree n takes the general form

$$Q(x,y) = \sum_{i=0}^{n} \binom{n}{i} a_i \, x^i y^{n-i}, \tag{3.3}$$

where the coefficients $a_0, \ldots, a_n \in \mathbb{R}$, and the binomial factors are included to conform with common conventions. The explicit action of $\mathrm{GL}(2)$ on the coefficients a_i is not hard to determine. For example, the coefficients of a general linear polynomial $Q(x,y) = ax + by$ will transform according to

$$\begin{pmatrix} \alpha & \gamma \\ \beta & \delta \end{pmatrix} \begin{pmatrix} \bar{a} \\ \bar{b} \end{pmatrix} = \begin{pmatrix} a \\ b \end{pmatrix}, \qquad A = \begin{pmatrix} \alpha & \beta \\ \gamma & \delta \end{pmatrix} \in \mathrm{GL}(2). \tag{3.4}$$

Therefore, the representation ρ_1 on the space $\mathcal{P}^{(1)}$ can be identified with the contragredient representation $A \mapsto A^{-T}$ of $\mathrm{GL}(2)$ on $\mathbb{R}^2$. On the space of quadratic polynomials

$$Q(x,y) = ax^2 + 2bxy + cy^2, \tag{3.5}$$

the induced representation ρ_2 acts on the coefficients via the second "symmetric power" of the contragredient representation:

$$\begin{pmatrix} \alpha^2 & 2\alpha\gamma & \gamma^2 \\ \alpha\beta & \alpha\delta+\beta\gamma & \gamma\delta \\ \beta^2 & 2\beta\delta & \delta^2 \end{pmatrix} \begin{pmatrix} \bar a \\ \bar b \\ \bar c \end{pmatrix} = \begin{pmatrix} a \\ b \\ c \end{pmatrix}, \qquad A = \begin{pmatrix} \alpha & \beta \\ \gamma & \delta \end{pmatrix} \in \mathrm{GL}(2). \tag{3.6}$$

A particularly important class of representations is obtained by tensoring the preceding representations with powers of the determinantal representation, so that the polynomial $Q(x,y)$ transforms to $\overline{Q}(\bar x, \bar y)$, which is given by

$$Q(x,y) = (\alpha\delta - \beta\gamma)^k \overline{Q}(\bar x, \bar y) = (\alpha\delta - \beta\gamma)^k \overline{Q}(\alpha x + \beta y, \gamma x + \delta y). \tag{3.7}$$

The restriction of the representation (3.7) to the space $\mathcal{P}^{(n)}$ of homogeneous polynomials serves to define the fundamental representation $\rho_{n,k}$ of *weight* k and *degree* n. For example, according to Exercise 3.3 above, the representation $\rho_{1,1}$ of weight 1 and degree 1 is isomorphic to the identity representation of GL(2).

Theorem 3.10. *The representation $\rho_{n,k}$ of* GL(2) *is irreducible.*

Proof: It suffices to prove that its restriction to the unimodular subgroup $\mathrm{SL}(2) \subset \mathrm{GL}(2)$ is irreducible. In view of Exercise 3.6, the easiest approach is to work infinitesimally, using the generators $\mathbf{v}_- = y\partial_x$, $\mathbf{v}_0 = x\partial_x - y\partial_y$, $\mathbf{v}_+ = x\partial_y$. Given a nonzero invariant subspace $\{0\} \neq W \subset \mathcal{P}^{(n)}$, let $Q(x,y) = c\,x^i y^{n-i} + d\,x^{i+1}y^{n-i-1} + \cdots \in W$, $c \neq 0$, be any nonzero element. Applying the "raising operator" $\mathbf{v}_+$ repeatedly, we find $(\mathbf{v}_+)^{n-i}Q$ is a nonzero multiple of x^n, and hence $x^n \in W$. Then, applying the "lowering operator" $\mathbf{v}_-$ repeatedly, we find $(\mathbf{v}_-)^j x^n$, which is a nonzero multiple of $x^{n-j}y^j$, must also lie in W. Therefore $W = \mathcal{P}^{(n)}$, which proves irreducibility. *Q.E.D.*

It is possible to prove that the representations $\rho_{n,k}$ provide a complete list of all irreducible finite-dimensional representations of GL(2); moreover, any tensor representation can be decomposed into a direct sum of the irreducible representations $\rho_{n,k}$, cf. [**223**].

If $n \geq 2$, the restriction of the representation $\rho_{n,k}$ to the rotation subgroup $\mathrm{SO}(2) \subset \mathrm{GL}(2)$ is no longer irreducible. For example, the representation of SO(2) on the space of quadratic polynomials $\mathcal{P}^{(2)}$ decomposes into the sum of two irreducible subrepresentations, a trivial one on the one-dimensional subspace spanned by $x^2 + y^2$, and a two-dimensional representation on the subspace spanned by $x^2 - y^2$ and xy.

Exercise 3.11. Decompose the restriction of $\rho_{n,k}$ to the subgroup $\mathrm{SO}(2) \subset \mathrm{GL}(2)$ into irreducible subrepresentations. *Hint*: Note that the subspaces $\{(x^2+y^2)^m Q(x,y) \mid Q \in \mathcal{P}^{(n-2m)}\}$ are rotationally invariant.

Multiplier Representations

Although induced actions of transformation groups on functions provide us with a wide variety of important representations, these are not quite general enough for our purposes. In order to appreciate this, consider the linear fractional action (2.4) of the group $\mathrm{GL}(2,\mathbb{R})$ on the projective line $\mathbb{RP}^1$. According to the preceding construction, cf. (3.1), this induces the representation

$$\overline{F}(\bar{p}) = \overline{F}\left(\frac{\alpha p + \beta}{\gamma p + \delta}\right) = F(p)$$

on the space of real-valued functions on $\mathbb{RP}^1$, written in terms of the projective coordinate $p = x/y$. However, with the exception of the constants, which are homogeneous of degree 0, this representation does not naturally contain any of the irreducible polynomial representations $\rho_{n,k}$ discussed above. Indeed, each homogeneous polynomial (3.3) has an inhomogeneous representative, given by

$$F(p) = Q(p,1) = \sum_{i=0}^{n} \binom{n}{i} a_i\, p^i. \tag{3.8}$$

The induced action of GL(2) corresponding to the representation $\rho_{n,k}$, cf. (3.7) is readily seen to be

$$\begin{aligned} F(p) &= (\alpha\delta - \beta\gamma)^k (\gamma p + \delta)^n \overline{F}(\bar{p}) \\ &= (\alpha\delta - \beta\gamma)^k (\gamma p + \delta)^n \overline{F}\left(\frac{\alpha p + \beta}{\gamma p + \delta}\right). \end{aligned} \tag{3.9}$$

For example, a quadratic polynomial $Q(x,y) = xy + 2y^2$ (of weight 0) has inhomogeneous representative $F(p) = p + 2$. The linear map $(x,y) \longmapsto (y,x)$ maps Q to $\overline{Q}(\bar{x},\bar{y}) = 2\bar{x}^2 + \bar{x}\bar{y}$, which has inhomogeneous form $\overline{F}(\bar{p}) = 2\bar{p}^2 + \bar{p}$. The corresponding projective map is the inversion $\bar{p} = 1/p$, and, in agreement with (3.9), $F(p) = p^2\overline{F}(1/p)$. Note, in particular, that even though F looks like a linear polynomial, its transformation properties are not linear, but rather coincide with those of a quadratic polynomial. Indeed, $F(p)$ should be viewed as a degenerate quadratic, whose second root is at $p = \infty$. Consequently, one cannot directly ascertain the transformation properties of an inhomogeneous polynomial, since it can serve as the representative of many inequivalent homogeneous forms.

The linear action (3.9) defines a more general kind of representation of the projective group on the space of functions over $\mathbb{RP}^1$. Such representations are known as multiplier representations, and the prefactor $(\alpha\delta - \beta\gamma)^k(\gamma p + \delta)^n$, or, rather, its reciprocal, is called the *multiplier*. The general definition follows,† cf. [**17**], [**169**], [**223**].

Definition 3.12. Let G be a group acting on a manifold M. A *multiplier representation* of G is a representation $\overline{F} = g \cdot F$ on the space of real-valued functions $\mathcal{F}(M)$ of the particular form

$$\overline{F}(\bar{x}) = \overline{F}(g \cdot x) = \mu(g, x)F(x), \qquad g \in G, \quad F \in \mathcal{F}. \tag{3.10}$$

The condition that (3.10) actually defines a representation of the group G requires that the multiplier μ satisfy a certain algebraic identity. A straightforward analysis of the group law $g \cdot (h \cdot F) = (g \cdot h) \cdot F$ demonstrates the following characterization of the multiplier.

Lemma 3.13. *A function $\mu: G \times M \to \mathbb{R}$ is a multiplier for a transformation group G acting on a manifold M if and only if it satisfies the* multiplier equation

$$\begin{aligned} \mu(g\cdot h, x) &= \mu(g, h\cdot x)\, \mu(h, x), \\ \mu(e, x) &= 1, \end{aligned} \qquad \text{for all} \qquad \begin{aligned} &g, h \in G, \\ &x \in M. \end{aligned} \tag{3.11}$$

If $\mu(g, x)$ and $\widehat{\mu}(g, x)$ are multipliers for a group G, so is their product $\mu(g, x) \cdot \widehat{\mu}(g, x)$, as is any power $\mu(g, x)^k$.

Example 3.14. In the case of the usual projective action (2.4) of the general linear group GL(2), the function

$$\mu_{n,k}\left(\begin{pmatrix} \alpha & \beta \\ \gamma & \delta \end{pmatrix}, p\right) = (\alpha\delta - \beta\gamma)^{-k}(\gamma p + \delta)^{-n} \tag{3.12}$$

† This definition of multiplier representation is *not* the same as that appearing in the work of Mackey, [**160**]; these are also known as projective representations, [**110**].

satisfies the multiplier equation (3.11), and hence defines a multiplier representation. For brevity, we shall say a scalar function $F(p)$ has *weight* (n, k) if it transforms according to the multiplier $\mu_{n,k}$, i.e., satisfies (3.9) under a linear fractional transformation. The restriction of this multiplier representation to the space $\mathcal{P}^n$ of polynomials of degree $\leq n$ coincides with the fundamental representation $\rho_{n,k}$ defined by (3.9).

There is a trivial way to obtain multipliers from the ordinary representation of a transformation group G on the function space $\mathcal{F}(M)$. Suppose we multiply every function by a fixed nonvanishing function $\eta\colon M \to \mathbb{R} \setminus \{0\}$, and set $F^*(x) = \eta(x)F(x)$ for each $F \in \mathcal{F}$. The function η is sometimes known as a *gauge factor*, and the operation of multiplying by η known as a *change of gauge.* This terminology comes from physics, and has its origins in Weyl's speculative attempt, [**225**], to unify electromagnetism and gravity. Our use of the term "gauge" is closer in spirit to Weyl's original coinage since, unlike the modern definition appearing in quantum electrodynamics, cf. [**224**], [**19**], our gauge factors $\eta(x) = e^{\varphi(x)}$ are not restricted to be of modulus 1, i.e., $\varphi(x)$ is not restricted to be purely imaginary. In terms of the new choice of gauge, then, the standard representation (3.1) takes the modified form

$$\overline{F}^*(\bar{x}) = \frac{\eta(\bar{x})}{\eta(x)}\, F^*(x), \qquad \text{where} \qquad \bar{x} = g\cdot x, \quad \overline{F}^* = g\cdot F^*.$$

The function $\mu(g, x) = \eta(g \cdot x)/\eta(x)$ is readily seen to satisfy the multiplier equation (3.11), and so defines a *trivial multiplier*, whose associated representation is equivalent to the usual representation of function space. More generally, we will call two multipliers μ and $\widetilde{\mu}$ *gauge equivalent* if they are related by a change of gauge:

$$\widetilde{\mu}(g, x) = \frac{\eta(g\cdot x)}{\eta(x)}\, \mu(g, x), \tag{3.13}$$

for some nonzero function $\eta(x)$, and thus prescribe the same representation, up to multiplication by a function.

Although the definition of multiplier representation was stated for scalar-valued functions, it can be readily extended to group actions on vector-valued functions, $F\colon M \to \mathbb{R}^n$. In this case, a multiplier $\mu(g, x)$ is a matrix-valued function on the group, $\mu\colon G \times M \to \mathrm{GL}(n)$, satisfying the matrix analogue of the multiplier equation (3.11).

Example 3.15. If G acts on M, then there is an induced action on the space of vector fields on M which takes the form of a multiplier representation. We identify a vector field $\mathbf{v} = \sum_{i=1}^{m} \xi^i(x)\partial_{x^i}$ with the vector-valued function $\xi(x) = (\xi^1(x), \ldots, \xi^m(x))$. According to (1.15) the action of the differential $\overline{\mathbf{v}} = dg(\mathbf{v})$ of a group transformation $g \cdot x = \chi_g(x)$ induces the Jacobian multiplier representation

$$\overline{\xi}(\bar{x}) = \mu_J(g, x)\xi(x), \qquad \text{where} \qquad \mu_J(g, x) = \left(\frac{\partial \chi_g^i}{\partial x^j}\right) \tag{3.14}$$

on the associated coefficient functions. The multiplier equation (3.11) in this case reduces to the usual chain rule formula for the Jacobian of the composition of two group transformations.

Exercise 3.16. What is the definition of a trivial matrix multiplier representation? Discuss gauge equivalence in this context.

There is an alternative, fully geometrical approach to the theory of multiplier representations, that allows us to use standard results on Lie group actions in this more general context.† Let $\mu: G \times M \to \mathrm{GL}(q)$ be a matrix-valued multiplier associated with a multiplier representation on the space of vector-valued functions $F: M \to U$, where $U \simeq \mathbb{R}^q$. We introduce the "extended space" $E = M \times U$, and consider the extended group action

$$g: (x, u) \longmapsto (g \cdot x, \mu(g, x)u), \qquad g \in G, \quad x \in M, \quad u \in U, \tag{3.15}$$

which is linear in the u coordinates.‡ The reader should verify that the condition that (3.15) define a local group action extending the action of G on M is equivalent to the condition that μ satisfy the multiplier equation (3.11). This means that the theory of multiplier representations is entirely equivalent to the theory of linearly extended local group actions (3.15). In particular, the Jacobian multiplier representation of Example 3.15 can be naturally identified with the extended action of the group G on the tangent bundle TM provided by the differentials $dg: TM \to TM$ of the group transformations $g \in G$.

† This construction is a special case of the general Lie theory of group actions on spaces of functions to be presented in Chapter 4.

‡ The action (3.15) can be easily generalized to the category of vector bundles over M; however, as all our considerations are local, we shall not lose any generality by restricting our attention to the Cartesian product case.

Exercise 3.17. Discuss the multiplier representation corresponding to the action of a transformation group on the space of differential one-forms on the manifold. Generalize to differential k-forms.

Infinitesimal Multipliers

As usual, for Lie groups, multiplier representations are most effectively treated using infinitesimal methods. Let G be a connected Lie group acting on a manifold M, and let $\mu(g,x)$ be a scalar multiplier. Consider first the geometrical version (3.15) of the multiplier action on the extended space $E = M \times U \simeq M \times \mathbb{R}$. The infinitesimal generators of this action are vector fields

$$\widetilde{\mathbf{v}} = \sum_{i=1}^{p} \xi^i(x) \frac{\partial}{\partial x^i} + h(x)u \frac{\partial}{\partial u}, \tag{3.16}$$

which are linear in the coordinate u. Here $\widehat{\mathbf{v}} = \sum_i \xi^i \partial_{x^i}$ is the infinitesimal generator of the action of G on M corresponding to the Lie algebra element $\mathbf{v} \in \mathfrak{g}$. The linear map $\sigma: \mathfrak{g} \to \mathcal{F}(M)$ which takes a Lie algebra element $\mathbf{v} \in \mathfrak{g}$ to the coefficient function $\sigma(\mathbf{v}) = h(x) = h_{\mathbf{v}}(x)$ in the extended infinitesimal generator (3.16) is called the *infinitesimal multiplier* for the multiplier representation. Applying equation (2.15) to the particular form (3.15) of the extended action, we find that the infinitesimal multiplier can be explicitly computed by the formula

$$\sigma(\mathbf{v}) = h_{\mathbf{v}}(x) = \frac{d}{dt}\, \mu(\exp t\mathbf{v}, x) \bigg|_{t=0}. \tag{3.17}$$

Exercise 3.18. Prove that a multiplier can be reconstructed from its infinitesimal form via the series

$$\begin{aligned} \mu(\exp t\mathbf{v}, x) &= 1 + t h_{\mathbf{v}}(x) + \tfrac{1}{2} t^2 [\mathbf{v}(h_{\mathbf{v}}) + h_{\mathbf{v}}^2] + \cdots \\ &= \exp\left[\int h_{\mathbf{v}}[\exp(t\mathbf{v})x]\, dt \right]. \end{aligned} \tag{3.18}$$

Note that the sum $\sigma + \widehat{\sigma}$ of two infinitesimal multipliers is also an infinitesimal multiplier, corresponding to the product $\mu \cdot \widehat{\mu}$ of the two multipliers. In particular, any constant multiple $k\sigma$ of an infinitesimal multiplier is also an infinitesimal multiplier, corresponding to the power μ^k of the associated multiplier.

It is convenient to identify the infinitesimal generator (3.16) of the extended multiplier action with the first order differential operator

$$\mathcal{D} = \mathcal{D}_{\mathbf{v}} = \widehat{\mathbf{v}} - \sigma(\mathbf{v}) = \sum_{i=1}^{p} \xi^i(x)\,\frac{\partial}{\partial x^i} - h(x), \tag{3.19}$$

on the original space M. (The reason for the apparently gratuitous change in sign will appear shortly.) In practical computations, this form of the infinitesimal multiplier representation is more convenient than the vector field version. It is readily verified that the Lie bracket between two such vector fields is mapped to the commutator

$$\mathcal{D}_{\mathbf{u}} = [\mathcal{D}_{\mathbf{v}}, \mathcal{D}_{\mathbf{w}}] = \mathcal{D}_{\mathbf{v}} \cdot \mathcal{D}_{\mathbf{w}} - \mathcal{D}_{\mathbf{w}} \cdot \mathcal{D}_{\mathbf{v}}, \qquad \widehat{\mathbf{u}} = [\widehat{\mathbf{v}}, \widehat{\mathbf{w}}], \tag{3.20}$$

between the corresponding differential operators. Since the infinitesimal generators of the extended group action form a Lie algebra of vector fields on E having the same commutation relations as those in the Lie algebra $\mathfrak{g}$, the corresponding differential operators (3.19) also form a Lie algebra of differential operators also obeying the same commutation relations. If G does not act locally effectively on M, the associated multiplier function may still be nontrivial, so G could act effectively on E. In such cases, some Lie algebra elements $\mathbf{v} \in \mathfrak{g}$ may be mapped to the zero vector field on M, so $\widehat{\mathbf{v}} = 0$, in which case the associated differential operator $\mathcal{D}_{\mathbf{v}} = -h(x)$ reduces to a pure *multiplication operator*, i.e., a differential operator of order 0. On the other hand, if G acts locally effectively, then we can identify $\mathbf{v} \in \mathfrak{g}$ with the vector field $\widehat{\mathbf{v}}$ on M, and the function h in (3.19) depends unambiguously on the vector field.

Example 3.19. The standard action of $\mathrm{GL}(2)$ on $\mathbb{R}^2$ has infinitesimal generators $x\partial_x$, $x\partial_y$, $y\partial_x$, and $y\partial_y$. The determinantal multiplier $\mu(A, \mathbf{x}) = \det A$, $A \in \mathrm{GL}(2)$, $\mathbf{x} \in \mathbb{R}^2$, has corresponding infinitesimal generators $x\partial_x - 1$, $x\partial_y$, $y\partial_x$, and $y\partial_y - 1$. In other words, the infinitesimal multiplier $\sigma\colon \mathfrak{gl}(2) \to \mathcal{F}(\mathbb{R}^2)$ is the linear map satisfying $\sigma(x\partial_x) = \sigma(y\partial_y) = 1$, $\sigma(x\partial_y) = \sigma(y\partial_x) = 0$. The differential operators obey the same $\mathfrak{gl}(2)$ commutation relations as the original vector fields.

The associated projective action of $\mathrm{GL}(2)$ on $\mathbb{RP}^1$ by linear fractional transformations, (2.4), is not effective. The infinitesimal generators are ∂_p, $p\partial_p$, $p^2\partial_p$, and 0, the latter coming from the trivial action of the diagonal subgroup $\{\lambda \mathbb{1}\} \subset \mathrm{GL}(2)$. The multiplier $\mu_{n,k}$ given in (3.12) is readily seen to have associated infinitesimal generators ∂_p, $p\partial_p - \frac{1}{2}n$, $p^2\partial_p - np$, and $2k$, the latter denoting the multiplication operator by the constant function $2k$. Again, these form a Lie algebra of differential operators isomorphic to $\mathfrak{gl}(2)$, having a locally effective action on the extended space when $k \neq 0$.

Theorem 3.20. *Let $\sigma: \mathfrak{g} \to \mathcal{F}(M)$ be a linear function. The following conditions are equivalent:*

(*i*) *σ is an infinitesimal multiplier.*

(*ii*) *σ satisfies*

$$\sigma([\mathbf{v}, \mathbf{w}]) = \widehat{\mathbf{v}}(\sigma(\mathbf{w})) - \widehat{\mathbf{w}}(\sigma(\mathbf{v})) \qquad \text{for all} \quad \mathbf{v}, \mathbf{w} \in \mathfrak{g}. \tag{3.21}$$

(*iii*) *The differential operators $\mathcal{D}_{\mathbf{v}} = \widehat{\mathbf{v}} - \sigma(\mathbf{v})$ form a Lie algebra of first order differential operators having the same commutation relations as $\mathfrak{g}$.*

Proof: Condition (3.21) arises from writing out the commutator relations (3.20) in detail. Using the vector field version (3.16) of the infinitesimal generators, Theorem 2.63 implies that G acts locally on the extended space E. Moreover, Exercise 1.26 implies that the action is linear in the u coordinate, i.e., of the form (3.15), and hence defines a multiplier representation. *Q.E.D.*

The infinitesimal characterization of trivial and equivalent multiplier representations is straightforward. The proof of the following result is left as an exercise for the reader.

Theorem 3.21. *Let $\mathcal{D}_{\mathbf{v}} = \widehat{\mathbf{v}} - \sigma(\mathbf{v})$ and $\widetilde{\mathcal{D}}_{\mathbf{v}} = \widehat{\mathbf{v}} - \widetilde{\sigma}(\mathbf{v})$ be two infinitesimal multiplier representations of the same connected Lie group action. The following conditions are equivalent:*

(*i*) *σ and $\widetilde{\sigma}$ generate equivalent multiplier representations.*

(*ii*) *The associated Lie algebras of differential operators are* gauge equivalent *meaning that there exists a nonvanishing function $\eta(x)$ satisfying*

$$\widetilde{\mathcal{D}}_{\mathbf{v}} = \eta \cdot \mathcal{D}_{\mathbf{v}} \cdot \frac{1}{\eta}, \qquad \text{for all} \qquad \mathbf{v} \in \mathfrak{g}. \tag{3.22}$$

(*iii*) *There exists a function $\varphi(x)$ such that the infinitesimal multipliers satisfy*

$$\widetilde{\sigma}(\mathbf{v}) = \sigma(\mathbf{v}) - \widehat{\mathbf{v}}(\varphi) \qquad \text{for all} \qquad \mathbf{v} \in \mathfrak{g}. \tag{3.23}$$

In particular, the infinitesimal generators of a trivial multiplier representation all have the form $\mathcal{D}_{\mathbf{v}} = \widehat{\mathbf{v}} - \widehat{\mathbf{v}}(\varphi)$ for some function $\varphi(x)$. In view of Theorems 3.20 and 3.21, the classification of scalar multiplier representations is the same as the classification of Lie algebras of first order differential operators up to gauge equivalence. More specifically, we shall call two Lie algebras of differential operators *gauge equivalent* if there exists a change of variables $\bar{x} = \chi(x)$ such that, under the change of variables, the two sets of generators are related as in (3.22) for some gauge factor $\eta(x) = e^{\varphi(x)}$. The problem of classifying Lie algebras of first order differential operators has important applications in the recent theory of "quasi-exactly solvable Schrödinger operators", which are differential operators that can be written as quadratic combinations of the generators of some finite-dimensional Lie algebra of first order differential operators, forming a "hidden" symmetry algebra of the operator. This remarkable theory is discussed at length in the survey paper [**97**] and the recent book [**217**].

Remark: The conditions (3.21), (3.23), have a convenient interpretation in terms of Lie algebra cohomology, cf. [**124**]. The first condition (3.21) says that a multiplier is a 1-cocyle, whereas (3.23) says that two equivalent multipliers differ by a coboundary. Therefore, the infinitesimal multipliers are classified by the Lie algebra cohomology space $\mathrm{H}^1(\mathfrak{g}, \mathcal{F}(M))$. See [**95**], [**97**], for more details.

Example 3.22. Let us apply Theorem 3.21 to classify the multiplier representations associated with the three inequivalent finite-dimensional group actions on the real line $\mathbb{R}$, as described in Theorem 2.70. The simplest is the translation group $x \mapsto x+t$, with infinitesimal generator ∂_x. Any multiplier representation will have infinitesimal generator $\partial_x - f(x)$, corresponding to the representation $u(x) \mapsto \mu(t,x)u(x-t)$, with $f(x) = \mu_t(0,x)$, cf. (3.17). In this case, the multiplier is always trivial since, according to (3.23), the trivial infinitesimal multipliers have the form $\partial_x - \varphi_x$, so that we can choose $\varphi(x) = \int f(x)\,dx$. Indeed, the multiplier equation (3.11) in this case is $\mu(s+t,x) = \mu(s,t+x)\mu(s,x)$ for $s,t,x \in \mathbb{R}$. If we set $x = 0$ and replace s by x we find $\mu(t,x) = \eta(t+x)/\eta(x)$, where $\eta(x) = \mu(x,0)$, thus re-proving the triviality of μ directly.

Any multiplier representation for the affine group $x \mapsto ax+b$ will have infinitesimal generators $\partial_x - f(x), x\partial_x - g(x)$, where, according to the Lie bracket condition (3.21), f and g must satisfy the differential equation $g_x - xf_x = f$. This implies that $g(x) = xf(x) + k$ for some constant k. On the other hand, by (3.23), a trivial infinitesimal multiplier has the form $\partial_x - \varphi_x$, $x\partial_x - x\varphi_x$. Choosing $\varphi(x) = \int f(x)dx$ as before, we see that the multiplier representation is equivalent to one with infinitesimal generators $\partial_x, x\partial_x - k$, corresponding to the multiplier $\mu((a,b),x) = a^k$.

For the projective action of the unimodular group SL(2) by linear fractional transformations, any multiplier representation will have infinitesimal generators

$$\partial_x - f(x), \qquad x\partial_x - g(x), \qquad x^2\partial_x - h(x),$$

where, according to (3.21), f, g, h, satisfy the differential equations

$$g_x - xf_x = f, \qquad h_x - x^2 f_x = 2g, \qquad xh_x - x^2 g_x = h. \tag{3.24}$$

On the other hand, a trivial infinitesimal multiplier has the form

$$\partial_x - \varphi_x, \qquad x\partial_x - x\varphi_x, \qquad x^2\partial_x - x^2\varphi_x.$$

If $\varphi(x) = \int f(x)dx$, then (3.24) implies that $g(x) = x\varphi_x + c$, and $h(x) = x^2\varphi_x + 2cx$, where c is a constant. Therefore, every infinitesimal multiplier is equivalent to one of the form

$$\partial_x, \qquad x\partial_x - c, \qquad x^2\partial_x - 2cx. \tag{3.25}$$

Example 3.19 shows that the associated multiplier is $(\gamma p+\delta)^{-2c}$, arising from the restriction of the multiplier (3.12), for $c = \frac{1}{2}n$, to the unimodular subgroup. Thus, Theorems 3.20 and 3.21 imply that *every* multiplier representation associated to the projective action of SL(2) is equivalent to the fundamental multiplier representation with multiplier $(\gamma p + \delta)^{-n}$ for some n (not necessarily integral).

Exercise 3.23. Prove that every multiplier for the projective action of the complex general linear group $\mathrm{GL}(2,\mathbb{C})$ is equivalent to one of the fundamental multipliers $\mu_{n,k}$ given in (3.12). The same holds for the connected component $\mathrm{GL}(2,\mathbb{R})^+$, while $\mathrm{GL}(2,\mathbb{R})$ admits an additional discrete multiplier $A \mapsto \operatorname{sign}\det A$.

The results of Example 3.22, coupled with the classification of possible Lie group actions on a one-dimensional manifold appearing in Theorem 2.70, provides a complete classification result for multiplier representations associated with one-dimensional locally effective group actions (without fixed points). In essence, they all arise as restrictions of the multiplier representations for $\mu_{n,k}$ of the projective group.

Exercise 3.24. Determine all the one-dimensional, noneffective multiplier representations without fixed points; see [**131**].

In [**95**], the same basic approach was used to completely classify multiplier representations for each of the complex planar transformation groups, as described in Tables 1–3 at the end of the book. The full results include some fairly complicated cases, and will not be presented here.

Exercise 3.25. Consider the standard action of SL(2) on $\mathbb{R}^2$. Let $k \in \mathbb{R}$. Prove that the differential operators $y\partial_x$, $x\partial_x - y\partial_y$, $x\partial_y - ky^{-2}$ define a nontrivial multiplier representation, with multiplier

$$\mu\left(\begin{pmatrix} \alpha & \beta \\ \gamma & \delta \end{pmatrix}; x, y\right) = \exp\left[\frac{k\gamma}{y(\gamma x + \delta y)}\right], \qquad \alpha\delta - \beta\gamma = 1.$$

Prove that, up to equivalence, these are the only nontrivial multiplier representations; see [**96**].

Exercise 3.26. Prove that the only nontrivial multiplier representations corresponding to the standard action of GL(2) on $\mathbb{R}^2$ have generators $x\partial_y$, $x\partial_x - k$, $y\partial_y - k$, $y\partial_x$, for $k \in \mathbb{R}$. Find the corresponding multiplier.

Exercise 3.27. Let $k \in \mathbb{R}$ be a constant. Let G be a Lie group acting on an open subset $M \subset \mathbb{R}^m$. Prove that the function that associates with every generator $\mathbf{v} = \sum_i \xi^i(x)\partial_{x^i}$ the differential operator $\mathbf{v} - k \operatorname{div} \xi$ defines a multiplier representation of G. Show that, for $k = -1$, one can identify this representation with the action of G on the space of differential m-forms on M. Find the corresponding multiplier. When is this multiplier trivial?

In the case of matrix multiplier representations, the infinitesimal generators of the associated extended action (3.15) take the form

$$\widetilde{\mathbf{v}} = \sum_{i=1}^{p} \xi^i(x)\frac{\partial}{\partial x^i} + \sum_{\alpha,\beta=1}^{q} h^\alpha_\beta(x) u^\beta \frac{\partial}{\partial u^\alpha}, \tag{3.26}$$

where $\sigma(\mathbf{v}) = \big(h^\alpha_\beta(x)\big)$ is the corresponding infinitesimal matrix multiplier. One can identify such vector fields with matrix differential operators of the form $\mathcal{D}_{\mathbf{v}} = \widehat{\mathbf{v}}\mathbb{1} - \sigma(\mathbf{v})$, i.e., the vector field part appears only on the diagonal of the matrix. The resulting matrix differential operators form a Lie algebra having the same commutation relations as that of G.

Exercise 3.28. Find the analogues of Theorems 3.20 and 3.21 in the matrix case.

Exercise 3.29. What are the infinitesimal multipliers corresponding to the extended actions of a transformation group on the tangent and cotangent bundles? (See Example 3.15 and Exercise 3.17.)

Relative Invariants

Note that an invariant I of a group of transformations can be regarded as a fixed point of the induced representation (3.1) on the space of functions: $g \cdot I = I$ for all $g \in G$. The analogue of an invariant for a multiplier representation is called a relative invariant.

Definition 3.30. Let $\mu: G \times M \to \mathbb{R}$ be a multiplier for a transformation group action. A *relative invariant* of *weight* μ is a function $R(x)$ which satisfies $R(g \cdot x) = \mu(g, x) R(x)$.

It is not hard to see that, as long as $R \not\equiv 0$, the weight function μ must necessarily be a multiplier for the group action. For clarity, ordinary invariants are, occasionally, referred to as *absolute invariants*. The values of absolute invariants of a group action have intrinsic geometric meaning since they are unaffected by the group transformations, and so equivalent points (i.e., points lying in the same group orbit) will have the same invariant values. In particular, if the group acts regularly, the orbits, and hence the canonical forms, are completely distinguished by the specification of sufficiently many invariants. On the other hand, general values of relative invariants usually have no intrinsic meaning, since they will change according to the multiplier. Only those points at which the relative invariant vanishes are necessarily mapped to each other by the group transformations. Such points are often found to play a distinguished role in the analysis of the group action.

As usual, if the Lie group is connected, the most effective method for determining relative invariants is by an associated infinitesimal condition. This condition explains our curious choice of sign in (3.19).

Theorem 3.31. *Let G be a connected group of transformations acting on M. A function $R(x)$ is a relative invariant for an associated multiplier representation if and only if it is annihilated by the differential operators spanning the infinitesimal multiplier:*

$$\mathcal{D}_{\mathbf{v}}(R) = \widehat{\mathbf{v}}(R) - \sigma(\mathbf{v})R = 0, \qquad \text{for every} \qquad \mathbf{v} \in \mathfrak{g}. \tag{3.27}$$

If R and S are relative invariants corresponding to the *same* multiplier μ, then any linear combination $c_1 R + c_2 S$ is also a relative invariant of weight μ. (However, this certainly does not hold if R and S are relative invariants corresponding to different multipliers!) If R has weight μ and S has weight ν, then their product $R \cdot S$ is a relative invariant for the product multiplier $\mu \cdot \nu$. In particular, the ratio R/S of two relative invariants having the *same* weight is an absolute invariant of the group. Therefore, once we know one relative invariant of a given weight μ, we can easily provide a complete list of all such relative invariants.

Proposition 3.32. *Let μ be a scalar multiplier for a transformation group G. If $R_0(x)$ is a nonvanishing relative invariant of weight μ, then every other relative invariant of weight μ has the form $R(x) = I(x)R_0(x)$, where I is any absolute invariant.*

According to Theorem 2.34, if G acts regularly on M with s-dimensional orbits, then there exist $m - s$ functionally independent absolute invariants, $I_1, \ldots, I_{m-r}$. Moreover, according to Proposition 3.32, every relative invariant for the scalar weight function μ can be written in the form $F(I_1(x), \ldots, I_{m-s}(x))R_0(x)$ for some particular relative invariant R_0 (if any exists).

Exercise 3.33. Discuss how the relative invariants of equivalent multipliers are related.

Exercise 3.34. Define a relative invariant for a matrix multiplier. Find an analogue of Proposition 3.32 in this case. (See Theorem 3.36.)

Exercise 3.35. Show that a relative invariant for the Jacobian multiplier representation (3.14) is the same as a G-invariant vector field. Similarly, a G-invariant one-form is a relative invariant for the multiplier representation of Exercise 3.17.

Given a multiplier representation, are there any relative invariants? In view of Exercise 3.35, the existence problem for relative invariants can be regarded as a generalization of the existence problems for invariant vector fields and differential forms. Now, in general, there may not exist any relative invariants of a given multiplier representation. For example, the infinitesimal criterion (3.27) implies that, in the scalar case, there are *no* relative invariants if the group action is not locally effective, but some trivially acting Lie subgroup has nontrivial multiplier. (This is equivalent to the existence of a nonzero element $\mathbf{v} \in \mathfrak{g}$ such that the associated vector field $\widehat{\mathbf{v}} \equiv 0$ is trivial, but $\mathcal{D}_{\mathbf{v}} = h_{\mathbf{v}}(x) \neq 0$ is a nonzero multiplication operator.)

In order to formulate a general theorem governing the existence of relative invariants for sufficiently regular group actions, we consider the extended group action (3.15) on the bundle $E = M \times U$. The key remark is that there is a one-to-one correspondence between relative invariants of weight μ and linear absolute invariants of the extended action. Specifically, a linear function $J(x,u) = \sum_{\alpha=1}^{n} R_\alpha(x) u^\alpha$ is an invariant of the extended group action (3.15) if and only if the vector-valued function $R(x) = (R_1(x), \dots, R_q(x))^T$ is a relative invariant of weight μ. Therefore, we need only produce a sufficient number of *linear* invariants of the extended action. Moreover, if $J(x,u)$ is any invariant of the extended group action, then it is not hard to prove that its linear Taylor polynomial is also an invariant, and hence provides a relative invariant for the multiplier representation. Thus, the only question is how many independent relative invariants can be constructed in this manner.

In the classification Theorem 2.34 for ordinary invariants, the dimension of the orbits of G plays the key role. In the present case, there are two fundamental orbit dimensions — that of the orbits of G on M, and that of the orbits of the extended action of G on E. Since the extended orbits project down to the orbits on M, the former must have dimension greater than the latter. If orbit dimensions are the *same*, then we can find a complete system of relative invariants.

Theorem 3.36. *Let G be an r-dimensional Lie group acting semi-regularly on the m-dimensional manifold M with s-dimensional orbits. Let $\mu: G \times M \to \mathrm{GL}(q)$ be a matrix multiplier representation on a q-dimensional space $U \simeq \mathbb{R}^q$. The maximal orbit dimension of the extended action of G on $M \times U$ is also equal to s if and only if there exists a complete system of q linearly independent relative invariants $R_1(x), \dots, R_q(x)$ taking values in U, in which case every other relative* *invariant can be written as a linear combination*

$$R(x) = \sum_{\nu=1}^{q} I_\nu(x) R_\nu(x), \tag{3.28}$$

whose coefficients I_ν are scalar absolute invariants for G.

Thus, a complete set of q independent relative invariants will exist if and only if the subspace of TM spanned by the infinitesimal generators $\widehat{\mathbf{v}}$ of G has the same dimension as the subspace of TE spanned by the extended infinitesimal generators $\mathcal{D}_{\mathbf{v}} = \widehat{\mathbf{v}} - \sigma(\mathbf{v})$. The proof of Theorem 3.36 will follow from a slight generalization of Frobenius' Theorem; see Example 14.6.

Corollary 3.37. *If G acts locally freely on M, then G admits a complete system of relative invariants.*

Proof: According to Proposition 2.22 the group orbits on M have the maximal possible dimension, $s = r = \dim G$. Therefore, the extended orbits on $M \times U$ must necessarily have the same dimension, and the corollary follows. *Q.E.D.*

Example 3.38. Consider the infinitesimal 2×2 matrix multiplier

$$\partial_x, \qquad x\partial_x - (\alpha u + \beta v)\partial_u - (\gamma u + \delta v)\partial_v$$

for the usual action of the affine group $\mathrm{A}(1)$ on $\mathbb{R}$. A relative invariant $R(x) = (f(x), h(x))^T$ can be identified with the invariant function $J(x, u, v) = f(x)u + h(x)v$. An easy computation shows that J is invariant if and only if f and h are constant, and, furthermore, $\alpha f + \gamma h = 0 = \beta f + \delta h$. Therefore, this multiplier representation admits a nonzero relative invariant if and only if $\alpha\delta - \beta\gamma = 0$. This demonstrates that, when the extended orbits have strictly larger dimension, the existence of relative invariants is fairly subtle — one cannot just count dimensions. A general theorem that counts the number of relative invariants of multiplier representations in all cases can be found in the recent paper by M. Fels and the author, "On relative invariants", *Math. Ann.*, to appear.

Exercise 3.39. Exercise 3.35 implies that we can identify invariant vector fields with relative invariants of the extended group action on TM. Prove that the extended group orbits have the same maximal dimension as the orbits of G on M if and only if G acts effectively freely. Use Corollary 3.37 to complete the proofs of Theorems 2.84 and 2.93.

Classical Invariant Theory

The classification of relative and absolute invariants of polynomials under the multiplier representations of the general linear group is a problem that achieved central importance in the mathematical literature of the last century. After a long fallow period, classical invariant theory, as the subject is now known, received new impetus from a variety of recent applications, and so is worth discussing in some detail. For brevity, we will only consider the planar case here. As is often the case, it is easier to look first for relative invariants, since they are polynomial functions, whereas absolute invariants are (typically) only rational functions. However, as remarked above, we can always produce absolute invariants by taking the ratio of two relative invariants of the same weight.

Consider the standard action of the general linear group GL(2) on the real (or complex) plane: $\mathbf{x} \mapsto A\mathbf{x}$, $\mathbf{x} \in \mathbb{R}^2$, $A \in \mathrm{GL}(2)$. According to Exercise 3.26, every associated multiplier is equivalent to the determinantal multiplier $\mu_k(A, \mathbf{x}) = (\det A)^{-k}$ for some k. The fundamental multiplier representation $\rho_{n,k}$ is obtained by restricting to the subspace $\mathcal{P}^{(n)}$ of homogeneous polynomials of degree n, so that a polynomial $Q(\mathbf{x}) \in \mathcal{P}^{(n)}$ of weight k transforms according to (3.7). A function $I[Q] = I(a_0, \ldots, a_n)$ depending on the coefficients $\mathbf{a} = (a_0, \ldots, a_n)$ of Q, as in (3.3), will define a *relative invariant* of *weight* m if it transforms according to

$$I[Q] = (\alpha\delta - \beta\gamma)^m \, I[\overline{Q}], \qquad A = \begin{pmatrix} \alpha & \beta \\ \gamma & \delta \end{pmatrix} \in \mathrm{GL}(2), \tag{3.29}$$

whenever $Q(\mathbf{x})$ and $\overline{Q}(\overline{\mathbf{x}})$ are related by the weight k transformation rule (3.7). In particular, each relative invariant defines an absolute invariant of the unimodular subgroup SL(2).

For example, suppose $Q = ax^2 + 2bxy + cy^2$ is a quadratic polynomial of weight zero, and hence its coefficients transform according to (3.6). In this case, the usual discriminant $\Delta = b^2 - ac$ is readily seen to obey the transformation rule $b^2 - ac = (\alpha\delta - \beta\gamma)^2 \cdot (\bar{b}^2 - \bar{a}\bar{c})$, and hence defines a relative invariant of weight 2. In fact, it is not difficult to see that the discriminant is the only invariant of a general quadratic polynomial, meaning that any other relative invariant is a power thereof; see Example 2.77. In fact, it was this elementary observation that inspired Boole to construct additional relative invariants of higher degree polynomials, thereby igniting the extensive development of classical invariant theory in the last century. Note that since the discriminant is only a relative invariant, its particular value carries no intrinsic significance. (Although for real polynomials, the sign of the discriminant does have significance since it has even weight.) The exceptions are the cases when the discriminant vanishes, which happens if and only if the quadratic form has a double root — a property that is clearly unaffected by linear transformations. Finally, we note that if Q has weight k rather than weight 0, then Δ is still a relative invariant, now of weight $2k+2$. This remark is a particular case of a general fact that relative invariants remain so under reweighting of the polynomial Q; see Exercise 3.41 below.

The infinitesimal invariance criterion of Theorem 3.31 provides a simple test for determining relative invariants in this case.

Theorem 3.40. *Let* GL(2) *act on the space* $\mathcal{P}^{(n)}$ *of homogeneous polynomials of degree* n *according to the weight* k *multiplier representation* (3.7). *Using the coefficients* $a_0, \ldots, a_n$ *as coordinates, cf.* (3.3), *we define the following vector fields on* $\mathcal{P}^{(n)}$:

$$\begin{aligned} \mathbf{v}_- &= a_1 \frac{\partial}{\partial a_0} + 2a_2 \frac{\partial}{\partial a_1} + \cdots + (n-1)a_{n-1} \frac{\partial}{\partial a_{n-2}} + na_n \frac{\partial}{\partial a_{n-1}}, \\ \mathbf{v}_0 &= -na_0 \frac{\partial}{\partial a_0} - (n-2)a_1 \frac{\partial}{\partial a_1} + \cdots + (n-2)a_{n-1} \frac{\partial}{\partial a_{n-1}} + na_n \frac{\partial}{\partial a_n}, \\ \mathbf{v}_+ &= na_0 \frac{\partial}{\partial a_1} + (n-1)a_1 \frac{\partial}{\partial a_2} + \cdots + 2a_{n-2} \frac{\partial}{\partial a_{n-1}} + a_{n-1} \frac{\partial}{\partial a_n}. \end{aligned} \tag{3.30}$$

A function $I(\mathbf{a})$ *is an invariant of the induced action of* SL(2) *if and only if* $\mathbf{v}_-(I) = \mathbf{v}_0(I) = \mathbf{v}_+(I) = 0$. *Moreover,* I *is a relative invariant of weight* m *for* GL(2) *if and only if* I *is homogeneous of degree* $l = 2m/(n+2k)$, *or, equivalently,* $\mathcal{D}(I) = 0$, *where*

$$\mathcal{D} = a_0 \frac{\partial}{\partial a_0} + a_1 \frac{\partial}{\partial a_1} + \cdots + a_{n-1} \frac{\partial}{\partial a_{n-1}} + a_n \frac{\partial}{\partial a_n} - \frac{2m}{n+2k} \tag{3.31}$$

is the infinitesimal generator of the scaling subgroup.

For example, in the case of a quadratic polynomial (3.5), the vector fields (3.30) are $\mathbf{v}_- = b\partial_a + 2c\partial_b$, $\mathbf{v}_0 = -2a\partial_a + 2b\partial_b$, $\mathbf{v}_+ = 2a\partial_b + b\partial_c$.

They are readily seen to annihilate the discriminant $\Delta = b^2 - 4ac$; see Example 2.77. The fact that the discriminant is a relative invariant of weight $m = 2 + 2k$ when Q has weight k is a simple consequence of the fact that it is a homogeneous polynomial of degree $l = 2$ in the coefficients of Q; alternatively, we note that it is annihilated by the scaling operator $a\partial_a + b\partial_b + c\partial_c - 2$.

A cubic polynomial

$$Q(\mathbf{x}) = a_0x^3 + 3a_1x^2y + 3a_2xy^2 + a_3y^3 \tag{3.32}$$

also has just one fundamental invariant — the *discriminant*

$$\Delta = 2a_0^2a_3^2 - 6a_1^2a_2^2 - 12a_0a_1a_2a_3 + 8a_0a_2^3 + 8a_1^3a_3. \tag{3.33}$$

If Q has weight 0, then Δ is an invariant of weight 6, as can be checked using the infinitesimal criterion of Theorem 3.40. As with quadratic polynomials, the vanishing of the discriminant implies the geometrical condition that Q has a multiple root.

Finally, a quartic polynomial

$$Q(\mathbf{x}) = a_0x^4 + 4a_1x^3y + 6a_2x^2y^2 + 4a_3xy^3 + a_4y^4 \tag{3.34}$$

has two fundamental invariants:

$$i = a_0a_4 - 4a_1a_3 + 3a_2^2, \qquad j = \det \begin{vmatrix} a_0 & a_1 & a_2 \\ a_1 & a_2 & a_3 \\ a_2 & a_3 & a_4 \end{vmatrix}, \tag{3.35}$$

which, if Q has weight 0, have respective weights 4 and 6. The geometric meaning of the vanishing of the invariants i and/or j is expressed in terms of the (complex) roots of the quartic. First, $i = j = 0$ if and only if Q has a triple or a quadruple root. If $i = 0$, $j \neq 0$, the roots are in "equi-anharmonic ratio", meaning that their cross-ratio is a complex cube root of -1, whereas $j = 0$ if and only if Q can be written as the sum of two fourth powers, $Q = (ap + b)^4 + (cp + d)^4$. Note further that since i has weight 4 and j has weight 6, both i^3 and j^2 are relative invariants of weight 12, and hence the ratio i^3/j^2 is an absolute invariant, and its value is fixed. Moreover, any linear combination of i^3 and j^2 is again a relative invariant of weight 12. The most important of these is the *discriminant* $\Delta = i^3 - 27j^2$, which vanishes if and only if the quartic has a multiple root. See **[107]** for the details.

Exercise 3.41. Suppose $Q(\mathbf{x})$ is a binary form of degree n and weight 0, as in (3.3). Let $I(a_0, \ldots, a_n)$ be a relative invariant which is a homogeneous polynomial of degree l in the coefficients a_i. Prove that I has weight $m = \frac{1}{2}ln$. Suppose now that Q has weight k. Prove that I remains a relative invariant, of revised weight $\widehat{m} = l(\frac{1}{2}n + k)$. For example, if Q is a binary quartic of weight -2, then both invariants (3.35) have weight 0, i.e., are absolute invariants.

The classification of homogeneous polynomials via their invariants has been one of the traditional problems of classical invariant theory. However, the (relative) invariants do not supply sufficient information to completely distinguish the different types or canonical forms. For example, in the case of a cubic polynomial, the only invariant is the discriminant, (3.33), but it vanishes for cubics having either double or triple roots, and therefore we cannot use it to distinguish between these two inequivalent types. Indeed, the invariants will, in accordance with Theorem 2.34, allow us to distinguish between different group orbits in the open subset of the space $\mathcal{P}^{(n)}$ where GL(2) acts regularly, but will not, in general, distinguish the different "singular orbits", which include, for example, the polynomials with multiple roots. This requires the introduction of more sophisticated invariant quantities.

Definition 3.42. A *covariant* of weight k of a binary form $Q(\mathbf{x}) = Q(x, y)$ is a function $J(\mathbf{a}, \mathbf{x}) = J(a_0, \ldots, a_n, x, y)$, depending both on the coefficients of Q, and on the independent variables $\mathbf{x} = (x, y)$, which, up to a determinantal factor, does not change under the action of the general linear group GL(2):

$$J(\mathbf{a}, \mathbf{x}) = (\alpha\delta - \beta\gamma)^k \, J(\bar{\mathbf{a}}, \bar{\mathbf{x}}), \qquad A = \begin{pmatrix} \alpha & \beta \\ \gamma & \delta \end{pmatrix} \in \mathrm{GL}(2). \tag{3.36}$$

In other words, a covariant is merely a relative invariant, of weight k, that also depends on the independent variables. One can similarly define joint covariants and invariants of any system of binary forms $Q_1, \ldots, Q_m$, which are functions depending on all the coefficients of the forms, and transform as in (3.36) when the forms are simultaneously subjected to a linear transformation.

Exercise 3.43. Formulate an infinitesimal criterion for covariants. (Compare Theorem 3.40.)

A fundamental fact from classical invariant theory is that all the polynomial covariants can be constructed using certain remarkable invariant differential operators.

Definition 3.44. Given a pair of functions $Q(x,y)$, $R(x,y)$, their r^{th} *transvectant* is the function

$$(Q,\, R)^{(r)} = \sum_{i=0}^{r} (-1)^i \binom{r}{i} \frac{\partial^r Q}{\partial x^{r-i} \partial y^i} \frac{\partial^r R}{\partial x^i \partial y^{r-i}}. \tag{3.37}$$

Theorem 3.45. *If Q and R are relative invariants having respective weights i and j, then their r^{th} transvectant $(Q,\, R)^{(r)}$ is a relative invariant of weight $i+j+r$. If, in addition, Q and R are homogeneous polynomial covariants of respective degrees n and m, then $(Q,\, R)^{(r)}$ is a polynomial covariant of degree $n+m-2r$. In particular, $(Q,\, R)^{(r)} \equiv 0$ if either $n < r$ or $m < r$.*

Proof: The simplest proof† of this theorem relies on an interesting device introduced by Cayley. Let $\mathbf{x}_1 = (x_1, y_1)$, $\mathbf{x}_2 = (x_2, y_2)$ be two pairs of independent variables. The second order differential operator

$$\Omega = \det \boldsymbol{\Omega} = \det \begin{vmatrix} \dfrac{\partial}{\partial x_1} & \dfrac{\partial}{\partial y_1} \\ \dfrac{\partial}{\partial x_2} & \dfrac{\partial}{\partial y_2} \end{vmatrix} = \frac{\partial^2}{\partial x_1 \partial y_2} - \frac{\partial^2}{\partial x_2 \partial y_1} \tag{3.38}$$

is known as Cayley's *omega process*. It is easy to see that we can write the rth transvectant in terms of the omega process as

$$(Q,\, R)^{(r)} = \Omega^r \{ Q(x_1, y_1) \cdot R(x_2, y_2) \} \Big|_{x_1 = x_2 = x,\; y_1 = y_2 = y}. \tag{3.39}$$

Now, if we subject the variables $\mathbf{x}_i$ to a simultaneous linear transformation, $\mathbf{x}_i \mapsto A\mathbf{x}_i$, $i = 1, 2$, then, by the usual chain rule, the matrix differential operator $\boldsymbol{\Omega}$ whose determinant defines the omega process is equivariantly transformed: $\boldsymbol{\Omega} \mapsto A^{-T}\boldsymbol{\Omega}$. Thus $\Omega \mapsto (\det A)^{-1}\,\Omega$, which, in view of (3.39), immediately implies the invariance of the corresponding transvectant. *Q.E.D.*

† One can also prove the theorem using the infinitesimal criterion of Exercise 3.43, although the combinatorics are a bit tricky.

Remark: The use of the omega process for constructing invariants and covariants is closely related to the symbolic method of classical invariant theory, cf. [**115**], [**99**], [**107**], as well as the transform method developed by Gel'fand and Dikii, [**83**], and Shakiban, [**203**], that allows one to apply powerful algebraic results to study questions in differential algebra and the formal calculus of variations. See [**180**] for a survey, and my forthcoming book on classical invariant theory for more details.

Remark: The omega process is, interestingly, related to the Hirota bilinear operator $\mathbb{D}_x$ which arises in the analysis of soliton equations, cf. [**177**]. Specifically, if we apply Ω to functions of the form $Q(x,y) = f(x)e^y$, $R(x,y) = g(x)e^y$, we recover the definition of the Hirota operator applied to $f(x)$, $g(x)$.

Let us look at the first few transvectants. The zeroth transvectant $(Q,\, R)^{(0)} = QR$ is just the product of Q and R, so Theorem 3.45 includes, as a special case, our earlier result that the product of two relative invariants is also a relative invariant. The first transvectant,

$$(Q,\, R)^{(1)} = Q_x R_y - Q_y R_x, \tag{3.40}$$

is just the Jacobian determinant of the functions, whose invariance under GL(2) is well known. The second transvectant is given by

$$(Q,\, R)^{(2)} = Q_{xx} R_{yy} - 2Q_{xy} R_{xy} + Q_{yy} R_{xx}, \tag{3.41}$$

a formula which appears in the von Karman equations of plate mechanics in elasticity, [**53**]. In particular, the important Hessian covariant is obtained as the second transvectant of a function with itself:

$$H[Q] = \tfrac{1}{2}(Q,Q)^{(2)} = Q_{xx}Q_{yy} - Q_{xy}^2. \tag{3.42}$$

The third and higher transvectants are less well known. Note that $(Q,R)^{(r)} = (-1)^r (R,Q)^{(r)}$, so any odd transvectant $(Q,\, Q)^{(2j+1)}$ of a function with itself is automatically 0.

The corresponding formulae for transvectants in terms of the projective coordinate $p = x/y$ are of interest. They are proved inductively by differentiating the basic formula

$$Q(x,y) = y^n F\left(\frac{x}{y}\right) \tag{3.43}$$

that reconstructs a homogeneous polynomial $Q(x,y)$ of degree n from its inhomogeneous counterpart $F(p) = Q(p,1)$.

Theorem 3.46. *Let $F(p)$ and $G(p)$ be polynomials of degrees n, m, respectively. The rth transvectant of F and G is the polynomial*

$$(F, G)^{(r)} = \sum_{k=0}^{r} (-1)^k \binom{r}{k} \frac{(m-k)!}{(m-r)!} \frac{(n-r+k)!}{(n-r)!} \frac{d^{r-k}F}{dp^{r-k}} \frac{d^k G}{dp^k}. \tag{3.44}$$

Theorem 3.45 holds unchanged in the projective context. Recall that a function transforming under the projective group multiplier $\mu_{n,k}$, as in (3.9), is said to have weight (n, k).

Theorem 3.47. *Suppose $F(p)$ and $G(p)$ are functions having weights (n, k) and (m, j), respectively. Then their r^{th} transvectant $(F, G)^{(r)}$ is a function of weight $(n + m - 2r, k + j + r)$.*

In particular, if $F \in \mathcal{P}^{(n)}$, $G \in \mathcal{P}^{(m)}$, are polynomial covariants of respective degrees n, m, then the rth transvectant $(F, G)^{(r)} \in \mathcal{P}^{(n+m-2r)}$ is a polynomial covariant of degree at most $n + m - 2r$, a fact that is *not* (at least to me) obvious from the projective formula (3.44).

Example 3.48. Let F have degree n and G degree m. The simplest cases of (3.44) are

$$\begin{aligned} &(F, G)^{(0)} = FG, \qquad (F, G)^{(1)} = mF_pG - nFG_p, \\ &(F, G)^{(2)} = m(m-1)F_{pp}G - 2(m-1)(n-1)F_pG_p + n(n-1)FG_{pp}. \end{aligned} \tag{3.45}$$

In particular, the Hessian of a polynomial F of degree n is

$$H[F] = \tfrac{1}{2}(F, F)^{(2)} = n(n-1)\left[FF_{pp} - \frac{n-1}{n}F_p^2\right]. \tag{3.46}$$

This formula provides a simple direct proof of the following classical result.

Theorem 3.49. *A homogeneous function (polynomial) $Q(x, y)$ of degree n has zero Hessian, $H[Q] \equiv 0$, if and only if $Q(x, y) = (ax + by)^n$ is the nth power of a linear form.*

Proof: Using projective coordinates, (3.46) implies that $H \equiv 0$ if and only if F satisfies the differential equation $FF_{pp} = \dfrac{n-1}{n}F_p^2$. It is not hard to prove that the general solution of this ordinary differential equation is $F(p) = (ap + b)^n$. (See Example 6.21 below.) *Q.E.D.*

Remark: Theorem 3.49 characterizes functions whose second transvectant with itself vanishes. The corresponding problem for the higher order transvectant equations $(F, F)^{(2j)} = 0$ is considerably more interesting. See Exercise 6.33 for the fourth order case, which has surprising connections with elliptic functions and Schwarz' theory of hypergeometric functions.

Remark: Transvectants play an important role in the theory of automorphic forms. By definition, an automorphic form $F(p)$ is a relative invariant for the restriction of a fundamental multiplier representation $\rho_{n,k}$ to a discrete subgroup $\Gamma \subset \mathrm{GL}(2)$; i.e., the polynomial satisfies (3.9) for all matrices $A \in \Gamma$. The particularly important case of modular forms occurs when $\Gamma = \mathrm{SL}(2, \mathbb{Z})$. Theorem 3.46 implies that *any* transvectant of two automorphic forms (corresponding to the same discrete subgroup) is also an automorphic form. For the Hessian covariant, this result appears in Rankin, [**196**]; it was generalized to higher transvectants by H. Cohen, [**55**].

Remark: In [**192**], transvectants, under the alias "Lie derivatives of order n", are applied to study Poisson brackets on the space of differential operators first introduced by Adler, [**4**], and Gel'fand and Dikii, [**84**]. These play a central role in the Hamiltonian approach to soliton systems and, more recently, the theory of quantum groups and the Virasoro algebra, cf. [**65**].

Given a binary form, or, more generally, a system of binary forms, the successive applications of the transvectant operations allow us to construct a large variety of relative invariants and covariants. For instance, starting with a single binary form $Q(\mathbf{x})$ of degree n, one can form a sequence of even transvectants $R_r = (Q, Q)^{(2r)}$, $0 \le 2r \le n$. (As remarked above, the odd transvectants all vanish.) Furthermore, each of these can be transvected with Q itself, or with each other, to produce yet more relative invariants, which can in turn be transvected. Does this process supply all possible covariants? The First Fundamental Theorem of Invariant Theory says that the answer is yes.

Theorem 3.50. *Let $Q_1, \ldots, Q_m$ be a system of binary forms. Then every polynomial covariant and invariant can be expressed as a linear combination of the successive transvectants between the forms.*

See, for example, [**99**], [**107**], [**211**], for more detailed statements and proofs of this important result. Theorem 3.50 has the consequence that every polynomial covariant of a system of binary forms can be written as a constant coefficient, homogeneous, differential polynomial in the forms. Of course, not all successive transvectants provide independent invariants, so an important problem is to construct a fundamental system of covariants; this requires the classification of syzygies (relations) among the covariants, which is covered by the Second Fundamental Theorem of classical invariant theory. In the algebraic approach, one looks for an algebraically independent set of invariants, known as a *Hilbert basis*, having the property that any other covariant can be expressed as a polynomial of the generating set. Hilbert's Basis Theorem, [**114**], [**115**], proves that a finite Hilbert basis of invariants or covariants always exists; see also [**211**]. However, the explicit determination of a Hilbert basis, even for forms of low degree, is an extremely difficult problem, and has only been solved if the degree is at most 8; see [**62**] for a recent survey of the history of this fundamental problem in algebra. If one relaxes the independence criterion, and just requires functional independence, or even rational independence, then matters are considerably simpler. We just state the following result, [**183**], which is based on the work of Stroh, [**209**].

Theorem 3.51. *Let Q be a binary form of degree n. Then Q, $R_r = (Q,\, Q)^{(2r)}$, $0 \le 2r \le n$, and $(Q,\, R_r)^{(1)}$, $0 \le 2r \le n-1$, provide a complete set of n rationally independent covariants of Q.*

Another approach, based on the theory of differential invariants, is discussed in Theorem 5.19. It is a general fact that binary forms can be completely classified by their covariants; in fact, in Theorem 10.10 we shall prove that, as a consequence of the Cartan equivalence method, one need only determine two absolute rational covariants in order to completely classify the equivalence class of a binary form!

Remark: The invariance of the transvection operations does not rely on the fact that Q and R are polynomials. Therefore, one can successfully apply these differential operators to develop a theory of differential polynomial covariants for systems of smooth or analytic functions. However, one needs a slight generalization, known as a "partial transvectant", in order to produce a complete system of invariant differential polynomials; see [**180**] for more details.

Example 3.52. In the case of a binary quartic, (3.1), the first two covariants are the Hessian, $H = \frac{1}{2}(Q,\, Q)^{(2)}$, which is a covariant of weight 2, and the quadratic invariant $i = \frac{1}{1372}(Q, Q)^{(4)}$, which has weight 4. Since i is an invariant, any transvectant with it produces zero, so further covariants arise only by transvecting with the Hessian covariant. The most important are the Jacobian covariant $T = (Q,\, H)^{(1)}$, which has weight 3, and the cubic invariant $j = \frac{1}{96}(Q,\, H)^{(4)}$, which has weight 6. It can be proven that every polynomial invariant or covariant of a binary quartic can be written in terms of the invariants i, j, and the covariants Q, H, T, cf. [**99**], [**107**]. Moreover, j can be expressed as a rational function of the other covariants, owing to the syzygy

$$jQ^3 = \frac{1}{72} iQ^2 H - \frac{1}{2^7 3^6} H^3 - \frac{1}{2^{12} 3^4} T^2. \tag{3.47}$$

The complete classification of complex quartic polynomials is readily effected using these basic covariants; see [**107**] for more details. A quartic has four simple roots if and only if $i^3 \neq 27j^2$; the different canonical forms are distinguished by the absolute invariant j^2/i^3 (or its reciprocal). A quartic has a double and two single roots if and only if $i^3 = 27j^2$ and $T \not\equiv 0$; if $i^3 = 27j^2$, $T \equiv 0$, but $i \neq 0$, it has two double roots. The case $i = j = 0, H \not\equiv 0$ characterizes quartics with a triple root, while those with quadruple roots are given by $H \equiv 0$.

Exercise 3.53. It can be shown that a binary cubic Q has basic covariants given by Q itself, the Hessian $H = \frac{1}{2}(Q,\, Q)^{(2)}$, the Jacobian $T = (Q,\, H)^{(1)}$, and the discriminant (3.33). Find a formula for the discriminant in terms of transvectants. Classify complex cubics using these covariants.

Exercise 3.54. Generalize the result of Exercise 3.41 to covariants. Show, in particular, that every polynomial covariant of a binary quartic of weight -2 has degree $n = 2i$ and weight $k = -i$ for some integer i. Find the analogous formula if the quartic has weight 0.

Chapter 4

Jets and Contact Transformations

The previous chapter was devoted to the analysis of linear actions of Lie groups on functions, provided by either ordinary or multiplier representations. Although of great importance, such actions are certainly not the most general one can imagine, and we will have occasion to make use of fully nonlinear group actions. Such general transformation groups figure prominently in Lie's theory of symmetry groups of differential equations, which we discuss in Chapter 6. They also reoccur in the Cartan equivalence method, to be discussed in the second part of the book. The transformation groups will act on the basic space coordinatized by the independent and dependent variables relevant to the system of differential equations under consideration, and we begin with a short discussion of the different types of transformation groups of interest. Since we are dealing with *differential equations* we must be able to handle the derivatives of the dependent variables on the same footing as the independent and dependent variables themselves. This chapter is devoted to a detailed study of the proper geometric context for these purposes —the so-called "jet spaces" or "jet bundles", well known to nineteenth century practitioners, but first formally defined by Ehresmann, [**66**]. After presenting a simplified version of the basic construction, we then discuss how group transformations are "prolonged" so that the derivative coordinates are appropriately acted upon, and, in the case of infinitesimal generators, deduce the explicit prolongation formula. Next we introduce the underlying contact structure on jet space, and present Lie's theory of contact transformations which, in the case of a single dependent variable, enlarge our repertoire of available group actions. Applications of these results to the study of differential equations, differential operators, and variational problems will be discussed in the following chapters.

Transformations and Functions

A general system of (partial) differential equations involves p independent variables $x = (x^1, \ldots, x^p)$, which we can view as local coordinates on the Euclidean space $X \simeq \mathbb{R}^p$, and q dependent variables $u = (u^1, \ldots, u^q)$, coordinates on $U \simeq \mathbb{R}^q$. The *total space* will be the Euclidean space $E = X \times U \simeq \mathbb{R}^{p+q}$ coordinatized by the independent and dependent variables.† The most basic (but not the only) type of symmetry we will discuss is provided by a (locally defined) diffeomorphism on the space of independent and dependent variables:

$$(\bar{x}, \bar{u}) = g \cdot (x, u) = (\chi(x, u), \psi(x, u)). \tag{4.1}$$

These are often referred to as *point transformations* since they act pointwise on the total space E. Aside from the contact transformations to be discussed below, point transformations form the most general class of symmetries to be used in the sequel. However, it *is* convenient to specialize, on occasion, to more restrictive classes of transformations. For example, *base transformations* are only allowed to act on the independent variables, and so have the form $\bar{x} = \chi(x)$, $\bar{u} = u$. If we wish to preserve the bundle structure of the space E, we must restrict to the class of *fiber-preserving transformations* in which the changes in the independent variable are unaffected by the dependent variable, and so take the form

$$(\bar{x}, \bar{u}) = g \cdot (x, u) = (\chi(x), \psi(x, u)). \tag{4.2}$$

Most, but not all, important group actions are fiber-preserving. If the function $\psi(x, u)$ depends linearly on the dependent variables, then (4.2) reduces to the geometric version (3.15) of a (matrix) multiplier representation, the base transformations corresponding to the standard representation on the space of functions.

† It should be remarked that all these considerations extend to the more general global context, in which the total space is replaced by a vector bundle E over the base X coordinatized by the independent variables, the fibers being coordinatized by the dependent variables. Since our primary considerations are local, we shall not lose anything by specializing to (open subsets of) Euclidean space, which has the added advantage of avoiding excessive abstraction. Moreover, the sophisticated reader can easily supply the necessary translations of the machinery if desired. Incidentally, an even more general setting for these techniques, which avoids any *a priori* distinction between independent and dependent variables, is provided by the extended jet bundles defined in [**186**; Chapter 3].

In the case of connected groups, the action of the group can be recovered from that of its associated infinitesimal generators. A general vector field

$$\mathbf{v} = \sum_{i=1}^{p} \xi^i(x,u)\,\frac{\partial}{\partial x^i} + \sum_{\alpha=1}^{q} \varphi^\alpha(x,u)\,\frac{\partial}{\partial u^\alpha}, \tag{4.3}$$

on the space of independent and dependent variables generates a flow $\exp(t\mathbf{v})$, which is a local one-parameter group of point transformations on E. The vector field will generate a one-parameter group of fiber-preserving transformations if and only if the coefficients $\xi^i = \xi^i(x)$ do not depend on the dependent variables. The group consists of base transformations if and only if the vector field is *horizontal*, meaning that all the vertical coefficients vanish: $\varphi^\alpha = 0$. Vice versa, *vertical* vector fields, which are characterized by the vanishing of the horizontal coefficients, $\xi^i = 0$, generate groups of vertical transformations; an important physical example is provided by the gauge transformations discussed at the end of Chapter 3.

A (classical) solution† to a system of differential equations will be described by a smooth function $u = f(x)$. (In the more general bundle-theoretic framework, solutions are described by *sections* of the bundle.) The graph of the function, $\Gamma_f = \{(x, f(x))\}$, determines a regular p-dimensional submanifold of the total space E. However, not every regular p-dimensional submanifold of E defines the graph of a smooth function. Globally, it can intersect each fiber $\{x_0\} \times U$ in at most one point. Locally, it must be *transverse*, meaning that its tangent space contains no vertical tangent directions. The Implicit Function Theorem demonstrates that, locally, the transversality condition is both necessary and sufficient for a submanifold to represent the graph of a smooth function.

Proposition 4.1. *A regular p-dimensional submanifold $\Gamma \subset E$ which is transverse at a point $z_0 = (x_0, u_0) \in \Gamma$ coincides locally, i.e., in a neighborhood of x_0, with the graph of a single-valued smooth function $u = f(x)$.*

† To avoid technical complications, we will only consider smooth or analytic solutions, although extensions of these results to more general types of solutions are certainly possible; see [**198**], [**199**].

Any point transformation (4.1) will act on a function $u = f(x)$ by pointwise transforming its graph. In other words if $\Gamma_f = \{(x, f(x))\}$ denotes the graph of f, then the transformed function $\bar{f} = g \cdot f$ will have graph

$$\Gamma_{\bar{f}} = \{(\bar{x}, \bar{f}(\bar{x}))\} = g \cdot \Gamma_f = \{g \cdot (x, f(x))\}. \tag{4.4}$$

In general, we can only assert that the transformed graph is another p-dimensional submanifold of E, and so the transformed function will not be well defined unless $g \cdot \Gamma_f$ is (at least) transverse. Transversality will, however, be ensured if the transformation g is sufficiently close to the identity transformation, and the domain of f is compact.

Example 4.2. Consider the one-parameter group of rotations

$$g_t \cdot (x, u) = (x \cos t - u \sin t,\ x \sin t + u \cos t), \tag{4.5}$$

acting on the space $E \simeq \mathbb{R}^2$ consisting of one independent and one dependent variable. Such a rotation transforms a function $u = f(x)$ by rotating its graph; therefore, the transformed graph $g_t \cdot \Gamma_f$ will be the graph of a well-defined function only if the rotation angle t is not too large. The equation for the transformed function $\bar{f} = g_t \cdot f$ is given in implicit form

$$\bar{x} = x \cos t - f(x) \sin t, \qquad \bar{u} = x \sin t + f(x) \cos t,$$

so that $\bar{u} = \bar{f}(\bar{x})$ is found by eliminating x from these two equations. For example, if $u = ax + b$ is affine, then the transformed function is also affine, and given explicitly by

$$\bar{u} = \frac{\sin t + a \cos t}{\cos t - a \sin t}\, \bar{x} + \frac{b}{\cos t - a \sin t}, \tag{4.6}$$

which is defined provided $\cot t \neq a$, i.e., provided the graph of f has not been rotated to be vertical.

In general, given a point transformation as in (4.1), the transformed function $\bar{u} = \bar{f}(\bar{x})$ is found by eliminating x from the parametric equations $\bar{u} = \psi(x, f(x))$, $\bar{x} = \chi(x, f(x))$, provided this is possible. If the transformation is fiber-preserving, then $\bar{x} = \chi(x)$ is a local diffeomorphism, and so the transformed function is always well defined, being given by $\bar{u} = \psi(\chi^{-1}(\bar{x}), f(\chi^{-1}(\bar{x})))$.

Exercise 4.3. Suppose the fiber-preserving transformation (4.2) depends linearly on the dependent variables, and so coincides with the geometric version of a multiplier representation. Prove that the definition (4.4) of the group action coincides with our earlier definition (3.10) of the action of the multiplier representation on functions $u = f(x)$.

Invariant Functions

Given a group of point transformations G acting on $E \simeq X \times U$, the characterization of all G-invariant functions $u = f(x)$ is of great importance.

Definition 4.4. A function $u = f(x)$ is said to be *invariant* under the transformation group G if its graph Γ_f is a (locally) G-invariant subset.

For example, the graph of any invariant function for the rotation group of Example 4.2 must be an arc of a circle centered at the origin, so $u = \pm\sqrt{c^2 - x^2}$. Note that there are no globally defined invariant functions in this case.

Exercise 4.5. Suppose the group action coincides with a multiplier representation on a space of functions with multiplier $\mu(x, g)$. Prove that a G-invariant function, as per Definition 4.4, is the same as a relative invariant of weight μ.

Remark: In this framework, it is important to distinguish between an "invariant function", which is a section $u = f(x)$ of the bundle E, and an "invariant", which, as in Definition 2.29, is a real-valued function $I(x, u)$ defined (locally) on E. It is hoped that this will not cause too much confusion in the sequel.

In general, since any invariant function's graph must, locally, be a union of orbits, the existence of invariant functions passing through a point $z = (x, u) \in E$ requires that the orbit $\mathcal{O}$ through z be of dimension at most p, the number of independent variables, and, furthermore, be transverse. Since the tangent space to the orbit $T\mathcal{O}|_z = \mathfrak{g}|_z$ agrees with the space spanned by the infinitesimal generators of G, the transversality condition requires that, at each point, $\mathfrak{g}|_z$ contain no vertical tangent vectors. For example, the infinitesimal generator of the rotation group is $\mathbf{v} = -u\partial_x + x\partial_u$, which is vertical at $u = 0$. Thus, the rotation group fails the transversality criterion on the x-axis, and, as we saw, there are no smooth, rotationally invariant functions $u = f(x)$ passing through such points. In the case the group acts both (semi-)regularly and transversally, then we can characterize the invariant functions by the use of the functionally independent (real-valued) invariants of the group action. The following result is a direct corollary of Theorem 2.38.

Theorem 4.6. *Let G be a transformation group acting semi-regularly and transversally on $E \simeq X \times U$ with r-dimensional orbits. Let $I_1(x,u), \ldots, I_{p-s}(x,u)$, $J_1(x,u), \ldots, J_q(x,u)$, be a complete set of functionally independent invariants for G. Then any G-invariant function $u = f(x)$, can, locally, be written in the implicit form*

$$w = h(y), \qquad \text{where} \qquad y = I(x,u), \quad w = J(x,u). \tag{4.7}$$

In the fiber-preserving case, if we assume that the orbits of the projected action of G on X also have dimension s, then we can take the first $p-s$ invariants $I_1(x), \ldots, I_{p-s}(x)$ to depend only on the independent variables, and forming a complete system of invariants on X.

Example 4.7. A "similarity solution" of a system of partial differential equations is just an invariant function for a group of scaling transformations. As a specific example, consider the one-parameter scaling group $(x, y, u) \mapsto (\lambda x, \lambda^\alpha y, \lambda^\beta u)$. The independent invariants are provided by the ratios $y = y/x^\alpha$, $w = u/x^\beta$, so any scale-invariant function can be written as $w = h(y)$, or, explicitly, $u = x^\beta h(y/x^\alpha)$.

As usual, the most convenient characterization of the invariant functions is based on an infinitesimal condition. Since the graph of a function is defined by the vanishing of its components $u^\alpha - f^\alpha(x)$, our general invariance Theorem 2.79 imposes the infinitesimal invariance conditions

$$0 = \mathbf{v}(u^\alpha - f^\alpha(x)) = \varphi^\alpha(x,u) - \sum_{i=1}^{p} \xi^i(x,u)\,\frac{\partial f^\alpha}{\partial x^i}\,, \qquad \alpha = 1, \ldots, q,$$

which must hold whenever $u = f(x)$, for every infinitesimal generator $\mathbf{v} \in \mathfrak{g}$, as in (4.3). These first order partial differential equations are known in the literature as the "invariant surface conditions" associated with the given transformation group, cf. **[22]**.

Definition 4.8. The *characteristic* of the vector field $\mathbf{v}$ given by (4.3) is the q-tuple of functions $Q(x, u^{(1)})$, depending on x, u and first order derivatives of u, defined by

$$Q^\alpha(x, u^{(1)}) = \varphi^\alpha(x, u) - \sum_{i=1}^{p} \xi^i(x, u) \frac{\partial u^\alpha}{\partial x^i}, \qquad \alpha = 1, \ldots, q. \tag{4.8}$$

Theorem 4.9. *A function $u = f(x)$ is invariant under a connected group of point transformations if and only if it is a solution to the first order system of partial differential equations*

$$Q^\alpha(x, u^{(1)}) = 0, \qquad \alpha = 1, \ldots, q, \tag{4.9}$$

determined by all the characteristics $Q(x, u^{(1)})$ of the set of infinitesimal generators $\mathbf{v} \in \mathfrak{g}$.

For example, the characteristic of the rotation vector field $-u\partial_x + x\partial_u$ is $Q = x + uu_x$. Any rotationally invariant function must satisfy the differential equation $x + uu_x = 0$. This equation can be readily integrated: $x^2 + u^2 = c$, and hence the graph is an arc of a circle. Similarly, the infinitesimal generator of the scaling group of Example 4.7 is the vector field $x\partial_x + \alpha y\partial_y + \beta u\partial_u$. The characteristic is $Q = \beta u - xu_x - \alpha yu_y$, and the scale-invariant functions constitute the general solution to the linear partial differential equation $xu_x + \alpha yu_y = \beta u$.

Jets and Prolongations

Since we are interested in studying the symmetries of differential equations, we need to know not only how the group transformations act on the independent and dependent variables, but also how they act on the derivatives of the dependent variables. In the last century, this was done automatically, without fretting about the precise mathematical foundations of the method; in more recent times, geometers have formalized this geometrical construction through the general definition of the "jet space" (or bundle) associated with the total space of independent and dependent variables. The jet space coordinates will represent the derivatives of the dependent variables. We describe a simple, direct formulation of these spaces, initially using local coordinates; the coordinate-free approach will be indicated a bit later on.

A smooth, scalar-valued function $f(x^1, \ldots, x^p)$ depending on p independent variables has $p_k = \binom{p+k-1}{k}$ different k^{th} order partial derivatives $\partial_J f(x) = \partial^k f / \partial x^{j_1} \partial x^{j_2} \cdots \partial x^{j_k}$, indexed by all unordered (symmetric) multi-indices $J = (j_1, \ldots, j_k)$, $1 \le j_\kappa \le p$, of order $k = \#J$. Therefore, if we have q dependent variables $(u^1, \ldots, u^q)$, we require $q_k = qp_k$ different coordinates u_J^α, $1 \le \alpha \le q$, $\#J = k$, to represent all the kth order derivatives $u_J^\alpha = \partial_J f^\alpha(x)$ of a function $u = f(x)$. For the total space $E = X \times U \simeq \mathbb{R}^p \times \mathbb{R}^q$, the nth *jet space* $\mathrm{J}^n = \mathrm{J}^n E = X \times U^{(n)}$ is the Euclidean space of dimension $p + q^{(n)} \equiv p + q\binom{p+n}{n}$, whose coordinates consist of the p independent variables x^i, the q dependent variables u^α, and the derivative coordinates u_J^α, $\alpha = 1, \ldots, q$, of orders $1 \le \#J \le n$. The points in the vertical space (fiber) $U^{(n)}$ are denoted by $u^{(n)}$, and consist of all the dependent variables and their derivatives up to order n; thus the coordinates of a typical point $z \in \mathrm{J}^n$ are denoted by $(x, u^{(n)})$. Since the derivative coordinates $u^{(n)}$ form a subset of the derivative coordinates $u^{(n+k)}$, there is a natural projection $\pi_n^{n+k} \colon \mathrm{J}^{n+k} \to \mathrm{J}^n$ on the jet spaces, with $\pi_n^{n+k}(x, u^{(n+k)}) = (x, u^{(n)})$. In particular, $\pi_0^n(x, u^{(n)}) = (x, u)$ is the projection from J^n to $E = \mathrm{J}^0$. If $M \subset E$ is an open subset, then $\mathrm{J}^n M = (\pi_0^n)^{-1} M \subset \mathrm{J}^n E$ is the open subset of the nth jet space which projects back down to M.

A smooth function $u = f(x)$ from X to U has n^{th} *prolongation* $u^{(n)} = f^{(n)}(x)$ (also known as the *n-jet* and denoted $\mathrm{j}_n f$), which is the function from X to $U^{(n)}$ defined by evaluating all the partial derivatives of f up to order n; thus the individual coordinate functions of $f^{(n)}$ are $u_J^\alpha = \partial_J f^\alpha(x)$. In particular, $f^{(0)} = f$. Note that the graph of the prolonged function $f^{(n)}$, namely $\Gamma_f^{(n)} = \{(x, f^{(n)}(x))\}$, will be a p-dimensional submanifold of J^n. At a point $x \in X$, two functions have the same n^{th} order prolongation, and so determine the same point of J^n, if and only if they have n^{th} *order contact*, meaning that they and their first n derivatives agree at the point, which is the same as requiring that they have the same n^{th} order Taylor polynomial at the point x. Thus, a more intrinsic way of defining the jet space J^n is to consider it as the set of equivalence classes of smooth functions using the equivalence relation of nth order contact. Note that the process of prolongation is compatible with the jet space projections, so $\pi_n^{n+k} \circ f^{(n+k)} = f^{(n)}$.

If g is a (local) point transformation (4.1), then g acts on functions by transforming their graphs, and hence also naturally acts on the derivatives of the functions. This allows us to define the induced *prolonged transformation* $(\bar{x}, \bar{u}^{(n)}) = g^{(n)} \cdot (x, u^{(n)})$ on the jet space J^n. More specifically, given a point $z_0 = (x_0, u_0^{(n)})$, choose a representative smooth function $u = f(x)$ whose n^{th} prolongation has the prescribed derivatives at x_0, so $z_0 = (x_0, f^{(n)}(x_0)) \in \mathrm{J}^n$. The transformed point $\bar{z}_0 = g^{(n)} \cdot z_0$ is found by evaluating the derivatives of the transformed function $\bar{f} = g \cdot f$ at the image point $\bar{x}_0$, defined so that $(\bar{x}_0, \bar{u}_0) = (\bar{x}_0, \bar{f}(\bar{x}_0)) = g \cdot (x_0, f(x_0))$; therefore $\bar{z}_0 = (\bar{x}_0, \bar{u}_0^{(n)}) = (x_0, \bar{f}(x_0))$. This definition assumes that $\bar{f}$ is smooth at $\bar{x}_0$ — otherwise the prolonged transformation is not defined at $(x_0, u_0^{(n)})$. Thus, the prolonged transformation $g^{(n)}$ maps the graph $\Gamma_f^{(n)}$ of the nth prolongation of a function $u = f(x)$ to the graph of the nth prolongation of its image $\bar{f} = g \cdot f$:

$$g^{(n)} \cdot \Gamma_f^{(n)} = \Gamma_{g \cdot f}^{(n)}. \tag{4.10}$$

A straightforward chain rule argument shows that the construction does not depend on the particular function f used to represent the point of J^n; in particular, in view of the identification of the points in J^n with Taylor polynomials of order n, it suffices to determine how the point transformations act on polynomials of degree at most n. Note that the prolongation process preserves compositions, $(g \circ h)^{(n)} = g^{(n)} \circ h^{(n)}$, and is compatible with the jet space projections, $\pi_n^{n+k} \circ g^{(n+k)} = g^{(n)}$. Given a (local) group of transformations acting on E, we define the prolonged group action, denoted by $G^{(n)}$, on the jet space $\mathrm{J}^n E$ by prolonging the individual transformations in G. In general, prolongation may only define a local action of the group G, although certain classes, e.g., global fiber-preserving actions, do have global prolongations.

Example 4.10. For a rotation in the one-parameter group considered in Example 4.2, the first prolongation $g_t^{(1)}$ will act on the space coordinatized by (x, u, p), where, in accordance with classical notation, we use p to represent the derivative coordinate u_x.† Given a point (x_0, u_0, p_0), we choose the linear polynomial $u = f(x) = p_0(x - x_0) + u_0$ as representative, noting that $f(x_0) = u_0$, $f'(x_0) = p_0$. The transformed

† Unfortunately, this notation conflicts with our earlier use of p to represent the number of independent variables in a bundle E, which, in the present case is fixed at $p = 1$. The different meanings of p should, however, be clear from context.

function is given by (4.6), so

$$\bar{f}(\bar{x}) = \frac{\sin t + p_0 \cos t}{\cos t - p_0 \sin t}\,\bar{x} + \frac{u_0 - p_0 x_0}{\cos t - p_0 \sin t}.$$

Then, $\bar{x}_0 = x_0 \cos t - u_0 \sin t$, so $\bar{f}(\bar{x}_0) = \bar{u}_0 = x_0 \sin t + u_0 \cos t$, as we already knew, and $\bar{p}_0 = \bar{f}'(\bar{x}_0) = (\sin t + p_0 \cos t)/(\cos t - p_0 \sin t)$, which is defined provided $p_0 \neq \cot t$. Therefore, dropping the 0 subscripts, the first prolongation of the rotation group is explicitly given by

$$g_t^{(1)} \cdot (x, u, p) = \left(x \cos t - u \sin t, x \sin t + u \cos t, \frac{\sin t + p \cos t}{\cos t - p \sin t} \right), \tag{4.11}$$

defined for $p \neq \cot t$. Note, in particular, that even though the original group action is globally defined, its first prolongation is only a local transformation group.

Example 4.11. Example 4.10 is a particular case of a general point transformation

$$\bar{x} = \chi(x, u), \qquad \bar{u} = \psi(x, u), \tag{4.12}$$

on the space $E \simeq \mathbb{R} \times \mathbb{R}$ coordinatized by a single independent and a single dependent variable. In view of the ordinary calculus chain rule, the derivative coordinate $p = u_x$ on the jet space J^1 will transform under the first prolongation of (4.12) according to a linear fractional transformation

$$\bar{p} = \frac{\alpha p + \beta}{\gamma p + \delta}, \tag{4.13}$$

whose coefficients

$$\alpha = \frac{\partial \psi}{\partial u}, \qquad \beta = \frac{\partial \psi}{\partial x}, \qquad \gamma = \frac{\partial \chi}{\partial u}, \qquad \delta = \frac{\partial \chi}{\partial x}, \tag{4.14}$$

are certain derivatives of the functions χ, ψ determining our change of variables. Further applications of the chain rule will yield the higher order prolongations in this case — see, for instance, (4.16) below.

Exercise 4.12. Before proceeding further, the reader should try to compute the second prolongation of the rotation group SO(2).

Total Derivatives

The chain rule computations used to compute prolongations are significantly simplified if we introduce the useful concept of a total derivative. We first define the functions that are to be differentiated.

Definition 4.13. A smooth, real-valued function $F\colon \mathrm{J}^n \to \mathbb{R}$, defined on an open subset of the nth jet space is called a *differential function* of *order* n.

Any n^{th} order *differential equation* will be determined by the vanishing of a differential function of order n. For example, the planar Laplace equation $u_{xx} + u_{yy} = 0$ is given by the second order differential function $F(x, u^{(2)}) = u_{xx} + u_{yy}$ defined on $\mathrm{J}^2 E$, where $E = \mathbb{R}^2 \times \mathbb{R}$ has coordinates x, y, u. Note that any differential function of order n automatically defines a differential function on any higher order jet space merely by treating the coordinates $(x, u^{(n)})$ of J^n as a subset of the coordinates $(x, u^{(n+k)})$ of J^{n+k} — this is the same as composing F with the projection $\pi^{n+k}_n\colon \mathrm{J}^{n+k} \to \mathrm{J}^n$. In the sequel, we will not distinguish between F and its compositions $F \circ \pi^{n+k}_n$. The *order* of a differential function will typically only refer to the minimal order jet space on which it is defined, which is the same as the maximal order derivative coordinate upon which it depends. Thus $u_{xx} + u_{yy}$ is a second order differential function, even though it also defines a function on any jet space J^k for $k \geq 2$.

Definition 4.14. Let $F(x, u^{(n)})$ be a differential function of order n. The *total derivative* F with respect to x^i is the $(n+1)^{\text{st}}$ order differential function $D_i F$ satisfying

$$D_i F(x, f^{(n+1)}(x)) = \frac{\partial}{\partial x^i}\, F(x, f^{(n)}(x)),$$

for any smooth function $u = f(x)$.

For example, in the case of one independent variable x and one dependent variable u, the total derivative of a given differential function $F(x, u^{(n)})$ with respect to x has the general formula

$$D_x F = \frac{\partial F}{\partial x} + u_x \frac{\partial F}{\partial u} + u_{xx} \frac{\partial F}{\partial u_x} + u_{xxx} \frac{\partial F}{\partial u_{xx}} + \cdots . \tag{4.15}$$

For example, $D_x(xuu_{xx}) = uu_{xx} + xuu_{xxx} + xu_x u_{xx}$.

As a first application, note that the chain rule prolongation formula (4.13) can now be simply written

$$\bar{p} = \frac{d\bar{u}}{d\bar{x}} = \frac{D_x\psi}{D_x\chi}.$$

Using this notation, the action of the second prolongation of the point transformation (4.12) on the second order derivative coordinate $q = u_{xx}$ can be compactly written as

$$\bar{q} = \frac{d^2\bar{u}}{d\bar{x}^2} = \frac{1}{D_x\chi}\, D_x\left(\frac{D_x\psi}{D_x\chi}\right) = \frac{D_x\chi \cdot D_x^2\psi - D_x\psi \cdot D_x^2\chi}{(D_x\chi)^3}. \tag{4.16}$$

For example, in the case of the rotation group discussed in Example 4.10, formula (4.16) implies the explicit form

$$\left(x\cos t - u\sin t,\; x\sin t + u\cos t,\; \frac{\sin t + p\cos t}{\cos t - p\sin t},\; \frac{q}{(\cos t - p\sin t)^3}\right), \tag{4.17}$$

for the second prolongation, thereby solving Exercise 4.12.

In the general framework, the total derivative with respect to the ith independent variable x^i is the first order differential operator

$$D_i = \frac{\partial}{\partial x^i} + \sum_{\alpha=1}^{q}\sum_J u^\alpha_{J,i}\,\frac{\partial}{\partial u^\alpha_J}, \tag{4.18}$$

where $u^\alpha_{J,i} = D_i(u^\alpha_J) = u^\alpha_{j_1\ldots j_k i}$. The sum in (4.18) is over all symmetric multi-indices J of arbitrary order. Even though D_i involves an infinite summation, when applying the total derivative to any particular differential function, only finitely many terms (namely, those for $\#J \le n$, where n is the order of F) are needed. In particular, each total derivative D_iF of an $(n+1)^{\text{st}}$ order differential function depends linearly on the $(n+1)^{\text{st}}$ derivative coordinates. Higher order total derivatives are defined in the obvious manner, so that $D_J = D_{j_1} \cdot \ldots \cdot D_{j_k}$ for any multi-index $J = (j_1, \ldots, j_k)$, $1 \le j_\nu \le p$.

Exercise 4.15. Prove that $F(x, u^{(n)})$ is a differential function all of whose total derivatives vanish, $D_iF = 0$, $i = 1, \ldots, p$, if and only if F is constant.

Prolongation of Vector Fields

Given a vector field $\mathbf{v}$ generating a one-parameter group of point transformations $\exp(t\mathbf{v})$ on $E \simeq X \times U$, the associated n^{th} order prolonged vector field $\mathbf{v}^{(n)}$ is the vector field on the jet space J^n which is the infinitesimal generator of the prolonged one-parameter group $\exp(t\mathbf{v})^{(n)}$. Thus, according to (1.7), at any point $(x, u^{(n)}) \in \mathrm{J}^n$,

$$\mathbf{v}^{(n)}\big|_{(x,u^{(n)})} = \frac{d}{dt}\exp(t\mathbf{v})^{(n)} \cdot (x, u^{(n)})\bigg|_{t=0}. \tag{4.19}$$

The explicit formula for the prolonged vector field is provided by the following "prolongation formula ". Although the formula can be proved by direct computation based on the definition (4.19), cf. [**186**], we will wait in order to present an alternative useful proof based on the contact structure of the jet space later in this chapter.

Theorem 4.16. *Let* $\mathbf{v}$ *be a vector field given by* (4.3), *and let* $Q = (Q^1, \ldots, Q^q)$ *be its characteristic, as in* (4.8). *The* n^{th} *prolongation of* $\mathbf{v}$ *is given explicitly by*

$$\mathbf{v}^{(n)} = \sum_{i=1}^{p} \xi^i(x,u)\,\frac{\partial}{\partial x^i} + \sum_{\alpha=1}^{q} \sum_{\#J=j=0}^{n} \varphi^\alpha_J(x,u^{(j)})\,\frac{\partial}{\partial u^\alpha_J}, \tag{4.20}$$

with coefficients

$$\varphi^\alpha_J = D_J Q^\alpha + \sum_{i=1}^{p} \xi^i\, u^\alpha_{J,i}. \tag{4.21}$$

Example 4.17. Suppose we have just one independent and dependent variable. A general vector field $\mathbf{v} = \xi(x,u)\partial_x + \varphi(x,u)\partial_u$ has characteristic

$$Q(x,u,u_x) = \varphi(x,u) - \xi(x,u)u_x. \tag{4.22}$$

The second prolongation of $\mathbf{v}$ is a vector field

$$\mathbf{v}^{(2)} = \xi(x,u)\,\frac{\partial}{\partial x} + \varphi(x,u)\,\frac{\partial}{\partial u} + \varphi^x(x,u^{(1)})\,\frac{\partial}{\partial u_x} + \varphi^{xx}(x,u^{(2)})\,\frac{\partial}{\partial u_{xx}}, \tag{4.23}$$

on J^2, whose coefficients φ^x, φ^{xx} are given by (4.21), hence

$$\begin{aligned} \varphi^x &= D_x Q + \xi u_{xx} = \varphi_x + (\varphi_u - \xi_x)u_x - \xi_u u_x^2, \\ \varphi^{xx} &= D_x^2 Q + \xi u_{xxx} = \varphi_{xx} + (2\varphi_{xu} - \xi_{xx})u_x + (\varphi_{uu} - 2\xi_{xu})u_x^2 - \\ &\qquad - \xi_{uu}u_x^3 + (\varphi_u - 2\xi_x)u_{xx} - 3\xi_u u_x u_{xx}. \end{aligned} \tag{4.24}$$

In (4.24) the subscripts on ξ and φ indicate partial derivatives. For example, the second prolongation of the generator $\mathbf{v} = -u\partial_x + x\partial_u$ of the rotation group is given by

$$\mathbf{v}^{(2)} = -u\,\frac{\partial}{\partial x} + x\,\frac{\partial}{\partial u} + (1+u_x^2)\,\frac{\partial}{\partial u_x} + 3u_x u_{xx}\,\frac{\partial}{\partial u_{xx}}\,. \tag{4.25}$$

The group transformations (4.17) can be readily recovered by integrating the system of ordinary differential equations governing the flow of $\mathbf{v}^{(2)}$, as in (1.5); these are

$$\frac{dx}{dt} = -u, \qquad \frac{du}{dt} = x, \qquad \frac{dp}{dt} = 1+p^2, \qquad \frac{dq}{dt} = 3pq,$$

where we have used p and q to stand for u_x and u_{xx} to avoid confusing derivatives with jet space coordinates. Note also that the first prolongation of $\mathbf{v}$ is obtained by omitting the second derivative terms in (4.23), or, equivalently, projecting back to J^1.

Example 4.18. As a second example, suppose we have two independent variables, x and t, and one dependent variable u. A vector field

$$\mathbf{v} = \xi(x,t,u)\partial_x + \tau(x,t,u)\partial_t + \varphi(x,t,u)\partial_u$$

has characteristic $Q = \varphi - \xi u_x - \tau u_t$. The second prolongation of $\mathbf{v}$ is the vector field

$$\mathbf{v}^{(2)} = \xi\partial_x + \tau\partial_t + \varphi\partial_u + \varphi^x\partial_{u_x} + \varphi^t\partial_{u_t} + \varphi^{xx}\partial_{u_{xx}} + \varphi^{xt}\partial_{u_{xt}} + \varphi^{tt}\partial_{u_{tt}}, \tag{4.26}$$

where, for example,

$$\begin{aligned}
\varphi^x &= D_xQ + \xi u_{xx} + \tau u_{xt} = \varphi_x + (\varphi_u - \xi_x)u_x - \tau_x u_t - \xi_u u_x^2 - \tau_u u_x u_t,\\
\varphi^t &= D_tQ + \xi u_{xt} + \tau u_{tt} \;= \varphi_t - \xi_t u_x + (\varphi_u - \tau_t)u_t - \xi_u u_x u_t - \tau_u u_t^2\\
\varphi^{xx} &= D_x^2Q + \xi u_{xxx} + \tau u_{xxt}\\
&= \varphi_{xx} + (2\varphi_{xu} - \xi_{xx})u_x - \tau_{xx}u_t + (\varphi_{uu} - 2\xi_{xu})u_x^2 - \\
&\quad - 2\tau_{xu}u_x u_t - \xi_{uu}u_x^3 - \tau_{uu}u_x^2 u_t + (\varphi_u - 2\xi_x)u_{xx} - \\
&\quad - 2\tau_x u_{xt} - 3\xi_u u_x u_{xx} - \tau_u u_t u_{xx} - 2\tau_u u_x u_{xt}.
\end{aligned} \tag{4.27}$$

See Chapter 6 for applications of these formulae to the study of symmetries of differential equations.

Exercise 4.19. Prove that the coefficients (4.21) of the prolonged vector field satisfy the classical recursive formula

$$\varphi^\alpha_{J,i} = D_i\varphi^\alpha_J - \sum_{j=1}^{p} D_i\xi^j\, u^\alpha_{J,j}. \tag{4.28}$$

In particular, the coefficients of the first prolongation of $\mathbf{v}$ are given by

$$\varphi^\alpha_i = D_i\varphi^\alpha - \sum_{j=1}^{p} (D_i\xi^j)\, u^\alpha_j. \tag{4.29}$$

For instance, formulae (4.27) can also be written in the form

$$\varphi^x = D_x\varphi - (D_x\xi)u_x - (D_x\tau)u_t, \qquad \varphi^t = D_t\varphi - (D_t\xi)u_x - (D_t\tau)u_t,$$
$$\varphi^{xx} = D_x\varphi^x - (D_x\xi)u_{xx} - (D_x\tau)u_{xt}.$$

An alternative approach to the prolongation formula for vector fields is to introduce the *evolutionary vector field*

$$\mathbf{v}_Q = \sum_{\alpha=1}^{q} Q^\alpha(x, u^{(1)})\, \frac{\partial}{\partial u^\alpha}\,, \tag{4.30}$$

based on the characteristic of our vector field $\mathbf{v}$, cf. (4.8); $\mathbf{v}_Q$ is an example of a "generalized" (or "Lie–Bäcklund") vector field in that it no longer depends on just the independent and dependent variables, but also on their derivatives, and so does not determine a well-defined geometrical transformation on the total space E; see [**11**], [**186**], for a survey of the general theory. Let

$$\mathbf{v}_Q^{(n)} = \sum_{\alpha=1}^{q} \sum_{\#J\geq 0} D_J Q^\alpha(x, u^{(1)})\, \frac{\partial}{\partial u^\alpha_J}\,, \tag{4.31}$$

be the corresponding formal prolongation of $\mathbf{v}_Q$. Then the nth prolongation of $\mathbf{v}$ can be written as

$$\mathbf{v}^{(n)} = \mathbf{v}_Q^{(n)} + \sum_{i=1}^{p} \xi^i D_i. \tag{4.32}$$

Since the prolongation process respects the composition of maps, the commutator formula (1.13) immediately proves that it preserves Lie brackets:

$$[\mathbf{v},\mathbf{w}]^{(n)} = [\mathbf{v}^{(n)},\mathbf{w}^{(n)}], \tag{4.33}$$

and therefore defines a Lie algebra homomorphism from the space of vector fields on E to the space of prolonged vector fields on $\mathrm{J}^n E$. Thus, the nth prolongation $\mathfrak{g}^{(n)}$ of a Lie algebra of vector fields on E defines an isomorphic Lie algebra of vector fields on J^n, generating the nth prolongation of the associated transformation group.

Example 4.20. According to the classification tables appearing at the end of the book, there are three distinct, locally inequivalent actions of the unimodular Lie group $\mathrm{SL}(2)$ on a two-dimensional complex manifold. Interestingly, the process of prolongation can be used to relate all three actions. Consider first the usual linear fractional action $(x,u) \mapsto ((\alpha x+\beta)/(\gamma x+\delta), u)$, whose infinitesimal generators

$$\mathbf{v}_- = \partial_x, \qquad \mathbf{v}_0 = x\partial_x, \qquad \mathbf{v}_+ = x^2\partial_x, \tag{4.34}$$

span the intransitive Lie algebra of Type 3.3; see also Example 2.64. Using $p = u_x$ to denote the derivative coordinate, we find that the first prolongation of this group action is generated by the vector fields

$$\mathbf{v}_-^{(1)} = \partial_x, \qquad \mathbf{v}_0^{(1)} = x\partial_x - p\partial_p, \qquad \mathbf{v}_+^{(1)} = x^2\partial_x - 2xp\partial_p, \tag{4.35}$$

which, in accordance with (4.33), form a Lie algebra having the same $\mathfrak{sl}(2)$ commutation relations. The Lie algebra (4.14) clearly projects to the (x,p)–plane, thereby defining the Lie algebra of vector fields of Type 1.1 in our tables. The corresponding unimodular group action is $(x,p) \mapsto ((\alpha x+\beta)/(\gamma x+\delta), (\gamma x+\delta)^2 p)$, and serves to define the fundamental multiplier representation with multiplier $\mu_{2,0} = (\gamma x+\delta)^2$, cf. (3.12). Further, setting $q = u_{xx}$, the second prolongation of the vector fields (4.34) yields the Lie algebra

$$\begin{gathered}\mathbf{v}_-^{(2)} = \partial_x, \qquad \mathbf{v}_0^{(2)} = x\partial_x - p\partial_p - 2q\partial_q, \\ \mathbf{v}_+^{(2)} = x^2\partial_x - 2xp\partial_p - (4xq+2p)\partial_q,\end{gathered} \tag{4.36}$$

again having the same $\mathfrak{sl}(2)$ commutation relations. Define

$$w = \frac{q}{2p} = \frac{p_x}{2p} = \frac{u_{xx}}{2u_x}, \tag{4.37}$$

and use (x, u, p, w) instead of (x, u, p, q) as coordinates on $\{p \neq 0\} \subset \mathrm{J}^2$. The vector fields (4.36) then have the form

$$\begin{aligned} \mathbf{v}_-^{(2)} &= \partial_x, \qquad \mathbf{v}_0^{(2)} = x\partial_x - p\partial_p - w\partial_w, \\ \mathbf{v}_+^{(2)} &= x^2\partial_x - 2xp\partial_p - (2xw+1)\partial_w. \end{aligned} \tag{4.38}$$

Again, we can project this action to the (x, w)–plane, on which the vector fields (4.38) span a Lie algebra of Type 1.2. The associated group action is $(x, w) \longmapsto ((\alpha x + \beta)/(\gamma x + \delta), (\gamma x + \delta)^2 w + \gamma(\gamma x + \delta))$, which is no longer a multiplier representation.

Thus, all three inequivalent actions of SL(2) on two-dimensional complex manifolds arise from a single source through the process of prolongation. This raises the interesting question of how general this phenomenon is: Are different (complex) actions of a given transformation group related by the processes of prolongation and projection? As for real actions, there is one additional inequivalent real action of $\mathrm{SL}(2, \mathbb{R})$, provided by Type 6.2. Does it appear in among the prolongations of the real intransitive action of Type 3.3?

Contact Forms

We have already remarked on how the nth order jet space can be abstractly constructed using the equivalence relation of nth order contact. We now investigate the contact structure of jet spaces in some detail, determining a remarkable system of differential forms — the contact forms — which serve to characterize precisely prolonged functions and transformations. One immediate consequence of this approach is a simple proof of the prolongation formula (4.20), which will, in turn, lead us straightforwardly to Lie's definition of a contact transformation. Contact forms also prove to be of fundamental importance in setting up and solving equivalence problems for differential equations and variational problems by the Cartan method; see Chapter 9 in particular.

To begin with, recall that each (smooth) function $u = f(x)$ determines a prolonged function $u^{(n)} = f^{(n)}(x)$ from X to the nth jet space J^n. The (easy) inverse problem is to characterize those sections of the jet space, meaning functions $F\colon X \to \mathrm{J}^n$ given by $u^{(n)} = F(x)$, which come from prolonging ordinary functions. This is essentially the same as the problem of determining which p-dimensional submanifolds of J^n are the graphs of the prolongations of functions. Although in local coordinates the solution to both problems is completely trivial — one just checks that the derivative coordinates match up properly — there is a more interesting and useful solution to these problems based on an intrinsically defined system of differential forms.

Definition 4.21. A differential one-form θ on the jet space J^n is called a *contact form* if it is annihilated by all prolonged functions. In other words, if $u = f(x)$ is any smooth function with n^{th} prolongation $f^{(n)}: X \to \mathrm{J}^n$, then the pull-back of θ to X via $f^{(n)}$ must vanish: $(f^{(n)})^* \theta = 0$.

Example 4.22. Consider the case of one independent and one dependent variable. On the first jet space J^1, with coordinates $x, u, p = u_x$, a general one-form takes the form $\theta = a\,dx + b\,du + c\,dp$, where a, b, c are functions of x, u, p. A function $u = f(x)$ has first prolongation $p = f'(x)$, hence $(f^{(1)})^*\theta$ equals

$$\big[a(x, f(x), f'(x)) + b(x, f(x), f'(x))f'(x) + c(x, f(x), f'(x))f''(x)\big]\,dx.$$

This will vanish for all functions f if and only if $c = 0$ and $a = -bp$; hence $\theta = b(x, u, p)\theta_0$ must necessarily be a multiple of the *basic contact form* $\theta_0 = du - p\,dx = du - u_x\,dx$. Proceeding to the second jet space J^2, with additional coordinate $q = u_{xx}$, a similar calculation shows that a one-form $\theta = a\,dx + b\,du + c\,dp + e\,dq$ is a contact form if and only if $\theta = b\,\theta_0 + c\,\theta_1$, where $\theta_1 = dp - q\,dx = du_x - u_{xx}\,dx$ is the next basic contact form. (Here, as we did earlier with differential functions, we are identifying the form θ_0 with its pull-back $(\pi_1^2)^*\theta_0$ to J^2.) In general, provided $x, u \in \mathbb{R}$, the general contact form can be written as a linear combination of the basic contact forms $\theta_k = du_k - u_{k+1}\,dx$, $k = 0, \ldots, n-1$, where $u_k = D_x^k u$ is the kth order derivative of u.

Similar elementary computations show that, in the case of two dependent variables u, v, every contact form is a linear combination of the basic contact forms for u and analogous ones for v, i.e., $dv - v_x\,dx$, $dv_x - v_{xx}\,dx$, etc. On the other hand, for two independent and one dependent variable, there is one basic contact form $\theta_0 = du - u_x\,dx - u_y\,dy$ on J^1, two basic contact forms $\theta_x = du_x - u_{xx}\,dx - u_{xy}\,dy$, $\theta_y = du_y - u_{xy}\,dx - u_{yy}\,dy$, on J^2, and so on. A similar argument provides us with the complete characterization of all contact forms.

Theorem 4.23. *Every contact form on* J^n *can be written as a linear combination,* $\theta = \sum_{J,\alpha} P_J^\alpha \theta_J^\alpha$, *with smooth coefficient functions* $P_J^\alpha(x, u^{(n)})$, *of the basic contact forms*

$$\theta_J^\alpha = du_J^\alpha - \sum_{i=1}^{p} u_{J,i}^\alpha \, dx^i, \qquad \alpha = 1, \ldots, q, \quad 0 \leq \#J < n. \tag{4.39}$$

In (4.39), we call $\#J$ the *order* of the contact form θ_J^α. Note especially that the contact forms on J^n have orders at most $n-1$.

Theorem 4.24. *A section* $u^{(n)} = F(x)$ *of the jet space* J^n *is the prolongation of some function* $u = f(x)$, *meaning* $F = f^{(n)}$, *if and only if* F *annihilates all the contact forms on* J^n:

$$F^* \theta_J^\alpha = 0, \qquad \alpha = 1, \ldots, q, \quad 0 \leq \#J < n. \tag{4.40}$$

A transverse p-dimensional submanifold $\Gamma^{(n)} \subset \mathrm{J}^n$ *is (locally) the graph of a prolonged function* $u = f^{(n)}(x)$ *if and only if all the contact forms vanish on it:* $\theta_J^\alpha \mid \Gamma^{(n)} = 0$.

The general argument will be clear from consideration of the following two special cases. Let $x, u \in \mathbb{R}$. A section F of J^1 is described by a pair of functions $u = f(x)$, $p = g(x)$. The section annihilates the single contact form $\theta_0 = du - p\,dx$ provided $0 = F^*\theta_0 = df - g\,dx = [f'(x) - g(x)]\,dx$, which vanishes if and only if $g(x) = f'(x)$, showing that F is just the first prolongation of f. Similarly, a section F of J^2 is given by $u = f(x)$, $p = g(x)$, $q = h(x)$. It annihilates the two basic contact forms provided $0 = F^*(du - p\,dx) = df - g\,dx = [f'(x) - g(x)]\,dx$, so $g(x) = f'(x)$, and $0 = F^*(dp - q\,dx) = dg - h\,dx = [g'(x) - h(x)]\,dx$, which implies $h(x) = g'(x) = f''(x)$. Therefore $F = f^{(2)}$ as desired.

The contact forms also provide us with a nice intrinsic characterization of the total derivative operators defined earlier. A one-form ω is called *horizontal* if it annihilates all the vertical tangent directions in J^n. Thus, in local coordinates, the horizontal one-forms are just linear combinations, $\omega = \sum Q_i(x, u^{(n)}) dx^i$, of the coordinate one-forms on the base X. To any one-form ω on J^n, there is an intrinsically defined horizontal form ω_H on J^{n+1}, called the *horizontal component* of ω, which is defined so that $\omega = \omega_H + \theta$ where θ is a contact form of order n. In coordinates, using the general formula (4.39), we find

$$\omega = \sum_{i=1}^{p} Q\,(x, u^{(n)})\,dx^i + \sum_{\alpha=1}^{q} \sum_{\#J\le n} P_\alpha^J(x, u^{(n)})\,du_J^\alpha$$
$$= \sum_{i=1}^{p} \left\{ Q_i + \sum_{\alpha=1}^{q} \sum_{\#J\le n} u_{J,i}^\alpha P_\alpha^J \right\} dx^i + \sum_{\alpha=1}^{q} \sum_{\#J\le n} P_\alpha^J \theta_J^\alpha.$$

The second summand is a contact form, and hence

$$\omega_H = \sum_{i=1}^{p} \left\{ Q_i(x, u^{(n)}) + \sum_{\alpha=1}^{q} \sum_{\#J\le n} u_{J,i}^\alpha P_\alpha^J(x, u^{(n)}) \right\} dx^i. \tag{4.41}$$

Warning: The horizontal component ω_H of a form ω on J^n depends on $(n+1)^{\mathrm{st}}$ order derivatives. This occurs because we are including the nth order contact forms, which are only defined on J^{n+1}, in the decomposition $\omega = \omega_H + \theta$. For example, if $p = q = 1$, the horizontal component of $\omega = Q\,dx + P\,du$ is $\omega_H = (Q + u_x P)\,dx$.

Definition 4.25. Let $F\colon \mathrm{J}^n \to \mathbb{R}$ be a differential function of order n. Then the *total differential* of F is the horizontal component of its ordinary differential dF, written $\mathrm{D}F = (dF)_H$.

To compute the total differential $\mathrm{D}F$, first note that

$$dF = \sum_{i=1}^{p} \frac{\partial F}{\partial x^i}\,dx^i + \sum_{\alpha=1}^{q} \sum_{\#J\le n} \frac{\partial F}{\partial u_J^\alpha}\,du_J^\alpha, \tag{4.42}$$

Thus, by (4.41) and (4.18), the total differential of F is given in terms of the total derivatives of F by the formula

$$\mathrm{D}F = \sum_{i=1}^{p} D_i F\,dx^i. \tag{4.43}$$

For example, if $F(x, u, p, q)$ is a second order differential function, with $x, u \in \mathbb{R}$, $p = u_x$, $q = u_{xx}$, then

$$dF = F_x\,dx + F_u\,du + F_p\,dp + F_q\,dq = (F_x + pF_u + qF_p + rF_q)\,dx +$$
$$+ F_u(du - p\,dx) + F_p(dp - q\,dx) + F_q(dq - r\,dx),$$

where $r = u_{xxx}$. The total differential $\mathrm{D}F = D_x F\, dx$ is the first component. Formula (4.42) proves the invariance of the total differential (and hence the "covariance" of the total derivatives) under general contact transformations, as defined below.

The decomposition $dF = \mathrm{D}F + \theta$ of an nth order differential function into horizontal and contact components does *not* work on J^n. However, on J^n we have the alternative decomposition

$$dF = \widehat{\mathrm{D}}F + d_n F + \vartheta. \tag{4.44}$$

Here ϑ is a contact form of order at most $n-1$, and

$$d_n F = \sum_{\alpha=1}^{q} \sum_{\#J=n} \frac{\partial F}{\partial u_J^\alpha}\, du_J^\alpha. \tag{4.45}$$

is the (intrinsically defined) n^{th} order *differential* of F. Furthermore,

$$\widehat{\mathrm{D}}F = \sum_{i=1}^{p} \widehat{D}_i F\, dx^i \tag{4.46}$$

is the "truncated total differential" of F. In (4.46), the operator $\widehat{D}_i$ is the nth order truncation of the total derivative (4.18), obtained by restricting the multi-index summation to run only over J's with $\#J \leq n-1$. For example, if $F = uu_{xx}$, then

$$\begin{aligned} dF &= u_{xx}\, du + u\, du_{xx} = u_x u_{xx}\, dx + u\, du_{xx} + u_{xx}(du - u_x\, dx) \\ &= (uu_{xxx} + u_x u_{xx})\, dx + u_{xx}(du - u_x\, dx) + u(du_{xx} - u_{xxx}\, dx). \end{aligned}$$

Therefore, $\mathrm{D}F = D_x F\, dx = (uu_{xxx} + u_x u_{xx})\, dx$, while $\widehat{\mathrm{D}}F = u_x u_{xx}\, dx$ and $d_2 F = u\, du_{xx}$.

Contact Transformations

The nth prolongation $g^{(n)}\colon \mathrm{J}^n \to \mathrm{J}^n$ of any point transformation g can be characterized by the property that it maps the prolonged graph of any function to the prolonged graph of the transformed function. Theorem 4.24 immediately implies that $g^{(n)}$ maps contact forms to contact forms. Indeed, this latter property, coupled with its compatibility with the projection $\pi_0^n\colon \mathrm{J}^n \to E$, serves to essentially define the prolongation of g.

Conversely, consider a local diffeomorphism $\Psi\colon \mathrm{J}^n \to \mathrm{J}^n$ which maps contact forms to contact forms. Since the graphs of prolonged functions are characterized by the vanishing of the contact forms thereon, the image $\Psi[\Gamma_f^{(n)}]$ of the prolonged graph of a function $u = f(x)$ is, provided it remains transverse, the graph of the nth prolongation of some transformed function $u = \bar{f}(x)$, so $\Psi[\Gamma_f^{(n)}] = \Gamma_{\bar{f}}^{(n)}$ serves to define an action of Ψ on functions $u = f(x)$. Consequently, maps which preserve the set of contact forms are candidates for symmetries of differential equations. The following definition dates back to Lie, [**157**].

Definition 4.26. A local diffeomorphism $\Psi\colon \mathrm{J}^n \to \mathrm{J}^n$ defines a *contact transformation* of order n if it preserves the space of contact forms, meaning that if θ is any contact form on J^n, then $\Psi^*\theta$ is also a contact form.

Example 4.27. The Legendre transformation, which is important in Hamiltonian mechanics, [**12**], is the first order contact transformation

$$\bar{x} = p, \qquad \bar{u} = u - xp, \qquad \bar{p} = -x, \tag{4.47}$$

where $p = u_x$, $\bar{p} = \bar{u}_{\bar{x}}$. In this case, the contact forms match up precisely, $d\bar{u} - \bar{p}\,d\bar{x} = du - p\,dx$, so (4.47) defines a contact transformation on J^1. If $u = f(x)$ is any function, its Legendre transform $\bar{u} = \bar{f}(\bar{x})$ is defined parametrically by the formulae $\bar{x} = f'(x)$, $\bar{u} = f(x) - xf'(x)$, and is locally well defined if and only if $f_{xx}(x) \neq 0$. A straightforward chain rule computation proves that $\bar{p} = -x = -(f')^{-1}(\bar{x})$, verifying the contact condition.

The Legendre transformation is a special case of a general first order contact transformation

$$\bar{x} = \chi(x,u,p), \qquad \bar{u} = \psi(x,u,p), \qquad \bar{p} = \pi(x,u,p), \tag{4.48}$$

in one independent and one dependent variable. The condition that equation (4.48) define a contact transformation is that

$$\Psi^*(d\bar{u} - \bar{p}\,d\bar{x}) = d\psi - \pi\,d\chi = \lambda(du - p\,dx), \tag{4.49}$$

for some function $\lambda(x,u,p)$, which means that

$$\psi_x - \pi\chi_x = -p\lambda, \qquad \psi_u - \pi\chi_u = \lambda, \qquad \psi_p - \pi\chi_p = 0.$$

Eliminating λ, we conclude that (4.48) must satisfy the pair of equations

$$\psi_x + p\psi_u = \pi(\chi_x + p\chi_u), \qquad \psi_p - \pi\chi_p = 0. \tag{4.50}$$

Equation (4.48) defines a prolonged point transformation if and only if χ and ψ are independent of the derivative coordinate p, in which case the first condition in (4.50) reduces to the first prolongation formula (4.13). On the other hand, if (4.48) is *not* a prolonged point transformation, we can use the Implicit Function Theorem to eliminate the variable p in (4.48), leading to a relation of the form $H(x, u, \bar{x}, \bar{u}) = 0$. Lie, [**157**], proves that *every* "genuine" contact transformation arises in this manner.

Theorem 4.28. *Every contact transformation which is not a prolonged point transformation can be determined by solving the following three implicit equations for* $\bar{x}, \bar{u}, \bar{p}$ *in terms of* x, u, p:

$$H(x, u, \bar{x}, \bar{u}) = 0, \qquad \frac{\partial H}{\partial x} + p\,\frac{\partial H}{\partial u} = 0, \qquad \frac{\partial H}{\partial \bar{x}} + \bar{p}\,\frac{\partial H}{\partial \bar{u}} = 0. \tag{4.51}$$

The generating function $H(x, u, \bar{x}, \bar{u})$ *is an arbitrary smooth function, subject only to the solvability of* (4.51) *for* $\bar{x}, \bar{u}, \bar{p}$.

For example, the function $H = \bar{u} - u - \bar{x}x$ generates the Legendre transformation (4.47). More generally, a generating function of the form $H = \bar{u} - u - S(\bar{x}, x)$ generates a contact transformation of the form $p = -S_x$, $\bar{p} = S_{\bar{x}}$, and $\bar{u} = u + S$. If we ignore the u-coordinate, then we have recovered the well-known generating function $S(\bar{x}, x)$ for a canonical transformation in Hamiltonian mechanics, [**12**], [**227**]. Thus, the theory of canonical transformations of Hamiltonian mechanics is a particular case of the general theory of contact transformations.

Exercise 4.29. Let $x \in \mathbb{R}^p$, $u \in \mathbb{R}$, and let $p_i = \partial u/\partial x^i$. Prove that if a map Ψ, of the form $\bar{x} = \xi(x, p)$, $\bar{u} = u + \psi(x, p)$, $\bar{p} = \pi(x, p)$, defines a first order contact transformation, then it is a symmetry of the canonical symplectic two-form $\Omega = \sum_i dp_i \wedge dx^i$, meaning $\Psi^*(\Omega) = \Omega$; see (1.30). Prove that the converse holds for a suitably defined function $\psi(x, p)$.

Just as we prolong point transformations, so also contact transformations on a jet space J^n can be prolonged to any higher order jet space J^{n+k}, $k > 0$. The following easy result includes point transformations as a special case (when $n = 0$) providing an alternative mechanism for defining their prolongation.

Proposition 4.30. *Let* $\Psi^{(n)}\colon \mathrm{J}^n \to \mathrm{J}^n$ *be any* n^{th} *order contact transformation. The* $(n+k)^{\text{th}}$ *order prolongation* $\Psi^{(n+k)}\colon \mathrm{J}^{n+k} \to \mathrm{J}^{n+k}$ *is the unique contact transformation on* J^{n+k} *which projects back to* $\Psi^{(n)}$*, meaning that* $\pi_n^{n+k} \circ \Psi^{(n+k)} = \Psi^{(n)} \circ \pi_n^{n+k}$.

The formulae for the prolongation of contact transformations are found by the same chain rule approach as was used with point transformations. For example, the prolongation of the first order contact transformation (4.48) to the second order jet space is governed by

$$\bar{q} = \frac{d\bar{p}}{d\bar{x}} = \frac{D_x \pi}{D_x \chi} = \frac{\pi_p q + \pi_u p + \pi_x}{\chi_p q + \chi_u p + \chi_x}.$$

Exercise 4.31. Suppose $\Psi\colon \mathrm{J}^1 \to \mathrm{J}^1$ is a contact transformation with local coordinate expression $\bar{x}^i = \chi^i(x, u^{(1)})$, $\bar{u}^\alpha = \psi^\alpha(x, u^{(1)})$, $\bar{u}_i^\alpha = \pi_i^\alpha(x, u^{(1)})$, where $u_i^\alpha = \partial u^\alpha / \partial x^i$. First show that

$$\sum_{i=1}^{p} \pi_i^\alpha(x, u^{(1)}) D_j \xi^i = D_j \psi^\alpha. \tag{4.52}$$

Then prove that the associated action on the basic contact forms is

$$\Psi^*(\bar{\theta}^\alpha) = \sum_{\beta=1}^{q} \left(\frac{\partial \psi^\alpha}{\partial u^\beta} - \sum_{i=1}^{p} \pi_i^\alpha \frac{\partial \chi^i}{\partial u^\beta} \right) \theta^\beta. \tag{4.53}$$

Can you extend this result to higher order contact forms?

Although the definition of contact transformation appears quite promising for significantly extending the types of symmetry transformations we are able to use in the study of differential equations, it turns out that, beyond prolonged point transformations, there are surprisingly few examples of contact transformations. The following remarkable theorem is due to Bäcklund, [**15**].

Theorem 4.32. *If the number of dependent variables is greater than one,* $q > 1$*, then every contact transformation is the prolongation of a point transformation. If* $q = 1$*, then there are first order contact transformations which do not come from point transformations, but every* n^{th} *order contact transformation is the prolongation of a first order contact transformation.*

Theorem 4.32 is true for completely general contact transformations. Bäcklund's original proof of Theorem 4.32 is of a geometrical nature and would take us too far afield to explain here; see [**15**], [**11**], for details. In the following subsection we will prove the infinitesimal version, which, via the usual exponentiation technique, will imply the result for any contact transformation which can be connected to the identity by a sequence of one-parameter groups of contact transformations. As alternatives, we shall present, in Chapter 11, a proof based on the Cartan equivalence method, which is yet further simplified using the theory of exterior differential systems in Chapter 15.

Remark: The restrictions of Bäcklund's Theorem 4.32 severely limit the range of applicability of contact transformations. However, it is possible that, on a submanifold of the jet space defined by a system of differential equations, there do exist additional higher order transformations which preserve the restriction of the contact ideal, known as internal symmetries. See [**9**] for details.

Infinitesimal Contact Transformations

Let us now turn to the analysis of the infinitesimal generators of groups of contact transformations. The infinitesimal criterion that a vector field $\mathbf{v}^{(n)}$ on J^n generate a one-parameter group of contact transformations follows immediately from Theorem 2.91.

Proposition 4.33. *The flow* $\exp(t\mathbf{v}^{(n)})$ *generated by a vector field* $\mathbf{v}^{(n)}$ *on the jet space* J^n *forms a one-parameter group of contact transformations if and only if the Lie derivative* $\mathbf{v}^{(n)}(\theta)$ *of any contact form is itself a contact form.*

Let us now use the infinitesimal contact conditions to provide a proof of Bäcklund's Theorem 4.32 for infinitesimal contact transformations. Let us begin with the first order case. Consider a general vector field $\mathbf{v}^{(1)}$ on J^1, given by

$$\sum_{i=1}^{p} \xi^i(x, u^{(1)}) \frac{\partial}{\partial x^i} + \sum_{\alpha=1}^{q} \varphi^\alpha(x, u^{(1)}) \frac{\partial}{\partial u^\alpha} + \sum_{\alpha=1}^{q} \sum_{i=1}^{p} \chi_i^\alpha(x, u^{(1)}) \frac{\partial}{\partial u_i^\alpha}. \tag{4.54}$$

In order to verify the infinitesimal contact condition of Proposition 4.33, we apply $\mathbf{v}^{(1)}$ to the basic contact form $\theta^\alpha = du^\alpha - \sum u_i^\alpha \, dx^i$, and obtain

$$\mathbf{v}^{(1)}(\theta^\alpha) = d\varphi^\alpha - \sum_{i=1}^{p} \left[\chi_i^\alpha \, dx^i + u_i^\alpha \, d\xi^i \right]$$

$$= \sum_{i=1}^{p} \left\{ \frac{\partial \varphi^\alpha}{\partial x^i} + \sum_{\beta=1}^{q} u_i^\beta \frac{\partial \varphi^\alpha}{\partial u^\beta} - \sum_{i=1}^{p} u_j^\alpha \left[\frac{\partial \xi^j}{\partial x^i} + \sum_{\beta=1}^{q} u_i^\beta \frac{\partial \xi^j}{\partial u^\beta} \right] - \chi_i^\alpha \right\} dx^i$$

$$+ \sum_{i=1}^{p} \sum_{\beta=1}^{q} \left\{ \frac{\partial \varphi^\alpha}{\partial u_i^\beta} - \sum_{j=1}^{p} u_j^\alpha \frac{\partial \xi^j}{\partial u_i^\beta} \right\} du_i^\beta + \sum_{\beta=1}^{q} \left\{ \frac{\partial \varphi^\alpha}{\partial u^\beta} - \sum_{j=1}^{p} u_j^\alpha \frac{\partial \xi^j}{\partial u^\beta} \right\} \theta^\beta .$$

Therefore, $\mathbf{v}^{(1)}$ determines an infinitesimal contact transformation if and only if the coefficients of dx^i and du_i^β in this formula vanish. The former requirements imply the *contact conditions*

$$\frac{\partial \varphi^\alpha}{\partial u_j^\beta} - \sum_{i=1}^{p} u_i^\alpha \frac{\partial \xi^i}{\partial u_j^\beta} = 0. \tag{4.55}$$

The latter requirements provide explicit formulae for the coefficients of the first derivative terms in $\mathbf{v}^{(1)}$:

$$\chi_i^\alpha = \widehat{D}_i \varphi^\alpha - \sum_{j=1}^{p} (\widehat{D}_i \xi^j) u_j^\alpha, \qquad \text{where} \qquad \widehat{D}_i = \frac{\partial}{\partial x^i} + \sum_{\alpha=1}^{q} u_i^\alpha \frac{\partial}{\partial u^\alpha} \tag{4.56}$$

denotes the zeroth order truncation of the total derivative operator D_i, cf. (4.46). In fact, the formulae (4.56) imply the usual prolongation formula,

$$\chi_i^\alpha = D_i \varphi^\alpha - \sum_{j=1}^{p} (D_i \xi^j) u_j^\alpha, \tag{4.57}$$

cf. (4.29), because the contact conditions (4.55) guarantee that all the terms in (4.57) involving second order derivatives of u cancel out. Thus the contact conditions uniquely prescribe the prolongation of any point or contact vector field.

Next, we recall the definition (4.8) of the characteristic of a vector field, which we adapt without change here. The contact conditions (4.55) can then be rewritten in the simpler form

$$\frac{\partial Q^\alpha}{\partial u_i^\beta} + \xi^i \delta_\beta^\alpha = 0, \qquad \alpha, \beta = 1, \ldots, q, \quad j = 1, \ldots, p, \tag{4.58}$$

where δ^α_β is the Kronecker delta. Conversely, if $Q = (Q^1, \ldots, Q^q)$ and $\xi = (\xi^1, \ldots, \xi^p)$ are any solutions to the contact conditions (4.58), then the vector field (4.54) with

$$\varphi^\alpha = Q^\alpha + \sum_{i=1}^{p} \xi^i u_i^\alpha, \qquad \chi_i^\alpha = \widehat{D}_i Q^\alpha = D_i Q^\alpha + \sum_{j=1}^{p} \xi^j u_{ij}^\alpha, \tag{4.59}$$

determines a solution to (4.55),(4.56), and hence is a contact vector field. In particular, if there is just one dependent variable, $q = 1$, we can drop the Greek indices, so there is just one characteristic $Q = \varphi - \sum_i \xi^i u_i$. In this case, (4.58), (4.59) reduce to

$$\xi^i = -\frac{\partial Q}{\partial u_i}, \quad \varphi = Q - \sum_{i=1}^{p} u_i \frac{\partial Q}{\partial u_i}, \quad \chi_i = \frac{\partial Q}{\partial x^i} + u_i \frac{\partial Q}{\partial u}, \quad i = 1, \ldots, p, \tag{4.60}$$

which serve to define the coefficients of $\mathbf{v}^{(1)}$ in terms of the characteristic Q. Formula (4.60) appears in Lie's original work on contact transformations, [**157**], where the function $W = -Q$ was called the "characteristic function" of the infinitesimal contact transformation determined by the vector field $\mathbf{v}$; this is the historical origin of our term "characteristic".

Proposition 4.34. *In the case of a single dependent variable, every first order differential function $Q(x, u^{(1)})$ is the characteristic of a uniquely defined contact vector field. The differential function Q is the characteristic of a point vector field if and only if it is an affine function of the first order derivative coordinates $\partial u/\partial x^i$.*

For example, in the case of one independent and one dependent variable, the first order contact vector field with characteristic $Q(x, u, u_x) = Q(x, u, p)$ has the explicit form

$$\mathbf{v} = -\frac{\partial Q}{\partial p}\frac{\partial}{\partial x} + \left(Q - p\frac{\partial Q}{\partial p}\right)\frac{\partial}{\partial u} + \left(\frac{\partial Q}{\partial x} + p\frac{\partial Q}{\partial u}\right)\frac{\partial}{\partial p}. \tag{4.61}$$

This defines the first prolongation of a point vector field if and only if $Q(x, u, p) = \varphi(x, u) - p\,\xi(x, u)$ is an affine function of p. As we shall see, the higher order prolongations of the vector field (4.61) are found using the standard prolongation formula (4.21) (which also applies to the coefficient of ∂_p).

Exercise 4.35. Suppose $q = 1$, and $\mathbf{v}$ is a contact vector field with characteristic $Q(x, u^{(1)})$. Prove that the Lie derivative of the basic contact form $\theta = du - \sum_{i=1}^{p} u_i \, dx^i$ with respect to $\mathbf{v}$ is

$$\mathbf{v}(\theta) = \frac{\partial Q}{\partial u}\,\theta. \tag{4.62}$$

Returning to the general case, if there is more than one dependent variable, then, choosing $\beta \neq \alpha$ in the contact conditions (4.58), we deduce that Q^α does not depend on u_i^β for $\alpha \neq \beta$. Moreover, $\xi^i = -\partial Q^\alpha / \partial u_i^\alpha$ for any choice of α. This in turn implies that ξ^i does not depend on any of the first order derivatives u_i^α, and, according to (4.59), neither does φ^α. This means that the vector field $\mathbf{v}$ is the first prolongation of a vector field on the space E defining an infinitesimal point transformation. Consequently, if there is more than one dependent variable, every (infinitesimal) first order contact transformation is a prolonged (infinitesimal) point transformation, which completes the proof of the infinitesimal form of Bäcklund's Theorem 4.32 in the first order case.

To prove the higher order version, consider a contact vector field

$$\mathbf{v}^{(n)} = \sum_{i=1}^{p} \xi^i(x, u^{(n)}) \frac{\partial}{\partial x^i} + \sum_{\alpha=1}^{q} \sum_{\#J=0}^{n} \varphi_J^\alpha(x, u^{(n)}) \frac{\partial}{\partial u_J^\alpha}, \tag{4.63}$$

on the nth order jet space J^n. Taking the Lie derivative of the contact form (4.39) with respect to (4.63), we find

$$\mathbf{v}^{(n)}(\theta_J^\alpha) = d\varphi_J^\alpha - \sum_{i=1}^{p} \left[\varphi_{J,i}^\alpha \, dx^i + u_{J,i}^\alpha \, d\xi^i \right], \qquad \#J < n. \tag{4.64}$$

Now, if this is to be a contact form on J^n, its horizontal component must vanish. Equations (4.41), (4.42) imply that the coefficients of $\mathbf{v}^{(n)}$ satisfy the inductive form of the prolongation formula (4.28). Since (4.28) serves to uniquely specify the higher order coefficients in terms of ξ^i and φ^α, we deduce that the coefficients of a general contact vector field on J^n are given by the same prolongation formula (4.21) as with a prolonged point transformation; in particular, this observation serves to complete the proof of the prolongation formula in Theorem 4.16. Thus, we have the following result, that includes our earlier version for point vector fields, Theorem 4.16, as a special case.

Proposition 4.36. *The prolongation of a vector field $\mathbf{v}^{(n)}$ on J^n to the jet space J^{n+k} is the unique contact vector field $\mathbf{v}^{(n+k)}$ on J^{n+k} whose projection to J^n coincides with $\mathbf{v}^{(n)} = d\pi_n^{n+k}(\mathbf{v}^{(n+k)})$. The coefficients of $\mathbf{v}^{(n+k)}$ are generated via the recursive prolongation formula* (4.28).

These are not the only conditions that $\mathbf{v}^{(n)}$ must satisfy. Indeed, the splitting of a one-form on J^n into its horizontal and contact components takes place, not on J^n, but on the next higher order jet space J^{n+1}, whereas Proposition 4.33 requires that the Lie derivative $\mathbf{v}^{(n)}(\theta_J^\alpha)$ be a contact form on J^n. Using the decomposition (4.44) in equation (4.64), we see that this requires that

$$d_n \varphi_J^\alpha - \sum_{i=1}^{p} u_{J,i}^\alpha \, d_n \xi^i = 0, \qquad \#J < n.$$

Thus, according to (4.45), the coefficient of du_I^α in the latter formula implies that

$$\frac{\partial \varphi_J^\alpha}{\partial u_I^\alpha} - \sum_{i=1}^{p} u_{J,i}^\alpha \frac{\partial \xi^i}{\partial u_I^\alpha} = 0, \qquad \#I = n, \quad \#J < n. \tag{4.65}$$

Note that (4.65) is the same as the condition that the coefficients $\varphi_{J,i}^\alpha$, as given by (4.28), do not depend on $(n+1)^{\mathrm{st}}$ order derivatives of the u's. A straightforward induction proves that the characteristic Q of the contact vector field $\mathbf{v}^{(n)}$ must depend on at most first order derivatives of u. Indeed, if $n > 2$, then, for $J = 0$, (4.65) is the same as $\partial Q^\alpha / \partial u_I^\alpha = 0$, and hence the characteristic Q of our contact vector field cannot depend on n^{th} order derivatives u_I^α, $\#I = n$. More generally, suppose $Q = Q(x, u^{(k)})$ only depends on k^{th} and lower order derivatives of u for some $k \geq 2$. In this case,

$$\frac{\partial}{\partial u_{I,J}^\alpha}\left[D_J Q^\alpha\right] = \frac{\partial Q^\alpha}{\partial u_I^\alpha}, \qquad \#I = k, \quad \#J = n - k.$$

Therefore, substituting the prolongation formula (4.21) into conditions (4.65), we deduce that $\partial Q^\alpha / \partial u_I^\alpha = 0$ for $\#I = k$, which proves that Q in fact only depends on at most $(k-1)^{\mathrm{st}}$ order derivatives of u. Therefore, we conclude that $Q = Q(x, u^{(1)})$ can depend on at most first order derivatives of u. Finally, analysis of (4.65) for $\#J = n - 1$ shows that Q must satisfy the first order contact conditions (4.58), and hence our first order analysis completes the proof of the infinitesimal version of Bäcklund's Theorem. *Q.E.D.*

Classification of Groups of Contact Transformations

In [**152**], Lie found the complete classification of all finite-dimensional Lie groups of contact transformations acting on a space of one independent and one dependent complex variable. (As far as I know, the corresponding real classification has never been done.) These include the groups of point transformations discussed in Chapter 2, and listed in Tables 1–3 at the end of the book. Lie found that, remarkably, beyond the previously classified groups of point transformations, there are, in fact, only three "genuine" contact transformation groups. These are listed in Table 4, which displays the characteristics of the infinitesimal generators of the group; the vector fields themselves are recovered via the formula (4.61). The precise theorem of Lie follows; see also Sokolov, [**205**], for a modern presentation, and [**212**] for further results.

Theorem 4.37. *Every connected, finite-dimensional Lie group of contact transformations acting on a complex two-dimensional manifold without fixed points is locally equivalent, under a contact transformation, to either one of the Lie groups of point transformations listed in Tables 1–3, or to one of the three contact transformation groups listed in Table 4.*

The structure of the three contact transformation groups is based on the following observations. The second of these groups, given by Case 4.2, is generated by the seven-dimensional algebra of contact vector fields

$$\partial_u,\, x\partial_u + \partial_p,\, x^2\partial_u + 2x\partial_p,\, u\partial_u + p\partial_p,\, \partial_x,\, x\partial_x - p\partial_p,\, 2p\partial_x + p^2\partial_u. \quad (4.66)$$

If, as in Example 4.20, we restrict the vector fields (4.66) to the (x,p)–plane, we obtain a six-dimensional subalgebra of vector fields which generates the planar affine group action $\mathrm{A}(2)$ given by Case 2.2. Similarly, restricting the six-dimensional contact group of Case 4.1 to the (x,p)–plane produces the five-dimensional special affine group action $\mathrm{SA}(2)$ given in Case 2.1. The largest contact group, which contains the other two as subgroups, is locally isomorphic to the group $\mathrm{SO}(5)$ of rotations in five-dimensional space, or, equivalently, the symplectic group $\mathrm{Sp}(4)$. (See [**119**] for a proof of the isomorphism $\mathfrak{so}(5) \simeq \mathfrak{sp}(4)$.) This group acts as a group of conformal transformations on the three-dimensional jet space J^1. Lie showed that (modulo a change of variables) this group is the largest group of contact transformations taking circles to circles, cf. [**151**].

As for the point transformation groups, they are not all distinct if we allow equivalence under contact transformations. In particular, of the three different actions of the unimodular group SL(2) discussed in Example 4.20, the two transitive cases (4.35), (4.36) can, in fact, be mapped to each other a contact transformation, even though there is no point transformation that will accomplish this! (The intransitive case, though, remains inequivalent since it remains intransitive when prolonged to J^1.) To demonstrate this fact, we rewrite the two Lie algebras in terms of x, u, as in Cases 1.1 and 1.2 of the tables. We begin by making the simple point transformation $(x, u) \longmapsto (1/u, x)$, in terms of which the first prolongation of the two Lie algebras is

$$\begin{aligned} &\mathbf{v}_- = \partial_u, \quad \mathbf{v}_0 = x\partial_x + u\partial_u, \quad \mathbf{v}_+ = 2xu\partial_x + u^2\partial_u - 2xp\partial_p, \\ &\mathbf{v}_- = \partial_u, \quad \mathbf{v}_0 = x\partial_x + u\partial_u, \quad \mathbf{v}_+ = (2xu + x^2)\partial_x + u^2\partial_u - 2x(p^2 + p)\partial_p. \end{aligned} \tag{4.67}$$

In terms of these new variables, the class of contact transformations

$$\bar{x} = \frac{x}{h'(p)}, \qquad \bar{u} = u + x\left(\frac{h(p)}{h'(p)} - p\right), \qquad \bar{p} = h(p), \tag{4.68}$$

where $h(p)$ is any smooth function with nonvanishing derivative, is readily seen to preserve the two-dimensional subalgebra spanned by $\mathbf{v}_-$, $\mathbf{v}_0$. Moreover, if $h(p) = \sqrt{\sqrt{p+1} + \sqrt{p}}$, so that $(p^2 + p)h'(p)^2 = h(p)^2$, then the resulting transformation (4.68) maps the second Lie algebra in (4.67) back to the first, rewritten in terms of $\bar{x}, \bar{u}, \bar{p}$, and thus provides the desired transformation. The details of the computation are left to the interested reader.

Chapter 5

Differential Invariants

Recall that an invariant of a group G acting on a manifold M is a function $I: M \to \mathbb{R}$ which is not affected by the group transformations. A *differential invariant* is merely an invariant, in the standard sense, for a prolonged group of transformations (or, more generally, a group of contact transformations) acting on the jet space J^n. Just as the ordinary invariants of a group action serve to characterize invariant equations, so differential invariants will completely characterize invariant systems of differential equations for the group, as well as invariant variational principles. As such they form the basic building blocks of many physical theories, where one begins by postulating the invariance of the differential equations, or the variational problem, under a prescribed symmetry group. Historically, the subject was initiated by Halphen, [**108**], and then developed in great detail, with numerous applications, by Lie, [**156**], and Tresse, [**214**]. In this chapter, we discuss the basic theory of differential invariants, and some fundamental methods for constructing them. Applications will appear here and in the following two chapters. The explicit examples of group actions that are presented to illustrate the general theory will be fairly elementary, in part because, rather surprisingly, the complete classification of differential invariants for many of the groups of physical importance, including the general linear, affine, Poincaré, and conformal groups, does not yet seem to be known!

Differential Invariants

The basic problem to be addressed in the first part of this chapter is the classification of the differential invariants of a given group action. We will be considering general (connected) transformation groups, by which we mean groups of contact transformations, which, according to Bäcklund's Theorem 4.32, are either prolonged point transformation groups or, in the case of a single dependent variable, prolonged first order contact transformation groups. If G denotes the transformation group, we let $G^{(n)}$ denote the corresponding prolonged action on the nth jet space $\mathrm{J}^n = \mathrm{J}^n E$ of the space of independent and dependent variables. We use the notation $g^{(n)}$ to denote the (prolonged) action of the individual group elements $g \in G$ on J^n, and $\mathbf{v}^{(n)}$ for the associated infinitesimal generators.

Definition 5.1. Let G be a group of point or contact transformations. A *differential invariant* for G is a differential function $I\colon \mathrm{J}^n \to \mathbb{R}$ which satisfies $I(g^{(n)} \cdot (x, u^{(n)})) = I(x, u^{(n)})$ for all $g \in G$ and all $(x, u^{(n)}) \in \mathrm{J}^n$ where $g^{(n)} \cdot (x, u^{(n)})$ is defined.

As usual, the differential invariant I may only be defined on an open subset of the jet space J^n, although, in accordance with our usual convention, we shall still write $I\colon \mathrm{J}^n \to \mathbb{R}$. Note that, as with general differential functions, any lower order differential invariant $I(x, u^{(k)})$, $k < n$ can also be viewed as an nth order differential invariant.

Example 5.2. Consider the usual action of the rotation group $\mathrm{SO}(2)$ on $E \simeq \mathbb{R}^2$, cf. Example 4.10. The radius $r = \sqrt{x^2 + u^2}$ is an (ordinary) invariant of $\mathrm{SO}(2)$ (and, *a fortiori*, a differential invariant too). The first prolongation $\mathrm{SO}(2)^{(1)}$ was given in (4.11), and has one-dimensional orbits at each point of J^1; therefore, by Theorem 2.34, besides the radius r, there is one additional first order differential invariant, which can be taken to be the function $w = (xu_x - u)/(x + uu_x)$, provided $x \neq -uu_x$. (Alternative pairs of differential invariants must be used near the points on J^1 where $x + uu_x = 0$.) Geometrically, $w = \tan\phi$, where ϕ is the angle between the line from the origin to the point $(x, u) = (x, f(x))$ and the tangent to the graph of $u = f(x)$ at that point; see Figure 2. The second prolongation $\mathrm{SO}(2)^{(2)}$ is given by (4.17), and also has one-dimensional orbits. (Indeed, the dimension of the orbits can never exceed that of the group.) The radius r and first order differential invariant w still provide two independent differential invariants on J^2, and the curvature $\kappa = (1+u_x^2)^{-3/2} u_{xx}$ is the additional second order differential invariant.

As with ordinary invariants, we shall classify differential invariants up to functional independence, since any function $H(I_1, \ldots, I_k)$ of a collection of differential invariants $I_1, \ldots, I_k$ is also a differential invariant. In accordance with Proposition 1.32, the differential functions

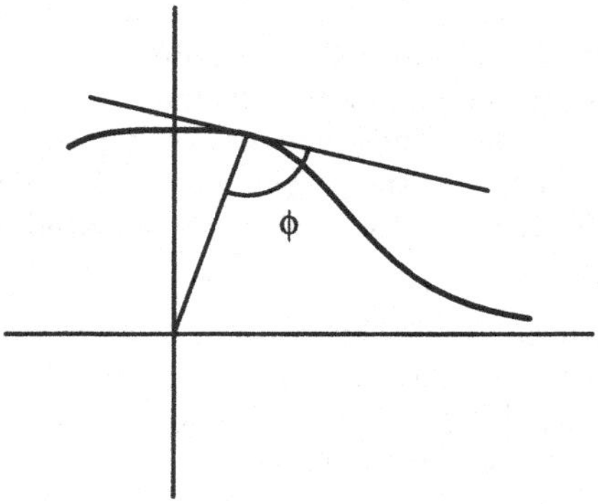

Figure 2. Rotation Group Invariant

$F_1, \ldots, F_k \colon \mathrm{J}^n \to \mathbb{R}$ are functionally independent if their differentials, cf. (4.42), are pointwise independent: $dF_1 \wedge \cdots \wedge dF_k \neq 0$. Since, as noted above, lower order differential invariants are also considered as n^{th} order differential invariants, it will be important to distinguish the differential invariants which genuinely depend on the nth order derivative coordinates. We will call a set of differential functions on J^n strictly independent if, as functions of the nth order derivative coordinates alone, they are functionally independent. More specifically, we make the following definition.

Definition 5.3. A collection of nth order differential functions $F_1(x, u^{(n)}), \ldots, F_k(x, u^{(n)})$ is called *strictly independent* if they and the derivative coordinate functions x, $u^{(n-1)}$, of order less than n, are all functionally independent.

For example, in the case of the rotation group discussed above, the second order differential invariants r, w, κ are functionally independent, but not strictly independent since r and w have order less than two. Indeed, there is only one strictly independent second order differential invariant — the curvature κ (or any function thereof).

Proposition 5.4. *The differential functions $F_1, \ldots, F_k$ are strictly independent if their n^{th} order differentials, as given in* (4.45), *are linearly independent at each point: $d_n F_1 \wedge d_n F_2 \wedge \cdots \wedge d_n F_k \neq 0$.*

In particular, strict independence implies that none of the F_ν's, or any function thereof, can be of order strictly less than n. In the regular case, then, the number of strictly independent functions in a collection $\{F_1, \ldots, F_k\}$ is equal to the rank of their $k \times q_n$ top order Jacobian matrix $\left(\partial F_\nu / \partial u_J^\alpha\right)$, $\#J = n$.

Dimensional Considerations

In order to study the differential invariants of a group of transformations, a more detailed knowledge of the structure of the prolonged group action is required. Of particular importance is an understanding of the dimension of the (generic) orbits of the different prolongations $G^{(n)}$ on J^n. Recall first that the dimension of the jet space J^n is denoted by $p + q^{(n)}$, where $q^{(n)} = q\binom{p+n}{n}$. The number of derivative coordinates of order exactly n is denoted by

$$q_n = \dim \mathrm{J}^n - \dim \mathrm{J}^{n-1} = q^{(n)} - q^{(n-1)} = q\binom{p+n-1}{n}. \tag{5.1}$$

We let s_n denote the maximal (generic) orbit dimension of $G^{(n)}$, so that $G^{(n)}$ acts semi-regularly on the open subset $V^n \subset \mathrm{J}^n$ which consists of all points contained in the orbits of maximal dimension. (If G is a group of point transformations, then s_n is well defined for each $n \geq 0$, whereas for a group of contact transformations, s_n is defined only for $n \geq 1$.) If G acts analytically, then the subset V^n is dense in J^n. For the time being, we restrict our attention to the subset V^n, thereby avoiding more delicate questions concerning singularities of the prolonged group action. Note that s_n equals the maximal dimension of the subspace $\mathfrak{g}^{(n)}|_z \subset T\mathrm{J}^n|_z$ spanned by the prolonged infinitesimal generators of the group action at points $z \in \mathrm{J}^n$. According to Proposition 2.67, the prolonged orbit dimensions can also be computed as $s_n = r - h_n$, where h_n denotes the dimension of the isotropy subgroup $H_z^{(n)} \subset G$ at any point $z \in V^n$.

According to Theorem 2.34, there are

$$i_n = p + q^{(n)} - s_n = p + q^{(n)} - r + h_n \tag{5.2}$$

functionally independent differential invariants of order at most n defined in a neighborhood of any point $z \in V^n$. Since each differential invariant of order less than n is included in this count, the integers i_n form a nondecreasing sequence: $i_0 \leq i_1 \leq i_2 \leq \cdots$. The difference

$$j_n = i_n - i_{n-1} = q_n - s_n + s_{n-1} = q_n + h_n - h_{n-1} \tag{5.3}$$

will count the number of strictly independent n^{th} order differential invariants. For groups of point transformations, we set $j_0 = i_0$ to be the number of ordinary invariants; for contact transformation groups, where i_0, s_0, and j_0 are not defined, we set $j_1 = i_1$. Note that j_n cannot exceed the number of independent derivative coordinates of order n, so $j_n \leq q_n$, which implies the elementary inequalities

$$i_{n-1} \leq i_n \leq i_{n-1} + q_n. \tag{5.4}$$

For example, in the case of the rotation group discussed in Example 5.2, each prolongation has one-dimensional orbits (indeed, SO(2) acts regularly on all of J^n for $n \geq 1$), and hence $s_0 = s_1 = s_2 = \cdots = 1$. Equation (5.2) implies that $i_0 = 1$, so there is one ordinary invariant — the radius r. Furthermore, $i_1 = 2$, so there is $j_1 = i_1 - i_0 = 1$ additional first order differential invariant, which can be chosen to be the angle ϕ. In general, $i_n = n+1$, so there is precisely $j_n = i_n - i_{n-1} = 1$ additional differential invariant at each order n. Beyond the curvature κ, the higher order differential invariants for the rotation group will be constructed in Example 5.17 below.

If $\mathcal{O}^{(n)} \subset \mathrm{J}^n$ is any orbit of $G^{(n)}$, then, for any $k < n$, its projection $\pi^n_k(\mathcal{O}^{(n)}) \subset \mathrm{J}^k$ is an orbit of the kth prolongation $G^{(k)}$. Therefore, the maximal orbit dimension s_n of $G^{(n)}$ is also a *nondecreasing* function of n, bounded by r, the dimension of G itself:

$$s_0 \leq s_1 \leq s_2 \leq \cdots \leq r. \tag{5.5}$$

On the other hand, since the orbits cannot increase in dimension any more than the increase in dimension of the jet spaces themselves, we have the the elementary inequalities

$$s_{n-1} \leq s_n \leq s_{n-1} + q_n, \tag{5.6}$$

governing the orbit dimensions. Note that, in view of equations (5.1) and (5.2), the inequalities (5.6) are equivalent to those in (5.4). Also note that (5.6) implies that the isotropy subgroups $H^{(n)}_z$, $z \in V^n$, have nonincreasing dimensions: $h_{n-1} \geq h_n = r - s_n$. This follows directly from the observation that the isotropy subgroup $H^{(k)}_w$ of the projection $w = \pi^n_k(z) \in \mathrm{J}^k$ of a point $z \in \mathrm{J}^n$ is always contained in the isotropy subgroup $H^{(n)}_z$.

The inequalities (5.5) imply that the maximal orbit dimension eventually stabilizes, so that there exists an integer s such that $s_m = s$ for all m sufficiently large. In particular, if the orbit dimension is ever the same as that of G, meaning $s_n = r$ for some n, then $s_m = r$ for all $m \geq n$. We shall call s the *stable orbit dimension*, and the minimal order n for which $s_n = s$ the *order of stabilization* of the group. Once we've reached the stabilization order, the number of higher order differential invariants is immediate.

Proposition 5.5. *Let n denote the order of stabilization of the group G. Then, for every $m > n$ there are precisely q_m strictly independent m^{th} order differential invariants.*

Consequently, any (finite-dimensional) group of transformations has an infinite number of differential invariants of arbitrarily large order. Thus the classification problem for differential invariants cannot be handled by purely algebraic techniques alone.

Infinitesimal Methods

As with ordinary invariants, it is easier to determine differential invariants (of connected groups) using an infinitesimal approach. The basic infinitesimal invariance condition for differential invariants is an immediate corollary of Theorem 2.74.

Proposition 5.6. *A function $I\colon \mathrm{J}^n \to \mathbb{R}$ is a differential invariant for a connected transformation group G if and only if it is annihilated by all the prolonged infinitesimal generators:*

$$\mathbf{v}^{(n)}(I) = 0 \qquad \text{for all} \qquad \mathbf{v} \in \mathfrak{g}. \tag{5.7}$$

Example 5.7. In the case of the rotation group $\mathrm{SO}(2)$, its second prolongation has infinitesimal generator $\mathbf{v}^{(2)}$ given by (4.25). Applying this vector field to the functions given in Example 5.2, we find $\mathbf{v}^{(2)}(r) = \mathbf{v}^{(2)}(w) = \mathbf{v}^{(2)}(\kappa) = 0$, re-proving the fact that r, w, κ are differential invariants. Note that these differential invariants can be deduced directly from the form of $\mathbf{v}^{(2)}$ using the method of characteristics, as was done in Example 2.75.

Example 5.8. Consider the three-parameter similarity group

$$(x, u) \longmapsto (\lambda x + a, \lambda u + b), \qquad (x, u) \in E \simeq \mathbb{R}^2,$$

consisting of translations and scalings, and generated by the vector fields $\partial_x, \partial_u, x\partial_x + u\partial_u$. There are no ordinary invariants since the group acts transitively on $E = \mathbb{R}^2$. Furthermore, all three vector fields happen to coincide with their first prolongations, and hence there is one independent first order differential invariant, namely u_x. The second prolongations are $\partial_x, \partial_u, x\partial_x + u\partial_u - u_{xx}\partial_{u_{xx}}$, and hence there are no differential invariants of (strictly) second order. There is a single third order differential invariant, namely $u_{xx}^{-2}u_{xxx}$, a single fourth order invariant, $u_{xx}^{-3}u_{xxxx}$, and, in general, a single n^{th} order differential invariant $u_{xx}^{1-n}D_x^n u$. Therefore, the number of strictly independent differential invariants is given by $j_0 = j_2 = 0$, $j_1 = j_3 = \cdots = j_n = 1$, $n \geq 3$. This implies that $i_0 = 0$, $i_1 = i_2 = 1$, $i_3 = 2, \ldots, i_n = n - 1$, and hence the maximal orbit dimensions are $s_0 = s_1 = 2$, $s_2 = s_3 = \cdots = 3 = \dim G$, a fact that can also be deduced by looking at the dimension of the space spanned by the prolonged infinitesimal generators. Note, in particular, that the orbit dimensions "pseudo-stabilized" at order 0 since $s_0 = s_1$, but that the true order of stabilization is $n = 2$. Thus, the fact that $s_k = s_{k+1}$ does *not*, in general, imply that k is the order of stabilization of a transformation group G. However, as we shall subsequently see, this pseudo-stabilization phenomenon is quite rare.

The preceding example corresponds to a particular case of our classification tables for Lie group actions in the plane — namely Case 1.7 with $\alpha = k = 1$. More generally, consider Case 1.7 with $\alpha = k = r - 2 \geq 1$, which is the r parameter group generated by the vector fields

$$\partial_x, \quad \partial_u, \quad x\partial_u, \quad \ldots, \quad x^{r-3}\partial_u, \quad x\partial_x + (r-2)u\partial_u.$$

(What is the associated group action?) Using the prolongation formula (4.21), it is not hard to see that the prolonged orbit dimensions are given by $s_0 = 2$, $s_1 = 3, \ldots, s_{r-3} = s_{r-2} = r - 1$, $s_{r-1} = s_r = \cdots = r$. In this case, the orbit dimensions pseudo-stabilize at order $r - 3$ and stabilize at order $r - 1$. Consequently, the prolonged orbit dimensions of a transformation group can pseudo-stabilize at an arbitrarily high order. Surprisingly, though, these are essentially the *only* examples in which a pseudo-stabilization occurs; see Theorem 5.24 below.

Exercise 5.9. Find the differential invariants of the latter case.

Exercise 5.10. Prove that the orbit dimensions for Case 1.7 with $\alpha \neq k$ do *not* pseudo-stabilize.

Stabilization and Effectiveness

As we already noted, cf. (5.5), the stable orbit dimension is bounded by the dimension of the transformation group G. Remarkably, these two dimensions are the same if and only if the group acts locally effectively, as defined in Chapter 2. This result is due to Ovsiannikov, [**191**].

Theorem 5.11. *Let G be an r-dimensional Lie group of point or contact transformations. The group G acts locally effectively if and only if its stable orbit dimension equals its dimension, so that $s_m = r$ for all m sufficiently large.*

Proof: We begin with the following elementary result.

Lemma 5.12. *A transformation group G acts locally effectively if and only if its prolongation $G^{(n)}$ acts locally effectively.*

Proof: This is an immediate consequence of the fact that the only transformation whose n^{th} prolongation is the identity map on J^n is the identity map on E itself. In particular, the global isotropy group of the prolonged action coincides with the global isotropy group G_E. *Q.E.D.*

Therefore, if G does not act (locally) effectively, neither does $G^{(n)}$, and hence, by Proposition 2.4, the orbit dimension of $G^{(n)}$ is strictly less than the dimension of G. Thus, if G does not act locally effectively, the stable orbit dimension is strictly less than the dimension of G, so $s < r$ in this case.

To prove the converse, suppose that $s < r$. Let $\mathbf{v}_1, \ldots, \mathbf{v}_r$ be a basis for $\mathfrak{g}$ such that $\mathbf{v}_1^{(n)}, \ldots, \mathbf{v}_s^{(n)}$ are linearly independent on an open subset $W^n \subset V^n \subset \mathrm{J}^n$. This implies that there exist *uniquely determined* n^{th} order differential functions $A_\mu^\nu(x, u^{(n)})$, $\nu = 1, \ldots, s$, $\mu = s+1, \ldots, r$ such that the remaining infinitesimal generators are written as

$$\mathbf{v}_\mu^{(n)} = \sum_{\nu=1}^{s} A_\mu^\nu \, \mathbf{v}_\nu^{(n)}, \qquad \mu = s+1, \ldots, r. \tag{5.8}$$

If we can prove that the coefficients A_μ^ν are, in fact, constants, then Theorem 5.11 follows directly from Theorem 2.62.

Since n is the order of stabilization, $s_m = s$ for all $m \geq n$, so the prolonged vector fields $\mathbf{v}_1^{(m)}, \ldots, \mathbf{v}_s^{(m)}$ must remain linearly independent on the open subset $W^m = (\pi_n^m)^{-1} W^n \subset \mathrm{J}^m$. Moreover, the linear identities (5.8) must also hold for all higher order prolongations of the generators:

$$\mathbf{v}_\mu^{(m)} = \sum_{\nu=1}^{s} A_\mu^\nu \, \mathbf{v}_\nu^{(m)}, \qquad \mu = s+1, \dots, r, \quad m \geq n, \tag{5.9}$$

for the *same* uniquely determined differential functions $A_\mu^\nu(x, u^{(n)})$. Indeed, if we apply $d\pi_n^m$ to (5.9), both sides will project to their counterparts in (5.8), and hence, by uniqueness, the coefficients must be the same. Let $Q_\mu = (Q_\mu^1, \dots, Q_\mu^q)$ be the characteristic of the vector field $\mathbf{v}_\mu$. According to the prolongation formula (4.21), equation (5.9) is equivalent to the identities

$$\xi_\mu^i = \sum_{\nu=1}^{s} a_\mu^\nu \, \xi_\nu^i, \qquad D_J Q_\mu^\alpha = \sum_{\nu=1}^{s} A_\mu^\nu \, D_J Q_\nu^\alpha, \qquad \mu = s+1, \dots, r, \tag{5.10}$$

valid for all $i = 1, \dots, p$, $\alpha = 1, \dots, q$, and *all* multi-indices J of order $\#J \leq m$, where $m \geq n$ can be taken to be arbitrarily large. If we apply the total derivative operators D_k to the second set of equations in (5.10), we find

$$D_k D_J Q_\mu^\alpha = \sum_{\nu=1}^{s} \big[(D_k A_\mu^\nu)\,(D_J Q_\nu^\alpha) + A_\mu^\nu \,(D_k D_J Q_\nu^\alpha) \big].$$

Subtracting the corresponding equation in (5.10), with J replaced by (J,k), we find

$$\sum_{\nu=1}^{s} (D_k A_\mu^\nu)\,(D_J Q_\nu^\alpha) = 0, \qquad \mu = s+1, \dots, r, \tag{5.11}$$

for all k, α, J. Now, A_μ^ν depends on at most n^{th} order derivatives of u, so its total derivative $D_k A_\mu^\nu$ depends on at most $(n+1)^{\text{st}}$ order derivatives. Moreover, since the characteristic Q_ν depends on at most first order derivatives, formula (4.8), or, in the case of contact transformations, formula (4.58), shows that its total derivatives have the form

$$D_J Q_\nu^\alpha = \sum_{\beta=1}^{q} \sum_{i=1}^{p} \frac{\partial Q_\nu^\alpha}{\partial u_i^\beta} \, u_{J,i}^\beta + \cdots = - \sum_{i=1}^{p} \xi_\nu^i \, u_{J,i}^\alpha + \cdots,$$

where the omitted terms depend on derivatives of order at most $\#J$. Therefore, if $\#J = m \geq n+1$, the only terms in (5.11) that depend on derivatives of order $m+1$ are

$$-\sum_{\nu=1}^{s}\sum_{i=1}^{p}(D_k A_\mu^\nu)\,\xi_\nu^i\,u_{J,i}^\alpha = 0, \qquad \mu = s+1,\ldots,r.$$

The independence of the derivative coordinates $u_{J,i}^\alpha$ implies that

$$\sum_{\nu=1}^{s}(D_k A_\mu^\nu)\,\xi_\nu^i = 0, \qquad \mu = s+1,\ldots,r, \quad i,k = 1,\ldots,p. \tag{5.12}$$

Again, by the prolongation formula (4.21), equations (5.11) and (5.12) imply that

$$\sum_{\nu=1}^{s}(D_k A_\mu^\nu)\,\mathbf{v}_\nu^{(m)} = 0, \qquad \mu = s+1,\ldots,r. \tag{5.13}$$

But the vector fields $\mathbf{v}_1^{(m)},\ldots,\mathbf{v}_s^{(m)}$ were assumed to be *linearly independent*; therefore equation (5.13) will hold if and only if $D_k A_\mu^\nu = 0$ for all k,μ,ν. Exercise 4.15 shows that the coefficients A_μ^ν must all be constant, which completes the proof of the theorem. *Q.E.D.*

Corollary 5.13. *If G is an r-dimensional transformation group with global isotropy subgroup G_E of dimension t, then the stable orbit dimension of G is $s = r - t = \dim G - \dim G_E$. Consequently, for each $m \geq n$, the order of stabilization, G acts effectively freely on the open subset $V^m \subset \mathrm{J}^m$ where the prolonged group orbits achieve their maximal dimension, namely s.*

If G does not act effectively, then, according to Proposition 2.12, we can replace G by the effectively acting quotient group G/G_E without any appreciable loss of generality. Therefore, to avoid unnecessary complications in our subsequent presentation, we can always assume that the transformation group G acts locally effectively, and so the stable orbit dimension s is the same as the dimension r of G itself.

Corollary 5.14. *If G acts (locally) effectively on an open subset $M \subset E$, then G acts (locally) freely on the open subset $V^m \subset \mathrm{J}^m$ where the prolonged group orbits achieve the stable orbit dimension $r = \dim G$.*

Invariant Differential Operators

Since any transformation group action has differential invariants of arbitrarily high order, it is incumbent upon us to find a more systematic method for determining them all. The basic tool is the use of certain "invariant" differential operators, introduced by Lie, **[156]**, and Tresse, **[214]**, which have the property of mapping n^{th} order differential invariants to $(n+1)^{\text{st}}$ order differential invariants, and thus, by iteration, produce hierarchies of differential invariants of arbitrarily large order. In fact, we can guarantee the existence of sufficiently many such differential operators and differential invariants so as to completely generate all the higher order independent differential invariants of the group by successively differentiating the lower order differential invariants. Thus, a complete description of all the differential invariants is provided by a collection of low order "fundamental" differential invariants along with the requisite invariant differential operators. To introduce the general method, we begin with the simplest case when there is only one independent variable, where the construction of higher order differential invariants is facilitated by the following result.

Proposition 5.15. *Suppose $X = \mathbb{R}$ and G is a transformation group acting on $E \simeq X \times U$. Let $s = I(x, u^{(n)})$ and $v = J(x, u^{(n)})$ be functionally independent differential invariants, at least one of which has order exactly n. Then the derivative $dv/ds = (D_x J)/(D_x I)$ is an $(n+1)^{\text{st}}$ order differential invariant.*

Proof: The statement can be verified directly, but we provide a more generally applicable proof based on the contact structure of J^n. According to Proposition 2.86, if $I(x, u^{(n)})$ is a differential invariant, its differential dI is an invariant one-form on J^n. As in Definition 4.25, we decompose dI into its horizontal and contact components, $dI = D_x I \, dx + \theta_I$, where θ_I is a contact form on J^{n+1}. Similarly, if J is any other n^{th} order differential invariant, then, on the open subset of J^{n+1} where $D_x I \neq 0$, we have $dJ = D_x J \, dx + \theta_J = [(D_x J)/(D_x I)] \, dI + \vartheta$ for some contact form ϑ. According to Proposition 4.30, the prolonged group transformations in $G^{(n+1)}$ map contact forms to contact forms. Therefore, the invariance of both dI and dJ immediately implies that the coefficient $D_x J / D_x I$ must be an invariant function for $G^{(n+1)}$. Finally, the functional independence of I and J is enough to guarantee that $(D_x J)/(D_x I)$ has order $n+1$ — see Lemma 5.34 below. *Q.E.D.*

Let us reinterpret this result. If $s = I(x, u^{(n)})$ is any given differential invariant, then $\mathcal{D} = d/ds = (D_x I)^{-1} D_x$ is an *invariant differential operator* for the prolonged group actions, since if J is any other differential invariant, so is $\mathcal{D}J$. Therefore, we can iterate $\mathcal{D}$, producing a sequence $\mathcal{D}^k J = d^k J/ds^k$, $k = 0, 1, 2, \ldots$, of higher and higher order differential invariants. Let us first apply this result when there is just one independent and one dependent variable. In this case, once we know two independent differential invariants, all higher order differential invariants can be calculated by successive differentiation with respect to the invariant differential operator $\mathcal{D}$.

Theorem 5.16. *Suppose G is a connected group of point or contact transformations acting on the jet spaces corresponding to $E \simeq \mathbb{R} \times \mathbb{R}$. Then, for some $n \geq 0$, there are precisely two functionally independent differential invariants I, J of order n (or less). Furthermore for any $k \geq 0$, a complete system of functionally independent differential invariants of order $n + k$ is provided by I, J, $\mathcal{D}J, \ldots, \mathcal{D}^k J$, where $\mathcal{D} = (D_x I)^{-1} D_x$ is the associated invariant differential operator.*

Proof: Since the dimension of J^{m+1} is one more than J^m, (5.4) implies that $i_m \leq i_{m+1} \leq i_m + 1$, meaning that, for each m, there is at most one strictly independent $(m+1)^{\mathrm{st}}$ order differential invariant. Moreover, if G is a group of point transformations, $i_0 \leq 2 = \dim E$, whereas if G is a group of contact transformations, $i_1 \leq 2$, since if $i_1 = 3$, then $s_1 = 0$, so G would have zero-dimensional orbits on J^1, which implies that it acts trivially, and is merely the first prolongation of a trivial point transformation group action. These considerations clearly imply that we can find an order n for which $i_n = 2$, and hence can always find the required local invariants; indeed, we can choose I to have order strictly less than n and J to have order exactly n. Then, using Proposition 5.15, the differential invariant $\mathcal{D}^k J$ is the required $(n+k)^{\mathrm{th}}$ order differential invariant. *Q.E.D.*

We will call the differential invariants I and J the *fundamental differential invariants* for the group G. Note that the differential invariants constructed by this method are well defined on the open subset of J^{n+k} where I and J are defined and where $D_x I \neq 0$. At the points where $D_x I = 0$, one must use an alternative differential invariant to avoid singularities.

Example 5.17. For the rotation group SO(2), as discussed in Example 5.2, we can apply Theorem 5.16 when $n = 1$, since r and w provide two independent first order differential invariants. The second order differential invariant resulting from Theorem 5.16, however, is not exactly the curvature, but the more complicated second order differential invariant

$$\frac{dw}{dr} = \frac{D_x w}{D_x r} = \frac{\sqrt{x^2+u^2}}{(x+uu_x)^3}\left[(x^2+u^2)u_{xx} - (1+u_x^2)(xu_x - u)\right].$$

However, since we know that there is only one independent second order differential invariant, we must be able to re-express the curvature in terms of this new differential invariant; we find

$$\kappa = (1+u_x^2)^{-3/2}u_{xx} = (1+w^2)^{-3/2}\left[w_r + r^{-1}(w+w^3)\right].$$

If we replace $w = \tan\phi$ by the angle ϕ described in Example 5.2, then we find the interesting SO(2)-invariant formula $\kappa = \phi_r \cos\phi + r^{-1}\sin\phi$ expressing the curvature of a curve in terms of the radial variation of the angle ϕ. Higher order differential invariants are given by successive derivatives $d^k w/dr^k$ (or, alternatively, $d^k\phi/dr^k$). These can be rewritten explicitly in terms of x and u using the invariant differential operator $D_r = r(x+uu_x)^{-1}D_x$, or, more simply, the alternative invariant differential operator $r^{-1}D_r = (x+uu_x)^{-1}D_x$.

Example 5.18. A particularly important example, since it forms the basis of classical invariant theory, is the linear fractional action

$$(x,u) \longmapsto \left(\frac{\alpha x+\beta}{\gamma x+\delta},\ (\gamma x+\delta)^{-n}u\right). \tag{5.14}$$

of the unimodular group SL(2) on $X \times U \simeq \mathbb{R}^2$. This action corresponds to the multiplier representation (3.9) (restricted to SL(2)). The infinitesimal generators are

$$\mathbf{v}_- = \frac{\partial}{\partial x}, \qquad \mathbf{v}_0 = x\frac{\partial}{\partial x} - \frac{n}{2}u\frac{\partial}{\partial u}, \qquad \mathbf{v}_+ = x^2\frac{\partial}{\partial x} - nxu\frac{\partial}{\partial u}. \tag{5.15}$$

All the actions for $n \neq 0$ are locally isomorphic to Case 1.1 in our classification tables, while if $n = 0$ we are in the intransitive Case 3.1. Note that for $n \neq 0$, the action is transitive on $\{u \neq 0\}$, and so there are no ordinary invariants, whereas if $n = 0$ the dependent variable u is invariant.

The differential invariants for these groups are most easily found by first looking for *relative differential invariants*, which are defined as relative invariants for the prolonged group action. According to (3.25), a relative m^{th} order differential invariant $R(x, u^{(m)})$ of weight k for the group SL(2) satisfies the infinitesimal invariance conditions

$$\mathbf{v}_-^{(m)}(R) = 0, \qquad \mathbf{v}_0^{(m)}(R) = \tfrac{1}{2}kR, \qquad \mathbf{v}_+^{(m)}(R) = kxR. \tag{5.16}$$

Absolute differential invariants, i.e., those of weight 0, can be found by taking the ratio of appropriate powers of relative differential invariants. The key remark is that the rth transvectant $(R,\, S)^{(r)}$ between two relative invariants of respective weights k and l, as given by (3.44), produces a relative differential invariant of weight $k + l - 2r$. Thus, for example, starting with u, which is a relative invariant of weight n, we can produce the Hessian

$$\widetilde{H} = \frac{1}{2(n-1)}(u,\, u)^{(2)} = nuu_{xx} - (n-1)u_x^2,$$

which is a relative invariant of weight $2n - 4$. (Here we have omitted the extra factor of $n - 1$ from the original definition (3.46).) For $n \neq 0$, the ratio

$$I = \frac{1}{n}u^{(4/n)-2}\widetilde{H} = u^{(4/n)-1}u_{xx} - \frac{n-1}{n}u^{(4/n)-2}u_x^2,$$

is therefore the first fundamental absolute differential invariant. (The reader should check that, for $n \neq 0$, there are no lower order differential invariants.) If R is any relative m^{th} order differential invariant of weight k, then the Jacobian covariant

$$(u,\, R)^{(1)} = nuD_xR - ku_xR \tag{5.17}$$

is a relative differential invariant of weight $k + n - 2$, which, provided $n \neq 0$, has order $m + 1$. In particular, if I is an absolute differential invariant of order m, then

$$\frac{1}{n}u^{(2/n)-1}(u,\, I)^{(1)} = u^{(2/n)}D_xI \tag{5.18}$$

is an absolute differential invariant of order $m + 1$. Thus, the differential operator $\mathcal{D} = u^{(2/n)}D_x$ has the property that it maps differential invariants to differential invariants. In particular, we can construct the fundamental third order differential invariant

$$J = u^{(2/n)} D_x I = u^{(6/n)-3}(u, H)^{(1)}$$
$$= u^{(6/n)-1} u_{xxx} - 3\frac{n-2}{n} u^{(6/n)-2} u_x u_{xx} + 2\frac{(n-1)(n-2)}{n^2} u^{(6/n)-3} u_x^3. \tag{5.19}$$

Thus, starting with the Hessian invariant, we can use the Jacobian differential operator (5.18) to construct a differential invariant of every order. This implies the following result that provides a functionally independent set of differential invariants for the basic representations of classical invariant theory.

Theorem 5.19. *Every relative differential invariant of the multiplier representation* (5.14) *is a homogeneous function of the differential invariants* $R_0 = u$, $R_2 = \widetilde{H}$, *and their successive Jacobians* $R_m = (u, R_{m-1})^{(1)}$, $m \geq 3$.

In particular, restricting to the case when u is a polynomial of degree n, the differential invariants constructed in Theorem 5.19 provide a complete system of functionally independent covariants for any binary form of degree n.

Exercise 5.20. According to Theorem 5.16, the higher order differential invariants can all be written as functions of the invariants I, J, and their derivatives $d^k J/d^k I$, $k > 0$. Find formulae for the fourth order invariants $u^{(2/n)} D_x J$ and $u^{-(2/n)}(u, u)^{(4)}$ in terms of the invariants I, J, dJ/dI.

The case when $n = 0$ is slightly different, since u is now an absolute invariant, and so cannot be paired with relative differential invariants to produce absolute differential invariants. However, the role played by u is now taken by u_x, which is a relative invariant of weight -2. The invariant process (5.17) used to construct higher order relative invariants is replaced by

$$-(u_x, R)^{(1)} = 2 u_x D_x R + k u_{xx} R, \tag{5.20}$$

which, when applied to a relative m^{th} order invariant of weight k produces a relative $(m+1)^{\text{st}}$ order invariant of weight $k-4$. In particular, given an absolute m^{th} order differential invariant I, the process

$$\frac{dI}{du} = u_x^{-1} D_x I = -u_x^{-2}(u_x, I)^{(1)}, \tag{5.21}$$

produces an absolute invariant of order $m + 1$. However, applying this to the zeroth order invariant u just produces the constant 1, so we use

$$S = \frac{2u_x u_{xxx} - 3u_{xx}^2}{u_x^4}, \qquad (5.22)$$

which is the ratio of relative invariants of weight 4 (the numerator is just a multiple of the Hessian of u_x), as our starting absolute differential invariant. The successive derivatives $d^k S/du^k$ provide a complete system of higher order absolute differential invariants.

Exercise 5.21. Prove that when $n = 0$ there are no (absolute or relative) differential invariants of second order.

Exercise 5.22. Show that if we interchange the role of independent and dependent variables, then the differential invariant (5.22) becomes $S = -2x_u^{-2}(x_u x_{uuu} - \frac{3}{2}x_{uu}^2)$. The latter equals minus twice the classical *Schwarzian derivative*, whose unimodular invariance is well known, cf. [**116**].

Exercise 5.23. Determine the fundamental differential invariants of the third action of SL(2), given by Case 1.2 in the tables. (*Hint:* Use the prolongation connection of Example 4.20. The answer appears in (6.50) below.)

The general construction of Theorem 5.16 allows us to obtain detailed information on the number of differential invariants and, consequently, the (generic) dimension of the prolonged group orbits in the particular case of one independent and one dependent variable. Suppose that the two fundamental differential invariants $I(x, u^{(k)})$ and $J(x, u^{(l)})$ have orders $0 \leq k < l$ respectively. The case $k = 0$ occurs if and only if G acts intransitively on E and hence has an ordinary invariant. (Here we are leaving aside the case when G acts completely trivially on E, where there are two independent zeroth order invariants, namely x and u, and every differential function is a differential invariant.) Under these assumptions, there is precisely one strictly independent differential invariant of each order $m \geq l$ obtained by invariantly differentiating J with respect to I. Therefore, the number of strictly independent differential invariants is given by $j_k = 1$, and $j_m = 1$ for every $m \geq l$, while $j_h = 0$ for $k \neq h < l$. This implies that, for each $m \geq l$, the number of differential invariants of order at most m is given by $i_m = m - l + 2$, and hence $s_m = m + 2 - i_m = l$. Therefore the orbit dimension has stabilized at order l (or less). However, since G was assumed to act locally effectively, Corollary 5.14 implies that the stable orbit dimension must be the same as the dimension r of G itself, and hence, in every case, the second fundamental differential invariant J has the *same* order as the dimension of G. The prolonged orbit dimensions and number of invariants are thus given by

$$s_m = \begin{cases} m+2, \\ m+1, \\ r, \end{cases} \quad i_m = \begin{cases} 0, \\ 1, \\ m-r+2, \end{cases} \qquad \text{when} \quad \begin{cases} m \leq k-1, \\ k \leq m \leq r-1, \\ m \geq r. \end{cases} \tag{5.23}$$

There are three subcases: If the first fundamental invariant I has order $k = 0$, the group acts intransitively on E, the orbit dimension stabilizes at order $r - 1$, and there is one ordinary invariant (of order 0) and one r^{th} order differential invariant. At the other extreme, if $k = r - 1$, then the group has fundamental differential invariants of orders $r - 1$ and r, and the orbit dimension stabilizes at order $r - 2$. The intermediate cases $0 < k < r - 1$ are when the orbit dimension pseudo-stabilizes at order k, and finally stabilizes at order $r - 1$. In particular, the group allows at most *one* such pseudo-stabilization; this is a particular case of the general result to be given in Theorem 5.37 below.

Thus, our methods provide rather detailed information on the possible orbit dimensions of prolonged group actions in the single variable case. However, these results can be made even more precise. As we remarked in Chapter 2, there is a complete classification of the Lie groups of both point and contact transformations acting on a single independent and a single dependent variable. Lie, [**156**], completely classified the differential invariants for each of the point transformation groups† — his results are summarized in Table 6. In most cases, the fundamental differential invariant of lowest order is displayed; the second fundamental differential invariant is obtained by invariant differentiation, to be discussed momentarily. Leaving aside the intransitive actions, we find that in almost every case, the two differential invariants have orders $r-1$ and r, where r is the dimension of the group. The only exceptions, in

† Apparently, Lie never published the differential invariants of the three contact transformation groups. These cases were analyzed with the help of R. Heredero and V. Sokolov.

which the orbit dimension pseudo-stabilizes, are the cases discussed in Example 5.8; they are subcases of Case 1.7 when $\alpha = k$, which is designated as Case 1.7a in Table 5. In this case, the fundamental differential invariants have orders $k = r - 2$ and r, the pseudo-stabilization occurs at order $r - 3$. Thus, remarkably, in the scalar case there are just three fundamentally different possibilities.

Theorem 5.24. *Let G be a connected r-dimensional Lie group of point transformations acting locally effective on $E \simeq \mathbb{R} \times \mathbb{R}$. Then G has two fundamental differential invariants $I(x, u^{(k)})$ and $J(x, u^{(r)})$ of orders $k < r$. Moreover, $k = r - 1$ and the group action stabilizes at order $r - 2$ unless either* a) *G acts intransitively, in which case $k = 0$, or* b) *the prolonged orbit dimensions pseudo-stabilize, in which case $k = r - 2$, and the pseudo-stabilization occurs at order $r - 3$; in both cases, the stabilization order is $r - 1$. The group pseudo-stabilizes if and only if it is equivalent, under a change of variables, to the r-dimensional Lie group action appearing in Case 1.7a, with $\alpha = k = r - 2$.*

Invariant Differential Forms

The construction of differential invariants described above admits an additional important simplification, based on the following observation. Note that the proof of Proposition 5.15 relies on a simple fact: if I is any differential invariant, its total differential $\mathrm{D}I = D_x I\, dx$, cf. Definition 4.25, is a "contact-invariant" one-form, in the following sense.

Definition 5.25. A differential one-form ω on J^n is called *contact-invariant* under a transformation group G if and only if, for every $g \in G$, we have $(g^{(n)})^* \omega = \omega + \theta$ for some contact form $\theta = \theta_g$.

Contact forms are trivially contact-invariant, so only the horizontal contact-invariant forms are of interest. In the scalar case, if $\omega = P\, dx$ is a horizontal contact-invariant one-form (e.g., $\omega = \mathrm{D}I$ is the total differential of a differential invariant I, in which case $P = D_x I$), then every other horizontal contact-invariant one-form is of the form $J\, \omega = JP\, dx$, where J is an arbitrary differential invariant. Thus, if we know two horizontal contact-invariant one-forms $P\, dx$, $\widetilde{P}\, dx$, their ratio $J = \widetilde{P}/P$ defines a differential invariant. A contact-invariant one-form serves to define an invariant differential operator.

Proposition 5.26. *Let G be a group of contact transformations, and let $\omega = P(x, u^{(n)})\, dx$ be a contact-invariant horizontal one-form* *on* J^n*. Then the associated differential operator* $\mathcal{D} = (1/P)D_x$ *is* G*-invariant, so that whenever* I *is a differential invariant, so is* $\mathcal{D}I$*.*

For example, in the case of the action (5.14) of the projective group, for $n \neq 0$ the one-form $u^{(2/n)}\,dx$ is readily seen to be contact-invariant. Proposition 5.26 immediately provides the invariant differential operator (5.18). If $n = 0$, the simplest contact-invariant one-form is just the total differential $\mathrm{D}u = u_x dx$ of the invariant function u, resulting in the invariant differential operator (5.21).

The infinitesimal criterion for contact-invariance is that the Lie derivative $\mathbf{v}^{(n)}(\omega)$ of the form with respect to the prolonged infinitesimal generators be a contact form for every infinitesimal generator $\mathbf{v} \in \mathfrak{g}$. If $\mathbf{v} = \xi\partial_x + \varphi\partial_u$, then the Lie derivative of a horizontal one-form $P(x, u^{(n)})\,dx$ with respect to the prolonged vector field is readily computed using the intrinsic formulation of the total derivative:

$$\mathbf{v}^{(n)}(P\,dx) = \mathbf{v}^{(n)}(P)\,dx + P\,d\xi = \left[\mathbf{v}^{(n)}(P) + PD_x\xi\right] dx + \theta, \qquad (5.24)$$

for some contact form θ. Therefore, $P\,dx$ is contact-invariant under the group G if and only if $\mathbf{v}^{(n)}(P) + PD_x\xi = 0$ for each infinitesimal generator. In other words, the differential function P is a relative differential invariant corresponding to the infinitesimal divergence multiplier $\mathbf{v}^{(n)} + D_x\xi$. Thus, the existence of contact-invariant one-forms is governed by our general result, Theorem 3.36, of the existence of relative invariants for general multiplier representations, coupled with Corollary 5.13 guaranteeing the freeness of the prolonged group action.

Theorem 5.27. *If* n *is the stabilization order of a transformation group, then there exists a nontrivial horizontal contact-invariant one-form* $\omega = P(x, u^{(n)})\,dx$ *of order at most* $\max\{1, n\}$*.*

Note that, in the ordinary cases, for an r-dimensional group action, the simplest differential invariant I has order $r - 1$, and produces the rth order contact-invariant one-form $\mathrm{D}I$, whereas the stabilization order is $n = r - 2$, and Theorem 5.27 shows that there is a contact-invariant of order $r - 2$. The formula for the simplest contact-invariant one-form for each of the transformation groups in Lie's classification of complex group actions is provided in Table 5. Note that in roughly half of the cases (specifically 1.2, 1.3, 1.7, 1.8, 1.9, 1.11, 2.2, and the three intransitive cases 3.1, 3.2, 3.3) the invariant one-form is of lower order than the order of stabilization of the group. It is not clear, though, how to recognize this phenomenon in advance. In all the cases except the pseudo-stabilization cases, the invariant one-form is of order strictly less than the order of the fundamental differential invariant, and a complete system of differential invariants is provided by successively applying the invariant differential operator to the fundamental differential invariant. The pseudo-stabilization Case 1.7a is unusual, in that it is the only one whose fundamental contact-invariant one-form is the total differential of the lowest order differential invariant, and hence requires a second fundamental differential invariant to generate all the higher order ones.

Example 5.28. The Euclidean group $\mathrm{SE}(2) = \mathrm{SO}(2) \ltimes \mathbb{R}^2$ acts via rotations and translations on $E \simeq \mathbb{R}^2$. Every (x, u)-independent rotational differential invariant, as given in Example 5.17, will provide a Euclidean differential invariant. In particular, the fundamental Euclidean invariant is the curvature $\kappa = (1 + u_x^2)^{-3/2} u_{xx}$. The simplest contact-invariant one-form is $\omega = \sqrt{1 + u_x^2}\, dx$, which is the Euclidean arc length element, often denoted ds. Proposition 5.26 implies the classical result that every Euclidean differential invariant is a function of the curvature and its derivatives $d^k \kappa / ds^k$ with respect to arc length.

The geometrical interpretation of the fundamental invariant one-form and differential invariant in the Euclidean case extends to other transformation groups, such as the special affine group SA(2), which is Case 2.1, and the projective group SL(3), which is Case 2.3, of importance in both differential geometry, [**106**], and, more recently, computer vision, [**189**]. In both cases, the simplest invariant one-form is identified with the group-invariant arc length element, while the fundamental differential invariant is identified with the group-invariant curvature. One is tempted to make a similar definition in all other cases (with the possible exceptions of the intransitive and pseudo-stabilization cases), so that a complete system of G-invariant differential invariants is provided by the G-invariant curvature and its derivatives with respect to the G-invariant arc length.

Let us consider an ordinary case, where G is r-dimensional, and acts effectively on $E \simeq \mathbb{R} \times \mathbb{R}$. Since G acts (locally) freely and transitively on (connected components of) $V^n \subset \mathrm{J}^n$, where $n = r - 2$ is the stabilization order, Theorem 2.93 guarantees the existence of a complete G-invariant coframe $\omega^1, \ldots, \omega^r$ on V^n, meaning that $\omega^1 \wedge \cdots \wedge \omega^r \neq 0$. The forms ω^i are strictly (meaning not just modulo contact forms) G-invariant, so

$(g^{(n)})^*\omega^i = \omega^i$. Indeed, we can identify V^n with a neighborhood of the identity in G, and use the corresponding *right-invariant* Maurer–Cartan forms as our coframe elements; see Example 2.85. Each of these one-forms can be written in the form $\omega^i = A_i(x, u^{(n)})\, dx + \theta^i + B_i(x, u^{(n)}) du_n$, where $u_n = D_x^n u$, for some contact form θ^i on J^n. Since the prolonged group transformations preserve contact forms, the horizontal component $\omega_H^i = (A_i + B_i u_{n+1})\, dx$ defines a contact-invariant one-form on J^{n+1}, which is linear in its highest derivative. Thus the ratio $I_{jk} = (A_j + B_j u_{n+1})/(A_k + B_k u_{n+1})$ provides an $(n+1)^{\text{st}}$ order differential invariant which is a linear fractional function of its top order derivative. Since the ω^i's are independent, at least one of these ratios is a nonconstant differential function, and the resulting invariant can be chosen as the fundamental differential invariant of the transformation group G. Thus, we have proved that every scalar transformation group has a fundamental differential invariant which is linear fractional in the top order derivative. (Actually, according to Table 6, the fundamental differential invariant can always be taken to be linear in u_{n+1}; see Exercise 5.31 below.) This method has been effectively used by Guggenheimer, [**106**], to construct differential invariants of geometric transformation groups. Guggenheimer uses a version of Cartan's moving frame method, [**34**], which has much in common with the equivalence method we discuss in the second part of the book, to construct the G-invariant coframe, and thus finds the differential invariants without having to integrate any differential equations. It would be of great interest to further develop this approach, particularly in multi-dimensional situations.

Example 5.29. Consider the Euclidean group $\mathrm{SE}(2) = \mathrm{SO}(2) \ltimes \mathbb{R}^2$ acting on the plane. The stabilization order is $n = 1$, and $\mathrm{SE}(2)$ acts transitively on J^1. Its first prolongation has infinitesimal generators ∂_x, ∂_u, $-u\partial_x + x\partial_u + (1 + u_x^2)\partial_{u_x}$. Fixing a point in J^1, we can locally identify J^1 with a neighborhood of the identity in $\mathrm{SE}(2)$, in such a way that the group acts on itself by left multiplication. As such, the Maurer-Cartan forms provide a G-invariant coframe, explicitly constructed (in two ways) in Example 2.94. Translating that result into the present notation, we find that a Euclidean invariant coframe is

$$\omega^1 = \frac{du - u_x\, dx}{\sqrt{1 + u_x^2}}, \qquad \omega^2 = \frac{du_x}{1 + u_x^2},$$
$$\omega^3 = \sqrt{1 + u_x^2}\, dx + \frac{u_x}{\sqrt{1 + u_x^2}}\, (du - u_x\, dx).$$

The first element is a contact form; the third is equivalent, modulo a contact form, to the contact-invariant Euclidean arc length form. The horizontal component of the second one-form is the second order contact-invariant one-form $\omega_H^2 = (1+u_x^2)^{-1} u_{xx}\, dx$. Dividing by the arc-length form produces the fundamental curvature invariant $\kappa = (1+u_x^2)^{-3/2} u_{xx}$ for the Euclidean group.

Exercise 5.30. Construct an invariant coframe and fundamental differential invariant for the similarity group, consisting of translations, rotations, and scalings $(x,u) \mapsto \lambda(x,u)$. A considerably more substantial exercise is to do this for the special affine, affine, and projective groups — Cases 2.1, 2.2, and 2.3 in the Tables; see [**106**].

Exercise 5.31. Prove that, in the ordinary case, if G has dimension $r = n+2$, then a G-invariant coframe on J^n can be found such that

$$\omega^1 = Q_0^1\, \theta_0, \quad \omega^2 = Q_0^2\, \theta_0 + Q_1^2\, \theta_1, \quad \ldots \quad \omega^n = Q_0^n\, \theta_0 + \cdots + Q_{n-1}^n\, \theta_{n-1},$$

are all contact forms, while

$$\begin{aligned} \omega^{n+1} &= P(x,u^{(n)})\, dx + Q_0^{n+1}\, \theta_0 + \cdots + Q_{n-1}^{n+1}\, \theta_{n-1}, \\ \omega^{n+2} &= A(x,u^{(n)})\, dx + B(x,u^{(n)})\, du_n + Q_0^{n+2}\, \theta_0 + \cdots + Q_{n-1}^{n+2}\, \theta_{n-1}, \end{aligned}$$

where $P\, dx$ is a contact-invariant one-form. Therefore, the fundamental differential invariant of order $r-1 = n+1$ is $I = (B u_{n+1} + A)/P$, and is an affine function of its highest derivative.

Several Dependent Variables

Similar constructions apply when there are several dependent variables and only one independent variable. We still require a single contact-invariant one-form $\omega = P(x,u^{(n)})\, dx$, which can be obtained as the total differential $\mathrm{D}I$ of a differential invariant I, and q, which is the number of dependent variables, additional differential invariants $J_1, \ldots, J_q$. A complete system of differential invariants is constructed by successively applying the invariant differential operator $\mathcal{D} = (1/P) D_x$ to the basic differential invariants J_ν.

Theorem 5.32. *Suppose that G is a group of point or contact transformations acting on a space $E = X \times U \simeq \mathbb{R} \times \mathbb{R}^q$ having one* *independent variable and q dependent variables. Then there exist $q+1$ fundamental independent differential invariants $I, J_1, \ldots, J_q$, such that, locally, every differential invariant of G can be written as a function of these differential invariants and their derivatives $\mathcal{D}^m J_\nu$, where $\mathcal{D} = (D_x I)^{-1} D_x$ is the invariant differential operator associated with the first differential invariant I.*

Example 5.33. Consider the case of SO(3) acting by rotations on the three-dimensional space $E \simeq \mathbb{R} \times \mathbb{R}^2$, with coordinates x, u, v. The orbit dimensions are $s_0 = 2$, $s_n = 3$, $n \geq 1$. Therefore, $i_0 = 1$, $i_1 = 2$, $i_2 = 4$, etc., so we have one ordinary invariant, which we can take to be the radius $r = \sqrt{x^2 + u^2 + v^2}$, and one first order differential invariant $w = (1 + u_x^2 + v_x^2)^{-1/2}(x + uu_x + vv_x)$. There are two second order differential invariants, namely

$$D_r w, \qquad \text{and} \qquad z = (1 + u_x^2 + v_x^2)^{-3/2}[(v - xv_x)u_{xx} + (xu_x - u)v_{xx}].$$

Every other differential invariant can be written as a function of the derivatives $D_r^n w, D_r^{n-1} z$, $n \geq 0$. Note that we can use the alternative contact-invariant one-form $\sqrt{1 + u_x^2 + v_x^2}\, dx = dr/w$, and associated invariant differential operator $(1 + u_x^2 + v_x^2)^{-3/2} D_x$.

The proof of Theorem 5.32 rests on the following independence result.

Lemma 5.34. *Let $K_1, \ldots, K_r$ be strictly independent n^{th} order differential invariants, such that $d_n K_1 \wedge \cdots \wedge d_n K_r \neq 0$. Suppose I is an additional differential invariant such that either* a) *I has order strictly less than n, or* b) *I has order n and $d_n I \wedge d_n K_1 \wedge \cdots \wedge d_n K_r \neq 0$, so I is strictly independent of $K_1, \ldots, K_r$. Let $\mathcal{D} = (D_x I)^{-1} D_x$ denote the invariant differential operator associated with I. Then the functions $\mathcal{D}K_1, \ldots, \mathcal{D}K_r$ are strictly independent $(n+1)^{\text{st}}$ order differential invariants on the subset of J^{n+1} where $D_x I \neq 0$.*

Proof: Recall that the total derivative of an nth order differential function $F(x, u^{(n)})$ has the form

$$D_x F = \sum_{\alpha=1}^{q} \frac{\partial F}{\partial u_n^\alpha}\, u_{n+1}^\alpha + \cdots, \tag{5.25}$$

where $u_n^\alpha = D_x^n u^\alpha$, and the omitted terms have coefficients depending on only n^{th} and lower order derivatives of the u's. Suppose first that the differential invariant I has order $< n$, so $D_x I$ has order $\leq n$. Then, using (5.25),

$$d_{n+1}[\mathcal{D}K_\nu] = \sum_{\alpha=1}^{q} \frac{\partial}{\partial u_{n+1}^\alpha}\left(\frac{D_x K_\nu}{D_x I}\right) du_{n+1}^\alpha = \frac{1}{D_x I}\sum_{\alpha=1}^{q} \frac{\partial K_\nu}{\partial u_n^\alpha}\, du_{n+1}^\alpha.$$

Therefore, the coefficient of the r-form $du_{n+1}^{\alpha_1} \wedge \cdots \wedge du_{n+1}^{\alpha_r}$ in the wedge product $d_{n+1}K_1 \wedge \cdots \wedge d_{n+1}K_r$ is the same as $(D_x I)^{-r}$ times the coefficient of the r-form $du_n^{\alpha_1} \wedge \cdots \wedge du_n^{\alpha_r}$ in $d_n K_1 \wedge \cdots \wedge d_n K_r$. This suffices to prove the lemma in this case.

The proof when I has order n is a bit more complicated. (Note that this case can only occur when $q \geq r + 1 \geq 2$.) As before, we compute

$$\begin{aligned} d_{n+1}[\mathcal{D}K_\nu] &= \frac{1}{(D_x I)^2}\sum_{\alpha=1}^{q}\left\{D_x I\, \frac{\partial K_\nu}{\partial u_n^\alpha} - D_x K_\nu\, \frac{\partial I}{\partial u_n^\alpha}\right\} du_{n+1}^\alpha \\ &= \frac{D_x I\, \widehat{d}_n K_\nu - D_x K_\nu\, \widehat{d}_n I}{(D_x I)^2}, \end{aligned}$$

where

$$\widehat{d}_n F = \sum_{\alpha=1}^{q} \frac{\partial F}{\partial u_n^\alpha}\, du_{n+1}^\alpha.$$

Since $\widehat{d}_n I \wedge \widehat{d}_n I = 0$, we find

$$d_{n+1}K_1 \wedge \cdots \wedge d_{n+1}K_r = (D_x I)^{-r-1}\Bigg[D_x I\, \widehat{d}_n K_1 \wedge \cdots \wedge \widehat{d}_n K_r +$$

$$\sum_{\nu=1}^{r}(-1)^\nu D_x K_\nu\, \widehat{d}_n K_1 \wedge \cdots \wedge \widehat{d}_n K_{\nu-1} \wedge \widehat{d}_n I \wedge \widehat{d}_n K_{\nu+1} \wedge \cdots \wedge \widehat{d}_n K_r\Bigg].$$

In view of (5.25), the coefficient of the r-form $u_{n+1}^{\alpha_1} du_{n+1}^{\alpha_2} \wedge \cdots \wedge du_{n+1}^{\alpha_{r+1}}$ in $d_{n+1}K_1 \wedge \cdots \wedge d_{n+1}K_r$ is $(D_x I)^{-r-1}$ times the coefficient of the $(r+1)$-form $du_n^{\alpha_1} \wedge du_n^{\alpha_2} \wedge \cdots \wedge du_n^{\alpha_{r+1}}$ in $d_n I \wedge d_n K_1 \wedge \cdots \wedge d_n K_r$. This completes the proof of the lemma. *Q.E.D.*

As a corollary of Lemma 5.34, the number of strictly independent invariants $j_k = i_k - i_{k-1}$ is found to satisfy the alternative inequalities

$$\begin{aligned} j_{k+1} &\geq j_k, & i_{k-1} &\geq 1, \\ j_{k+1} &\geq j_k - 1, & i_{k-1} &= 0. \end{aligned} \tag{5.26}$$

By (5.3), this implies that the prolonged orbit dimensions are bounded by

$$\begin{aligned} s_{k+1} &\le 2s_k - s_{k-1}, & & & s_{k-1} &\le kq, \\ & & &\text{if} & & \\ s_{k+1} &\le 2s_k - s_{k-1} + 1 = 2s_k - kq, & & & s_{k-1} &= 1 + kq. \end{aligned} \tag{5.27}$$

For example, if $s_0 \le q$ (so that G acts intransitively on E), then (5.27) implies that $s_2 \le 2s_1$ whereas our earlier inequality (5.6) implies that $s_2 \le s_1 + q$, which is a less restrictive inequality when $s_0 \le s_1 < q$.

Proof of Theorem 5.32: Let k denote the minimal order at which $i_k \ge 2$, meaning that the group admits at least two independent k^{th} order differential invariants. The functionally independent differential invariants on J^k will be denoted by I, which has order $\le k$, and $J_1, \ldots, J_{i_k-1}$ which all have order exactly k, and are strictly independent. (If $i_{k-1} = 1$, then I is the unique differential invariant of order $< k$; otherwise, if $i_{k-1} = 0$, then the choice of the distinguished k^{th} order differential invariant I is more or less arbitrary.) Let $\mathcal{D}$ denote the invariant differential operator associated with I. According to Lemma 5.34, the $(k+1)^{\text{st}}$ order differential invariants $\mathcal{D}J_1, \ldots, \mathcal{D}J_{i_k-1}$ are all strictly independent. If $i_{k+1} = 2i_k - 1$, then this implies that we have determined a complete set of functionally independent differential invariants of order $k+1$ or less; otherwise, we can choose $i_{k+1} - 2i_k + 1$ new $(k+1)^{\text{st}}$ order differential invariants $J_{i_k}, \ldots, J_{i_{k+1}-i_k+1}$, which are strictly independent of the previously differentiated invariants $\mathcal{D}J_1, \ldots, \mathcal{D}J_{i_k-1}$. According to Lemma 5.34, the $(k+2)^{\text{nd}}$ order differentiated invariants $\mathcal{D}^2 J_1, \ldots, \mathcal{D}^2 J_{i_k-1}, \mathcal{D}J_{i_k}, \ldots, \mathcal{D}J_{i_{k+1}-i_k+1}$ are also strictly independent. If $i_{k+2} = 2i_{k+1} - i_k$, then there are no further independent $(k+2)^{\text{nd}}$ order differentiated invariants; otherwise, we include $i_{k+2} - 2i_{k+1} + i_k$ further strictly independent differential invariants of order $k+2$. The process now continues by induction until we reach an order n at which there are exactly q (the maximum possible) strictly independent n^{th} order differential invariants among the invariant derivatives $\mathcal{D}^i J_\mu$ of the successively introduced differential invariants. At this point, Lemma 5.34 implies that the further derivatives of these differential invariants remain strictly independent, and hence the orbit dimension has stabilized at this point. *Q.E.D.*

As a consequence of the proof, we deduce that there are essentially only two possibilities for the orders of the fundamental differential invariants of an r-dimensional group of transformations acting on a space with just one independent variable. Let n denote the order of stabilization. Then either

(a) The fundamental differential invariants have order at most $n+1$. In this case, there exist $q+1$ differential invariants: I, of order $\leq n$, $J_1, \ldots, J_{q-1}$ of order $\leq n+1$, and J_q of order $= n+1$. In this case $\dim G = r \leq 1 + (n+1)q$.

(b) The fundamental differential invariants have order at most $n+2$. In this case, the fundamental differential invariants are $I, J_1, \ldots, J_{q-1}$ of order $= n+1$, and J_q of order $= n+2$. This case can *only* occur if $\dim G = r = 1 + (n+1)q$.

Consequently, the stabilization order n of an r-dimensional locally effective group action obeys the inequalities

$$\frac{r-1}{q} - 1 \leq n \leq r - 1. \tag{5.28}$$

Exercise 5.35. Carefully state and prove a generalization of Theorem 5.32 that replaces the differential invariant I by a contact-invariant one-form $\omega = P(x, u^{(n)})\, dx$.

Lemma 5.34 has some important consequences governing the order of stabilization of a group action.

Theorem 5.36. *Let G be a transformation group, and let s_n denote the maximal orbit dimension of its nth prolongation $G^{(n)}$. Suppose that $s_{n-1} < s_n = s_{n+1} \leq q^{(n)}$. Then n is the order of stabilization of G, so $s_m = s_n$ for all $m \geq n$.*

Proof: Since $p = 1$, formula (5.2), combined with our hypothesis, implies that $i_n = 1 + q^{(n)} - s_n \geq 1$. Therefore, there exists at least one differential invariant I of order $\leq n$. Moreover, since $q^{(n+1)} = q^{(n)} + q$,

$$i_{n+1} = p + q^{(n+1)} - s_{n+1} = i_n + q \geq 1 + q,$$

which shows that there exist q strictly independent differential invariants $J_1, \ldots, J_q$ of order $n+1$. Lemma 5.34 shows that, for $m \geq n+1$, there are also q strictly independent differential invariants of order m, namely $\mathcal{D}^{m-n-1} J_1, \ldots, \mathcal{D}^{m-n-1} J_q$. Consequently, $i_{m+1} = i_m + q$ for $m \geq n+1$, and hence $s_{m+1} = s_m$, which completes the proof. *Q.E.D.*

As we saw in Example 5.8, it is possible for the orbit dimension to pseudo-stabilize, meaning that $s_k = s_{k+1} < s_{k+2}$. The next result shows that there can be at most one such pseudo-stabilization.

Theorem 5.37. *Suppose the maximal orbit dimensions of the prolonged group action satisfy $s_k = s_{k+1}$ and, also, $s_n = s_{n+1}$ for some $n > k$. Then $s_m = s_n$ for all $m \geq n$.*

Proof: According to Theorem 5.36, if $s_k \leq q^{(k)} = (k+1)q$, then $s_n = s_k$, and the orbit dimension has already stabilized at order k. Otherwise, $s_k = 1 + (k+1)q$, and, using the inequality (5.6), we find

$$s_n \leq s_{k+1} + (n-k-1)q = s_k + (n-k-1)q = 1 + nq \leq (n+1)q = q^{(n)}.$$

Thus, again by Theorem 5.36, the orbit dimension stabilizes at order n (or less). *Q.E.D.*

Both these results are valid as stated in the general case of several independent variables and several dependent variables — see below. Theorem 5.37 provides a significant strengthening of Ovsiannikov's stabilization theorem, [**191**; p. 313], which states that if $s_{n-1} < s_n = s_{n+1} = s_{n+2}$, then the orbit dimension stabilizes at order n. Indeed, even in this case, the proof of Theorem 5.37 is new and much more direct than that of Ovsiannikov, which requires a more detailed investigation of the methods used to prove Theorem 5.11.

In his study of the differential invariants of curves in a homogeneous space, [**102**], M. Green discovered a striking formula relating the number of fundamental differential invariants to the dimensions of the isotropy subgroups of the prolonged group action. Our Theorem 5.32 implies that Green's results are, in fact, valid for completely general transformation groups on spaces with one independent variable! Let k_n denote the number of strictly independent fundamental differential invariants of order n, i.e., those differential invariants which are not expressed in terms of differentiated invariants of any lower order. Since, according to Lemma 5.34, the differentiated invariants coming from strictly independent differential invariants are themselves strictly independent, these numbers satisfy $k_n = j_n - j_{n-1}$ provided $i_{n-2} \geq 1$, so there is at least one lower order invariant to provide the required invariant differential operator.

Theorem 5.38. *Given a transformation group G, the number k_n of fundamental differential invariants of order n is given in terms of the* *minimal dimension* h_n *of the isotropy subgroups of* $G^{(n)}$ *according to*

$$k_n = \begin{cases} h_n - 2h_{n-1} + h_{n-2} + 1, & \textit{if } k_0 = \cdots = k_{n-2} = 0,\ k_{n-1} > 0, \\ h_n - 2h_{n-1} + h_{n-2}, & \textit{otherwise.} \end{cases} \tag{5.29}$$

Equation (5.29) *is valid for all* $n \geq 0$ *provided we set* $i_n = 0$ *and* $h_n = r - 1 - (n+1)q$ *when the action* $G^{(n)}$ *is not defined, i.e., for* $n = -2, -1$, *and, in the case of a contact transformation group,* $n = 0$.

Proof: For brevity, we just do the case of a group of point transformations. (The contact case only applies if there is one dependent variable, and is proved by similar arguments.) First suppose $n \geq 2$. According to (5.3), $j_n = q + h_n - h_{n-1}$. If $i_{n-2} > 0$, then $k_n = j_n - j_{n-1} = h_n - 2h_{n-1} + h_{n-2}$ as desired. If $i_{n-2} = 0$ and $i_{n-1} \geq 1$, so that $k_0 = \cdots = k_{n-2} = 0$, $k_{n-1} > 0$, then one of the $(n-1)^{\text{st}}$ order differential invariants must be used to define the invariant differential operator, and hence there are only $i_{n-1} - 1$ independent n^{th} order differentiated invariants. Therefore, $k_n = j_n - i_{n-1} + 1 = i_n - 2i_{n-1} + 1 = h_n - 2h_{n-1} + h_{n-2} + 1$. Finally, if $i_{n-2} = i_{n-1} = 0$, then $h_{n-2} = h_{n-1} = r$, $h_n = r - i_n$, hence $k_n = i_n = i_n - 2i_{n-1} + i_{n-2} = h_n - 2h_{n-1} + h_{n-2}$.

Next, suppose $n = 1$. If $i_0 = 1 + q - r + h_0 \geq 1$, then $j_1 = q + h_1 - h_0$, hence $k_1 = j_1 - i_0 + 1 = h_1 - 2h_0 + r = h_1 - 2h_0 + h_{-1} + 1$, since $h_{-1} = r - 1$, proving (5.29) in this case. On the other hand, if $i_0 = 0$, then $h_0 = r - q - 1$, and hence $k_1 = i_1 = 1 + 2q - r + h_1 = h_1 - 2h_0 + r - 1 = h_1 - 2h_0 + h_{-1}$, again verifying (5.29). Finally, for $n = 0$, we have $k_0 = i_0 = 1 + q - r + h_0 = h_0 - 2h_{-1} + h_{-2}$, since $h_{-1} = r - 1$ and $h_{-2} = q + r - 1$. This completes the proof of (5.29) in the point transformation case. *Q.E.D.*

A particularly important special case occurs when G is an r-dimensional Lie group and $H \subset G$ a closed subgroup of dimension s, so that G/H is a homogeneous space of dimension $r - s$. Let $E = \mathbb{R} \times G/H$, so that the functions $u = f(x)$ parametrize curves in the homogeneous space G/H. The group G acts on E by the Cartesian product of the trivial action on the independent variable $x \in \mathbb{R}$ and its usual action via left multiplication on G/H. In this case, $h_0 = s = \dim H$, and $i_0 = j_0 = k_0 = 1$, since there is a single ordinary invariant x, with consequential invariant differential operator D_x. Formula (5.29) implies Green's main result, [**102**], that there are $k_n = h_n - 2h_{n-1} + h_{n-2}$ fundamental differential invariants of order $n \geq 2$. For $n = 1$, we have

$k_1 = h_1 - 2h_0 + h_{-1} + 1 = h_1 - 2s + r$ since $i_{-1} = 0$, $i_0 = 1$, while $h_{-1} = r - 1$ according to our convention. (Green sets $h_{-1} = r$, but this choice does not agree with our general formula.) See **[102]** for a wide variety of applications and explicit examples, including affine, projective, and conformal geometry.

Example 5.39. Consider the differential invariants of curves in Euclidean space. Here we realize $\mathbb{R}^q = \mathrm{SE}(q)/\mathrm{SO}(q)$ as the quotient of the Euclidean group by the rotation subgroup, and are looking for Euclidean differential invariants of maps $u: \mathbb{R} \to \mathbb{R}^q$. The nth order isotropy group of a curve at a point $u_0 = u(x_0)$ is the subgroup of $\mathrm{SE}(q)$ which maps the curve to a curve $\bar{u} = g \cdot u$, $g \in \mathrm{SE}(q)$, having n^{th} order contact at the same point $u_0 = \bar{u}(x_0)$. The particular curve $u(x) = (x, x^2, \ldots, x^{n+1})$ is sufficiently general, and it is not hard to see that its nth order isotropy subgroup is $H_n = \mathrm{SO}(q - n)$, $n = 0, \ldots, q$. Therefore, $h_n = \dim H_n = \frac{1}{2}(q-n)(q-n-1)$, a formula that also works for $n = -1$. According to Theorem 5.38, for each $1 \leq n \leq q$, the number of fundamental n^{th} order differential invariants for Euclidean curves is $k_n = h_n - 2h_{n-1} + h_{n-2} = 1$. Thus, for each $n = 1, \ldots, q$, there is precisely one fundamental differential invariant of order n, which can be taken to be $|u_n|^2$, where $u_n = D_x^n u$. All other differential invariants are obtained as functions of them and their derivatives $D_x^m |u_n|^2$.

Several Independent Variables

The method of invariant differential operators can be extended to include transformation groups involving several independent and dependent variables. The basic requirement is the existence of a sufficient number of invariant differential forms on some jet bundle J^n. Let $E \simeq X \times U \simeq \mathbb{R}^p \times \mathbb{R}^q$. As in Definition 5.25, a one-form on the associated jet space J^n is called *contact-invariant* under a group of contact transformations $G^{(n)}$ if its horizontal component is unaffected by the group elements. In order to construct suitably many invariant differential operators, we need to determine not just one, but as many contact-invariant one-forms on the jet space as the number of independent variables. In anticipation of later terminology in connection with the Cartan equivalence method (see Chapter 8) we formalize the requirement in the following definition.

Definition 5.40. Let G be a transformation group acting on a space having p independent variables. An nth order *contact-invariant* *coframe* for G is a collection of p linearly independent horizontal one-forms which are contact-invariant under the prolonged action of $G^{(n)}$ on the jet space J^n.

As before, if $I(x, u^{(n)})$ is any n^{th} order differential invariant, its total differential $\mathrm{D}I = \sum D_j I \, dx^j$ is a contact-invariant form. Thus, if we know p independent n^{th} order differential invariants, we can always produce an $(n+1)^{\text{st}}$ order contact-invariant coframe by taking their total differentials. However, as we saw above, this is not the only way in which contact-invariant one-forms arise. Indeed, a contact-invariant coframe can be identified with a relative invariant for the matrix multiplier representation given by the total Jacobian multiplier, $\sigma(\mathbf{v}) = J = (D_j \xi^i)$, for infinitesimal generators $\mathbf{v}$ given by (4.3). Thus, Theorem 3.36 and Corollary 5.13 imply that we can always find a contact-invariant coframe whose order is at most the stabilization order of the group.

Theorem 5.41. *Let G be a transformation group, and n the stabilization order of G. Then there exists a contact-invariant coframe of order at most* $\max\{1, n\}$.

Given a contact-invariant coframe $\omega^1, \ldots, \omega^p$, the associated *invariant differential operators* $\mathcal{D}_1, \ldots, \mathcal{D}_p$ are constructed so that the formula

$$\mathrm{D}I = \sum_{j=1}^{p} \mathcal{D}_j I \, \omega^j, \tag{5.30}$$

expressing the total differential I in terms of the coframe is valid.† It is not hard to see that this uniquely defines the differential operators $\mathcal{D}_j$. Moreover, they map differential invariants to differential invariants:

Proposition 5.42. *Let G be a transformation group, and let $\mathcal{D}_1, \ldots, \mathcal{D}_p$ be the differential operators associated with a contact-invariant coframe on J^n. If $I(x, u^{(m)})$ is any m^{th} order differential invariant, then $\mathcal{D}_j I$ is a differential invariant of order at most* $\max\{n, m+1\}$.

Proof: This follows directly from the identity (5.30). Indeed, since I is a differential invariant, the left hand side $\mathrm{D}I$ is a contact-invariant

† Formula (5.30) shows that the differential operators $\mathcal{D}_i$ are the jet space versions of the usual coframe derivatives; see Chapter 8.

one-form. Furthermore, the one-forms ω^j are contact-invariant by hypothesis. Since the ω^j are assumed to be independent, their individual coefficients $\mathcal{D}_j I$ are also necessarily invariant under G. Thus formula (5.30) immediately implies that the operators $\mathcal{D}_j$ are invariant differential operators for the (prolonged) group action. *Q.E.D.*

A collection of one-forms $\omega^i = \sum_j J^i_j(x, u^{(n)})dx^j$, $i = 1, \ldots, p$, forms a contact-invariant coframe provided $\omega^1 \wedge \cdots \wedge \omega^p \neq 0$ on (an open subset of) J^n, so that the associated $p \times p$ matrix of differential functions $\mathbf{J} = \big(J^i_j(x, u^{(n)})\big)$ is invertible. The associated invariant differential operators are $\mathcal{D}_j = \sum_i K^i_j(x, u^{(n)}) D_i$, $j = 1, \ldots, p$, where $\mathbf{K} = \big(K^i_j(x, u^{(n)})\big) = \mathbf{J}^{-T}$ is the inverse transpose of $\mathbf{J}$. In particular, if the $\omega^i = \mathrm{D}I_i$ are obtained from p functionally independent differential invariants $I_1, \ldots, I_p$, then $\mathbf{J} = \big(D_j I_i(x, u^{(n)})\big)$ is their total Jacobian matrix, and the invariant differential operators are given by

$$\mathcal{D}_j J = \frac{D(I_1, \ldots, I_{j-1}, J, I_{j+1}, \ldots, I_p)}{D(I_1, \ldots, I_p)}, \qquad j = 1, \ldots, p, \tag{5.31}$$

where $D(I_1, \ldots, I_p) = \det\big(D_i I_j\big)$ denotes the total Jacobian determinant. (See Tresse, [**214**].)

Exercise 5.43. Prove that if $\mathcal{D}_1, \ldots, \mathcal{D}_p$ and $\overline{\mathcal{D}}_1, \ldots, \overline{\mathcal{D}}_p$ are two complete sets of invariant differential operators for the same group action, then they are related by $\mathcal{D}_j = \sum_i A^i_j \overline{\mathcal{D}}_i$, where the A^i_j are differential invariants.

Exercise 5.44. Prove that the invariant differential operators (5.31) commute: $[\mathcal{D}_i, \mathcal{D}_j] = 0$. Discuss how, in the general case when the contact-invariant coframe is not constructed from differential invariants, the commutators $[\mathcal{D}_1, \mathcal{D}_2] = \mathcal{D}_1\mathcal{D}_2 - \mathcal{D}_2\mathcal{D}_1$ of invariant differential operators act on differential invariants. Do the invariant differential operators necessarily form a Lie algebra?

The next result we require is a version of the independence result for the differentiated invariants similar to that in Lemma 5.34. The multi-variable case, though, is more complicated since the strict independence of nth order differential invariants $J_1, \ldots, J_r$ does not necessarily imply the independence of the differentiated invariants $\mathcal{D}_j J_\kappa$. For instance, if $J_1 = \mathcal{D}_1 J$, $J_2 = \mathcal{D}_2 J$, where J is an $(n-1)^{\text{st}}$ order differential invariant, then we would not expect $\mathcal{D}_2 J_1 = \mathcal{D}_2 \mathcal{D}_1 J$ and $\mathcal{D}_1 J_2 = \mathcal{D}_1 \mathcal{D}_2 J$ to be strictly independent $(n+1)^{\text{st}}$ order differential invariants — see Exercise 5.44. Thus the problem is more difficult in the multi-variable case. In general, it only appears possible to provide inequalities governing the number of independent differentiated invariants. Our approach relies on a combinatorial result used by Macaulay, [**159**], in his study of the Hilbert function of an algebraic polynomial ideal.† The method rests on the following elementary result on decomposing integers into sums of binomial coefficients.

Lemma 5.45. *Let $p \geq 1$ be an integer. Then any nonnegative integer $l \in \mathbb{N}$ can be uniquely written in the form*

$$l = \binom{k_1 + p - 1}{p} + \binom{k_2 + p - 2}{p-1} + \cdots + \binom{k_s + p - s}{p - s + 1}, \tag{5.32}$$

where $k_1 \geq k_2 \geq \cdots \geq k_s \geq 1$ form a nonincreasing sequence of positive integers and $1 \leq s \leq p$.

Write $k_p(l) = (k_1, \ldots, k_s)$ for the integer sequence associated with $l \in \mathbb{N}$. Define the function $\mu_p : \mathbb{N} \to \mathbb{N}$ which takes an integer l, represented by the sequence $k_p(l) = (k_1, \ldots, k_s)$, to the integer $\mu_p(l)$ which satisfies $k_p(\mu_p(l)) = (k_1 + 1, k_2 + 1, \ldots, k_s + 1)$; thus

$$\mu_p(l) = \binom{k_1 + p}{p} + \binom{k_2 + p - 1}{p-1} + \cdots + \binom{k_s + p - s + 1}{p - s + 1}. \tag{5.33}$$

For example, if $p = 2$ and $l = 4$, then $4 = \binom{3}{2} + \binom{1}{1}$, hence $\mu_2(4) = \binom{4}{2} + \binom{2}{1} = 8$. The function μ_p satisfies the following convexity condition.

Lemma 5.46. *Suppose $l = i_1 + \cdots + i_q = j_1 + \cdots + j_r$, where $i_1 \geq i_2 \geq \cdots \geq i_q > 0$, and $j_1 \geq j_2 \geq \cdots \geq j_r > 0$. If $i_1 > j_1$, then*

$$\mu_p(l) \leq \mu_p(i_1) + \cdots + \mu_p(i_q) < \mu_p(j_1) + \cdots + \mu_p(j_r). \tag{5.34}$$

The importance of the functions μ_p is underlined by the following theorem due to Macaulay, [**159**], which determines lower bounds for the dimensions of the homogeneous‡ components of polynomial ideals:

† In more recent years, Macaulay's theorem has been extensively generalized in the combinatorial theory of extremal multi-sets and forms a special case of the Kruskal–Katona Theorem, cf. [**103**].

‡ Macaulay also extends his result to nonhomogeneous ideals.

Theorem 5.47. *Let $\mathcal{I} \subset \mathbb{R}[x_1, \ldots, x_p]$ be a homogeneous polynomial ideal in p variables. Let $d_n = \dim \mathcal{I}^{(n)}$ be the dimension of the set $\mathcal{I}^{(n)} = \{P \in \mathcal{I} \mid P(\lambda x) = \lambda^n P(x)\}$ of polynomials of degree n in the ideal $\mathcal{I} = \bigoplus_{n \geq 0} \mathcal{I}^{(n)}$. Then $d_{n+1} \geq \mu_p(d_n)$.*

Given $n \geq 1$, we define $\mu_p^n : \mathbb{N} \to \mathbb{N}$ as follows: For $l \in \mathbb{N}$, we write $l = i p_n + j$, where i is the quotient, and j the remainder, when l is divided by $p_n = \binom{p+n-1}{n}$. Then $\mu_p^n(l) = i p_{n+1} + \mu_p(j)$. Note, in particular, that $\mu_p^n(l) = \mu_p(l)$ if $l \leq p_n$. Also $\mu_p^n(q_n) = q_{n+1}$. With this notation, we can now state a fundamental inequality for the number of differential invariants.

Theorem 5.48. *Let G be a transformation group acting on a space with p independent variables and q dependent variables. Suppose $\mathcal{D}_1, \ldots, \mathcal{D}_p$ form a complete set of invariant differential operators coming from a contact-invariant coframe of order n or less. Suppose $J_1, \ldots, J_l$ are strictly independent n^{th} order differential invariants. Then the set of differentiated invariants $\mathcal{D}_i J_\nu$, $i = 1, \ldots, p$, $\nu = 1, \ldots, l$, contains at least $\mu_p^n(l)$ strictly independent $(n+1)^{\text{st}}$ order differential invariants. In particular, if there are a maximal number of strictly independent n^{th} order differential invariants, $J_1, \ldots, J_{q_n}$, then the set of differentiated invariants $\mathcal{D}_i J_\nu$, $i = 1, \ldots, p$, $\nu = 1, \ldots, q_n$, contains a complete set of q_{n+1} strictly independent $(n+1)^{\text{st}}$ order differential invariants.*

Proof: The second statement follows from the first since $\mu_p^n(q_n) = q_{n+1}$. Suppose first that $q = 1$, so there is just one independent variable, and we can drop the α superscript. Given a differential function $F(x, u^{(n)})$, we define its *symbol* at a point $(x, u^{(n)}) \in \mathrm{J}^n$ to be the homogeneous polynomial

$$P(\zeta) = \sum_{\#J=n} \frac{\partial F}{\partial u_J}(x, u^{(n)})\, \zeta_J, \tag{5.35}$$

depending on the auxiliary variables $\zeta = (\zeta_1, \ldots, \zeta_p)$, where, given a symmetric multi-index $J = (j_1, \ldots, j_n)$, we let $\zeta_J = \zeta_{j_1} \zeta_{j_2} \cdots \zeta_{j_n}$ be the associated monomial. (The polynomial (5.35) is the same as the symbol of the partial differential equation $F(x, u^{(n)}) = 0$. The complex roots of the symbol polynomial determine the characteristic directions, cf. [**186**] — see also Definition 14.8.) If $J_1, \ldots, J_r$ are any regular set of nth order differential functions, then, at each point $(x, u^{(n)})$, the dimension of the space spanned by their n^{th} order differentials $d_n J_1, \ldots, d_n J_r$ is clearly the same as the dimension of the space spanned by their symbol polynomials $P_1(\zeta), \ldots, P_r(\zeta)$, which, if constant, equals the number of strictly independent differential functions among them. Let $\mathcal{I} = \mathcal{I}(x, u^{(n)})$ denote the homogeneous polynomial ideal generated by the symbol polynomials associated with our n^{th} order differential invariants at the point $(x, u^{(n)})$. I claim that the homogeneous component $\mathcal{I}^{(n+k)}$, $k \geq 0$, of $\mathcal{I}$ coincides with the space spanned by the symbol polynomials associated with the k-fold differentiated invariants $\mathcal{D}_{i_1} \cdot \ldots \cdot \mathcal{D}_{i_k} J_\nu$. In particular, the number of strictly independent differentiated invariants $\mathcal{D}_i J_\nu$ equals the dimension of $\mathcal{I}^{(n+1)}$, which, by Macaulay's Theorem 5.47, is at least $\mu_p(l)$, where $l = \dim \mathcal{I}^{(n)}$. This suffices to prove the theorem in the case of one dependent variable. (One technical detail: The differentiated invariants may not form a regular family; however, it is not hard to see that one can choose a regular subfamily which still satisfies the required dimension bound.)

To prove the claim, let $\mathcal{D}_i = \sum_k K_i^k(x, u^{(n)}) D_k$, $i = 1, \ldots, p$, be the invariant differential operators. Then, by (4.18), the $(n+1)^{\text{st}}$ order differential invariants are given by

$$\mathcal{D}_i J_\nu = \sum_{\#J=n} \sum_{k=1}^{p} K_i^k \frac{\partial J_\nu}{\partial u_J} u_{J,k} + \cdots, \qquad \begin{array}{l} i = 1, \ldots, p, \\ \nu = 1, \ldots, l \end{array}, \tag{5.36}$$

where we have only indicated terms depending on $(n+1)^{\text{st}}$ order derivatives of u. Therefore, the symbols of the differentiated invariants are the homogeneous polynomials

$$Q_{\nu,i}(\zeta) = \sum_{\#J=n} \sum_{k=1}^{p} K_i^k \frac{\partial J_\nu}{\partial u_J} (x, u^{(n)}) \, \zeta_J \zeta_k = L_i(\zeta) P_\nu(\zeta),$$

where $L_i(\zeta) = \sum_{k=1}^{p} K_i^k \zeta_k$. The linear polynomials $L_1(\zeta), \ldots, L_p(\zeta)$ are linearly independent since they are the symbols of the invariant differential operators $\mathcal{D}_1, \ldots, \mathcal{D}_p$. The independence of the L_i implies that the symbols $Q_{\nu,i}(\zeta) = L_i(\zeta) P_\nu(\zeta)$ span the space $\mathcal{I}^{(n+1)}$ of homogeneous polynomials of degree $n+1$ in the polynomial ideal generated by the P_ν's. This proves the claim, and completes the proof of Theorem 5.48 in the single dependent variable case.

The proof proceeds along similar lines in the case of several dependent variables. We now introduce an additional set of auxiliary variables $\eta = (\eta^1, \ldots, \eta^q)$, and consider the associated symbol polynomials

$$P_\nu(\eta;\zeta) = \sum_{\alpha=1}^{q} \sum_{\#J=n} \frac{\partial J_\nu}{\partial u_J^\alpha}(x, u^{(n)})\, \eta^\alpha \zeta^J,$$

which are linear in the η's and of degree n in the ζ's. As before, the number l of strictly independent invariants is equal to the dimension of the space $\mathcal{I}^{(n)}$ spanned by these polynomials. The differentiated invariants $\mathcal{D}_i J_\nu$ have corresponding symbol polynomials $L_i(\zeta) P_\nu(\eta;\zeta)$, spanning the space $\mathcal{I}^{(n+1)}$, which, by definition, is the set of polynomials in the ideal generated by the P_ν which are linear in the η's and of degree $n+1$ in the ζ's. The claim is that the dimension of this space is at least $\mu_p^n(l)$. In order to prove this, let us define the projections $\pi_\alpha \colon P(\eta,\zeta) \mapsto P(e_\alpha,\zeta)$, $\alpha = 1, \ldots, q$, where e_α denotes the αth standard basis vector of $\mathbb{R}^q$. Define $l_\alpha = \dim\, \mathcal{I}_\alpha^{(n)}$, where

$$\mathcal{I}_\alpha^{(n)} = \left\{ \pi_\alpha(P) \;\middle|\; P \in \mathcal{I}^{(n)}, \pi_\beta(P) = 0 \text{ for } \beta < \alpha \right\}.$$

Clearly $l = l_1 + \cdots + l_q$, and, moreover, $0 \le l_\alpha \le p_n$. Moreover, applying the same construction to $\mathcal{I}^{(n+1)}$, we find that each $\mathcal{I}_\alpha^{(n+1)}$ is contained in the polynomial ideal generated by $\mathcal{I}_\alpha^{(n)}$. Therefore,

$$\dim\, \mathcal{I}^{(n+1)} = \sum_{\alpha=1}^{q} \dim\, \mathcal{I}_\alpha^{(n+1)} \ge \sum_{\alpha=1}^{q} \mu_p(l_\alpha).$$

Moreover, the convexity Lemma 5.46 implies that the right hand side has minimal value when the nonzero l_α's are as large as possible. For $l = ip_n + j$, $0 \le j < p_n$, this happens when i of them equals p_n, one equals j, and the remainder are 0, in which case the right hand side equals $s\mu_p(p_n) + \mu_p(j) = sp_{n+1} + \mu_p(j) = \mu_p^n(l)$. This completes the proof of the theorem. *Q.E.D.*

We now apply this result to our study of differential invariants. Theorems 5.36 and 5.37, governing the order of stabilization and the number of possible pseudo-stabilizations of the orbit dimensions, hold as in the single variable case (although, of course, the formula for $q^{(n)}$ is different). Indeed, according to (5.2), if the maximal orbit dimension of $G^{(n)}$ satisfies $s_n \leq q^{(n)}$, then there are at least p functionally independent n^{th} order differential invariants, $I_1, \ldots, I_p$. Moreover, if $s_{n+1} = s_n$, then there are $j_{n+1} = i_{n+1} - i_n = q_{n+1} - s_{n+1} + s_n = q_{n+1}$ strictly independent $(n+1)^{\text{st}}$ order differential invariants. Therefore, we can apply Theorem 5.48 (with $n+1$ replacing n) to construct a complete system of q_m strictly independent differential invariants of any order $m \geq n+1$ by successively differentiating the $(n+1)^{\text{st}}$ order invariants using the invariant differential operators associated with the I_ν's. This implies that $s_m = s_n$, and hence the orbit dimension has stabilized at order n. The proof of Theorem 5.37 in this context is straightforward. The main result now states that every differential invariant can be constructed by successively applying the invariant differential operators to a finite collection of fundamental differential invariants. Note that, as with Theorem 5.32, this theorem holds locally on the jet spaces J^n.

Theorem 5.49. *Suppose that G is a group of point or contact transformations. Let n denote the order of stabilization of the group action. Then there exists a contact-invariant coframe $\omega^1, \ldots, \omega^p$ on J^n, with corresponding invariant differential operators $\mathcal{D}_1, \ldots, \mathcal{D}_p$, and fundamental differential invariants $J_1, \ldots, J_m$, of order at most $n+2$, such that, locally, every differential invariant can be written as a function of these differential invariants and their derivatives $\mathcal{D}_{j_1} \cdots \mathcal{D}_{j_\kappa} J_\nu$, $\kappa \geq 0$, $\nu = 1, \ldots, m$.*

Note that it is not asserted (in contrast to the single variable case in Theorem 5.32) that the differentiated invariants are necessarily functionally independent. Indeed, the classification of syzygies, or functional dependencies, among the differentiated invariants is an interesting problem that, as far as I know, has not been investigated in any degree of generality. Theorem 5.49 states that if the orbit dimension stabilizes at order n, then all the differential invariants can be obtained from those of order at most $n+2$ by applying the invariant differential operators. Moreover, if the stable orbit dimension satisfies $r = s_n \leq q^{(n)}$, and so there are at least p independent differential invariants $I_1, \ldots, I_p$ of order n, then the contact-invariant coframe $\mathrm{D}I_1, \ldots, \mathrm{D}I_p$ lives on J^{n+1}, and the fundamental differential invariants have order at most $n+1$. Finally, Theorem 5.49 also implies that our earlier stabilization and pseudo-stabilization Theorems 5.36 and 5.37 remain valid in the general multi-dimensional context. However, I am unaware of any interesting examples of multi-dimensional transformation groups whose orbit dimensions pseudo-stabilize beyond elementary generalizations of the one-dimensional examples of Case 1.7a. Moreover, except in specific examples, I do not know precisely how many fundamental differential invariants are required, although the dimension bounds of Theorem 5.48 should provide some useful estimates. Indeed, as remarked in [**102**], the generalization of Theorem 5.38 to the multi-dimensional case would be of great importance for studying, for example, the differential invariants of surfaces and higher dimensional submanifolds of homogeneous spaces.

Example 5.50. Let us look at the case of the rotation group $\mathrm{SO}(p)$ acting on $E \simeq \mathbb{R}^p \times \mathbb{R}$ by rotating the independent variables. There are two ordinary invariants: the radius $r = |\mathbf{x}|$ and the dependent variable u itself. Their total differentials $\frac{1}{2}d(r^2) = \mathbf{x} \cdot d\mathbf{x} = \sum_i x^i \, dx^i$ and $\mathrm{D}u = \nabla u \cdot d\mathbf{x} = \sum_i u_{x^i} \, dx^i$ provide contact-invariant one-forms. In the two-dimensional case, $p = 2$, these two forms provide a differential invariant coframe, leading to the two invariant differential operators $-yD_x + xD_y$ and $-u_yD_x + u_xD_y$. (Here $\mathbf{x} = (x, y)$, and we have omitted the determinantal factor $xu_y - yu_x$ since it is a differential invariant.) Another invariant differential operator is the scaling process $xD_x + yD_y$ which can be found by using the invariant one-form $y\,dx - x\,dy$ instead of $\mathrm{D}u$ in the construction. A complete system of first order differential invariants is then provided by $r, u, -yu_x + xu_y, xu_x + yu_y$. For example, we can write the norm of the gradient in terms of these basic differential invariants: $u_x^2 + u_y^2 = \frac{1}{2}r^{-2}[(-yu_x + xu_y)^2 + (xu_x + yu_y)^2]$. In this case, the orbit dimension has already stabilized at order 0, and, moreover, $s_0 = 1 \leq q_0 = 1$. All second order differential invariants are found by differentiating the first order invariants, and we can take

$$
\begin{aligned}
U &= x^2 u_{xx} + 2xyu_{xy} + y^2 u_{yy}, \\
V &= -xyu_{xx} + (x^2 - y^2)u_{xy} + xyu_{yy}, \\
W &= y^2 u_{xx} - 2xyu_{xy} + x^2 u_{yy},
\end{aligned} \tag{5.37}
$$

as fundamental second order differential invariants. In particular, the Laplacian can be written as $\Delta u = u_{xx} + u_{yy} = r^{-2}(U + W)$.

In the three-dimensional case, we need one further contact-invariant one-form in addition to the two we already know. The vector triple product $(\mathbf{x} \wedge \nabla u) \cdot d\mathbf{x}$ serves the purpose. The corresponding invariant differential operators are $\mathbf{J}^{-T}\mathbf{D}$, where $\mathbf{D} = (D_x, D_y, D_z)$, and $\mathbf{J}$ is the matrix of coefficients of the contact-invariant coframe. However, note that, under SO(3), the matrices $\mathbf{J}$ and $\mathbf{J}^{-T}$ have identical transformation properties, and hence simpler invariant differential operators are given by $\mathbf{J} \cdot \mathbf{D}$, or, explicitly, $\mathbf{x} \cdot \mathbf{D}$, $\nabla u \cdot \mathbf{D}$, $(\mathbf{x} \wedge \nabla u) \cdot \mathbf{D}$. Applying these operators to u yields first order invariants, $\mathbf{x} \cdot \nabla u = x u_x + y u_y + z u_z$, and $|\nabla u|^2 = u_x^2 + u_y^2 + u_z^2$. Five second order invariants can be found by applying these operators to the first order invariants; if $\mathbf{H} = \nabla^2 u$ denotes the Hessian matrix, these are $\mathbf{x}^T \mathbf{H} \mathbf{x}$, $\mathbf{x}^T \mathbf{H} \nabla u$, $\nabla u^T \mathbf{H} \nabla u$, $\mathbf{x}^T \mathbf{H}(\mathbf{x} \wedge \nabla u)$, $\nabla u^T \mathbf{H}(\mathbf{x} \wedge \nabla u)$. Then $(\mathbf{x} \wedge \nabla u)^T \mathbf{H}(\mathbf{x} \wedge \nabla u)$ provides remaining second order invariant. In this case, $s_0 = 2$, $s_1 = 3$, and the orbit dimension stabilizes at order 1. Theorem 5.49 implies that a complete system of higher order differential invariants of SO(3) is provided by successively applying the three invariant differential operators to the second order invariants.

Exercise 5.51. Prove that for SO(p), $p \geq 4$, the orbit dimension stabilizes at order 2. Find a complete system of second order differential invariants, and prove that all the higher order differential invariants can be found by successively differentiating the second order invariants.

See [**75**] for a more cumbersome analysis of the rotation group, as well as computations of the second order differential invariants of the Euclidean, Poincaré, and conformal groups. The complete analysis of the higher order differential invariants of these physically important groups is apparently unknown.

Example 5.52. For the special linear group SL(2) acting on $E \simeq \mathbb{R}^2 \times \mathbb{R}$ via the standard representation, first order differential invariants are provided by

$$u, \qquad \text{and} \qquad Z = x u_x + y u_y.$$

A contact-invariant coframe is provided by the pair of one-forms $\mathrm{D}u = u_x\, dx + u_y\, dy$ and $-y\, dx + x\, dy$, leading to the invariant differential operators $x D_x + y D_y$, $-u_y D_x + u_x D_y$. Applying the operators to this first order invariant leads to a pair of second order differential invariants:

$$A = x^2 u_{xx} + 2xy u_{xy} + y^2 u_{yy},$$
$$B = x u_y u_{xx} + (y u_y - x u_x) u_{xy} - y u_x u_{yy}.$$

There are two additional second order differential invariants:

$$H = u_{xx} u_{yy} - u_{xy}^2, \qquad J = u_y^2 u_{xx} - 2 u_x u_y u_{xy} + u_x^2 u_{yy}.$$

which agree, up to a factor, with the Gaussian curvature of the graph of u and the curvature of the level sets of u, cf. [**106**]. Since $\dim \mathrm{J}^2 = 8$, and SL(2) acts effectively on (an open subset of) J^2, since it already acts effectively on J^1, which has dimension 5, and just 2 invariants, there can be at most 5 functionally independent differential invariants of order at most 2. The reader can check that the invariants satisfy the "syzygy"

$$B^2 - JA + Z^2 H = 0. \tag{5.38}$$

Exercise 5.53. Find the differential invariants and invariant differential operators for GL(2) acting via the standard representation on $E \simeq \mathbb{R}^2 \times \mathbb{R}$. Do the same for the special affine group $\mathrm{SA}(2) = \mathrm{SL}(2) \ltimes \mathbb{R}^2$ and the full affine group $\mathrm{A}(2) = \mathrm{GL}(2) \ltimes \mathbb{R}^2$.

Exercise 5.54. Find the differential invariants and invariant differential operators for SO(3) acting via rotations on $E \simeq \mathbb{R}^2 \times \mathbb{R}$.

Chapter 6

Symmetries of Differential Equations

In this chapter we discuss the foundations and some applications of Lie's theory of symmetry groups of differential equations. The basic infinitesimal method for calculating symmetry groups is presented, and used to determine the general symmetry group of some particular differential equations of interest. We include a number of important applications, including integration of ordinary differential equations, classification of invariant differential equations based on their symmetry groups, and linearization theorems for partial differential equations. Additional applications appear in the following chapter on variational methods. The reader interested in pursuing these matters in greater depth should consult the texts [**22**], [**186**], [**191**], for a wide variety of additional examples, theoretical developments, and applications to both physical and mathematical problems. Extensive tables of symmetry groups of particular differential equations can be found in the recent handbook, [**122**].

Symmetry Groups and Differential Equations

Consider a general n^{th} order system of differential equations

$$\Delta_\nu(x, u^{(n)}) = 0, \qquad \nu = 1, \ldots, m, \tag{6.1}$$

in p independent variables $x = (x^1, \ldots, x^p)$, and q dependent variables $u = (u^1, \ldots, u^q)$, with $u^{(n)}$ denoting the derivatives of the u's with respect to the x's up to order n. The system of differential equations (6.1), which we often abbreviate as $\Delta = 0$, is thus defined by the vanishing of a collection of differential functions $\Delta_\nu \colon \mathrm{J}^n \to \mathbb{R}$ defined on the nth jet space J^n. (For simplicity, we shall restrict our attention to systems defined by smooth functions, although extensions of these methods to more general systems are possible.) The system (6.1) can therefore be viewed as defining (or defined by) a variety

$$\mathcal{S}_\Delta = \left\{ (x, u^{(n)}) \;\middle|\; \Delta_\nu(x, u^{(n)}) = 0, \quad \nu = 1, \ldots, m \right\}, \tag{6.2}$$

contained in the nth order jet space, consisting of all points $(x, u^{(n)}) \in \mathrm{J}^n$ satisfying the system. The defining functions Δ_ν are assumed to be regular, as per Definition 1.8, in a neighborhood of $\mathcal{S}_\Delta$; in particular, this is the case if the Jacobian matrix of the functions Δ_ν with respect to the jet variables $(x, u^{(n)})$ has maximal rank m everywhere on $\mathcal{S}_\Delta$. Theorem 1.21 implies that $\mathcal{S}_\Delta$ is a submanifold of J^n of dimension $p + q\binom{p+n}{n} - r$, where r is the rank of the system. Essentially all systems of differential equations arising in applications satisfy this condition (at least away from singularities). We also assume that the projection $\pi^n_X \colon \mathrm{J}^n \to X$ maps $\mathcal{S}_\Delta$ onto an open subset of the space of independent variables, since, otherwise, the system would include constraints on the independent variables, and therefore fall outside the traditional realm of differential equations.

A (smooth) function $u = f(x)$ will define a *solution* to the system of differential equations (6.1) if and only if its nth prolongation $f^{(n)}(x)$ satisfies the system, i.e., $\Delta_\nu(x, f^{(n)}(x)) = 0$, $\nu = 1, \ldots, m$. This is equivalent to the requirement that the graph $\Gamma_f^{(n)} = \{(x, f^{(n)}(x))\}$ of the nth prolongation of f be entirely contained in the variety defined by the system: $\Gamma_f^{(n)} \subset \mathcal{S}_\Delta$. The reader should unravel the relevant definitions so as to be convinced that this is merely a geometric reformulation of the classical notion of solution to a system of differential equations.

As with systems of algebraic equations, a *symmetry* of the system of differential equations (6.1) means a transformation which maps (smooth) solutions to solutions. The most basic type of symmetry is a group of point transformations on the space of independent and dependent variables, as discussed in Chapter 4, although, for either physical or mathematical reasons, one may wish to restrict attention to fiber-preserving transformations, since they do not mix up independent and dependent variables, or other more restrictive classes, e.g., volume-preserving, symplectic, etc. (Indeed, many systems arising in applications only have fiber-preserving symmetries.) Alternatively, if there is just one dependent variable, it might be advantageous to expand our repertoire of symmetries to include contact transformations. However, to begin with we will primarily concentrate on studying the basic point transformation symmetry groups.

Definition 6.1. A point transformation $g: E \to E$ acting on the space $E \simeq X \times U$ of independent and dependent variables is called a *symmetry* of the system of partial differential equations (6.1) if, whenever $u = f(x)$ is a solution to (6.1), and the transformed function $\bar{f} = g \cdot f$ is well defined, then $\bar{u} = \bar{f}(\bar{x})$ is also a solution to the system (6.1).

Definition 6.1 formulates the notion of symmetry of a system of differential equations in terms of its action on the space of solutions. A key geometrical reformulation of this property is obtained by considering the action of the prolonged group transformation on the associated variety (6.2).

Proposition 6.2. *Consider a system of nth order differential equations* (6.1) *with associated variety $\mathcal{S}_\Delta \subset \mathrm{J}^n$ as in* (6.2). *If the nth prolongation $g^{(n)}: \mathrm{J}^n \to \mathrm{J}^n$ of a point transformation $g: E \to E$ leaves $\mathcal{S}_\Delta$ invariant, so that $g^{(n)}(\mathcal{S}_\Delta) \subset \mathcal{S}_\Delta$, then g is a symmetry of the system.*

Proof: The proof follows directly from the basic definitions. As remarked above, a function $u = f(x)$ is a solution to the system if and only if the graph $\Gamma_f^{(n)}$ of $f^{(n)}$ is a subset of $\mathcal{S}_\Delta$. Now, invariance of $\mathcal{S}_\Delta$ implies that $g^{(n)} \cdot \Gamma_f^{(n)}$, which is the graph $\Gamma_{g \cdot f}^{(n)}$ of the prolongation of the transformed function $g \cdot f$, is also contained in $\mathcal{S}_\Delta$, and hence is also a solution. *Q.E.D.*

Note that Proposition 6.2 holds without change in the case of (prolonged) contact transformations: A contact map $g^{(n)}: \mathrm{J}^n \to \mathrm{J}^n$ will act on a solution $u = f(x)$ by pointwise transforming its prolonged graph $\Gamma_f^{(n)}$ which, in view of the contact conditions, forms the prolonged graph of the transformed function $\bar{u} = \bar{f}(\bar{x})$. The map $g^{(n)}$ will therefore determine a symmetry to the system (6.1) if it leaves the variety $\mathcal{S}_\Delta$ invariant.

The converse to Proposition 6.2 is, perhaps surprisingly, not necessarily true. Whereas for a system of algebraic equations, every point in the corresponding variety is (tautologically) a solution, this is no longer necessarily the case for systems of differential equations. Only those points in $\mathcal{S}_\Delta$ which correspond to actual solutions are required to be mapped to each other by the prolonged symmetry transformation. We formalize the required concept with the following definition.

Definition 6.3. A system of differential equations is called *locally solvable* at a point $(x_0, u_0^{(n)}) \in \mathcal{S}_\Delta$ if there exists a smooth solution

$u = f(x)$, defined in a neighborhood of x_0, which achieves the values of the indicated derivatives there: $u_0^{(n)} = f^{(n)}(x_0)$. A system of differential equations which satisfies both the regularity and local solvability conditions at each of its points is called *fully regular*.

Most of the systems arising in practice are fully regular. For instance, local solvability of a system of ordinary differential equations is the same as the property of existence of solutions to the standard initial value problem. For a system of partial differential equations, the local solvability problem is of a different character than the more usual Cauchy or boundary value problems, since the initial data are only specified at a single point. Nevertheless, the Cauchy–Kovalevskaya Existence Theorem, cf. Chapter 15, proves that any normal, analytic system of partial differential equations is locally solvable. The Frobenius and Cartan–Kähler Theorems imply local solvability of more general involutive systems of partial differential equations. The principal types of systems which do not satisfy the local solvability criterion are systems of differential equations with nontrivial integrability conditions, and certain smooth, non-analytic systems of partial differential equations, first discovered by H. Lewy, [**149**], which have no solutions. See [**186**; Chapter 2] for a more detailed discussion of regularity and local solvability.

Theorem 6.4. *A transformation g is a symmetry of a locally solvable system of differential equations* (6.1) *if and only if the corresponding variety $\mathcal{S}_\Delta$ is invariant under the prolonged transformation: $g^{(n)}(\mathcal{S}_\Delta) \subset \mathcal{S}_\Delta$.*

Proof: Given $z_0 = (x_0, u_0^{(n)}) \in \mathcal{S}_\Delta$, such that $\bar{z}_0 = g^{(n)} \cdot z_0$ is defined, we use the local solvability of $\mathcal{S}_\Delta$ to choose a solution $u = f(x)$ defined near x_0 with $u_0^{(n)} = f^{(n)}(x_0)$. Then $\bar{f} = g \cdot f$ is also a solution (at least near the image point $(\bar{x}_0, \bar{u}_0) = g \cdot (x_0, u_0)$), and hence $\bar{z}_0 = (\bar{x}_0, \bar{f}^{(n)}(\bar{x}_0)) \in \mathcal{S}_\Delta$, as desired. *Q.E.D.*

Infinitesimal Methods

We will henceforth assume that we are dealing with a connected group of point (or contact) transformations G. In the case of point transformations, the infinitesimal generators form a Lie algebra $\mathfrak{g}$ consisting of vector fields

$$\mathbf{v} = \sum_{i=1}^{p} \xi^i(x,u) \frac{\partial}{\partial x^i} + \sum_{\alpha=1}^{q} \varphi^\alpha(x,u) \frac{\partial}{\partial u^\alpha}, \tag{6.3}$$

on the space of independent and dependent variables. Let $\mathbf{v}^{(n)}$ denote the nth prolongation of the vector field to the jet space J^n — the explicit formula appears in Theorem 4.16. Applying our basic infinitesimal symmetry criterion Theorem 2.79 and the geometric reformulation of symmetry in Theorem 6.4, we deduce the fundamental infinitesimal symmetry criterion for a system of differential equations.

Theorem 6.5. *A connected group of transformations G is a symmetry group of the fully regular system of differential equations $\Delta = 0$ if and only if the classical infinitesimal symmetry conditions*

$$\mathbf{v}^{(n)}(\Delta_\nu) = 0, \qquad \nu = 1, \ldots, r, \qquad \textit{whenever} \qquad \Delta = 0, \tag{6.4}$$

hold for every infinitesimal generator $\mathbf{v} \in \mathfrak{g}$ of G.

The conditions (6.4) are known as the *determining equations* of the symmetry group for the system; note in particular that they are only required to hold on points $(x, u^{(n)}) \in \mathcal{S}_\Delta$ satisfying the equations. If we substitute the explicit formulas (4.20), (4.21) for the coefficients of the prolongation $\mathbf{v}^{(n)}$, we find that the determining equations form a large, over-determined, linear system of partial differential equations for the coefficients ξ^i, φ^α of $\mathbf{v}$. In almost every example of interest, the determining equations are sufficiently elementary that they can be explicitly solved, and thereby one can determine the complete (connected) symmetry group of the system (6.1). However, the required calculations can become very lengthy and time consuming. Fortunately, there are now a wide variety of computer algebra packages available which will automate all the routine steps in the calculation of the symmetry group of a given system of partial differential equations; see [**44**], [**112**] for surveys of the different packages available, including a discussion of their strengths and weaknesses.

One simplification, which is especially important in the case of contact symmetries, is to use the characteristic form (4.32) of the prolongation formula. Since any (smooth) solution to an equation $\Delta_\nu(x, u^{(n)}) = 0$ also satisfies all the equations $D_i\Delta_\nu = 0$, $i = 1, \ldots, p$, obtained by applying the total derivative operators (4.18), the infinitesimal symmetry conditions (6.4) have the alternative reformulation

$$\mathbf{v}_Q^{(n)}(\Delta_\nu) = 0, \qquad \nu = 1, \ldots, r, \qquad \text{on solutions to } \Delta = 0. \tag{6.5}$$

Note that in (6.5), the system *and* its derivatives must be taken into account. In particular, since, in the case of a single dependent variable, there is a one-to-one correspondence between infinitesimal contact transformations and first order characteristic functions $Q(x, u^{(1)})$, cf. Proposition 4.34, the determination of contact symmetry groups is most easily effected using (6.5); it can also be advantageously used for ordinary point symmetries.

Remark: The symmetry criterion (6.5) can be generalized to higher order "generalized symmetries", in which one allows the characteristic $Q = Q(x, u^{(k)})$ to depend on derivatives of a specified (finite) order. Although such higher order infinitesimal symmetries do *not* have a geometric, group-theoretic counterpart, they do play an essential role in the study of integrable soliton equations; we refer the interested reader to [**186**] for a complete development.

We now illustrate the practical use of the infinitesimal symmetry criterion (6.4) for determining the full (connected) symmetry group of several concrete differential equations of interest.

Example 6.6. We begin with an elementary example illustrating the basic techniques. Consider the second order ordinary differential equation

$$u_{xx} = 0. \tag{6.6}$$

(The reader may try to guess in advance the symmetry group of this equation based on the fact that its general solution is an affine function $u = ax + b$; thus every symmetry of (6.6) will map straight lines to straight lines.) An infinitesimal point symmetry of equation (6.6) will be a vector field of the general form $\mathbf{v} = \xi(x,u)\partial_x + \varphi(x,u)\partial_u$ on $E = \mathbb{R}^2$; our task is to determine which particular coefficient functions ξ, φ will produce infinitesimal symmetries. In order to apply Theorem 6.5, we must compute the second prolongation of $\mathbf{v}$, which is the vector field

$$\mathbf{v}^{(2)} = \xi(x,u)\,\frac{\partial}{\partial x} + \varphi(x,u)\,\frac{\partial}{\partial u} + \varphi^x(x,u^{(1)})\,\frac{\partial}{\partial u_x} + \varphi^{xx}(x,u^{(2)})\,\frac{\partial}{\partial u_{xx}}, \tag{6.7}$$

on J^2 whose coefficients φ^x, φ^{xx} can be found in (4.24). The infinitesimal symmetry criterion (6.4) in this case is simply

$$\varphi^{xx} = 0 \qquad \text{whenever} \qquad u_{xx} = 0. \tag{6.8}$$

Substituting the formula (4.24) for φ^{xx} into (6.8), and setting the second order derivative u_{xx} to zero, results in the explicit symmetry condition

$$\varphi_{xx} + (2\varphi_{xu} - \xi_{xx})u_x + (\varphi_{uu} - 2\xi_{xu})u_x^2 - \xi_{uu}u_x^3 = 0. \tag{6.9}$$

Since we can prescribe x, u, and u_x arbitrarily, and ξ and φ only depend on x, u, equation (6.9) will be satisfied if and only if the individual coefficients of the powers of u_x vanish. The resulting overdetermined system of partial differential equations

$$\varphi_{xx} = 0, \qquad 2\varphi_{xu} = \xi_{xx}, \qquad \varphi_{uu} = 2\xi_{xu}, \qquad \xi_{uu} = 0, \tag{6.10}$$

forms the complete set of *determining equations* for the symmetry group of the original ordinary differential equation (6.6). The general solution to the determining equations (6.10) is readily found:

$$\begin{aligned} \xi(x,u) &= c_1x^2 + c_2xu + c_3x + c_4u + c_5, \\ \varphi(x,u) &= c_1xu + c_2u^2 + c_6x + c_7u + c_8, \end{aligned}$$

where the c_i are arbitrary constants. Therefore, equation (6.6) admits an eight-dimensional Lie algebra of infinitesimal symmetries, spanned by the vector fields

$$\partial_x,\ \partial_u,\ x\partial_x,\ x\partial_u,\ u\partial_x,\ u\partial_u,\ x^2\partial_x + xu\partial_u, \qquad xu\partial_x + u^2\partial_u. \tag{6.11}$$

Note that (6.11) appear as the Lie algebra in Case 2.3 of our classification tables. The corresponding Lie group is the projective group SL(3), acting via linear fractional transformations

$$(x,u) \longmapsto \left(\frac{ax+bu+c}{hx+ju+k}, \frac{dx+eu+f}{hx+ju+k}\right), \qquad \det\begin{vmatrix} a & b & c \\ d & e & f \\ h & j & k \end{vmatrix} \neq 0, \tag{6.12}$$

which forms the most general point symmetry of the elementary ordinary differential equation (6.6).

The situation is markedly different if we try to find all contact symmetries of (6.6). As noted above, the calculations are simplified if we first place the infinitesimal generator of the contact symmetry group into characteristic form $\mathbf{v}_Q = Q(x,u,u_x)\partial_u$, cf. (4.30). The infinitesimal invariance condition (6.5) is simply $D_x^2 Q = 0$, which must hold whenever $u_{xx} = u_{xxx} = 0$. Evaluating the second total derivative, we deduce that any solution $Q(x,u,p)$ to the linear second order partial differential equation

$$\frac{\partial^2 Q}{\partial x^2} + 2p\,\frac{\partial^2 Q}{\partial x \partial u} + p^2\,\frac{\partial^2 Q}{\partial u^2} = 0 \tag{6.13}$$

determines an infinitesimal contact symmetry of (6.6). Note that (6.13) can be simply written as $(\partial_x + p\partial_u)^2 Q = 0$, and hence its general solution

$$Q(x,u,p) = A(u - xp, p) + xB(u - xp, p), \qquad p = u_x, \tag{6.14}$$

depends on two arbitrary functions of two variables — the characteristic variables for the vector field $\partial_x + p\partial_u$. Therefore, in contrast to the point transformation symmetry group, the contact transformation symmetry group is infinite-dimensional. In fact, the same argument proves that *every* second order ordinary differential equation has an infinite-dimensional contact symmetry group, which reflects the fact, to be proved later, that every (nonsingular) second order ordinary differential equation can be mapped by a suitable contact transformation to the elementary equation $u_{xx} = 0$. On the other hand, as we will see, the point symmetry group of a second order ordinary differential equation has dimension at most eight, and, moreover, only those equations which can be mapped to (6.6) have a symmetry group of the maximal dimension.

Exercise 6.7. Prove that any first order differential equation $u_x = P(x,u)$ has an infinite-dimensional point symmetry group. In particular, any vector field of the form $\xi(x,u)\big(\partial_x + P(x,u)\partial_u\big)$ is automatically a symmetry. Interpret this fact geometrically.

Exercise 6.8. Show that the second order ordinary differential equation $u_{xx} = [(x + x^2)e^u]_x$ has no (continuous) symmetries. Find the general solution to this equation. (See [**120**; p. 20].)

Exercise 6.9. Determine the contact symmetry group of the third order ordinary differential equation $u_{xxx} = 0$, cf. [**121**; p. 105].

Example 6.10. A classic example illustrating the basic techniques for computing symmetry groups of partial differential equations is the linear heat equation

$$u_t = u_{xx}. \tag{6.15}$$

An infinitesimal point symmetry of the heat equation will be given by a vector field of the form $\mathbf{v} = \xi\partial_x + \tau\partial_t + \varphi\partial_u$, whose coefficients ξ, τ, φ are functions of x, t, u. The infinitesimal symmetry criterion (6.4) is

$$\varphi^t = \varphi^{xx} \qquad \text{whenever} \qquad u_t = u_{xx}. \tag{6.16}$$

Here, φ^t and φ^{xx} are the coefficients of the terms $\partial_{u_t}, \partial_{u_{xx}}$ in the second prolongation of $\mathbf{v}$, cf. (4.26). Substituting the formulas (4.27) into (6.16), and replacing u_t by u_{xx} wherever it occurs, we are left with a polynomial equation involving the various derivatives of u whose coefficients are certain derivatives of ξ, τ, φ. Since ξ, τ, φ only depend on x, t, u we can equate the individual coefficients to zero, leading to the complete set of *determining equations*:

Coefficient	*Monomial*
$0 = -2\tau_u$	$u_x u_{xt}$
$0 = -2\tau_x$	u_{xt}
$0 = -\tau_{uu}$	$u_x^2 u_{xx}$
$-\xi_u = -2\tau_{xu} - 3\xi_u$	$u_x u_{xx}$
$\varphi_u - \tau_t = -\tau_{xx} + \varphi_u - 2\xi_x$	u_{xx}
$0 = -\xi_{uu}$	u_x^3
$0 = \varphi_{uu} - 2\xi_{xu}$	u_x^2
$-\xi_t = 2\varphi_{xu} - \xi_{xx}$	u_x
$\varphi_t = \varphi_{xx}$	1

The general solution to these elementary differential equations is

$$\xi = c_1 + c_4 x + 2c_5 t + 4c_6 xt, \qquad \tau = c_2 + 2c_4 t + 4c_6 t^2,$$
$$\varphi = (c_3 - c_5 x - 2c_6 t - c_6 x^2)u + \alpha(x,t),$$

where c_i are arbitrary constants and $\alpha_t = \alpha_{xx}$ is an arbitrary solution to the heat equation. Therefore, the symmetry algebra of the heat equation is spanned by the vector fields

$$\begin{aligned} &\mathbf{v}_1 = \partial_x, \qquad \mathbf{v}_2 = \partial_t, \qquad \mathbf{v}_3 = u\partial_u, \qquad \mathbf{v}_4 = x\partial_x + 2t\partial_t, \\ &\mathbf{v}_5 = 2t\partial_x - xu\partial_u, \qquad \mathbf{v}_6 = 4xt\partial_x + 4t^2\partial_t - (x^2 + 2t)u\partial_u, \\ &\mathbf{v}_\alpha = \alpha(x,t)\partial_u, \qquad \text{where} \qquad \alpha_t = \alpha_{xx}. \end{aligned} \tag{6.17}$$

The corresponding one-parameter groups are, respectively, x and t translations, scaling in u, the combined scaling $(x,t) \mapsto (\lambda x, \lambda^2 t)$, Galilean boosts, an "inversional symmetry", and the addition of solutions stemming from the linearity of the equation. Each of these groups has the property of mapping solutions of the heat equation to other solutions. For example, the inversional group has the explicit form

$$(x,t,u) \longmapsto \left(\frac{x}{1-4\varepsilon t}, \frac{t}{1-4\varepsilon t}, u\,\sqrt{1-4\varepsilon t}\, \exp\left\{ \frac{-x\varepsilon x^2}{1-4\varepsilon t} \right\} \right),$$

where ε is the group parameter. Note that this only defines a local transformation group. Computing the induced action on graphs of functions, we conclude that if $u = f(x,t)$ is any solution to the heat equation, so is

$$u = \frac{1}{1+4\varepsilon t} \exp\left\{ \frac{-\varepsilon x^2}{1+4\varepsilon t} \right\} f\left(\frac{x}{1+4\varepsilon t}, \frac{t}{1+4\varepsilon t} \right),$$

a fact that can, of course, be checked directly. In particular, starting with the trivial constant solution $u = c$, we produce the non-trivial solution $u = c(1+4\varepsilon t)^{-1/2} \exp[-\varepsilon(1+4\varepsilon t)^{-1} x^2]$, which is a multiple of the fundamental solution at the point $(0, -(4\varepsilon)^{-1})$. The reader should construct the actions of the other one-parameter groups and their combinations, and then verify that they all do take solutions to solutions. See [**186**] for more details.

Exercise 6.11. Does the heat equation have any contact symmetries which are not point symmetries?

Example 6.12. A similar computation can be used to determine the symmetry group of the nonlinear diffusion equation

$$u_t = u_{xx} + u_x^2, \tag{6.18}$$

known as the potential form of Burgers' equation. Infinitesimal symmetries have the same form as in the heat equation example, and the symmetry criterion (6.4) is

$$\varphi^t = \varphi^{xx} + 2u_x \varphi^x, \tag{6.19}$$

which must hold whenever (6.18) is satisfied. Again, we substitute the formulas (4.27) into (6.19), and replace u_t by $u_{xx} + u_x^2$. The general solution to the resulting determining equations is

$$\xi = c_1 + c_4 x + 2c_5 t + 4c_6 xt, \qquad \tau = c_2 + 2c_4 t + 4c_6 t^2,$$
$$\varphi = c_3 - c_5 x - 2c_6 t - c_6 x^2 + \alpha(x,t) e^{-u},$$

where c_i are arbitrary constants and $\alpha_t = \alpha_{xx}$ is an arbitrary solution to the *heat equation*: $\alpha_t = \alpha_{xx}$. Therefore, the symmetry algebra of Burgers' equation is spanned by the vector fields

$$\begin{aligned} &\mathbf{v}_1 = \partial_x, \qquad \mathbf{v}_2 = \partial_t, \qquad \mathbf{v}_3 = \partial_u, \qquad \mathbf{v}_4 = x\partial_x + 2t\partial_t, \\ &\mathbf{v}_5 = 2t\partial_x - x\partial_u, \qquad \mathbf{v}_6 = 4xt\partial_x + 4t^2\partial_t - (x^2 + 2t)\partial_u, \\ &\mathbf{v}_\alpha = \alpha(x,t)e^{-u}\partial_u, \qquad \text{where} \qquad \alpha_t = \alpha_{xx}. \end{aligned} \tag{6.20}$$

The symmetry algebra (6.20) is isomorphic to that of the heat equation (6.17); indeed, the transformation $u \mapsto e^u$ maps the Burgers' symmetry generators to those of the heat equation, and, in fact, linearizes the potential Burgers' equation by mapping it to the heat equation. We have thus, based on a symmetry analysis, rediscovered the well-known Hopf–Cole transformation — see [**226**] for applications.

The preceding observation is, in fact, a particular case of a general linearization theorem for a system of partial differential equations — see Theorem 6.46 below. It also serves to illustrate the following simple general fact.

Proposition 6.13. *Two equivalent differential equations have isomorphic symmetry groups.*

Indeed, if Φ is a transformation mapping one differential equation to another and g is a symmetry of the first equation, then $\bar{g} = \Phi \circ g \circ \Phi^{-1}$ is clearly a symmetry of the second equation. Of course, to interpret Proposition 6.13, one must be careful that the symmetries are included in the class of diffeomorphisms (i.e., fiber-preserving transformations, point transformations, contact transformations, etc.) which define equivalences. If we make a change of variables outside the class, then there is no guarantee that the symmetry groups will have anything to do with each other. For example, consider the more usual form

$$v_t = v_{xx} + 2vv_x \tag{6.21}$$

of Burgers' equation, which is obtained by differentiating (6.18) and setting $v = u_x$, so that u plays the role of a potential function for v. The symmetry group of (6.21) is only five-dimensional; nevertheless we can map it to the linear heat equation $w_t = w_{xx}$, which has an infinite dimensional symmetry group, by the *nonlocal* Hopf–Cole map

$v = (\log w)_x$. Thus, the symmetry group of a physical equation and that of the equation for the potential can be quite different, because local transformations in one case become nonlocal in the other (and vice versa). See [**22**] for the theory of *potential symmetries*, and [**168**], [**186**] for an approach to such more general linearization questions based on higher order symmetries.

Example 6.14. The Boussinesq equation

$$u_{tt} + uu_{xx} + u_x^2 + u_{xxxx} = 0 \tag{6.22}$$

is a well-known soliton equation, and arises as a model equation for the unidirectional propagation of solitary waves in shallow water, [**177**]. The basic symmetry condition (6.4) now takes the form

$$\varphi^{tt} + u\varphi^{xx} + u_{xx}\varphi + 2u_x\varphi^x + \varphi^{xxxx} = 0,$$

which must hold whenever (6.22) is satisfied. Here $\varphi^{tt}, \varphi^x, \varphi^{xx}, \varphi^{xxxx}$ are the coefficients of the vector fields $\partial_{u_{tt}}, \partial_{u_x}, \partial_{u_{xx}}, \partial_{u_{xxxx}}$ in the fourth prolongation of $\mathbf{v}$, with formulae similar to (4.27). A straightforward (but quite lengthy) calculation eventually yields the complete symmetry algebra, which is spanned by

$$\mathbf{v}_1 = \partial_x, \qquad \mathbf{v}_2 = \partial_t, \qquad \mathbf{v}_3 = x\partial_x + 2t\partial_t - 2u\partial_u. \tag{6.23}$$

In this example, the classical symmetry group is disappointingly trivial, consisting of easily guessed translations and scaling symmetries. Theorem 6.5 guarantees that these are the only continuous classical symmetries of the equation. There are, however, higher order generalized symmetries, [**186**], which account for the infinity of conservation laws of this equation, and, thus, its remarkable integrability properties.

Thus, the complicated calculation of the symmetry group of a system of differential equations sometimes yields only rather trivial symmetries; however, there are numerous examples where this is not the case and new and physically and/or mathematically important symmetries have arisen from a complete group analysis. See [**122**], [**186**], [**191**] for examples and references.

Exercise 6.15. Determine the form of the symmetry group of a nonlinear diffusion equation of the form

$$u_t = D_x\big(h(u)u_x\big). \tag{6.24}$$

The answer will depend on the form of $h(u)$; see [**122**; §10.2].

Exercise 6.16. Prove that the Korteweg–deVries equation

$$u_t = u_{xxx} + uu_x, \tag{6.25}$$

has a four-parameter symmetry group. Find the symmetry transformations explicitly, cf. [**186**; Example 2.44].

Exercise 6.17. Let $x \in \mathbb{R}^p$, $u \in \mathbb{R}$. Prove that the contact vector field (4.60) with characteristic $Q(x, u^{(1)})$ defines a symmetry of the first order partial differential equation $Q(x, u^{(1)}) = 0$.

Integration of Ordinary Differential Equations

Lie made the remarkable observation that virtually all the classical methods for solving specific types of ordinary differential equations (separable, homogeneous, exact, etc.) are particular examples of a general method for integrating ordinary differential equations that admit a group of symmetries. In particular, knowledge of a one-parameter group of symmetries of an ordinary differential equation allows us to reduce its order by one. Before beginning, though, we must remark that the method cannot be used to find every solution to the equation.

Definition 6.18. Let $\mathbf{v}$ be a vector field on the space of independent and dependent variables. A function $u = f(x)$ is called *nontangential* provided $\mathbf{v}$ is nowhere tangent to the graph of f.

Theorem 6.19. *Let $\Delta(x, u^{(n)}) = 0$ be an nth order scalar ordinary differential equation admitting a regular one-parameter symmetry group G. Then all nontangential solutions can be found by quadrature from the solutions to an ordinary differential equation $(\Delta/G)(x, u^{(n-1)}) = 0$ of order $n-1$, called the symmetry reduced equation.*

Remark: Note that a solution is everywhere tangential if and only if it is invariant under G, so the method will not, in particular, produce invariant solutions. However, in the scalar case, the graph of an invariant function must locally coincide with a one-dimensional orbit of the group, and hence the invariant solutions can be determined by inspection. See [**22**; §3.6] for applications to envelopes and separatrices.

Proof: Let us introduce rectifying coordinates $y = \chi(x, u)$, $v = \psi(x, u)$, in terms of which the infinitesimal generator of G is the vertical translation field ∂_v. If $\mathbf{v}$ is not tangent to the graph Γ of a solution

$u = f(x)$, then, in terms of the new y, v coordinates, Γ remains transverse, and therefore will locally coincide with the graph of a smooth function $v = h(y)$. In the new coordinates, the group transformations, and their prolongations, are simply given by translation $v \mapsto v + \varepsilon$ in the v coordinate alone, the derivative coordinates remaining fixed. (The infinitesimal form is $\mathbf{v}^{(n)} = \partial_v$, $n \geq 0$.) Therefore the variety $\mathcal{S}_\Delta = \{\Delta(y, v^{(n)}) = 0\}$ is invariant if and only if it does not depend on the variable v, and hence the equation is equivalent to one that does not depend explicitly on v itself (although it does depend on the derivatives of v with respect to y). Therefore, replacing v by $w = v_y$ reduces the equation to one of order $n - 1$ for $w = w(y)$; moreover, we recover the solution to our original equation by quadrature: $v = \int w(y)dy$. *Q.E.D.*

In particular, a first order equation $u_x = P(x, u)$ admitting a one-parameter symmetry group can be solved by quadrature. However, the symmetry must be nontangential: The trivial symmetries of Exercise 6.7, $\mathbf{v} = \xi(x, u)\big(\partial_x + P(x, u)\partial_u\big)$, are everywhere tangential to solutions; moreover, the characteristic method for finding the rectifying coordinates of such a vector field is the essentially same problem as integrating the equation itself, so the reduction method is of no help. Indeed, the problem of determining the most general symmetry group of a first order equation is more complicated than solving the equation itself, so we can only successfully apply Lie's method if, by inspection (perhaps motivated by geometry or physics), we can detect a relatively simple symmetry group.

Example 6.20. A classical example is provided by the homogeneous equation

$$\frac{du}{dx} = x^{m-1} F\left(\frac{u}{x^m}\right), \tag{6.26}$$

which admits the scaling group $(x, u) \mapsto (\lambda x, \lambda^m u)$ with infinitesimal generator $x\partial_x + mu\partial_u$. For $x \neq 0$, rectifying coordinates are given by $y = u/x^m$, $v = \log x$, in terms of which the equation reduces to

$$\frac{dv}{dy} = \frac{1}{F(y) - my}.$$

This can clearly be integrated, $v = h(y) = \int dy/\{F(y) - my\}$, thereby defining u implicitly: $\log x = h(u/x)$. (The reader might enjoy comparing this method with the one taught in elementary ordinary differential equation texts for solving homogeneous equations.) In this case, the nontangentiality condition requires that $du/dx \neq mu/x$, meaning that $F(y) \neq my$, and this method (and the standard one) break down at such singularities. In particular, the scale-invariant function $u = cx^m$, which will be a solution provided $F(c) = mc$, cannot be recovered by this approach, and constitutes a singular solution to the equation.

Example 6.21. Consider the second order ordinary differential equation

$$uu_{xx} = \alpha u_x^2, \tag{6.27}$$

where α is a nonzero constant. The equation admits three obvious symmetries: a translation $x \mapsto x + \varepsilon$ in the independent variable, reflecting the fact that the equation is autonomous, and two independent scaling transformations $(x, u) \mapsto (\lambda x, \mu u)$. To reduce with respect to the translation group, we set $y = u$, $v = x$, so that $u_x = 1/v_y$, $u_{xx} = -v_{yy}/v_y^3$, and the equation reduces to a linear equation $yv_{yy} + \alpha v_y = 0$, with solution $v = cy^{1-\alpha} + d$ for $\alpha \neq 1$, or $v = c \log y + d$ for $\alpha = 1$. The corresponding solution of (6.27) is then $u = (ax + b)^{1/(1-\alpha)}$, or $u = \exp(ax + b)$. (Note that the translationally invariant solutions $u =$ constant are recovered as limiting cases of these solutions.) Alternatively, if we use the scaling symmetry in u, then the appropriate coordinates are x and $v = \log u$, in terms of which the equation becomes $v_{yy} = (\alpha - 1)v_y^2$, which reduces to a homogeneous (separable) equation for $w = v_y$. The reader may enjoy seeing what happens if we reduce with respect to the other scaling symmetry $x \mapsto \lambda x$ instead.

Example 6.22. Finally, consider a general homogeneous second order linear equation

$$u_{xx} + f(x)u_x + g(x)u = 0. \tag{6.28}$$

This clearly admits the scaling symmetry generated by $u\partial_u$. According to the general reduction procedure, as long as $u \neq 0$, we can introduce the new variable $v = \log u$, in terms of which the equation becomes $v_{xx} + v_x^2 + f(x)v_x + g(x) = 0$, which is a first order Riccati equation for $w = v_x = u_x/u$. We have thus recovered the well-known correspondence between second order linear equations and first order Riccati equations.

If a higher order equation admits several symmetries, then it may be reducible in order more than once. However, unless the symmetry group has additional structure, we may not be able to make a full reduction since the reduced equations may not inherit the full symmetry group of the original equation. (On the other hand, they may admit additional symmetries not shared by the original system!) See [**186**], [**22**] for full details of the reduction techniques available.

Theorem 6.23. *Suppose $\Delta = 0$ is an nth order ordinary differential equation admitting a symmetry group G. Let $H \subset G$ be a one-parameter subgroup. Then the H-reduced equation $\Delta/H = 0$ admits the quotient group G_H/G, where $G_H = \{g \mid g \cdot H \cdot g^{-1} \subset H\}$ is the* normalizer subgroup*, as a symmetry group.*

Proof: Let $y = \chi(x,u)$, $v = \psi(x,u)$ be the rectifying coordinates for the infinitesimal generator $\mathbf{v} = \partial_v$ of the subgroup H. According to Theorem 6.19, the original differential equation reduces to an $(n-1)^{\text{st}}$ order equation for $w = v_y = \omega(x,u,u_x)$. Consider an infinitesimal symmetry $\mathbf{w} \in \mathfrak{g}$ of the original equation, which we re-express in terms of the rectifying coordinates: $\mathbf{w} = \eta(y,v)\partial_y + \zeta(y,v)\partial_v$. Clearly, the vector field $\mathbf{w}$ will induce a point symmetry of the reduced equation if and only if its first prolongation $\mathbf{w}^{(1)} = \eta\partial_y + \zeta\partial_u + \zeta^y \partial_{v_y}$ can be reduced to a local vector field $\widehat{\mathbf{w}} = \eta(y,w)\partial_y + \theta(y,w)\partial_w$ depending on just y and w. This will happen if and only if η and ζ^y are independent of v. According to the prolongation formula (4.24), this occurs if and only if $\zeta = \alpha(y) + cv$ for c constant. Since $\mathbf{v} = \partial_y$, this condition is equivalent to the requirement that $[\mathbf{v},\mathbf{w}] = c\mathbf{v}$. Exercise 2.59 demonstrates that $\mathbf{w} \in \mathfrak{g}_H$, the subalgebra of $\mathfrak{g}$ corresponding to the normalizer subgroup G_H, thereby completing the proof. *Q.E.D.*

Example 6.24. Consider a second order equation of the form

$$x^2 u_{xx} = F(xu_x - u), \tag{6.29}$$

which admits the two-parameter symmetry group $(x,u) \mapsto (\lambda x, u + \mu x)$ with infinitesimal generators $\mathbf{v} = x\partial_x$, $\mathbf{w} = x\partial_u$. Since $[\mathbf{v},\mathbf{w}] = \mathbf{w}$, if we reduce with respect to $\mathbf{w}$, then the resulting first order equation will retain a symmetry corresponding to $\mathbf{v}$, and hence Theorem 6.23 guarantees that it can be integrated. In this case, we set $v = u/x$, $w = v_x = x^{-2}(xu_x - u)$, so that (6.29) reduces to $x^3 w_x = F(x^2 w) + 2x^2 w$. This equation admits a scaling symmetry generated by the reduced vector field $\widehat{\mathbf{v}} = x\partial_x - 2w\partial_w$, which means that it is of homogeneous form

(6.26) and can be integrated as in Example 6.20. On the other hand, if we try to reduce the original equation with respect to $\mathbf{v}$ initially, using the variables $\widetilde{y} = u$, $\widetilde{v} = \log x$, equation (6.29) reduces to a first order equation $w_y = -w[1 + F(w^{-1} - y)]$ having no obvious symmetry. It is worth remarking that the latter equation *can* be solved — just reverse the procedure so as to replace it by the original second order equation, and use the first reduction method. Building on the original suggestions in [**186**], a detailed investigation into this enhanced reduction technique was recently undertaken by Abraham–Shrauner and Guo, [**1**]; see also the extensive tables in [**231**].

An r-dimensional Lie group G is called *solvable* if there exists a sequence of subgroups $\{e\} = G_0 \subset G_1 \subset \cdots \subset G_{r-1} \subset G_r = G$ such that each G_i is a normal subgroup of G_{i+1}. This is equivalent to the requirement that the corresponding subalgebras of $\mathfrak{g}$ satisfy $[\mathfrak{g}_i, \mathfrak{g}_{i+1}] \subset \mathfrak{g}_i$. A Theorem of Bianchi, [**18**], states that if an ordinary differential equation admits a (sufficiently regular) r-dimensional solvable symmetry group, then its solutions can be determined, by quadrature, from those to a reduced equation of order $n - r$; see also [**186**; Theorem 2.64].

Finally, it is worth reiterating the fact that not every integration method for ordinary differential equations is based on symmetry. Indeed, the equation appearing in Exercise 6.8 provides a simple example of an equation with no symmetries, but which can, nevertheless, be explicitly solved.

Characterization of Invariant Differential Equations

One of the most important uses of differential invariants is in the construction of general systems of differential equations (and variational problems) which admit a prescribed symmetry group. This is especially important in modern physical theories, where one begins by postulating the basic "symmetry group" of the theory, and then determines which field equations are admissible. The basic result that allows us to immediately write down the most general system of differential equations which is invariant under a prescribed transformation group is a direct corollary of Theorem 2.38 characterizing invariant systems of algebraic equations. Note that this result is valid for both ordinary and partial differential equations.

Theorem 6.25. *Let G be a Lie group acting on E. Assume that the nth prolongation of G acts regularly and has a complete set of* *functionally independent n^{th} order differential invariants $I_1, \ldots, I_k$ on an open subset $V^n \subset \mathrm{J}^n$. A system of nth order differential equations admits G as a symmetry group if and only if, on V^n, it can be rewritten in terms of the differential invariants:*

$$\Delta_\nu(x, u^{(n)}) = F_\nu(I_1(x, u^{(n)}), \ldots, I_k(x, u^{(n)})) = 0, \qquad \nu = 1, \ldots, l. \tag{6.30}$$

Example 6.26. Suppose we have just one independent variable and one dependent variable, and consider the usual rotation group SO(2) acting on $E = \mathbb{R} \times \mathbb{R}$. According to Example 5.2, there are two first order differential invariants — the radius $r = \sqrt{x^2 + u^2}$ and the angular invariant $w = (xu_x - u)/(x + uu_x)$. Therefore, the most general first order ordinary differential equation admitting SO(2) as a symmetry group can be written in the form $F(r, w) = 0$. Solving for w, we deduce that the equation has the explicit form

$$\frac{xu_x - u}{x + uu_x} = H\left(\sqrt{x^2 + u^2}\right) \qquad \text{or} \qquad u_x = \frac{u + xH(\sqrt{x^2 + u^2})}{x - uH(\sqrt{x^2 + u^2})}. \tag{6.31}$$

In terms of the polar coordinates r, θ, equation (6.31) takes the separable form $\theta_r = r^{-1}H(r)$ and can thus be integrated.

The most general second order differential equation admitting a rotational symmetry group can be written in the form $F(r, w, \kappa) = 0$, where $\kappa = (1 + u_x^2)^{-3/2} u_{xx}$ is the curvature. Solving for u_{xx}, we find

$$u_{xx} = (1 + u_x)^{3/2} H\left(\sqrt{x^2 + u^2}, \ \frac{xu_x - u}{x + uu_x}\right). \tag{6.32}$$

This second order equation can also be rewritten in terms of polar coordinates:

$$r\theta_{rr} = (1 + \theta_r^2) H(r, r\theta_r) - r^2\theta_r^3 - 2\theta_r.$$

Since the latter equation does not explicitly depend on θ, it can be reduced to a first order equation by introducing the variable $v = d\theta/dr$.

Example 6.27. Let $E = \mathbb{R}^2 \times \mathbb{R}$, and consider the action of the rotation group SO(2) acting on the independent variables x, y only. According to Example 5.50, every first order SO(2)-invariant partial differential equation has the form $H(xu_x + yu_y, yu_x - xu_y, u, r) = 0$. Similarly, every second order SO(2)-invariant partial differential equation can be written in terms of the second (and lower) order differential invariants derived in (5.37):

$$H(x^2 u_{xx} + 2xyu_{xy} + y^2 u_{yy}, -xyu_{xx} + (x^2 - y^2)u_{xy} + xyu_{yy},$$
$$y^2 u_{xx} - 2xyu_{xy} + x^2 u_{yy}, yu_x - xu_y, u, r) = 0.$$

In particular, the rotational invariance of $\Delta u + \lambda u = 0$, the Helmholtz equation, follows because of the identity for the Laplacian found in Example 5.50.

The construction of Theorem 6.25 suggests an alternative reduction method for ordinary differential equations invariant under a symmetry group. Under the assumptions of Theorem 6.25, we can rewrite any n^{th} order ordinary differential equation admitting an r parameter symmetry group in the form

$$F\left(y, w, \frac{dw}{dy}, \ldots, \frac{d^{n-r}w}{dy^{n-r}}\right) = 0 \tag{6.33}$$

involving only the two fundamental differential invariants: $y = I(x, u^{(s)})$, $w = J(x, u^{(r)})$, which, by Theorem 5.24, have orders $s < r = \dim G$, with $s = r - 1$ unless G is either intransitive, in which case $s = 0$, or pseudo-stabilizes, in which case $s = r-2$. Therefore, we have reduced the original n^{th} order differential equation to an $(n - r)^{\text{th}}$ order differential equation in the differential invariants. However, once we have solved (6.33) for $w = h(y)$, we then must solve an auxiliary r^{th} order differential equation

$$J(x, u^{(r)}) = h[I(x, u^{(s)})], \tag{6.34}$$

in order to recover u as a function of x. (If (6.33) only involves the first differential invariant y, then (6.34) has the form $I(x, u^{(s)}) = c$.) The Lie reduction method discussed above can be applied to equation (6.34) although, unless the Lie group is solvable, we will not in general be able to integrate it by quadrature alone.

A particularly interesting class of applications is provided by the three inequivalent planar actions of the special linear group SL(2); these appear as Cases 3.3, 1.1, and 1.2 in the tables at the end of the book. As noted in Example 4.20, all three actions are related via a process of prolongation and projection; this provides a ready means of simultaneously classifying and reducing the corresponding invariant ordinary differential equations.

For the first unimodular group action, any n^{th} order ordinary differential equation admitting the symmetry group generated by ∂_x, $x\partial_x$, $x^2\partial_x$ can be written in the form†

$$\frac{d^{n-3}s}{du^{n-3}} = H\left(u, s, \frac{ds}{du}, \ldots, \frac{d^{n-4}s}{du^{n-4}}\right), \tag{6.35}$$

where

$$u \qquad \text{and} \qquad s = \frac{u_x u_{xxx} - \frac{3}{2}u_{xx}^2}{u_x^4}. \tag{6.36}$$

are the two fundamental differential invariants, cf. (5.22). Once we know the solution $s = F(u)$ to the reduced ordinary differential equation (6.35), we recover the solution to the original equation by solving the invariant third order equation $s = F(u)$, or, explicitly,

$$u_x u_{xxx} - \tfrac{3}{2}u_{xx}^2 = u_x^4 F(u). \tag{6.37}$$

We can reduce (6.37) to a first order differential equation by applying the Lie reduction method associated with the two-dimensional solvable subgroup generated by ∂_x and $x\partial_x$. The two consequent reductions can be combined by setting

$$z = \frac{u_{xx}}{u_x^2}, \qquad \text{in terms of which} \qquad s = \frac{dz}{du} + \tfrac{1}{2}z^2. \tag{6.38}$$

Equation (6.37) then reduces to the Riccati equation

$$\frac{dz}{du} + \tfrac{1}{2}z^2 = F(u). \tag{6.39}$$

Once we solve equation (6.39) for $z = z(u)$, we use (6.38) to recover the solution $u = f(x)$ to equation (6.37) by a pair of quadratures. Alternatively, the reduction procedure of Example 6.22 can be reversed to

† For simplicity, we treat only the regular equations here, leaving the singular differential equation defined by the vanishing of the corresponding Lie determinant, as discussed below, as an exercise for the reader.

linearize the Riccati equation (6.39). The function $\psi(u) = \sqrt{u_x}$ is a solution to the second order, homogeneous, linear Schrödinger equation

$$\frac{d^2\psi}{du^2} - \tfrac{1}{2}F(u)\psi = 0. \tag{6.40}$$

We can recover $u(x)$ by a single quadrature:

$$\int^u \frac{d\hat{u}}{\psi(\hat{u})^2} = x + k. \tag{6.41}$$

Note that, according to the method of variation of parameters, cf. [**123**; p. 122], if $\psi(u)$ is one solution to the linear ordinary differential equation (6.40), then a second, linearly independent solution, is given by

$$\varphi(u) = \psi(u) \int^u \frac{d\hat{u}}{\psi(\hat{u})^2}. \tag{6.42}$$

Comparing with (6.41), and absorbing the integration constant, we conclude that the general solution to the invariant equation (6.37) is given, parametrically, as a ratio $x = \varphi(u)/\psi(u)$ of two arbitrary linearly independent solutions to the linear Schrödinger equation (6.40). According to Exercise 5.22, the differential invariant s can be identified with the negative of the Schwarzian derivative of $x = x(u)$; therefore, our symmetry reduction of (6.37) provides a direct proof of a classical theorem due to Schwarz; see [**116**; Theorem 10.1.1].

Theorem 6.28. *The general solution to the Schwarzian equation*

$$\frac{x_u x_{uuu} - \frac{3}{2}x_{uu}^2}{x_u^2} = \widehat{F}(u) \tag{6.43}$$

has the form $x = \varphi(u)/\psi(u)$, where $\varphi(u)$ and $\psi(u)$ form two linearly independent, but otherwise arbitrary, solutions to the linear Schrödinger equation $\psi_{uu} + \frac{1}{2}\widehat{F}(u)\psi = 0$. Alternatively, $x = w_{uu}/w_u$, where w is an arbitrary solution to the Riccati equation $w_u = \frac{1}{2}w^2 + \widehat{F}(u)$.

Corollary 6.29. *The equation $x_u x_{uuu} = \frac{3}{2}x_{uu}^2$ has, as general solution, the linear fractional functions $x = (\alpha u + \beta)/(\gamma u + \delta)$.*

Consider the next unimodular group action, generated by the vector fields ∂_x, $x\partial_x - v\partial_v$, $x^2\partial_x - 2xv\partial_v$. According to Example 4.20, the group coincides with the first prolongation of the preceding action under the identification $v = u_x$. Any invariant n^{th} order ordinary differential equation has the form

$$\frac{d^{n-3}r}{ds^{n-3}} = H\left(s, r, \frac{dr}{ds}, \ldots, \frac{d^{n-4}r}{ds^{n-4}}\right), \tag{6.44}$$

where

$$s = \frac{vv_{xx} - \frac{3}{2}v_x^2}{v^4}, \qquad r = \frac{ds}{du} = \frac{v^2v_{xxx} - 6vv_xv_{xx} + 6v_x^3}{v^6}, \tag{6.45}$$

are the two fundamental differential invariants. Once we know the solution $r = G(s)$ to the reduced ordinary differential equation (6.44), we recover the solution to the original ordinary differential equation by solving the invariant third order equation $r = G(s)$, or, explicitly,

$$v^2v_{xxx} - 6vv_xv_{xx} + 6v_x^3 = v^6G\left(\frac{vv_{xx} - \frac{3}{2}v_x^2}{v^4}\right). \tag{6.46}$$

Since $r = ds/du$, we can integrate equation (6.46) once to yield

$$H(s) = \int^s \frac{d\hat{s}}{G(\hat{s})} = u + k, \tag{6.47}$$

which, once solved for $s = F(u)$, reduces to the invariant third order equation (6.37), an equation we know how to solve in terms of a pair of quadratures and a Riccati equation. Alternatively, the solution based on the linear Schrödinger equation (6.40) can also be effectively employed. Note that $v = u_x = \psi(u)^2/\omega$, where $\omega = \varphi\psi_u - \varphi_u\psi$ denotes the Wronskian of the two solutions $\varphi(u)$ and $\psi(u)$ to (6.40), which, by Abel's formula, is constant. Therefore, the general solution to the invariant equation (6.46) can be written in parametric form as

$$x = \frac{\varphi(u)}{\psi(u)}, \qquad v = \frac{\psi(u)^2}{\omega}. \tag{6.48}$$

As for the third unimodular group action, the generators are ∂_x, $x\partial_x - w\partial_w$, $x^2\partial_x - (2xw+1)\partial_w$, where we identify $w = v_x/2v = u_{xx}/2u_x$. Any invariant n^{th} order ordinary differential equation has the form

$$\frac{d^{n-3}y}{dt^{n-3}} = H\left(t, y, \frac{dy}{dt}, \ldots, \frac{d^{n-4}y}{dt^{n-4}}\right), \tag{6.49}$$

where the fundamental differential invariants are

$$\begin{aligned} t &= \frac{r}{s^{3/2}} = \frac{w_{xx} - 6ww_x + 4w^3}{\sqrt{2}\,(w_x - w^2)^{3/2}}, \\ y &= \frac{2}{s^2}\frac{dr}{du} = 24 + \frac{w_{xxx} - 12ww_{xx} + 18w_x^2}{(w_x - w^2)^2}. \end{aligned} \tag{6.50}$$

Once we know the solution $y = 24 + K(t)$ to the reduced ordinary differential equation (6.49), we recover the solution to the original ordinary differential equation by solving the invariant third order equation

$$w_{xxx} - 12ww_{xx} + 18w_x^2 = (w_x - w^2)^2\, K\left(\frac{w_{xx} - 6ww_x + 4w^3}{\sqrt{2}(w_x - w^2)^{3/2}}\right). \tag{6.51}$$

Using (6.50), equation (6.51) takes the first order separable form

$$s\,\frac{dt}{ds} = \frac{K(t)}{t} - \frac{3t}{2}, \tag{6.52}$$

hence $t = M(s)$ can be found by quadrature. Then $r = s^{3/2}M(s) = G(s)$ has the form (6.46), and so can be solved as before. The general solution to the invariant equation (6.51) can thus be expressed in the parametric form

$$x = \frac{\varphi(u)}{\psi(u)}, \qquad w = \frac{\psi_x}{\psi} = \frac{\psi\psi_u}{\omega}, \tag{6.53}$$

where $\varphi(u)$ and $\psi(u)$ form two independent solutions to the second order linear equation (6.40), whose coefficient function $s = F(u)$ is found by inverting the integral (6.47).

The simplest class of invariant equations for the third unimodular group action is when the function K in (6.51) is constant. The resulting family of equations

$$w_{xxx} - 12ww_{xx} + 18w_x^2 = \alpha(w_x - w^2)^2, \tag{6.54}$$

are equivalent to the equations found by Chazy, [**45**], in his deep study of third order ordinary differential equations having the "Painlevé property", cf. [**123**; Chapter 14]. The Chazy equation appears as a similarity reduction of the Yang–Mills equations from particle physics, [**43**], and the Prandtl boundary layer equations from fluid mechanics, [**54**]. It also has deep connections to number theory and automorphic forms, [**213**]. Chazy showed that when

$$\alpha = 0, \qquad \text{or} \qquad \alpha = \frac{864}{36 - k^2}, \quad \text{where} \quad 6 < k \in \mathbb{N}, \tag{6.55}$$

then the nontrivial solutions $w = f(x)$ to (6.54) have a moveable natural boundary which forms a circle in the complex plane whose position depends on the initial data. Chazy's results, in fact, follow from the representation of the solution provided by the above analysis. First, in terms of the fundamental differential invariants r, s, the Chazy equation (6.54) takes the form

$$\frac{r}{s^2}\frac{dr}{ds} = \beta = \tfrac{1}{2}(\alpha - 24).$$

According to the general procedure, we solve for r, and then introduce a parameter u so that $r = ds/du$. The result is $r^2 = (ds/du)^2 = \frac{2}{3}\beta s^3 + c$, for c constant. If $\beta \neq 0$, i.e., $\alpha \neq 24$, then $s = F(u) = (6/\beta)\wp(u + k)$, where $\wp$ denotes the Weierstrass elliptic function with parameters $g_2 = 0$, $g_3 = -\frac{1}{36}\beta^2 c$. Therefore, we have the following result.

Theorem 6.30. *The general solution to the Chazy equation* (6.54) *for* $\alpha \neq 24$ *has the parametric form*

$$x = \frac{\varphi(u)}{\psi(u)}, \qquad y = \frac{1}{\psi}\frac{d\psi}{dx} = \frac{\psi\psi_u}{\omega}, \tag{6.56}$$

where $\varphi(u)$, $\psi(u)$, *are two linearly independent solutions of the Lamé equation*

$$\frac{d^2\psi}{du^2} - \frac{3\wp(u+k)}{\beta}\psi = 0, \tag{6.57}$$

and $\omega = \varphi\psi_u - \varphi_u\psi$ *is their Wronskian.*

Exercise 6.31. Show how, in the special case $\alpha = 24$ the solution to the Chazy equation can be recovered from that of the Airy equation $\psi_{uu} + cu\psi = 0$.

Exercise 6.32. Chazy, [**45**], expresses the general solution to (6.54) in the form

$$x = \frac{\varphi(y)}{\psi(y)}, \qquad w = \frac{1}{\psi(y)}\,\frac{d\psi}{dx}, \tag{6.58}$$

where $\varphi(y)$ and $\psi(y)$ are two arbitrary linearly independent solutions of the hypergeometric equation

$$y(1-y)\,\frac{d^2\psi}{dy^2} + \left(\frac{1}{2} - \frac{7}{6}y\right)\frac{d\psi}{dy} + \frac{1}{6(\alpha-24)}\,\psi = 0. \tag{6.59}$$

Show how to transform (6.59) into the Lamé equation (6.57) and thereby connect the two solutions. (See Example 6.50 below, and [**54**] for more details.) It is interesting that equation (6.59) arises in Schwarz's theory of algebraic hypergeometric functions; in fact, for α given by (6.55), with $k = 2, 3, 4$, and 5, all of the solutions are algebraic functions; these parameter values correspond to hypergeometric equations admitting the discrete symmetry groups of dihedral triangle, tetrahedral, octahedral, and icosahedral class. See Hille, [**116**; §10.3], for details of Schwarz's theory.

Exercise 6.33. Prove that $w(x)$ is a solution to the Chazy equation (6.54) if and only if $w = \frac{1}{12}(k-6)(\log z)_x$, where k is related to α via (6.55), and $z(x)$ is a relative invariant of weight $n = 12/(6-k)$ of SL(2) that solves the fourth transvectant equation $(z, z)^{(4)} = 0$. See (3.44), with $F = G = z$, $m = n = 12/(6-k)$ and p replaced by x. Thus one can express the general solution to the transvectant equation $(z, z)^{(4)} = 0$ in terms of hypergeometric functions! See Chazy, [**45**], for more details.

Lie Determinants

The characterization of invariant differential equations using the method of differential invariants is based, ultimately, on the characterization of invariant equations in Theorem 2.38. The latter result is valid only on the open subset where the group acts (semi-)regularly and, thus, a complete set of functionally independent invariants exists; indeed, it is not necessarily true that, on the singular variety where the orbit dimension drops, the invariant equations are characterized by the vanishing of one or more invariants. Similarly, given a transformation group G, if we are interested in a complete classification of all invariant systems of differential equations, and not just those in the open subset $V^n \subset \mathrm{J}^n$ where the orbits achieve their maximal dimension, then we must understand this singular variety in more detail. According to Proposition 2.65, it can be characterized as the set where the prolonged infinitesimal generators span a subspace of the tangent space to J^n of less than maximal dimension. Specifically, if s_n denotes the maximal orbit dimension of the prolonged group action $G^{(n)}$, then the singular variety is given by

$$\mathcal{S}^n = \mathrm{J}^n \setminus V^n = \left\{ z \in \mathrm{J}^n \;\middle|\; \dim \mathfrak{g}^{(n)}|_z < s_n \right\}.$$

In particular, assuming G acts locally effectively, if n is greater than or equal to the stabilization order of G, then $\mathcal{S}^n$ is just the subset of J^n where the prolonged infinitesimal generators of G are linearly dependent. Therefore, once n is greater than the stabilization order, the singular variety $\mathcal{S}^{n+1}$ is contained in $\mathcal{S}^n$ (or, more specifically, $(\pi_n^{n+1})^{-1}\mathcal{S}^n$). In other words, if we regard $\mathcal{S}^n \subset \mathrm{J}^n$ as a system of differential equations, then, for sufficiently large n, the "solution space" to $\mathcal{S}^{n+1}$ is always contained in that of $\mathcal{S}^n$.

Any G-invariant system of differential equations can then be decomposed into the union of a regular component, which is a subset of V^n and, as in Theorem 6.25, can be (locally) characterized by the vanishing of a system of differential invariants, and a singular component, which is a subset of $\mathcal{S}^n$, and hence is characterized by the linear dependence of the infinitesimal generators, possibly combined with additional constraints.

In the scalar case — one independent and one dependent variable — the singular variety can be readily characterized using the method of Lie determinants introduced by Lie, [**150**]. Suppose G is an r-dimensional Lie group acting locally effectively on $E \simeq \mathbb{R} \times \mathbb{R}$, with infinitesimal generators $\mathbf{v}_1, \ldots, \mathbf{v}_r$. The infinitesimal generators of the prolongation $G^{(r-2)}$ take the form

$$\mathbf{v}_\mu^{(r-2)} = \xi_\mu \frac{\partial}{\partial x} + \sum_{k=0}^{r-2} \varphi_\mu^k \frac{\partial}{\partial u_k}, \qquad \mu = 1, \ldots, r, \tag{6.60}$$

where $u_k = D_x^k u$. In the standard case, meaning that G acts transitively and the orbit dimensions do not pseudo-stabilize, cf. Theorem 5.24, the stabilization order of G is $r-2$, so the maximal orbit dimension is $s_{r-2} = r = \dim \mathrm{J}^{r-2}$. In this case, the singular variety is characterized as the subset where the prolonged infinitesimal generators (6.60) are linearly dependent, and hence is given by the determinantal condition

$$\det \begin{vmatrix} \xi_1 & \varphi_1 & \varphi_1^1 & \cdots & \varphi_1^{r-1} \\ \xi_2 & \varphi_2 & \varphi_2^1 & \cdots & \varphi_2^{r-1} \\ \vdots & \vdots & \vdots & \ddots & \vdots \\ \xi_r & \varphi_r & \varphi_r^1 & \cdots & \varphi_r^{r-1} \end{vmatrix} = 0. \tag{6.61}$$

Note that (6.61) is a single ordinary differential equation for u as a function of x. Its left hand side is known as the *Lie determinant* associated with the given transformation group. Any invariant differential equation of order $r-2$ or less must be contained in the singular variety, and hence is described by the vanishing of the Lie determinant. Moreover, as we remarked above, since all the higher order singular subvarieties are contained therein, we conclude that *all* invariant differential equations can be written either in terms of the fundamental differential invariants, or by the vanishing of the Lie determinant.

Example 6.34. Consider the four parameter group generated by $\partial_x, x\partial_x, \partial_u, u\partial_u$, which is Case 1.9 with $k=1$ in the classification tables. According to Table 5, the fundamental differential invariants of this group are $I = u_x u_{xxx}/u_{xx}^2$ and $J = u_x^2 u_{xxxx}/u_{xx}^3$. The second prolongations of these vector fields are ∂_x, $x\partial_x - u_x\partial_{u_x} - 2u_{xx}\partial_{u_{xx}}$, ∂_u, and $u\partial_u + u_x\partial_{u_x} + u_{xx}\partial_{u_{xx}}$, hence the Lie determinant is

$$\det \begin{vmatrix} 1 & 0 & 0 & 0 \\ x & 0 & -u_x & -2u_{xx} \\ 0 & 1 & 0 & 0 \\ 0 & u & u_x & u_{xx} \end{vmatrix} = -u_x u_{xx}.$$

Therefore, the singular invariant differential equations are $u_x = 0$ and $u_{xx} = 0$. Every other invariant differential equation can be written in terms of the fundamental differential invariants I, J, and their derivatives. For example, the invariant third order equations are all of the form $u_x u_{xxx} = c u_{xx}^2$.

In the anomalous cases (intransitive or pseudo-stabilization), the Lie determinant (6.61) vanishes identically since $s_{r-2} < r$. However, according to Theorem 5.24, in such cases the stabilization order is $r-1$, and hence the singular variety is given as the subset of J^{r-1} where the prolonged infinitesimal generators $\mathbf{v}_1^{(r-1)}, \ldots, \mathbf{v}_r^{(r-1)}$ are linearly dependent. This condition can be checked by forming an $r \times (r+1)$ matrix having the form (6.61) but whose columns go up to order $r-1$, and computing the determinant of a suitable maximal $r \times r$ minor. By a slight abuse of terminology, we shall call this maximal minor the *Lie determinant* in this case. In this case, any invariant differential equation of order $r-1$ or less is given either by the vanishing of the modified Lie determinant, or, in the pseudo-stabilization cases, by the condition $I(x, u^{(r-2)}) = c$ for some constant c, where I is the fundamental differential invariant of lowest order. In the intransitive case, the lowest order differential invariant is a function $I(x,u)$ of x and u alone, and so the condition $I(x,u) = c$ describes a variety of E — a "zeroth order" differential equation.

Example 6.35. Consider the three parameter group generated by $\partial_x, \partial_u, x\partial_x + u\partial_u$. According to Table 5, the fundamental differential invariants of this group are $I = u_x$ and $J = u_{xxx}/u_{xx}^2$. The second prolongations of these vector fields, ∂_x, ∂_u, $x\partial_x + u\partial_u - u_{xx}\partial_{u_{xx}}$, are linearly dependent if and only if $u_{xx} = 0$, hence the modified Lie determinant is merely u_{xx}. Consequently, every invariant differential equation of order ≤ 3 has one of the following three forms: $u_x = c$, $u_{xx} = 0$, or $u_{xxx} = u_{xx}^2 H(u_x)$.

Theorem 6.36. *Suppose G is a transformation group acting on $E \simeq \mathbb{R} \times \mathbb{R}$. Then every invariant differential equation can be written in terms of either the fundamental differential invariants and their derivatives, or the vanishing of the associated Lie determinant.*

The full list of Lie determinants for all complex Lie groups of point and contact transformations can be found in Table 5. In this table, we have omitted any inessential constant factors.

Exercise 6.37. Does the method of Lie determinants generalize to the case of several dependent variables?

Symmetry Classification of Ordinary Differential Equations

Lie's complete classification of all possible finite-dimensional Lie groups acting on the plane constitutes a powerful tool that enables one to conduct a detailed analysis of the symmetry groups of scalar ordinary differential equations. As a consequence of Theorem 6.36, the differential invariants and Lie determinants associated with each of the point or contact transformation groups provide complete lists of canonical forms for equations admitting symmetry groups, and also allow one to completely describe all possible symmetry reductions available for scalar ordinary differential equations. Unfortunately, since the classification is only available in two dimensions, corresponding results for systems of ordinary differential equations, or for partial differential equations, are not available, and, as a consequence, other tools must be employed.

Before applying our classification tables, one first need be sure that the symmetry group of the differential equation of interest is, in fact, finite-dimensional, and hence is equivalent to one of the groups appearing in the tables. The underlying result that allows this approach to work is the following theorem, valid near any regular point of the differential equation.

Theorem 6.38. *The point transformation symmetry group of a normal system of ordinary differential equations of order two or more is finite-dimensional, without fixed points. The contact transformation symmetry group of a normal ordinary differential equation of order three or more is finite-dimensional, without fixed points.*

By *normal* we mean that the ordinary differential equation can be solved for the highest derivative $u_n = H(x, u^{(n-1)})$. See [**156**], [**89**], for a geometrical proof of this result; an alternative approach will be discussed later in this chapter. For second order equations, the finite-dimensionality of the point transformational symmetry group can also be proved using the Cartan equivalence method, as described in Chapter 12; see Dickson, [**61**], for yet another approach.

Consider first the point symmetry group of a scalar ordinary differential equation of order two or more. Since the symmetry group is finite-dimensional, it must be equivalent, under a complex change of variables, to one of the groups of point transformations appearing in Tables 1, 2, or 3. We shall now illustrate the range of classification results which are available, beginning with the case of a single second order ordinary differential equation.

Theorem 6.39. *The point transformation symmetry group of a second order ordinary differential equation has dimension at most eight. Moreover, the equation admits an eight-dimensional symmetry group if* *and only if it can be mapped, by a point transformation, to the linear equation* $u_{xx} = 0$, *which has symmetry group* SL(3).

Proof: This follows by inspection of Table 5. The maximally symmetric ordinary differential equations appear among the Lie determinants of the various point transformation groups. In particular, the projective group SL(3), which is Case 2.3, has the highest dimension of all groups having a second order Lie determinant. *Q.E.D.*

In Chapter 12 we will present a completely different proof of Theorem 6.39, which is based on the Cartan equivalence method. This proof does not rely on the complete classification of transformation groups, and is therefore more amenable to generalization.

Exercise 6.40. Prove that any linear second order ordinary differential equation can be mapped to the equation $u_{xx} = 0$. What are the eight independent point symmetries of a general linear equation?

Exercise 6.41. Show that $u_{xx} - 3uu_x + u^3 = 0$ — the modified Emden equation arising in stellar dynamics — admits an eight-dimensional symmetry group. Find the linearizing transformation. (See [**163**].)

Exercise 6.42. Prove that a second order equation admits a two-dimensional intransitive symmetry group if and only if it is equivalent to $u_{xx} = 0$. (See [**121**].)

Indeed, with the classification tables in hand, it is not hard to conduct a complete symmetry classification of all second order ordinary differential equations. The results appear in Table 7. It is interesting that, if the differential equation is not equivalent to $u_{xx} = 0$, and so has a symmetry group of dimension less than eight, then it admits at most a three-dimensional symmetry group. Table 7 lists the canonical forms for all possible second order equations admitting one-, two- and three-dimensional symmetry groups. See [**142**], [**121**] for more details.

The corresponding bound on the number of independent symmetries admitted by a higher order ordinary differential equation was also found by Lie, [**155**]. First we state the result for point transformations.

Theorem 6.43. *Let* $n \geq 3$. *Any* n^{th} *order scalar ordinary differential equation admits at most an* $(n+4)$-*parameter symmetry group of point transformations. Moreover, the symmetry group is* $(n+4)$-*dimensional if and only if the equation is equivalent to the linear equation* $D_x^n u = u_n = 0$, *with symmetry group of type 1.11 for* $k = n$.

Using the additional classification of the differential invariants and Lie determinants of contact transformation groups, we are able to prove a similar result for contact symmetries. Since a second order equation admits an infinite-dimensional contact symmetry group, the classification tables only apply to equations of order at least three.

Theorem 6.44. *Any third order ordinary differential equation admits at most a ten-parameter symmetry group of contact transformations. Moreover, the symmetry group is ten-dimensional if and only if the equation can be mapped, by a contact transformation, to the linear equation* $u_{xxx} = 0$*, with symmetry group of type 4.3. An ordinary differential equation of order* $n \geq 4$ *admits at most an* $(n+4)$*-parameter symmetry group of contact transformations, with maximal symmetry occurring if and only if the equation is equivalent to the linear equation* $u_n = 0$*, with symmetry group of type 1.11 for* $k = n$*.*

In principle, the classification tables allow one to construct a complete symmetry classification of all scalar ordinary differential equations, similar to those appearing in Table 7 in the second order case and, consequently, all possible symmetry reductions available. The most extensive results in this direction appear in Lie, [**150**]; see also [**82**] for the third order case. For reasons of space, we shall just describe the classification of higher order ordinary differential equations with submaximal symmetry groups, meaning those having dimension as large as possible without being maximal. For $n = 3$, the submaximal point symmetry group has dimension six, and there is just one invariant differential equation,

$$2u_x u_{xxx} - 3u_{xx}^2 = 0, \tag{6.62}$$

which has a symmetry group $\mathrm{SL}(2) \times \mathrm{SL}(2)$ of type 1.4. According to Corollary 6.29, the general solution of this equation is a linear fractional function. For $n = 5$, the submaximal point or contact symmetry group has dimension eight; the invariant differential equation is

$$9u_{xx}^2 u_{xxxxx} - 45u_{xx}u_{xxx}u_{xxxx} + 40u_{xxx}^3 = 0, \tag{6.63}$$

which has a symmetry group $\mathrm{SL}(3)$ of type 2.3. The solutions $u = f(x)$ of (6.63) are all graphs of conic sections. For $n = 7$, the submaximal contact symmetry group has dimension ten; the invariant differential equation is

$$10u_3^3 u_7 - 70u_3^2 u_4 u_6 - 49u_3^2 u_5^2 + 280u_3 u_4^2 u_5 - 175u_4^4 = 0, \tag{6.64}$$

which has a contact symmetry group SO(5) of type 4.3. In all other cases, the submaximal symmetry group has dimension $n+2$. The equation is either linear (but not equivalent to $u_n = 0$), or equivalent to either

$$(n-1)u_{n-2}u_n - nu_{n-1}^2 = 0, \qquad \text{or} \qquad 3u_{xx}u_{xxxx} - 5u_{xxx}^2 = 0,$$

having respective symmetry groups of type 1.11, for $k = n-2$, or 2.2.

The extension of the preceding classification results to systems of ordinary differential equations is not so apparent. According to Theorem 6.43, the only scalar ordinary differential equations admitting a maximal symmetry group are those equivalent to an (elementary) linear equation. It is known, [**91**], that the symmetry group of a system of second order ordinary differential equations in q dependent variables has the maximal dimension q^2+4q+3, which is also the dimension of the symmetry group $\mathrm{SL}(q+2,\mathbb{R})$ of the free particle system $d^2u^\alpha/dx^2 = 0$, $\alpha = 1,\ldots,q$. Recently, Fels, [**69**], [**70**], used the Cartan equivalence method to prove that the free particle system is, up to equivalence, the only system of second order possessing a symmetry group of maximal dimension. Fels extends this result to third order systems, proving that the dimension of the symmetry group is at most q^2+3q+3, and, moreover, a system admits a symmetry group of this dimension if and only if it is equivalent to the system $d^3u^\alpha/dx^3 = 0$. However, for a general system of nth order ordinary differential equations in q variables, the best known bound on the dimension of the symmetry group is $q^2+(n+1)q+2$ when $n>3$, but this is greater than the dimension of the symmetry group of the trivial system $d^nu^\alpha/dx^n = 0$, which is q^2+nq+3, cf. [**90**]. Perhaps the new method of proving Theorem 6.44 that appears in the next section may shed some further light on this problem.

A Proof of Finite Dimensionality

Here we discuss a new, alternative proof of Theorem 6.44, based on a method originally introduced by Sokolov, [**205**], to bound the dimension of the symmetry group of an nth order evolution equation. In particular, this proof implies the finite-dimensionality of the symmetry group of the differential equation, proving Theorem 6.38 when $n \geq 4$. A key lemma states that the *kernel* of an nth order total differential operator $\mathcal{D}$, meaning all differential functions annihilated by $\mathcal{D}$, has dimension at most n. In fact, this is a consequence of the fact that the ring of ordinary differential operators admits a (non-commutative) version of the Euclidean division algorithm.

Lemma 6.45. *Let $\mathcal{D} = \sum_{i=0}^{n} P_i[u] D_x^i$ be an nth order total differential operator. Then* $\dim\ker \mathcal{D} \le n$.

Proof: Note that any differential function Q lies in the kernel of the first order differential operator $QD_x - (D_x Q)$. Therefore, if $Q \in \ker \mathcal{D}$, we can factor $\mathcal{D} = \mathcal{E}\cdot[QD_x - (D_x Q)]$, where $\mathcal{E}$ is a differential operator of order $n-1$. Moreover, $R \in \ker \mathcal{D}$ if and only if $S = [QD_x - (D_x Q)]R \in \ker \mathcal{E}$. Let $\mathcal{I}_Q = \{QD_x R - RD_x Q\}$ denote the image of this operator in the space of differential functions, so that $S \in \mathcal{R}_Q = \ker \mathcal{E} \cap \mathcal{I}_Q$. Now, by induction (the case $n = 0$ being trivial) $\dim\ker \mathcal{E} \le n-1$, hence the subspace $\mathcal{R}_Q$ has dimension $\le n-1$ also. Let $S_1, \ldots, S_k$, $k \le n-1$, be a basis for the subspace $\mathcal{R}_Q$, so each $\mathcal{S}_i = [QD_x - (D_x Q)]R_i$ for some R_i. Then the general solution to the equation $\mathcal{D}R = 0$ is readily seen to be $R = c_0 Q + c_1 R_1 + \cdots + c_k R_k$, for constants $c_0, \ldots, c_k$. Therefore, $\dim\ker \mathcal{D} = k + 1 \le n$. *Q.E.D.*

We use this lemma to provide a proof of Theorem 6.44 when $n \ge 4$. (The case $n = 3$ can also be verified in this way, but the calculations are more complicated.) Consider a normal n^{th} order equation

$$K(x, u^{(n)}) = u_n - F(x, u^{(n-1)}) = 0, \tag{6.65}$$

where we have solved for the nth order derivative $u_n = d^n u/dx^n$. Define the nth order differential operator

$$\mathcal{D}_n = D_x^n - F_{n-1} D_x^{n-1} - F_{n-2} D_x^{n-2} - \cdots - F_0,$$

where $F_k = \partial F/\partial u_k$. (The differential operator can be identified with the Fréchet derivative of the differential function K, cf. [**186**].) The infinitesimal symmetry condition (6.5) for the characteristic $Q(x, u, p)$, $p = u_x$, of a contact symmetry of (6.65) is then $\mathcal{D}_n Q = 0$, which must be satisfied whenever (6.65) and its derivatives are satisfied. Since $\mathcal{D}_n Q$ depends on at most $(n+1)^{\text{st}}$ order derivatives of u, we deduce that Q must satisfy

$$\mathcal{D}_n Q = [AD_x + B]K, \tag{6.66}$$

for some A, B. The next step is to determine the form of A, B in (6.66). Let $\partial_n = \partial/\partial u_n$. We make use of the commutation formula

$$\partial_k D_x^n = \sum_{i=0}^{\min\{k,n\}} \binom{n}{i} D_x^{n-i} \partial_{k-i}, \tag{6.67}$$

which is easily proved by induction starting with the simplest case $\partial_n D_x = D_x \partial_n + \partial_{n-1}$. Now, using the fact that Q only depends on x, u, and p, applying ∂_{n+1} to (6.66) and using (6.67), we find $A = Q_p$. Similarly, applying ∂_n, and using the formula for A, we find $B = nD_x Q_p + Q_u$. Therefore, the symmetry condition reduces to

$$\mathcal{D}_n Q = [nKD_x + (D_x K)]Q_p + KQ_u. \tag{6.68}$$

Our goal is to eliminate the partial derivatives Q_p and Q_u from (6.68) so as to produce a total differential operator which annihilates Q. This is done by further differentiating (6.68). If $n \geq 3$, applying ∂_{n-1} to (6.68) and canceling a few terms produces

$$\mathcal{D}_{n-1} Q = \frac{n(n-1)}{2} D_x^2 Q_p + F_{n-1} D_x Q_p + (D_x F_{n-1}) Q_p + nD_x Q_u, \tag{6.69}$$

where

$$\mathcal{D}_{n-1} = F_{n-1,n-1} D_x^{n-1} + F_{n-1,n-2} D_x^{n-2} + \cdots F_{n-1,0}$$

is an $(n-1)^{\text{st}}$ order total differential operator. Using equation (6.68) to substitute for Q_u in (6.69), we find an equation of the form

$$\mathcal{D}_{n+1} Q = \mathcal{E}_2 Q_p. \tag{6.70}$$

Here $\mathcal{D}_{n+1} = nD_x^{n+1} + \cdots$, and $\mathcal{E}_2 = \frac{1}{2}n(n+1)KD_x^2 + \cdots$, are total differential operators of orders $n+1$ and 2, respectively, with indicated leading terms. Assuming $n \geq 4$, we can apply ∂_{n-2} to (6.68); a similar calculation produces an equation of the form

$$\mathcal{D}_{n+2} Q = \mathcal{E}_3 Q_p, \tag{6.71}$$

where $\mathcal{D}_{n+2} = \frac{1}{2}n(n+1)D_x^{n+2} + \cdots$ and $\mathcal{E}_3 = \frac{1}{2}n(n^2-1)KD_x^3 + \cdots$. The final step is to eliminate Q_p from (6.70), (6.71). Let $\mathcal{G} = \mathcal{F}_{k+1}\mathcal{E}_2 = \mathcal{F}_k \mathcal{E}_3$ be the least common multiple of the differential operators $\mathcal{E}_2, \mathcal{E}_3$; we can assume that $k \leq 2$. Set

$$\mathcal{D}_{n+k+2} = \mathcal{F}_{k+1}\mathcal{D}_{n+1} - \mathcal{F}_k \mathcal{D}_{n+2} = \tfrac{n(n-1)}{12} D_x^{n+k+2} + \cdots,$$

which is a total differential operator of order at most $n+4$. A straightforward calculation using (6.70), (6.71) proves that $\mathcal{D}_{n+k+2} Q = 0$. Therefore, by Lemma 6.45, provided $n \geq 4$, the space of contact symmetries of (6.65) has dimension at most $n+k+2 \leq n+4$. *Q.E.D.*

Linearization of Partial Differential Equations

Because of the preceding results, trivial linearizable ordinary differential equations are (as far as we know) uniquely characterized by the property that they admit a symmetry group of the maximal, finite dimension. Symmetry groups can also be used effectively to characterize linearizable systems of partial differential equations. The key remark is that a linear system of partial differential equations $\mathcal{D}[u] = 0$, where $\mathcal{D}$ is an nth order linear differential operator, has (assuming the system is not overdetermined) an infinite-dimensional symmetry group since we can add any other solution to a given solution. The infinitesimal generators of the relevant infinite-dimensional symmetry group take the form

$$\mathbf{v}_\psi = \sum_{\alpha=1}^{q} \psi^\alpha(x) \frac{\partial}{\partial u^\alpha}, \tag{6.72}$$

where $\psi = (\psi^1(x), \ldots, \psi^q(x))$ is any solution to the linear system $\mathcal{D}[\psi] = 0$. Note that the vector fields (6.72) commute, and hence generate an infinite-dimensional abelian symmetry group. Assuming the system $\mathcal{D}[\psi] = 0$ is locally solvable, the operators (6.72) span a q-dimensional subspace, namely the space of vertical tangent directions in TE. According to Proposition 6.13, any equivalent system of partial differential equations must also admit such an infinite-dimensional symmetry group, and hence any system of partial differential equations which has only a finite-dimensional symmetry group is certainly not linearizable (at least by a local transformation). Versions of the following theorem appear in [**20**], [**21**]; see also [**125**], [**126**].

Theorem 6.46. *Let $\Delta(x, u^{(n)}) = 0$ be an nth order system of q independent partial differential equations in $p \geq 2$ independent variables and q unknowns. If the system admits an infinite-dimensional abelian symmetry group, having q-dimensional orbits, and infinitesimal generators depending linearly on the general solution to an nth order system of q independent linear partial differential equations $\mathcal{D}[\psi] = 0$, then it can, by a change of variables, be mapped to an inhomogeneous form of the linear system $\mathcal{D}[u] = f$.*

Proof: According to Frobenius' Theorem (see Exercise 14.4 below), any abelian transformation group with q-dimensional orbits can, through a change of variables, be mapped to a group generated by vector fields of the form (6.72). The additional hypothesis implies that the coefficient functions $\psi(x)$ form the general solution to a linear n^{th} order system of partial differential equations $\mathcal{D}[\psi] = 0$. Note that, in terms of the original coordinates, the generators take the form

$$\mathbf{v}_\psi = \sum_{\alpha=1}^{q} \psi^\alpha\big(\eta^1(x,u), \ldots, \eta^p(x,u)\big)\mathbf{w}_\alpha, \tag{6.73}$$

where the vector fields $\mathbf{w}_1, \ldots, \mathbf{w}_q$ are linearly independent, commute, and satisfy $\mathbf{w}_\alpha(\eta^i) = 0$ for all α, i. Now, in the new coordinates, the system must be equivalent to the inhomogeneous form of the linear system. To see this, it suffices to note that the only n^{th} order differential invariants of the infinite-dimensional group generated by the vector fields (6.72) are the components of $\mathcal{D}[u]$ and the variables x. Therefore, according to Theorem 6.25, any invariant system of differential equations must be isomorphic to one of the form $H(\mathcal{D}[u], x) = 0$. But, since the system consists of q independent equations, we can solve the system for the components of $\mathcal{D}[u]$, placing the system into the desired inhomogeneous form. *Q.E.D.*

Example 6.47. The nonlinear diffusion equation

$$u_t = u_x^{-2} u_{xx} \tag{6.74}$$

admits the following six symmetry generators:

$$\partial_t, \qquad \partial_u, \qquad x\partial_x, \qquad 2t\partial_t + u\partial_u,$$
$$xu\partial_x - 2t\partial_u, \qquad (2xt + xu^2)\partial_x - 4t^2\partial_t - 4tu\partial_u,$$

as well as the infinite dimensional abelian subalgebra

$$\alpha(t,u)\partial_x, \qquad \text{where} \qquad \alpha_t = \alpha_{uu}.$$

According to Example 6.34, the hodograph transformation $v = x$, $y = u$, linearizes (6.74) to the (homogeneous) heat equation $v_t = v_{yy}$.

Example 6.48. The equation

$$u_t u_{xx} + u_x^m = 0, \tag{6.75}$$

admits only a finite dimensional algebra of point symmetries, but has an infinite dimensional abelian algebra of contact symmetries generated by $F(t, u_x)\partial_u$ where $F(t, p)$ satisfies the linear parabolic equation $F_t = p^m F_{pp}$. Theorem 6.46 can clearly be generalized to include contact symmetries, where the linearization maps are now allowed to be contact transformations. In the present case, the invertible contact transformation $y = u_x, v = u - xu_x, v_y = -x, v_t = u_t$ (which leaves t unchanged) linearizes (6.75) to the diffusion equation $v_t = y^m v_{yy}$. See [**20**], [**21**], [**144**] for further examples and applications.

Differential Operators

The classification of linear differential equations is a special case of the general problem of classifying differential operators, which has a variety of important applications, including quantum mechanics and the projective geometry of curves. The theory, dating back to the work of Halphen, [**109**], Laguerre, [**148**], and Forsyth, [**71**], provides a striking area of application of our results from classical invariant theory. In this section we shall solve several types of equivalence problems for higher order differential operators. For simplicity, we shall only deal with the local equivalence problem for scalar differential operators in a single independent variable, although these problems are important for matrix-valued and partial-differential operators as well. The classical reference for these results is the book of Wilczynski, [**228**]; see also Se-ashi, [**202**], for a modern approach, and [**176**] for an analysis of the global equivalence problem.

Consider an nth order ordinary differential operator

$$\mathcal{D} = f^n(x)D_x^n + f^{n-1}(x)D_x^{n-1} + \cdots + f^1(x)D_x + f^0(x), \tag{6.76}$$

whose coefficients $f^i(x)$ are smooth functions of $x \in \mathbb{R}$. We assume that $f^n(x) \neq 0$ on the domain of interest, so that the operator determines a nonsingular, homogeneous linear ordinary differential equation

$$\mathcal{D}[u] = f^n(x)u_n + f^{n-1}(x)u_{n-1} + \cdots + f^1(x)u_x + f^0(x)u = 0; \tag{6.77}$$

as usual $u_n = D_x^n u$. We are interested in finding out when two operators, or two linear differential equations, can be mapped to each other by a suitable change of variables. In order to preserve linearity, we restrict to those of the form

$$\bar{x} = \chi(x), \qquad \bar{u} = \eta(x)u, \tag{6.78}$$

combining a change of independent variables, which induces the chain rule action $D_{\bar{x}} = (1/\chi_x)D_x$ on the total derivative operator, together with a rescaling of the dependent variable by a nonvanishing function $\eta(x) = e^{\varphi(x)}$, called, as in Chapter 3, the *gauge factor*.

Several different equivalence problems present themselves. Two differential operators $\mathcal{D}$ and $\overline{\mathcal{D}}$ are said to be *gauge equivalent* if they satisfy

$$\overline{\mathcal{D}} = \eta \cdot \mathcal{D} \cdot \frac{1}{\eta} \qquad \text{when} \qquad \bar{x} = \chi(x). \tag{6.79}$$

We have already encountered the problem of gauge equivalence in our classification of multiplier representations and Lie algebras of differential operators; see (3.22). In quantum mechanics, gauge equivalence plays an important role since it preserves the solution set to the associated Schrödinger equation $iu_t = \mathcal{D}[u]$, or its stationary counterpart, the eigenvalue problem $\mathcal{D}[u] = \lambda u$. On the other hand, if our principal concern is the associated homogeneous differential equations, (6.77), then we have the additional freedom to multiply the differential operator by a nonvanishing function $\zeta(x)$; the combined action

$$\overline{\mathcal{D}} = \zeta \cdot \eta \cdot \mathcal{D} \cdot \frac{1}{\eta} \qquad \text{when} \qquad \bar{x} = \chi(x), \tag{6.80}$$

will, in view of later applications, be referred to as *projective equivalence* of differential operators. Finally, the *direct equivalence problem* requires that the two linear differential polynomials agree, $\overline{\mathcal{D}}[\bar{u}] = \mathcal{D}[u]$, so the operators satisfy

$$\overline{\mathcal{D}} = \mathcal{D} \cdot \frac{1}{\eta} \qquad \text{when} \qquad \bar{x} = \chi(x). \tag{6.81}$$

The first order of business is to use the various equivalence transformations in a systematic manner to produce a set of (semi-)canonical forms for differential operators. We begin with the gauge equivalence problem.

Proposition 6.49. *A nonsingular n^{th} order differential operator (6.76) is, locally, gauge equivalent to an operator of the form*

$$\mathcal{D} = \pm D_x^n + p^{n-2}(x) D_x^{n-2} + \cdots + p^0(x). \tag{6.82}$$

Indeed, a straightforward calculation demonstrates that the required change of variables (6.78) is given by

$$\bar{x} = \chi(x) = \int \frac{dx}{\sqrt[n]{|f^n(x)|}}, \tag{6.83}$$

with associated gauge factor

$$\eta(x) = \left|p^n(x)\right|^{(1-n)/(2n)} \exp\left\{\int^x \frac{p^{n-1}(y)}{np^n(y)}\, dy\right\}. \tag{6.84}$$

Example 6.50. Consider a second order differential operator

$$\mathcal{D} = f(x)D_x^2 + g(x)D_x + h(x), \tag{6.85}$$

with $f(x) > 0$. According to (6.83), (6.84), the change of variables

$$\bar{x} = \chi(x) = \int \frac{dx}{\sqrt{f(x)}}, \qquad \eta(x) = \frac{1}{\sqrt[4]{f(x)}} \exp\left\{\int^x \frac{g(y)}{2f(y)}\, dy\right\}, \tag{6.86}$$

will place the operator in the form of a Schrödinger operator

$$\overline{\mathcal{D}} = D_{\bar{x}}^2 + V(\bar{x}), \tag{6.87}$$

with potential

$$V(\bar{x}) = \frac{8gf_x - 3f_x^2 - 4g^2}{16f} + h - \tfrac{1}{2}g_x + \tfrac{1}{4}f_{xx}, \tag{6.88}$$

where the right hand side is evaluated at $x = \chi^{-1}(\bar{x})$.

Exercise 6.51. Show that any second order operator is directly equivalent to a Schrödinger operator (6.87). Find the formula for the potential $V(\bar{x})$ in this case.

Example 6.52. Consider a third order operator of the special form

$$\mathcal{D} = D_x^3 + 3f(x)D_x^2 + 3g(x)D_x + h(x). \tag{6.89}$$

The gauge factor $\eta(x) = \exp \int f(x)\, dx$ will reduce it to the form

$$\overline{\mathcal{D}} = D_x^3 + 3p(x)D_x + q(x), \quad \text{where} \quad \begin{cases} p = g - f^2 - f_x, \\ q = h - 3fg + 2f^3 - f_{xx}. \end{cases} \tag{6.90}$$

Exercise 6.53. Show that any nonsingular n^{th} order operator is directly equivalent to one of the form (6.82), where the sign in front of D_x^n can always be taken positive.

In the projective equivalence problem, we can use the additional freedom to further normalize the coefficient of D_x^{n-2} to zero.

Theorem 6.54. *A nonsingular* n^{th} *order linear differential equation is projectively equivalent to one in* Laguerre–Forsyth normal form

$$u_n + r^{n-3}(x)u_{n-3} + \cdots + r^0(x)u = 0. \tag{6.91}$$

Proof: We first gauge the operator into the normal form (6.82) with positive leading coefficient. An orientation-preserving change of variables (6.80) will maintain this normal form provided

$$\bar{x} = \chi(x), \qquad \bar{u} = \chi_x^{(n-1)/2}\, u, \qquad \zeta = \chi_{\bar{x}}^{-n}. \tag{6.92}$$

The coefficient of u_{n-2} vanishes if and only if χ satisfies the Schwarzian equation

$$\frac{n(n^2-1)}{12}\,\frac{\chi_x\chi_{xxx} - \frac{3}{2}\chi_{xx}^2}{\chi_x^2} = p^{n-2}(x). \tag{6.93}$$

According to Theorem 6.28, the required change of variables has the form $\bar{x} = \chi(x) = \varphi(x)/\psi(x)$, where φ and ψ are any two linearly independent solutions of the auxiliary second order equation

$$v_{xx} + \frac{6}{n(n^2-1)}p^{n-2}(x)v = 0. \tag{6.94}$$

Thus, modulo solving the auxiliary Schrödinger equation (6.94), we have successfully produced the desired transformation to the Laguerre–Forsyth normal form. *Q.E.D.*

Remark: Theorem 6.54 proves that any linear homogeneous second order ordinary differential equation is projectively equivalent to the elementary equation $u_{xx} = 0$. However, this fact is not of much use in effecting the solution to the original equation, since the auxiliary equation (6.94) is the *same* as the associated Schrödinger equation.

Example 6.55. Consider the case of a third order differential equation, whose Laguerre–Forsyth normal form is

$$u_{\bar{x}\bar{x}\bar{x}} + r(\bar{x})u = 0. \tag{6.95}$$

To find the formula for r, we divide through by the leading coefficient, so the original equation is

$$u_{xxx} + 3f(x)u_{xx} + 3g(x)u_x + h(x)u = 0, \tag{6.96}$$

corresponding to the operator (6.89). The gauge transformation of (6.90) reduces the equation to the semi-normal form

$$u_{xxx} + 3p(x)u_x + q(x)u = 0. \tag{6.97}$$

To obtain the normal form (6.95), we invoke the change of variables $\bar{x} = \chi(x) = \varphi(x)/\psi(x)$, where φ and ψ are linearly independent solutions to the auxiliary Schrödinger equation $v_{xx} + \frac{3}{4}p(x)v = 0$. The remaining coefficient function is explicitly given by

$$r(\bar{x}) = r(\chi(x)) = \frac{\hat{r}(x)}{\chi_x^3}, \qquad \text{where} \qquad \hat{r} = q - \tfrac{3}{2}p_x. \tag{6.98}$$

Exercise 6.56. Find a formula for the Laguerre–Forsyth normal form for a fourth order equation. (See [**228**].)

The Laguerre–Forsyth normal form is not a uniquely defined canonical form for the differential operator, since there is a residual group of transformations that preserves such operators.

Proposition 6.57. *A change of variables* (6.80) *will preserve the Laguerre–Forsyth normal form* (6.91) *if and only if it has the linear fractional form*

$$\bar{x} = \frac{\alpha x + \beta}{\gamma x + \delta}, \qquad \bar{u} = \frac{(\alpha\delta - \beta\gamma)^{(n-1)/2}}{(\gamma x + \delta)^{n-1}}\, u. \tag{6.99}$$

This result follows from the solution to (6.93), with $p^{n-2} = 0$, given in Corollary 6.29. Note that the symmetry transformation (6.99) corresponds to the fundamental multiplier representation of GL(2) with multiplier $\mu_{n-1,(1-n)/2}$; see equation (3.9). In other words, the function $u(x)$ has weight $\left(n-1, \frac{1}{2}(1-n)\right)$ under the projective group action. The action (6.99) induces a representation of GL(2) on the coefficients of the Laguerre–Forsyth normal form, which we first analyze in the special case of a third order differential equation. Consider first an equation in Laguerre–Forsyth normal form

$$u_{xxx} + r(x)u = 0. \tag{6.100}$$

According to (6.99), the solution u has weight $(2, -1)$. A simple chain rule calculation proves that

$$\bar{u} = \frac{\alpha\delta - \beta\gamma}{(\gamma x + \delta)^2}\, u, \qquad \frac{d^3\bar{u}}{d\bar{x}^3} = \frac{(\gamma x + \delta)^4}{(\alpha\delta - \beta\gamma)^2}\, \frac{d^3 u}{dx^3}, \tag{6.101}$$

so u_{xxx} has weight $(-4, 2)$. (It is worth noting that, although both u and u_{xxx} transform according to multiplier representations of GL(2), neither the first nor second derivative of u has this property.) Therefore the coefficient $r(x)$ transforms with weight $(-6, 3)$:

$$r(x) = \frac{(\alpha\delta - \beta\gamma)^3}{(\gamma x + \delta)^6}\, \bar{r}(\bar{x}) = \frac{(\alpha\delta - \beta\gamma)^3}{(\gamma x + \delta)^6}\, \bar{r}\left(\frac{\alpha x + \beta}{\gamma x + \delta}\right). \tag{6.102}$$

According to Theorem 3.47, and the further discussion in Example 5.18, the successive transvectants of the function r will provide us with a wide array of relative differential invariants associated with the original linear differential equation. The simplest example is the Hessian, cf. (3.46), which, up to multiple, is given by $H[r] = r r_{xx} - \frac{7}{6} r_x^2$, and has weight $(-16, 8)$. Therefore, the function

$$I = \frac{H}{r^{8/3}} = \frac{r r_{xx} - \frac{7}{6} r_x^2}{r^{8/3}} \tag{6.103}$$

is the simplest absolute differential invariant associated with (6.100). Note that we can rewrite the differential invariant I in terms of the coefficients of a general third order equation (6.96), by using our previous formulas for changing it into Laguerre–Forsyth form. We find that

$$I = \frac{6\hat{r}\hat{r}_{xx} - 7\hat{r}_x^2 - 27p\hat{r}^2}{6\hat{r}^{8/3}}, \tag{6.104}$$

where $\hat{r}$, p are given by (6.98), (6.90).

Exercise 6.58. Consider a fourth order equation, which we place in normal form

$$u_{xxxx} + r(x)u_x + s(x)u = 0. \tag{6.105}$$

Prove that the functions r and $v = s - \frac{1}{2}r_x$ have respective weights $(-6,3)$, $(-8,4)$. Note that the coefficient s does *not* transform according to a multiplier representation of $\mathrm{GL}(2)$. Find formulas for these two invariants in terms of the original coefficients of the equation.

Theorem 6.59. *Consider the action of the group (6.99) on an nth order differential equation in Laguerre–Forsyth normal form (6.91). Then, for each $k = 0, \ldots, n-3$, the function*

$$v^k(x) = \sum_{k=i}^{n-3} (-1)^{i+k} \frac{k!(n+1-i)!(2n+2-i-k)!}{i!(k-i)!(n+1-k)!(2n+2-2i)!} \frac{d^{k-i}r^k}{dx^{k-i}} \tag{6.106}$$

is a relative invariant of weight $(2k-2n, n-k)$ under the projective action of $\mathrm{GL}(2)$.

See [**228**] for the proof. As in the construction of joint covariants of binary forms, one can readily produce a large variety of further relative invariants from the basic relative invariants v^k using successive transvectants. See [**228**] and [**170**] for geometric applications, and [**158**] for applications to the theory of orthogonal polynomials. These invariants can be used effectively to characterize the symmetries and particular normal forms of linear ordinary differential equations.

Theorem 6.60. *Consider an nth order linear homogeneous ordinary differential equation, where $n \geq 3$. The following conditions are equivalent:*

(i) *The equation is equivalent to the equation $u_n = 0$.*
(ii) *All the invariants of the equation vanish: $v^k \equiv 0$.*
(iii) *The point symmetry group of the equation has dimension $n+4$.*
(iv) *The general solution to the equation can be written in the form $u(x) = \sum_i c_i w_1(x)^i w_2(x)^{n-i}$, where $w_1(x)$ and $w_2(x)$ form a basis for the solution space to a second order linear, homogeneous ordinary differential equation.*

Theorem 6.61. *Consider an nth order linear homogeneous ordinary differential equation, where $n \geq 3$. The following conditions are equivalent:*

(*i*) *The equation is equivalent to a constant coefficient equation.*
(*ii*) *All the invariants of the equation are constant.*
(*iii*) *The symmetry group of the equation has dimension* $n+2$ *or* $n+4$.

As remarked earlier, no n^{th} order differential equation (linear or nonlinear) has an $(n+3)$-dimensional symmetry group. For linear equations, the possible dimensions are $n+4$, if the equation is equivalent to $D_x^n u = 0$, or $n+2$, if the equation is equivalent to a constant coefficient equation, or, otherwise, $n+1$, since we can always add in any solution or multiply by a constant.

Exercise 6.62. The (formal) *adjoint* of a differential operator $\mathcal{D} = \sum_k f^k(x) D_x^k$ is the operator $\mathcal{D}^* = \sum_k (-1)^k D_x^k \cdot f^k$. The operator is called self-adjoint (or skew-adjoint) if $\mathcal{D}^* = \mathcal{D}$ (or $\mathcal{D}^* = -\mathcal{D}$). Prove that an even (odd) order linear homogeneous ordinary differential equation is equivalent to a self-adjoint (skew-adjoint) equation if and only if all its odd order linear invariants v^{2j+1} vanish.

Applications to the Geometry of Curves

As noted by Wilczynski, [**228**], there is an intimate connection between the theory of curves in projective space and linear homogeneous ordinary differential equations. Consider a smooth curve $C \subset \mathbb{R}^n$, which we parametrize by an n-tuple of functions $u: \mathbb{R} \to \mathbb{R}^n$. We will always assume that the curve C is "truly" n-dimensional, meaning that it is not contained in any proper subspace of $\mathbb{R}^n$. This means that the functions $u^1(x), \ldots, u^n(x)$ are linearly independent, i.e., there is no nontrivial relation of the form $\sum c_\alpha u^\alpha(x) = 0$ for constant c_α not all zero. In this case, it is well known that the coordinate functions $u^\alpha(x)$ form a basis for the solutions to a linear, homogeneous ordinary differential equation.

Proposition 6.63. *Any set of* n *linearly independent functions* $u^1(x), \ldots, u^n(x)$ *forms a basis for the solution space to a homogeneous linear* n^{th} *order ordinary differential equation* $\mathcal{D}[u] = 0$.

Proof: If each of the functions $u^\alpha(x)$ is to be a solution, then the quotients $h^j(x) = f^j(x)/f^n(x)$ of the coefficients in (6.77) will be solutions to the linear system

$$h^{n-1}(x) D_x^{n-1} u^\alpha + \cdots + h^1(x) D_x u^\alpha + h^0(x) u^\alpha = -D_x^n u^\alpha, \quad j = 1, \ldots, n.$$

Cramer's rule implies that we can set $f^k(x) = W^k(x)$, where $W^k(x)$ is the $n \times n$ determinant obtained by omitting the kth column from the

$n \times (n+1)$ matrix whose entries are $D_x^i u^\alpha$, $i = 0, \ldots, n$, $\alpha = 1, \ldots, n$. Singular points of the differential equation, which include the origin $u = 0$, will correspond to points where the ordinary Wronskian $W^n(x)$ of the functions $u^1, \ldots, u^n$ vanishes. *Q.E.D.*

Two curves will correspond to the same linear differential operator (6.76) if and only if they can be mapped to each other by a (constant coefficient) linear transformation: $\overline{u}(x) = A\,u(x)$. An invertible change of variables $\bar{x} = \chi(x)$ will describe a reparametrization of the curve. Finally, two curves will project to the same curve in projective space $\mathbb{RP}^{n-1}$ if and only if their components are related by a common scalar gauge transformation: $\bar{u}(x) = \eta(x)u(x)$. The upshot is that the projective equivalence of linear differential operators can be reinterpreted as the equivalence of curves in projective space.

Proposition 6.64. *The preceding construction defines a one-to-one correspondence between projectively equivalent curves and projectively equivalent differential operators.*

As a specific example, let us consider curves in the projective plane $\mathbb{RP}^2$. In homogeneous coordinates, they are parametrized by a map $u: \mathbb{R} \to \mathbb{R}^3$, whose components form the solution space to a third order homogeneous equation (6.96). We can assume, without loss of generality, that the curve does not intersect the (x, y)-plane, so that we can rescale to use the parametrization $u(t) = (x(t), y(t), 1)$. The coefficients of (6.96) are (with t replacing x)

$$f(t) = \frac{1}{3}\frac{\mathbf{x}_t \wedge \mathbf{x}_{ttt}}{\mathbf{x}_t \wedge \mathbf{x}_{tt}}, \qquad g(t) = -\frac{1}{3}\frac{\mathbf{x}_{tt} \wedge \mathbf{x}_{ttt}}{\mathbf{x}_t \wedge \mathbf{x}_{tt}}, \qquad h = 0, \tag{6.107}$$

where $\mathbf{x}(t) = (x(t), y(t))$, and $\wedge$ denotes the cross product between vectors in the plane. The fundamental differential invariant (6.103) is the *projective curvature* of the curve. In particular, if we eliminate the parameter t by treating $y = u(x)$ as a function of x (assuming the curve is never vertical), then we recover the formula for the fundamental differential invariant of the projective group SL(3) listed in Table 5. The projective curvature is just the simplest projective invariant; one can similarly compute higher order transvectants to produce a wide range of further invariants.

Exercise 6.65. The curve described by the adjoint of a differential equation, cf. Exercise 6.62, is known as the *dual curve*. Geometrically it corresponds to the curve described by the osculating hyperplane, using the well-known duality between points and hyperplanes in projective space. Prove that a planar curve is projectively equivalent to a conic section if and only if it is self-dual.

Exercise 6.66. Adapt the preceding constructions to the determination of affine differential invariants of curves in $\mathbb{R}^n$.

Chapter 7

Symmetries of Variational Problems

The applications of symmetry groups to problems arising in the calculus of variations have their origins in the late papers of Lie, e.g., [**154**], which introduced the subject of "integral invariants". Lie showed how the symmetry group of a variational problem can be readily computed based on an adaptation of the infinitesimal method used to compute symmetry groups of differential equations. Moreover, for a given symmetry group, the associated invariant variational problems are completely characterized using the fundamental differential invariants and contact-invariant coframe, as presented in Chapter 5. This result lies at the foundation of modern physical theories, such as string theory and conformal field theory, which are constructed using a variational approach and postulating the existence of certain physical symmetries. The applications of Lie groups to the calculus of variations gained added importance with the discovery of Noether's fundamental theorem, [**178**], relating symmetry groups of variational problems to conservation laws of the associated Euler–Lagrange equations. We should also mention the applications of integral invariants to Hamiltonian mechanics, developed by Cartan, [**33**], which led to the modern symplectic approach to Hamiltonian systems, cf. [**165**]; unfortunately, space precludes us from pursuing this important theory here. More details on the applications of symmetry groups to variational problems and to Hamiltonian systems, along with many additional physical and mathematical examples, can be found in my earlier text [**186**]. Included in our presentation is a discussion of the Cartan form for scalar variational problems, and some new applications to the classification of invariant evolution equations, which have recently arisen in applications to computer vision.

The Calculus of Variations

The starting point will be a discussion of some of the foundational results in the calculus of variations. The presentation here is rather brief, and the interested reader can consult one of the standard textbooks, e.g., [**85**], for a more detailed development. As usual, we work over an open subset of the total space $E = X \times U \simeq \mathbb{R}^p \times \mathbb{R}^q$ coordinatized by independent variables $x = (x^1, \ldots, x^p)$ and dependent variables $u = (u^1, \ldots, u^q)$. The associated n^{th} jet space J^n is coordinatized by the derivatives $u^{(n)}$ of the dependent variables. Let $\Omega \subset X$ denote a connected open set with smooth boundary $\partial\Omega$. By an nth order *variational problem*, we mean the problem of finding the extremals (maxima and/or minima) of a *functional*

$$\mathcal{L}[u] = \mathcal{L}_\Omega[u] = \int_\Omega L(x, u^{(n)})\, dx \tag{7.1}$$

over some space of functions $u = f(x)$, $x \in \Omega$. The integrand $L(x, u^{(n)})$, which is a smooth differential function on the jet space J^n, is referred to as the *Lagrangian* of the variational problem (7.1); the horizontal p-form $L\, dx = L\, dx^1 \wedge \cdots \wedge dx^p$ is the *Lagrangian form*. The precise space of functions upon which the functional (7.1) is to be extremized will depend on any boundary conditions which may be imposed — e.g., the Dirichlet conditions $u = 0$ on $\partial\Omega$ — as well as smoothness requirements. More generally, although this is beyond our scope, one may also impose additional constraints — holonomic, non-holonomic, integral, etc. In our applications, the precise nature of the boundary conditions will, by and large, be irrelevant. Moreover, as in our discussion of differential equations, we shall always restrict our attention to smooth extremals, leaving aside important, but technically more complicated problems for more general extremals.

The most basic result in the calculus of variations is the construction of the fundamental differential equations — the Euler–Lagrange equations — which must be satisfied by any smooth extremal.† The Euler–Lagrange equations constitute the infinite-dimensional version of the basic theorem from calculus that the maxima and minima of a smooth function $f(x)$ occur at the point where the gradient vanishes: $\nabla f = 0$.

† See [**16**] for examples of variational problems with nonsmooth extremals which do *not* satisfy the Euler–Lagrange equations!

In the functional context, the gradient's role is played by the "variational derivative", whose components, in concrete form, are found by applying the fundamental Euler operators.

Definition 7.1. Let $1 \leq \alpha \leq q$. The differential operator $\mathrm{E} = (\mathrm{E}_1, \ldots, \mathrm{E}_q)$, whose components are

$$\mathrm{E}_\alpha = \sum_J (-D)_J \frac{\partial}{\partial u_J^\alpha}, \qquad \alpha = 1, \ldots, q, \tag{7.2}$$

is known as the *Euler operator*. In (7.2), the sum is over all symmetric multi-indices $J = (j_1, \ldots, j_k)$, $1 \leq j_\nu \leq p$, and $(-D)_J = (-1)^k D_J$ denotes the corresponding signed higher order total derivative.

Theorem 7.2. *The smooth extremals $u = f(x)$ of a variational problem with Lagrangian $L(x, u^{(n)})$ must satisfy the system of* Euler–Lagrange equations

$$\mathrm{E}_\alpha(L) = \sum_J (-D)_J \frac{\partial L}{\partial u_J^\alpha} = 0, \qquad \alpha = 1, \ldots, q. \tag{7.3}$$

Note that, as with the total derivatives, even though the Euler operator (7.2) is defined using an infinite sum, for any given Lagrangian only finitely many summands are needed to compute the corresponding Euler–Lagrange expressions $\mathrm{E}(L)$.

Proof: The proof of Theorem 7.2 relies on the analysis of variations of the extremal u. In general, a one-parameter family of functions $u(x, \varepsilon)$ is called a *family of variations*† of a fixed function $u(x) = u(x, 0)$ provided that, outside a compact subset $K \subset \Omega$, the functions coincide: $u(x, \varepsilon) = u(x)$ for $x \in \Omega \setminus K$. In particular, all the functions in the family satisfy the same boundary conditions as u. Therefore, if u is, say, a minimum of the variational problem, then, for any family of variations $u(x, \varepsilon)$, the scalar function $h(\varepsilon) = \mathcal{L}[u(x, \varepsilon)]$, must have a minimum at $\varepsilon = 0$, and so, by elementary calculus, satisfies $h'(0) = 0$. In view of

† In the usual approach, one employs a family of linear variations $u(x, \varepsilon) = u(x) + \varepsilon v(x)$, where $v(x)$ has compact support, since the inclusion of higher order terms in ε has no effect on the method.

our smoothness assumptions, we can interchange the integration and differentiation to evaluate this derivative:

$$0 = \frac{d}{d\varepsilon} \mathcal{L}[u(x,\varepsilon)] \bigg|_{\varepsilon=0} = \int_\Omega \left\{ \sum_{\alpha=1}^{q} \sum_J \frac{\partial L}{\partial u_J^\alpha} (x, u^{(n)}) \, D_J \, v^\alpha \right\} dx, \qquad (7.4)$$

where $v(x) = u_\varepsilon(x,0)$. The method now is to integrate (7.4) by parts. The Leibniz rule

$$PD_iQ = -QD_iP + D_i[PQ], \qquad i = 1, \ldots, p, \qquad (7.5)$$

for total derivatives implies, using the Divergence Theorem, the general integration by parts formula

$$\int_\Omega PD_iQ \, dx = - \int_\Omega QD_iP \, dx + \\ + \int_{\partial\Omega} (-1)^{i-1} PQ dx^1 \wedge \cdots \wedge dx^{i-1} \wedge dx^{i+1} \wedge \cdots \wedge dx^p, \qquad (7.6)$$

which holds for any smooth function $u = f(x)$. Applying (7.6) repeatedly to integral on the right hand side of (7.4), and using the fact that v and its derivatives vanish on $\partial\Omega$, we find

$$0 = \int_\Omega \left\{ \sum_{\alpha=1}^{q} \sum_J (-D)_J \left(\frac{\partial L}{\partial u_J^\alpha} \right) v^\alpha \right\} dx \\ = \int_\Omega \left\{ \sum_{\alpha=1}^{q} E_\alpha(L) v^\alpha \right\} dx = \int_\Omega [\mathrm{E}(L) \cdot v] \, dx.$$

Since the resulting integrand must vanish for *every* smooth function with compact support $v(x)$, the Euler–Lagrange expression $\mathrm{E}(L)$ must vanish everywhere in Ω, completing the proof. *Q.E.D.*

Let us specialize to the scalar case, when there is one independent and one dependent variable. Here, the Euler–Lagrange equation associated with an nth order Lagrangian $L(x, u^{(n)})$ is the ordinary differential equation

$$\frac{\partial L}{\partial u} - D_x \left(\frac{\partial L}{\partial u_x} \right) + D_x^2 \left(\frac{\partial L}{\partial u_{xx}} \right) - \cdots + (-1)^n D_x^n \left(\frac{\partial L}{\partial u_n} \right) = 0. \qquad (7.7)$$

For example, the Euler–Lagrange equation associated with the classical Newtonian variational problem $\mathcal{L}[u] = \int \left\{ \frac{1}{2} u_x^2 - V(u) \right\} dx$ (which equals kinetic energy minus potential energy) is the second order differential equation $-u_{xx} - V'(u) = 0$ governing motion in a potential force field.

In general, the Euler–Lagrange equation (7.7) associated with an nth order Lagrangian is an ordinary differential equation of order $2n$ provided the Lagrangian satisfies the classical *nondegeneracy condition*

$$\frac{\partial^2 L}{\partial u_n^2} \neq 0. \tag{7.8}$$

Isolated points at which the nondegeneracy condition (7.8) fails constitute singular points of the Euler–Lagrange equation. Note that, in this context, Lagrangians which are affine functions of the highest order derivative, $L(x, u^{(n)}) = A(x, u^{(n-1)}) u_n + B(x, u^{(n-1)})$, are degenerate everywhere. However, a straightforward integration by parts will reduce such a Lagrangian to a nondegenerate one of lower order, and so the exclusion of such Lagrangians is not essential; see Exercise 7.6 below.

Of particular interest are the *null Lagrangians*, which, by definition, are Lagrangians whose Euler–Lagrange expression vanishes identically: $\mathrm{E}(L) \equiv 0$. The associated variational problem is completely trivial, since $\mathcal{L}[u]$ depends only on the boundary values of u, and hence every function provides an extremal.

Theorem 7.3. *A differential function $L(x, u^{(n)})$ defines a null Lagrangian, $\mathrm{E}(L) \equiv 0$, if and only if it is a total divergence, so $L = \operatorname{Div} P = D_1 P_1 + \cdots D_p P_p$, for some p-tuple $P = (P_1, \ldots, P_p)$ of differential functions.*

Proof: Clearly, if $L = \operatorname{Div} P$, the Divergence Theorem implies that the integral (7.1) only depends on the boundary values of u. Therefore, the functional is unaffected by any variations, and so $\mathrm{E}(L) \equiv 0$. Conversely, suppose $L(x, u^{(n)})$ is a null Lagrangian. Consider the expression

$$\frac{d}{dt} L(x, t\, u^{(n)}) = \sum_{\alpha, J} u_J^\alpha \frac{\partial L}{\partial u_J^\alpha} (x, t\, u^{(n)}).$$

Each term in this formula can be integrated by parts, using (7.5) repeatedly. The net result is, as in the proof of Theorem 7.2,

$$\frac{d}{dt} L(x, t\, u^{(n)}) = \sum_{\alpha=1}^{q} u^\alpha \mathrm{E}_\alpha(L)(x, t\, u^{(2n)}) + \operatorname{Div} R(t, x, u^{(2n)}), \tag{7.9}$$

for some well-defined p-tuple of differential functions $R = (R_1, \ldots, R_p)$ depending on L and its derivatives. Since $\mathrm{E}(L) = 0$ by assumption, we can integrate (7.9) with respect to t from $t = 0$ to $t = 1$, producing the desired divergence identity,

$$L(x, u^{(n)}) = L(x, 0) + \operatorname{Div} \widehat{P} = \operatorname{Div} P.$$

Here $\widehat{P}(x, u^{(2n)}) = \int_0^1 P(t, x, u^{(2n)})\, dt$, and $P = P_0 + \widehat{P}$, where $P_0(x)$ is any p-tuple such that $\operatorname{div} P_0 = L(x, 0)$. *Q.E.D.*

Remark: The proof of Theorem 7.3 assumes that $L(x, u^{(n)})$ is defined everywhere on the jet space J^n. A more general result will, as in the deRham theory, depend on the underlying topology of the domain of definition of L, cf. [**6**], [**186**].

Corollary 7.4. *Two Lagrangians define the same Euler–Lagrange expressions if and only if they differ by a divergence:* $\widetilde{L} = L + \operatorname{Div} P$.

Remark: It is possible for two Lagrangians to give rise to the same Euler–Lagrange equations even though they do not differ by a divergence. For instance, both of the scalar variational problems $\int u_x^2\, dx$ and $\int \sqrt{1 + u_x^2}\, dx$ lead to the same Euler–Lagrange equation $u_{xx} = 0$, even though their Euler–Lagrange expressions are not identical. The characterization of such "inequivalent Lagrangians" is a problem of importance in the theory of integrable systems; see [**10**] and the references therein.

Example 7.5. In the case of one independent variable and one dependent variable, every null Lagrangian is a total derivative,

$$L(x, u^{(n)}) = D_x P(x, u^{(n-1)}) = u_n \frac{\partial P}{\partial u_{n-1}} + \widetilde{L}(x, u^{(n-1)}),$$

and hence an affine function of the top order derivative u_n. This is no longer the case for several dependent variables; for example the Hessian covariant

$$H = u_{xx} u_{yy} - u_{xy}^2 = D_x[u_x u_{yy}] - D_y[u_x u_{xy}],$$

is a divergence, and hence a null Lagrangian, even though it is quadratic in the top order derivatives. In fact, it can be shown that any null Lagrangian is a "total Jacobian polynomial" function of the top order derivatives of u, [**179**].

Remark: The problem of characterizing systems of differential equations which are the Euler–Lagrange equations for some variational problem is known as the "inverse problem" of the calculus of variations, and has been studied by many authors. See **[10]**, **[186]** for a discussion of results and the history of this problem.

Exercise 7.6. Suppose $p = 1$. Prove that any nontrivial Lagrangian is equivalent to a nondegenerate Lagrangian (not necessarily of the same order). Can this result be extended to Lagrangians involving several independent variables?

Equivalence of Functionals

The basic equivalence problem in the calculus of variations is to determine when two variational problems can be transformed into each other by a suitable change of variables. In order to properly formulate the equivalence problem, we must specify the type of change of variables which is to be admitted, e.g., fiber-preserving, point, or contact transformations. Let us first consider the "standard equivalence problem" for a first order scalar variational problem

$$\mathcal{L}[u] = \int L(x, u, u_x)\, dx, \qquad x, u \in \mathbb{R}. \tag{7.10}$$

Consider a fiber-preserving transformation g, whose first prolongation is given explicitly by

$$\bar{x} = \chi(x), \quad \bar{u} = \psi(x, u), \quad \bar{p} = \frac{\alpha p + \beta}{\delta}, \quad \begin{array}{c} \alpha = \psi_u, \ \beta = \psi_x, \\ \delta = \chi_x. \end{array} \tag{7.11}$$

A function $u = f(x)$ defined on a domain $\Omega \subset X$ will be mapped under (7.11) to the function $\bar{u} = \bar{f}(\bar{x})$ defined on the transformed domain $\overline{\Omega} = g \cdot \Omega$. We require that the two functionals have the same values, $\mathcal{L}_\Omega[f] = \overline{\mathcal{L}}_{\overline{\Omega}}[\bar{f}]$, or, more explicitly,

$$\int_\Omega L(x, f(x), f'(x))\, dx = \int_{\overline{\Omega}} \overline{L}(\bar{x}, \bar{f}(\bar{x}), \bar{f}'(\bar{x}))\, d\bar{x}. \tag{7.12}$$

Since (7.12) is required to hold for all functions f and subdomains Ω where f is defined, the usual change of variables formula for integrals implies that the associated integrands must satisfy

$$L(x, u, p) = \overline{L}(\bar{x}, \bar{u}, \bar{p})\, \chi_x. \tag{7.13}$$

Thus, two functionals are equivalent under a fiber-preserving transformation if and only if their Lagrangian forms are equivalent under the prolonged transformation: $(g^{(1)})^*(\overline{L}\,d\bar{x}) = L\,dx$.

Extending this notion of equivalence to point or contact transformations is slightly trickier. Consider a prolonged point transformation

$$\bar{x} = \chi(x,u), \quad \bar{u} = \psi(x,u), \quad \bar{p} = \frac{\alpha p + \beta}{\gamma p + \delta}, \quad \begin{matrix} \alpha = \psi_u, & \beta = \psi_x, \\ \gamma = \chi_u, & \delta = \chi_x. \end{matrix} \tag{7.14}$$

The basic equivalence condition (7.12) holds as stated above, although one should note that the transformed domain $\overline{\Omega}$ now depends on both the original domain Ω and the function f itself. Thus, given f, the new independent variables are given by $\bar{x} = \chi(x, f(x))$, and so the factor χ_x in the fiber-preserving transformation rule (7.13) should be replaced by the total derivative $D_x\chi = \chi_x + p\chi_u = \gamma p + \delta$, leading to the basic equivalence condition

$$L(x,u,p) = (\gamma p + \delta)\overline{L}(\bar{x},\bar{u},\bar{p}) = (\gamma p + \delta)\overline{L}\left(\chi(x,u), \psi(x,u), \frac{\alpha p + \beta}{\gamma p + \delta}\right), \tag{7.15}$$

for Lagrangians under point transformations (7.14). Unlike the fiber-preserving case, this does not imply that the associated Lagrangian forms are equivalent, since now $\overline{L}\,d\bar{x}$ will be changed into a linear combination of dx and du, which cannot directly match up with the Lagrangian form $L\,dx$. However, recall the formula

$$d\bar{x} = d\chi = \chi_x\,dx + \chi_u\,du = D_x\chi\,dx + \chi_u(du - p\,dx),$$

which provided the intrinsic definition of the total derivative, cf. Definition 4.25. The two Lagrangians will satisfy the equivalence condition (7.15) if and only if

$$\overline{L}\,d\bar{x} = L\,dx + \mu(du - p\,dx), \tag{7.16}$$

for some function $\mu(x,u,p)$, which, for the point transformation (7.14), equals $\chi_u\overline{L}$. Thus, two functionals are equivalent under a point transformation if and only if the Lagrangian forms are *contact-equivalent*, in the sense of Definition 5.25, under the prolonged transformation.

Under a point transformation, the derivative coordinate p transforms via a linear fractional transformation (albeit with (x,u) dependent coefficients). This implies that there is a remarkable connection between first order scalar Lagrangians and the representation theory of the general linear group GL(2), discussed in Chapter 3. Comparing (7.15) with (3.9), we see that a Lagrangian $L(p)$ which depends only on the derivative coordinate p is a relative invariant for the fundamental multiplier $\mu_{1,0} = (\gamma p + \delta)^{-1}$ associated with the projective action of GL(2). This implies that the equivalence problem for such Lagrangians includes, as a special case, the equivalence problem for binary forms. Given a homogeneous polynomial† $Q(x, y)$ of degree n (and weight 0), its inhomogeneous representative $R(p) = Q(p, 1)$ will transform according to the representation $\rho_{n,0}$ of GL(2). Therefore, the "Lagrangian"

$$L(p) = \sqrt[n]{R(p)} = \sqrt[n]{Q(p, 1)} \tag{7.17}$$

will transform according to $\rho_{1,0}$. In (7.17), we can either treat L as a multiply-valued Lagrangian, or can choose any convenient branch of the nth root, all of which are related by a simple rescaling of L by an nth root of unity. The preceding observation implies that there is a direct connection between the two equivalence problems, a result which will prove to be of profound importance in applying the solution to the equivalence problem for Lagrangians to classical invariant theory.

Proposition 7.7. *Two homogeneous polynomials $Q(x, y)$ and $\overline{Q}(\bar{x}, \bar{y})$ are equivalent under a linear transformation $A \in$ GL(2) if and only if the Lagrangians $L(p) = \sqrt[n]{Q(p, 1)}$ and $\overline{L}(\bar{p}) = \sqrt[n]{\overline{Q}(\bar{p}, 1)}$ are equivalent under a point transformation.*

Remark: One might object that the linear fractional transformation (7.14) can have coefficients which depend on x and u; however, two functions depending only on p which are equivalent under an (x, u)-dependent transformation are automatically equivalent under any constant coefficient transformation given by fixing the coefficients.

Let us now consider more general variational problems. The standard notion of equivalence can be readily extended to multiple and higher order variational problems.

† The x coordinate in Q has nothing to do with the x coordinate in the variational problem.

Definition 7.8. Two functionals are said to be *equivalent* under a contact transformation $(\bar{x}, \bar{u}^{(n)}) = g^{(n)} \cdot (x, u^{(n)})$ if and only if

$$\mathcal{L}_\Omega[u] = \overline{\mathcal{L}}_{\overline{\Omega}}[\bar{u}], \qquad \text{i.e.,} \qquad \int_\Omega L(x, u^{(n)})\, dx = \int_{\overline{\Omega}} \overline{L}(\bar{x}, \bar{u}^{(n)})\, d\bar{x} \tag{7.18}$$

for every smooth function $u = f(x)$, $x \in \Omega$, with transformed counterpart $\bar{u} = \bar{f}(\bar{x})$, $\bar{x} \in \overline{\Omega}$.

Proposition 7.9. *Two n^{th} order variational problems are equivalent under a contact transformation $g^{(n)}\colon \mathrm{J}^n \to \mathrm{J}^n$ if and only if their Lagrangian forms are contact-equivalent:*

$$(g^{(n)})^*(\overline{L}\, d\bar{x}) = L\, dx + \Theta, \tag{7.19}$$

where Θ is a contact form.

In coordinates, this requires that the Lagrangians obey the transformation rule

$$L(x, u^{(n)}) = \overline{L}(\bar{x}, \bar{u}^{(n)}) \det J = \overline{L}(g^{(n)} \cdot (x, u^{(n)})) \det J, \tag{7.20}$$

where $J = J(x, u^{(2)}) = (D_j \chi^i)$ is the total Jacobian matrix of the base transformation $\bar{x} = \chi(x, u^{(1)})$; the determinantal factor comes from the basic pull-back formula for volume forms, cf. Exercise 1.34. Note that, for point transformations, the total Jacobian depends on first order derivatives of u, whereas for contact transformations, J has order 2. Thus, a first order Lagrangian does *not*, in general, map to a first order Lagrangian under a contact transformation. On the other hand, contact transformations do take second order Lagrangians to second order Lagrangians, and so the standard equivalence problem under contact transformations makes sense only for higher order variational problems.

Exercise 7.10. Show how to formulate the equivalence problem for homogeneous polynomials in n variables under the general linear group $\mathrm{GL}(n)$ as a Lagrangian equivalence problem involving one independent variable and $n - 1$ dependent variables.

Invariance of the Euler–Lagrange Equations

As we demonstrated in Theorem 7.2, the extremals of a variational problem satisfy the Euler–Lagrange equations. Since the property of extremizing the problem does not depend on the particular coordinate system, the fact that the Euler–Lagrange equations are invariantly associated with the variational problem under general changes of variables is not surprising. However, the precise formula relating the two Euler–Lagrange expressions is a little more complicated to establish. This result does not follow immediately from the fact that the extremals are solutions, because there can be many other solutions to the Euler–Lagrange equations whose invariance is not so obvious.

Theorem 7.11. *Let $L(x, u^{(n)})$ be an nth order Lagrangian, with Euler–Lagrange expression $\mathrm{E}(L) = (\mathrm{E}_1(L), \ldots, \mathrm{E}_q(L))$. Consider a contact transformation $\bar{x} = \chi(x, u^{(1)})$, $\bar{u} = \psi(x, u^{(1)})$, and let $\bar{L}(\bar{x}, \bar{u}^{(n)})$ be the transformed Lagrangian, as given in (7.20). Then the Euler–Lagrange expression $\overline{\mathrm{E}}(\bar{L})$ for $\bar{L}$ is related to that of L via the formula*

$$\mathrm{E}(L) = \mathrm{F} \cdot \overline{\mathrm{E}}(\bar{L}), \tag{7.21}$$

where $\mathrm{F} = (\mathrm{F}_{\alpha\beta})$ is the $q \times q$ matrix of differential functions

$$\mathrm{F}_{\alpha\beta} = \det \begin{vmatrix} D_1\chi^1 & \cdots & D_p\chi^1 & \partial\chi^1/\partial u^\alpha \\ \vdots & & \vdots & \vdots \\ D_1\chi^p & \cdots & D_p\chi^p & \partial\chi^p/\partial u^\alpha \\ D_1\psi^\beta & \cdots & D_p\psi^\beta & \partial\psi^\beta/\partial u^\alpha \end{vmatrix}, \qquad \alpha, \beta = 1, \ldots, q. \tag{7.22}$$

Example 7.12. Let $x, u \in \mathbb{R}$. Under a point transformation $\bar{x} = \chi(x, u)$, $\bar{u} = \psi(x, u)$, the transformation matrix (7.22) is the scalar differential function

$$\begin{aligned} \mathrm{F} = \det \begin{vmatrix} D_x\chi & \chi_u \\ D_x\psi & \psi_u \end{vmatrix} &= \det \begin{vmatrix} \chi_x + u_x\chi_u & \chi_u \\ \psi_x + u_x\psi_u & \psi_u \end{vmatrix} \\ = \det \begin{vmatrix} \chi_x & \chi_u \\ \psi_x & \psi_u \end{vmatrix} &= \chi_x\psi_u - \psi_x\chi_u. \end{aligned}$$

Thus, if $L(x, u^{(n)})$ is transformed into $\bar{L}(\bar{x}, \bar{u}^{(n)})$, the Euler–Lagrange expressions are related by $\mathrm{E}(\bar{L}) = (\chi_x\psi_u - \psi_x\chi_u)^{-1}\mathrm{E}(L)$. For example, under the hodograph transformation $\bar{x} = u$, $\bar{u} = x$, we have $\mathrm{F} = -1$. If $L = u_x^2$, then $\bar{L} = 1/\bar{u}_{\bar{x}}$. The Euler–Lagrange expressions are $\mathrm{E}(L) = -2u_{xx} = -2\bar{u}_{\bar{x}}^{-3}\bar{u}_{\bar{x}\bar{x}} = -\mathrm{E}(\bar{L})$, verifying (7.21) in this case.

Exercise 7.13. Suppose there are one independent and two dependent variables. Determine the effect of the "hodograph" change of variables $\bar{x} = v$, $\bar{u} = u$, $\bar{v} = x$ on a variational problem. Prove that the Euler–Lagrange expressions for the two Lagrangians are related by the formulae $E_u(L) = v_x E_{\bar{u}}(\bar{L})$, $E_v(L) = -u_x E_{\bar{u}}(\bar{L}) - E_{\bar{v}}(\bar{L})$.

Exercise 7.14. In the scalar case discussed in Example 7.12, we were able to replace the total derivatives in the matrix (7.22) by partial derivatives. Is this permissible in general?

Proof of Theorem 7.11: A direct proof based on the variational formulation of the Euler–Lagrange equation can be found in [**186**; Theorem 4.8]. Here we provide an alternative proof that relies on the contact structure of the jet space. We begin by extending the total derivative operators so that they act on arbitrary differential forms on J^n as Lie derivatives. Specifically, we define

$$D_i(du^\alpha_J) = du^\alpha_{J,i}, \qquad D_i(dx^j) = 0, \tag{7.23}$$

for all i, j, α, J. Thus, in general, the total derivative operator maps differential forms on J^n to differential forms on J^{n+1}. Equation (7.23) allows us to extend the total differential, as in Definition 4.25, using the analogue of (4.43) for forms:

$$\mathrm{D}\omega = \sum_{i=1}^{p} dx^i \wedge D_i\omega; \tag{7.24}$$

see also Exercise 2.89. In particular, the total differential of a contact form is

$$\begin{aligned} \mathrm{D}\theta^\alpha_J &= \mathrm{D}\left(du^\alpha_J - \sum_{i=1}^{p} u^\alpha_{J,i}\,dx^i\right) = \sum_{j=1}^{p} dx^j \wedge D_j\left(du^\alpha_J - \sum_{i=1}^{p} u^\alpha_{J,i}\,dx^i\right) \\ &= \sum_{j=1}^{p} dx^j \wedge \left(du^\alpha_{J,j} - \sum_{i=1}^{p} u^\alpha_{J,i,j}\,dx^i\right) = \sum_{j=1}^{p} dx^j \wedge \theta^\alpha_{J,j}, \end{aligned}$$

which is also a contact form. Consequently, the ordinary and total differentials of the basic contact forms agree up to sign:

$$d\theta^\alpha_J = -\mathrm{D}\theta^\alpha_J. \tag{7.25}$$

This relationship clearly generalizes to arbitrary contact forms.

Lemma 7.15. *If* Θ *is any contact form on* J^n, *then, on* J^{n+1},

$$d\Theta = -\mathrm{D}\Theta + \Upsilon,$$

where Υ *is a "quadratic contact form", i.e., a sum of terms each of which contains the wedge product of two or more contact forms.*

Lemma 7.16. *Let* $L\,dx$ *be a Lagrangian form of degree* n. *Then the Euler–Lagrange expression* $\mathrm{E}(L)$ *is the uniquely defined* p*-tuple of differential functions satisfying*

$$d[L\,dx] = [\mathrm{E}(L)\cdot\theta]\wedge dx + \mathrm{D}\Omega = \sum_{\alpha=1}^{q} \mathrm{E}_\alpha(L)\theta^\alpha \wedge dx^1 \wedge \cdots \wedge dx^p + \mathrm{D}\Omega. \quad (7.26)$$

for some p*-form* Ω *on* J^{2n}.

Proof: We compute

$$d[L\,dx] = \sum_{\alpha=1}^{q} \sum_{\#J=1}^{n} \frac{\partial L}{\partial u_J^\alpha}\, du_J^\alpha \wedge dx = \sum_{\alpha=1}^{q} \sum_{\#J=1}^{n} \frac{\partial L}{\partial u_J^\alpha}\, \theta_J^\alpha \wedge dx$$
$$= \sum_{\alpha=1}^{q} \sum_{\#J=1}^{n} \frac{\partial L}{\partial u_J^\alpha}\, D_J\big[\theta^\alpha \wedge dx\big].$$

A straightforward integration by parts, using the differential form analogue of the basic formula (7.5), completes the demonstration of (7.26). The fact that the Euler–Lagrange expression is uniquely determined by such a formula is demonstrated by noting that a differential form of the form $\sum_\alpha P_\alpha\, \theta^\alpha \wedge dx$ lies in the image of the total differential D if and only if it is identically zero. *Q.E.D.*

Let $g^{(n)}$ denote the nth prolongation of a contact transformation to J^n. Then, according to (7.19),

$$(g^{(2n)})^*\Big[(\overline{\mathrm{E}}(\bar L)\cdot\bar\theta)\wedge d\bar x + \mathrm{D}\overline{\Omega}\Big] = (g^{(n)})^*[d(\bar L\,d\bar x)] = d\{(g^{(n)})^*[\bar L\,d\bar x]\}$$
$$= d[L\,dx + \Theta] = (\mathrm{E}(L)\cdot\theta)\wedge dx + \mathrm{D}\Omega + d\Theta,$$

where $\overline{\Omega}$ and Ω are p-forms, and Θ is a contact form. Lemma 7.15 and the invariance of the total differential imply that

$$(g^{(2n)})^*\Big[(\overline{\mathrm{E}}(\bar L)\cdot\bar\theta)\wedge d\bar x\Big] = (\mathrm{E}(L)\cdot\theta)\wedge dx + \mathrm{D}\widehat{\Omega} + \Upsilon,$$

where Υ is a quadratic contact form. The local coordinate formula for the behavior of the Euler–Lagrange expressions under the change of variables follows immediately from the formulae (4.52), (4.53), showing how the contact forms behave under a contact transformation. The remaining details are left to the reader. *Q.E.D.*

Corollary 7.17. *If two variational problems are equivalent, then their associated Euler–Lagrange equations are equivalent.*

The converse to Corollary 7.17 is not necessarily true. For instance, one can always add in a null Lagrangian without affecting the Euler–Lagrange equation. On the other hand, if we are primarily interested in transformations which preserve the solutions to the Euler–Lagrange equations, then this remark motivates the following relaxation of our equivalence condition.

Definition 7.18. Two Lagrangians $\bar{L}(\bar{x}, \bar{u}^{(n)})$ and $L(x, u^{(n)})$ are said to be *divergence equivalent* if there is a contact transformation $g^{(n)}$ such that

$$(g^{(n)})^*[\bar{L}\, d\bar{x}] = L\, dx + \mathrm{D}\Omega + \Theta, \tag{7.27}$$

for some p-form Ω and some contact form Θ. In terms of local coordinates, this is equivalent to the requirement that

$$L(x, u^{(n)}) = \bar{L}(\bar{x}, \bar{u}^{(n)}) \det J + \operatorname{Div} P, \quad \text{when} \quad (\bar{x}, \bar{u}^{(n)}) = g^{(n)}(x, u^{(n)}), \tag{7.28}$$

for some p-tuple of differential functions $P(x, u^{(n)})$.

For example, let $\bar{L}(\bar{x}, \bar{u}, \bar{p})$ and $L(x, u, p)$ be first order Lagrangians in one independent and one dependent variable. Under a point transformation $\bar{x} = \chi(x, u)$, $\bar{u} = \psi(x, u)$ the divergence equivalence condition (7.27) takes the explicit form

$$\bar{L}(\bar{x}, \bar{u}, \bar{p}) D_x \chi = L(x, u, p) + D_x A, \tag{7.29}$$

where $D_x A = A_x + p A_u$ is the total derivative (divergence) of A. Note that, in the case of point transformations, the function A cannot depend on the derivative coordinate p since otherwise the transformed Lagrangian would be of second order. However, if we are considering contact transformations, so $\bar{x} = \chi(x, u, p)$ depends on p also, we can allow $A(x, u, p)$ to also depend on p and thereby eliminate the second order term $q\chi_p \overline{L}$, so as to map the first order Lagrangian $\overline{L}$ to a divergence equivalent first order Lagrangian L. In fact, in Chapter 11, we shall prove that any two nondegenerate first order Lagrangians are locally divergence equivalent under a contact transformation.

Exercise 7.19. Prove that the change of variables formula (7.21) for the Euler–Lagrange equations holds without change for divergence equivalent Lagrangians. Consequently, divergence equivalent Lagrangians have equivalent Euler–Lagrange equations.

Symmetries of Variational Problems

Maps that preserve variational problems serve to define variational symmetries. The precise definition is as follows.

Definition 7.20. A point or contact transformation g is called a *variational symmetry* of the functional (7.1) if and only if the transformed functional agrees with the original one, which means that for every smooth function f defined on a domain Ω, with transformed counterpart $\bar{f} = g \cdot f$ defined on $\overline{\Omega}$, we have

$$\int_\Omega L(x, f^{(n)}(x))\, dx = \int_{\overline{\Omega}} L(\bar{x}, \bar{f}^{(n)}(\bar{x}))\, d\bar{x}. \tag{7.30}$$

In other words, a variational symmetry is merely a standard self-equivalence of a functional. Setting $\overline{L} = L$ in the equivalence condition (7.19), we see that a transformation group G is a variational symmetry group if and only if the Lagrangian form $L(x, u^{(n)})\, dx$ is a contact-invariant p-form, so

$$(g^{(n)})^* \big[L(\bar{x}, \bar{u}^{(n)})\, d\bar{x} \big] = L(x, u^{(n)})\, dx + \Theta, \qquad g \in G, \tag{7.31}$$

for some contact form $\Theta = \Theta_g$, possibly depending on the group element g. In particular, if the group transformation g is fiber-preserving, then $\Theta = 0$, and the Lagrangian form is strictly invariant. In local coordinates, the contact invariance condition (7.31) takes the form

$$L(x, u^{(n)}) = L(\bar{x}, \bar{u}^{(n)}) \det J, \qquad \text{when} \qquad (\bar{x}, \bar{u}^{(n)}) = g^{(n)} \cdot (x, u^{(n)}), \tag{7.32}$$

where $J = (D_i \chi^j)$ is the total Jacobian matrix, cf. (7.20). In other words, contact invariance of the p-form $L\, dx$ is the same as the condition that the Lagrangian L defines a relative G-invariant of weight $\det J$. Since, according to Theorem 7.11, the Euler–Lagrange equations are correspondingly transformed under an equivalence map, we immediately deduce the following useful result.

Theorem 7.21. *Every variational symmetry group of a variational problem is a symmetry group of the associated Euler–Lagrange equations.*

Note, though, that the converse to Theorem 7.21 is not true. The most common examples of symmetries which fail to be variational are those generating groups of scaling transformations. For example, the variational problem $\int \frac{1}{2} u_x^2 \, dx$ has Euler–Lagrange equation $u_{xx} = 0$, admitting the two-parameter scaling symmetry group $(x, u) \mapsto (\lambda x, \mu u)$. However, this group does not leave the variational problem invariant, but, rather scales it by the factor μ^2/λ. Note, however, that the one-parameter subgroup $(x, u) \mapsto (\lambda^2 x, \lambda u)$ *is* a variational symmetry group. Therefore, to determine variational symmetries, one can proceed by first determining the complete symmetry group of the Euler–Lagrange equations using Lie's method, and then analyzing which of the usual symmetries satisfy the additional criterion of being variational. See also [**186**; Proposition 5.55] for an alternative approach based directly on the infinitesimal invariance of the Euler–Lagrange equations.

In the case of a connected transformation group, we let g in (7.31) belong to a one-parameter subgroup, and differentiate to retrieve the basic infinitesimal invariance criterion for variational symmetry groups. This requires that the Lie derivative of the Lagrangian p-form with respect to the prolonged vector field $\mathbf{v}^{(n)}$ be a contact form. (See also (5.24).) In particular, as in Example 2.92, the horizontal component of the Lie derivative of the volume form with respect to $\mathbf{v}^{(n)}$ is

$$\mathbf{v}^{(n)}(dx^1 \wedge \cdots \wedge dx^p) = (\operatorname{Div} \xi) \, dx^1 \wedge \cdots \wedge dx^p, \tag{7.33}$$

where $\operatorname{Div} \xi = D_1 \xi^1 + \cdots D_p \xi^p$ denotes the total divergence of the base coefficients of $\mathbf{v}$. This suffices to prove the basic invariance criterion.

Theorem 7.22. *A connected transformation group G is a variational symmetry group of the Lagrangian $L(x, u^{(n)})$ if and only if the infinitesimal variational symmetry condition*

$$\mathbf{v}^{(n)}(L) + L \operatorname{Div} \xi = 0, \tag{7.34}$$

holds for every infinitesimal generator $\mathbf{v} \in \mathfrak{g}$.

Note that $\operatorname{Div} \xi$ is the infinitesimal multiplier associated with the total Jacobian multiplier representation, of weight $\det J$, on the space of differential functions, so that (7.34) is just the infinitesimal condition that L define a relative invariant thereof.

Example 7.23. The Boussinesq equation (6.22) is not the Euler–Lagrange equation for any variational problem. However, replacing u by u_{xx}, we form the "potential Boussinesq equation"

$$u_{xxtt} + \tfrac{1}{2}D_x^2(u_{xx}^2) + u_{xxxxxx} = 0, \tag{7.35}$$

which is the Euler–Lagrange equation for the variational problem

$$\mathcal{L}[u] = \int\!\!\int \left\{ \tfrac{1}{2}u_{xt}^2 + \tfrac{1}{6}u_{xx}^3 - \tfrac{1}{2}u_{xxx}^2 \right\} dx \wedge dt. \tag{7.36}$$

The symmetry group of the potential form (7.35) is spanned by the translation and scaling vector fields

$$\mathbf{v}_1 = \partial_x, \qquad \mathbf{v}_2 = \partial_t, \qquad \mathbf{v}_3 = x\partial_x + 2t\partial_t, \tag{7.37}$$

and the two infinite families of vector fields

$$\mathbf{v}_f = f(t)\partial_u, \qquad \mathbf{v}_h = h(t)x\partial_u, \tag{7.38}$$

where $f(t)$ and $h(t)$ are arbitrary functions of t; the corresponding group action $u \mapsto u + f(t) + h(t)x$ indicates the ambiguity in our choice of potential. (Compare with the symmetry group of the usual form of the Boussinesq equation found in Example 6.14.) The most general variational symmetry is found by substituting a general symmetry vector field $\mathbf{v} = c_1\mathbf{v}_1 + c_2\mathbf{v}_2 + c_3\mathbf{v}_3 + \mathbf{v}_f + \mathbf{v}_h$ into the infinitesimal criterion (7.34), which requires that

$$\tfrac{1}{4}c_3(-3u_{xt}^2 + u_{xx}^3 - 3u_{xxx}^2) + h'(t)u_{xt} = 0.$$

Therefore $c_3 = 0$ and h is constant, hence the two translations, and the group $u \mapsto u + cx + f(t)$, with c constant, are variational, whereas the scalings and the more general fields $\mathbf{v}_h$, h not constant, are not ordinary variational symmetries, but do define divergence symmetries, in the following sense.

Definition 7.24. A vector field $\mathbf{v}$ is a *divergence symmetry* of a variational problem with Lagrangian L if and only if it satisfies

$$\mathbf{v}^{(n)}(L) + L\,\mathrm{Div}\,\xi = \mathrm{Div}\,B, \tag{7.39}$$

for some p-tuple of functions $B = (B_1, \ldots, B_p)$.

A divergence symmetry is a divergence self-equivalence of the Lagrangian form, so that (7.31) holds modulo an exact p-form $d\,\Xi$. The divergence symmetry groups form the most general class of symmetries related to conservation laws. Indeed, Noether's Theorem provides a one-to-one correspondence between generalized divergence symmetries of a variational problem and conservation laws of the associated Euler-Lagrange equations; see [**186**] for a detailed discussion.

Exercise 7.25. Prove that every divergence symmetry of a variational problem is an (ordinary) symmetry of its Euler–Lagrange equations.

Exercise 7.26. Show that the Korteweg–deVries equation $u_t = u_{xxx} + uu_x$ can be placed into variational form through the introduction of a potential $u = v_x$. Determine the conservation laws associated with the point symmetries found in Exercise 6.16.

Invariant Variational Problems

As with differential equations, the most general variational problem admitting a given symmetry group can be readily characterized using the differential invariants of the prolonged group action. The key additional requirement is the existence of a suitable contact-invariant p-form, where p is the number of independent variables. The following theorem is a straightforward consequence of the infinitesimal variational symmetry criterion (7.34), and dates back to Lie, [**154**].

Theorem 7.27. *Let G be a transformation group, and assume that the nth prolongation of G acts regularly on (an open subset of) J^n. Assume further that there exists a nonzero contact-invariant horizontal p-form $\Omega_0 = L_0(x, u^{(n)})\, dx$ on J^n. A variational problem admits G as a variational symmetry group if and only if it can be written in the form $\int I\,\Omega_0 = \int I L_0\, dx$, where I is a differential invariant of G.*

In particular, any contact-invariant coframe $\omega^1, \ldots, \omega^p$ provides a contact-invariant p-form $\Omega_0 = \omega^1 \wedge \cdots \wedge \omega^p$. Hence every G-invariant variational problem has the form

$$\mathcal{L}[u] = \int L(x, u^{(n)})\, dx = \int F\big(I_1(x, u^{(n)}), \ldots, I_k(x, u^{(n)})\big)\, \omega^1 \wedge \cdots \wedge \omega^p, \tag{7.40}$$

where $I_1, \ldots, I_k$ are a complete set of functionally independent n^{th} order differential invariants. Theorem 5.41 implies that there exists a contact-invariant coframe of order n or less, where n denotes the stabilization order of G. I do not know useful conditions guaranteeing the existence of a lower order contact-invariant p-form, although they do occur fairly frequently.

Example 7.28. Consider the rotation group $\mathrm{SO}(2)$ acting on $E \simeq \mathbb{R} \times \mathbb{R}$, cf. Example 4.2. Since the radius $r = \sqrt{x^2 + u^2}$ is an invariant, its total differential provides a contact-invariant one-form, which we slightly modify: $\omega = (x + uu_x)\, dx$. Therefore, according to Theorem 7.27, any first order variational problem admitting the rotation group as a variational symmetry group has the form

$$\int F(r, w)(x + uu_x)\, dx = \int (x + uu_x) F\left(\sqrt{x^2 + u^2}\,, \frac{xu_x - u}{x + uu_x}\right) dx.$$

According to Theorem 7.21, the Euler–Lagrange equation for such a variational problem is a second order ordinary differential equation admitting $\mathrm{SO}(2)$ as a symmetry group, and hence of the form (6.32). In polar coordinates, the variational problem becomes $\frac{1}{2} \int F(r, r\theta_r) r\, dr$, with Euler–Lagrange equation $D_r\left[\frac{1}{2} r^2 F(r, r\theta_r)\right] = 0$. The latter can be immediately integrated once, leading to $r^2 F(r, r\theta_r) = c$, an equation defining θ_r implicitly as a function of r, and thereby soluble by quadrature. This fact is, as we shall see, no accident.

Example 7.29. Consider the rotation group $\mathrm{SO}(2)$ acting on $E \simeq \mathbb{R}^2 \times \mathbb{R}$ by rotating the independent variables. The differential invariants were computed in Example 5.50. The area form $dx \wedge dy$ is invariant, hence any first order variational problem admitting $\mathrm{SO}(2)$ as a variational symmetry group has the form

$$\mathcal{L}[u] = \int F\left(\sqrt{x^2 + y^2}\,, u, -yu_x + xu_y, xu_x + yu_y\right) dx \wedge dy.$$

Note that the one-forms $\omega^1 = x\, dx + y\, dy$, $\omega^2 = -y\, dx + x\, dy$ form a contact-invariant coframe, so $\omega^1 \wedge \omega^2 = (x^2 + y^2) dx \wedge dy$ provides an alternative contact-invariant two-form which, in view of Theorem 7.27, is an invariant multiple of the area form. Again, the Euler–Lagrange equation is rotationally invariant. For example, the Dirichlet variational problem has Lagrangian

$$u_x^2 + u_y^2 = \frac{(-yu_x + xu_y)^2 + (xu_x + yu_y)^2}{x^2 + y^2},$$

and is rotationally invariant; its Euler–Lagrange equation is the Laplace equation.

Remark: If G is a given transformation group, then, as we have seen, any G-invariant variational problem, and its associated Euler–Lagrange equations, can both be written in terms of the differential invariants of G. However, I do not know a general formula for calculating the invariant formulation of the Euler–Lagrange equations directly from the invariant formula for the Lagrangian. Some interesting special cases have, however, been examined by I. Anderson, [**7**].

Symmetry Classification of Variational Problems

The classification of the differential invariants and invariant one-forms for all finite-dimensional groups of point and contact transformations acting on the plane described in Table 5 allows us to classify the possible symmetry groups of scalar variational problems, depending upon just one independent and one dependent variable. We will be particularly interested in those variational problems admitting symmetry groups of maximal dimension. Note that the dimension of the point transformation symmetry group of a nondegenerate variational problem, of order at least one, is finite because it must be a subgroup of the symmetry group of the associated Euler–Lagrange equation, which was observed to be finite-dimensional in Theorem 6.38.

Theorem 7.30. *The variational symmetry group admitted by a nonsingular first order Lagrangian has dimension at most three. Moreover, any first order Lagrangian having a three-dimensional variational symmetry group is equivalent, under a complex point transformation, to a constant multiple of one of the following three Lagrangians:*

$$u_x^\alpha, \qquad \sqrt{u_x - u^2}, \qquad e^{-u_x}. \tag{7.41}$$

The symmetry groups of these inequivalent Lagrangians are, respectively, Cases 1.7 (with $k = 1$), 1.2, and 1.8 of the classification tables.

The constant multiple can be absorbed in the first and third cases by a suitable change of variables; however, it is essential in the second case — see Chapter 10. The first two types of Lagrangians admit solvable symmetry groups, whereas the symmetry group of the third is isomorphic to the simple Lie group SL(2). Not every three-dimensional Lie group can be realized as a symmetry group for a nonsingular first order variational problem — in particular, a first order Lagrangian never has an abelian three-parameter symmetry group of fiber-preserving transformations.

Theorem 7.31. *For $n \geq 2$, a nonsingular n^{th} order Lagrangian admits a variational symmetry group of point transformations of dimension at most $n+3$. Such a Lagrangian admits a variational symmetry group of dimension $n+3$ if and only if it is equivalent, under a complex point transformation, to a constant multiple of one of the following:*

$$u_{xx}^{1/3}, \qquad \frac{\sqrt{2u_x u_{xxx} - 3u_{xx}^2}}{u_x}, \qquad u_n^{2/(n+1)}, \quad n \geq 2, \tag{7.42}$$
$$\frac{\left(9u_{xx}^2 u_{xxxxx} - 45u_{xx}u_{xxx}u_{xxxx} + 40u_{xxx}^3\right)^{1/3}}{u_{xx}}.$$

The symmetry group of $L = u_n^{2/(n+1)}$ is of type 1.10. The symmetry groups of the three "anomalous" Lagrangians, of orders 2, 3, 5, are the special affine group SA(2), which is Case 2.1, the product $\mathrm{SL}(2) \times \mathrm{SL}(2)$ of two projective groups, Case 1.4, and the full planar projective group SL(3), Case 2.3, each of which has fundamental geometric significance. As noted in Chapter 5, in each case, the associated Lagrangian one-form can be identified with the group-invariant arc length element. Theorem 7.31 implies that the invariant arc length functionals of these three distinguished groups — the special affine, projective product, and full projective groups — play a distinguished role among scalar variational problems. Remarkably, the simple quadratic Lagrangian $L = u_n^2$, which has linear Euler–Lagrange equation $u_{2n} = 0$, has only an $(n+2)$-dimensional symmetry group, also of type 1.7, spanned by

$$\partial_x, \quad x\partial_x + (n - \tfrac{1}{2})u\partial_u, \quad \partial_u, \quad x\partial_u, \quad \ldots, \quad x^{n-1}\partial_u.$$

Thus, quadratic Lagrangians do *not* have the most symmetry, providing an explicit counterexample to the "meta-theorem" that linear objects are the ones with the highest degree of symmetry. It should be remarked, however, that for divergence symmetries, the quadratic Lagrangians regain their role as those having the maximal symmetry group, which is of dimension $2n+3$. See González–López, [**93**], for further details, including the corresponding theorem for real-valued variational problems. The analogous theorem for contact transformation groups relies on Table 4.

Theorem 7.32. *For $n \geq 2$, a nonsingular n^{th} order Lagrangian admits a variational symmetry group of contact transformations of dimension at most $n+3$. The symmetry group has maximal dimension if and only if the Lagrangian is equivalent, under a complex contact transformation, to a constant multiple of one of either the maximally symmetric Lagrangians in* (7.42), *or the following two Lagrangians:*

$$u_{xxx}^{1/3}, \qquad \frac{\left(10u_3^3u_7 - 70u_3^2u_4u_6 - 49u_3^2u_5^2 + 280u_3u_4^2u_5 - 175u_4^4\right)^{1/4}}{u_3},$$

whose symmetry groups are of type 4.1, 4.3, respectively.

Exercise 7.33. Prove that the Lagrangians $u_{xx}^{1/3}$ and $u_{xx}^{2/3}$ are equivalent under a contact transformation.

Exercise 7.34. Determine the Lagrangians which have submaximal symmetry groups.

First Integrals

For systems of "conservative" ordinary differential equations — meaning systems of Euler–Lagrange equations — the power of the Lie's symmetry method for reducing the order is effectively doubled. This is due to a fundamental result of E. Noether, [**178**], that relates symmetries and first integrals of Euler–Lagrange equations.

Definition 7.35. For a system of ordinary differential equations $\Delta(x, u^{(n)}) = 0$, a *first integral* is a function $P(x, u^{(m)})$ which is constant on solutions.

Physical examples of first integrals include the usual conservation laws of linear and angular momentum and energy. The fact that P is constant is equivalent to the statement that its total derivative vanishes on solutions, so $D_xP = 0$ whenever $u = f(x)$ is a solution to $\Delta = 0$. In the regular case, this requires that P satisfy an identity of the form $D_xP = \sum_i Q_i D_x^i \Delta$ for some differential functions Q_i.

Theorem 7.36. *If* $\mathbf{v}$ *is an infinitesimal variational symmetry with characteristic* Q*, then the product* $Q\,\mathrm{E}(L) = D_x P$ *is a total derivative, and thus a first integral of the Euler–Lagrange equation* $\mathrm{E}(L) = 0$.

Proof: If $\mathbf{v}$ defines a variational symmetry, then according to (7.34) and (4.32),

$$0 = \mathbf{v}^{(n)}(L) + L\, D_x \xi = \mathbf{v}_Q^{(n)}(L) + D_x(L\xi) = \sum_{i=0}^{n} (D_x^i Q) \cdot \frac{\partial L}{\partial u_i} + D_x(L\xi).$$

Now, applying the basic integration by parts formula (7.5) repeatedly, we find

$$Q \cdot \mathrm{E}(L) - D_x P = Q \cdot \left[\sum_{i=0}^{n} (-D_x)^i \left(\frac{\partial L}{\partial u_i} \right) \right] - D_x P = 0$$

for some function P depending on Q, L, and their derivatives. Thus $D_x P$ is a multiple of the Euler–Lagrange equation, which suffices to prove the result. *Q.E.D.*

Corollary 7.37. *If a variational problem admits a one-parameter variational symmetry group, then its Euler–Lagrange equation can be reduced in order by 2.*

Proof: In terms of the rectifying coordinates y, v introduced in the proof of Theorem 6.19, the infinitesimal generator $\mathbf{v} = \partial_v$ is a variational symmetry if and only if the Lagrangian $L(x, v_y, v_{yy}, \ldots)$ is independent of v. Therefore, the Euler–Lagrange equations have the form $D_y P(x, v_y, v_{yy}, \ldots, v_{n-1}) = 0$, which can be integrated once. If $w = v_y$, then the reduced ordinary differential equation $P(y, w^{(n-2)}) = c$, for c constant, forms the Euler–Lagrange equation of order $n-2$ for the "reduced" Lagrangian $L(y, w, w_y, \ldots) - cw$. *Q.E.D.*

Thus, the fact that we could integrate the rotationally invariant Euler–Lagrange equation in Example 7.28 was no accident. For multi-dimensional groups, the reduced variational problem will not, in general, admit the original variational symmetries *unless* they commute with the reducing symmetry. Therefore, only abelian r-dimensional symmetry groups will yield complete reductions of the Euler–Lagrange equation by order $2r$. This fact is closely connected with the reduction theory for

Hamiltonian systems, as described by Marsden and Weinstein, [**165**]; see also [**186**]. Noether's Theorem is considerably more general than the ordinary differential equation version of Theorem 7.36. In full generality, it prescribes a one-to-one correspondence between generalized variational symmetries and conservation laws of the Euler–Lagrange equations. Unfortunately, space precludes us from going into any more details here, and we refer the interested reader to [**186**] for the complete picture, including numerous applications.

The Cartan Form

As we have seen, if we restrict our attention to fiber-preserving transformation groups, then the Lagrangian form associated with an invariant variational problem defines an invariant differential form on the jet space. This result does not extend to more general point or contact transformation groups owing to the presence of contact forms in the general change of variables formula (7.19). We now describe, in the simplest case of a first order Lagrangian in one independent variable and one dependent variable, an important modification of the Lagrangian form that does remain invariant under general point transformations. This one-form, commonly known as the Cartan form, plays a significant role in the geometric theory of the calculus of variations, including the connection between symmetries and conservation laws. Hilbert's invariant integral, which is of central importance in the classical field theory governing the existence of extremals, cf. [**85**], is found by integrating the Cartan form along a solution to the Euler–Lagrange equation. The utility of the Cartan form was emphasized by Cartan, [**33**], and Goldschmidt and Sternberg, [**86**]. See [**98**] for more details on the history and development of this important variational concept.

Definition 7.38. Let $L(x,u,p)$, $p = u_x$, be a first order Lagrangian in one independent variable and one dependent variable. The *Cartan form* associated with L is the one-form

$$\eta_L = L\,dx + \frac{\partial L}{\partial p}\,\theta = L\,dx + \frac{\partial L}{\partial p}\,(du - p\,dx). \tag{7.43}$$

Theorem 7.39. *If* $\bar{L}$ *and* L *are two Lagrangians which are mapped to each other by a point transformation* g, *then the two Cartan forms are equivalent:* $(g^{(1)})^*(\bar{\eta}_{\bar{L}}) = \eta_L$.

Proof: First note that if we differentiate the basic formula (7.15) with respect to p, we find, using the notation of equation (7.14),

$$\frac{\partial L}{\partial p} = \frac{\alpha\delta - \beta\gamma}{\gamma p + \delta}\,\frac{\partial \overline{L}}{\partial \overline{p}} + \gamma \overline{L}. \tag{7.44}$$

On the other hand, according to (4.53)

$$(g^{(1)})^*\,\overline{\theta} = \left(\frac{\partial \psi}{\partial u} - \overline{p}\,\frac{\partial \chi}{\partial u}\right)\theta = \frac{\alpha\delta - \beta\gamma}{\gamma p + \delta}\,\theta. \tag{7.45}$$

The invariance of the Cartan form is now a simple consequence of (7.16), (7.44) and (7.45):

$$\begin{aligned}(g^{(1)})^*(\overline{\eta}_{\overline{L}}) &= (g^{(1)})^*(\overline{L}\,d\overline{x} + \overline{L}_{\overline{p}}\theta)\\ &= L\,dx + \gamma \overline{L}\,\theta + (g^{(1)})^*(\overline{L}_{\overline{p}}\theta) = L\,dx + L_p\theta = \eta_L.\end{aligned}$$

This completes the proof of the theorem. *Q.E.D.*

An alternative proof of the invariance of the Cartan form, based on the Cartan equivalence method, will be presented in Chapter 10.

Exercise 7.40. Prove that the Cartan forms associated with two divergence equivalent first order Lagrangians are equivalent modulo an exact one-form, meaning that

$$(g^{(1)})^*(\overline{\eta}_{\overline{L}}) = \eta_L + dR, \tag{7.46}$$

for some first order differential function $R(x,u,p)$. Prove that this result also extends to the divergence equivalence of first order Lagrangians under contact transformations; see the discussion following (7.29).

Corollary 7.41. *If G is a variational symmetry group of the Lagrangian $L(x,u,p)$, then the associated Cartan form η_L is a G-invariant one-form on* J^1.

The Cartan form is closely connected with the Euler–Lagrange equation. This observation is based on the following formula for its differential:

$$\begin{aligned} d\eta_L &= (L_u - L_{px} - pL_{pu})\,du\wedge dx + L_{pp}\,dp\wedge\theta\\ &= \mathrm{E}(L)\,\theta\wedge dx + L_{pp}\,\theta_x\wedge\theta,\end{aligned} \tag{7.47}$$

where $\theta_x = dp - q\,dx$ is the next contact form on J^2.

Proposition 7.42. *A parametrized curve $\phi(t)$ contained in J^2 will define the graph of the second prolongation of a solution to the Euler–Lagrange equation if and only if it is transverse and its tangent vector satisfies $\phi'(t) \lrcorner\, d\eta_L = 0$.*

(See equation (2.32) for the definition of the interior product $\lrcorner$.) In the language of Chapter 14, $\phi(t)$ defines a characteristic curve for the differential system generated by the Cartan form η_L.

As a consequence of Corollary 7.41 and (2.33), we find that if $\mathbf{v}$ is any variational symmetry, then

$$0 = \mathbf{v}^{(1)}(\eta_L) = d\langle \mathbf{v}^{(1)} ; \eta_L \rangle + \mathbf{v}^{(1)} \lrcorner\, d\eta_L .$$

Equation (7.47) shows that the second term vanishes on solutions to the Euler–Lagrange equation, and hence the first term defines a first integral. Thus our earlier version of Noether's Theorem relating symmetry groups of variational problems and first integrals of the associated Euler–Lagrange equations has the following alternative formulation.

Proposition 7.43. *If $\mathbf{v}$ is any variational symmetry of the variational problem with Lagrangian $L(x, u, p)$ and Cartan form η_L, then the function $\langle \mathbf{v}^{(1)} ; \eta_L \rangle$ is a first integral of the Euler–Lagrange equations.*

The extension of the Cartan form to multi-dimensional variational problems is not so straightforward. There is a commonly accepted definition of a Cartan form for higher order Lagrangians in one independent and several dependent variables, cf. [**118**]. However, for variational problems involving multiple integrals, there is a good deal of controversy surrounding the definition of the Cartan form, cf. [**98**]. We shall return to this question in Chapter 10, and produce the invariant Cartan form for a general first order variational problem — see (10.110). Invariant Cartan forms for second order variational problems are found in [**187**]. However, there is still no proper definition of a Cartan form for a third or higher order multi-dimensional Lagrangian!

Invariant Contact Forms and Evolution Equations

Recent applications to solitons and computer vision have required the classification of evolution equations which are invariant under a prescribed group action. This task was begun by Sokolov, [**205**], in a study of symmetry and integrability. The classification has received added impetus from recent work, done in collaboration with G. Sapiro and A. Tannenbaum, [**201**], [**189**], [**190**], on applications to image processing based on geometric diffusion. Suppose we have a transformation group G acting on a space $E \simeq X \times U$ coordinatized by independent variables x and dependent variables u. We consider a system of evolution equations

$$u_t = K(x, u^{(n)}), \tag{7.48}$$

in which t is an additional independent variable (the time), and the right hand side depends only on the spatial derivatives of u. The group action of G is extended to the dynamical space $\mathbb{R} \times E$ by letting it act trivially on the time variable $t \in \mathbb{R}$. There is an intimate connection between invariant evolution equations and invariant contact forms. To begin with, we consider the scalar case when there is just one dependent variable $u \in \mathbb{R}$. Our presentation relies on the theory of relative invariants for multiplier representations developed in Chapter 3.

Proposition 7.44. *Let G act on $E \simeq \mathbb{R} \times \mathbb{R}$. An evolution equation* (7.48) *admits G as a symmetry group if and only if the contact form*

$$\frac{\theta}{K} = \frac{1}{K(x, u^{(n)})} \left\{ du - \sum_{i=1}^{p} u_i \, dx^i \right\}, \tag{7.49}$$

is a G-invariant one-form on J^n. (Note: *We require strict invariance of the form under the prolonged group transformations, not mere contact-invariance.*)

Proof: Let $\mathbf{v}$ be any infinitesimal generator, with characteristic $Q(x, u^{(1)})$. On the one hand, according to (4.62), $\mathbf{v}^{(1)}(\theta) = Q_u \, \theta$, so that the contact-form θ plays the role of a relative G-invariant corresponding to the infinitesimal multiplier $\sigma(\mathbf{v}) = Q_u$; see Theorem 3.31. The contact form θ / K will therefore be G-invariant if and only if K is also a relative invariant of weight σ. On the other hand, since G acts trivially on the time variable t, the prolongation formula (4.59) implies that $\mathbf{v}^{(1)}(u_t) = Q_u u_t$, so that u_t is also a relative invariant of weight σ. Thus, (7.48) will be G-invariant if and only if K is also a relative invariant of weight σ. *Q.E.D.*

Remark: According to Exercise 5.31, if n denotes the stabilization order of G, then there exists an invariant contact form (7.49), and hence an invariant evolution equation (7.48), where $K(x, u^{(n)})$ has order (at most) n.

Theorem 7.45. *Let G be as above. Suppose $L(x, u^{(n)})$ is a G-invariant Lagrangian such that $\mathrm{E}(L) \neq 0$. Then every G-invariant evolution equation has the form*

$$u_t = \frac{L}{\mathrm{E}(L)}\, I, \tag{7.50}$$

where I is an arbitrary differential invariant of G.

Proof: In view of Proposition 7.44, this result is equivalent to the statement that $L\,dx$ defines a contact-invariant p-form if and only if $(\mathrm{E}(L)/L)\theta$ is an invariant contact form.† We begin by formulating the infinitesimal analog of the change of variables formula (7.21) for the Euler-Lagrange expression.

Lemma 7.46. *Let $\mathcal{L}[u] = \int L\,dx$ be a variational problem. Then*

$$\mathrm{E}[\mathrm{pr}\,\mathbf{v}(L) + L\,\mathrm{Div}\,\xi] = \mathrm{pr}\,\mathbf{v}(\mathrm{E}[L]) + (Q_u + \mathrm{Div}\,\xi)\mathrm{E}[L]. \tag{7.51}$$

Proof: Formula (7.51) can be deduced by differentiating the change of variables formula formula (7.21), for a one parameter group of contact transformations, with respect to the group parameter. Alternatively, a direct proof proceeds as follows: Let $u(x, \varepsilon)$ be a family of variations of the function $u(x)$. Let $u(x, \varepsilon, t) = \exp(t\mathbf{v})u(x, \varepsilon)$ be the corresponding transformed functions. The fact that the variations have compact support implies that, for t sufficiently small, the family $u(x, t, \varepsilon)$ also satisfies the relevant boundary conditions. As we shall see, (7.51) is just a statement of the equality of mixed partials of $\mathcal{L}[u(x, \varepsilon, t)]$ with respect to ε and t. The Taylor series for $u(x, \varepsilon, t)$ in ε and t is

$$u(x, \varepsilon, t) = u(x) + \varepsilon v(x) + tQ(x, u(x)) + \varepsilon t Q_u(x, u(x))v(x) + \cdots. \tag{7.52}$$

Therefore,

$$\left.\frac{\partial}{\partial \varepsilon}\mathcal{L}[u(x, \varepsilon, t)]\right|_{\varepsilon=0} = \int \mathrm{E}(L)[u(x, t)] \cdot v(x, t)dx$$

† Readers familiar with the variational bicomplex, cf. [**6**], [**7**], can thus justify Theorem 7.45 in another, more direct manner, pointed out to me by I. Anderson and J. Pohjanpelto.

where $u(x,t) = u(x,0,t)$, $v(x,t) = u_\varepsilon(x,0,t)$. Note that, by the preceding expansion (7.52), $v(x,t) = v(x) + tQ_u(x,u(x))v(x) + \cdots$. Consequently, using (7.33), we find

$$\frac{\partial^2}{\partial\varepsilon\partial t}\mathcal{L}[u(x,\varepsilon,t)]\bigg|_{\varepsilon=t=0} = \int \{\operatorname{pr}\mathbf{v}(\mathrm{E}(L)) + (Q_u + \operatorname{Div}\xi)\mathrm{E}(L)\}\, v\, dx. \tag{7.53}$$

On the other hand, if we first differentiate with respect to t, we find

$$\frac{\partial}{\partial t}\mathcal{L}[u(x,\varepsilon,t)]\bigg|_{t=0} = \int \{\operatorname{pr}\mathbf{v}(L) + L\operatorname{Div}\xi\} dx,$$

the integrand being evaluated at $u(x,\varepsilon) = u(x,\varepsilon,0)$. Therefore, using the original definition of the Euler operator,

$$\frac{\partial^2}{\partial\varepsilon\partial t}\mathcal{L}[u(x,\varepsilon,t)]\bigg|_{\varepsilon=t=0} = \int \mathrm{E}[\operatorname{pr}\mathbf{v}(L) + L\operatorname{Div}\xi]\, v\, dx. \tag{7.54}$$

Since the two integrals (7.53), (7.54) must agree for arbitrary variations v, we conclude the validity of the identity (7.51). *Q.E.D.*

Returning to the proof of Theorem 7.45, if the p-form $P\,dx$ is G-invariant, so $\operatorname{pr}\mathbf{v}(P) + P\operatorname{Div}\xi = 0$, then (7.51) implies the identity

$$\operatorname{pr}\mathbf{v}[\mathrm{E}(P)] + (\operatorname{Div}\xi + Q_u)\mathrm{E}(P) = 0.$$

Therefore, $\mathrm{E}(P)$ is a relative invariant of weight $-\operatorname{Div}\xi - Q_u$. On the other hand, P itself is a relative invariant of weight $-\operatorname{Div}\xi$. Therefore, the ratio $\mathrm{E}(P)/P$ is a relative invariant of weight $-Q_u$. We now invoke the proof of Proposition 7.44 to complete our argument. *Q.E.D.*

Remark: Theorem 7.45 can be extended to several dependent variables. If there are q dependent variables, we require q distinct G-invariant Lagrangians $L_1, \ldots, L_q$ with the property that their "Euler–Lagrange matrix" $\mathbf{E} = \big(\mathrm{E}_\alpha(L_\beta)\big)$ is invertible. The most general G-invariant evolution equation then has the form

$$u_t = L_1\,\mathbf{E}^{-1}\,\mathbf{I}, \tag{7.55}$$

where $\mathbf{I} = (I_1, \ldots, I_q)^T$ is a column vector of differential invariants. (Note that each $L_\beta = I_\beta L_1$ for some differential invariant I_β, so L_1 in (7.55) can be replaced by any other L_β by suitably modifying the invariant vector $\mathbf{I}$.)

Exercise 7.47. Suppose G is a group of point transformations on $E \simeq \mathbb{R}^p \times \mathbb{R}$ that preserves the $(p+1)$-form $dx^1 \wedge \cdots \wedge dx^p \wedge du$. (In other words, G is a group of volume-preserving transformations on E.) Prove that if $L(x, u^{(n)})$ is a G-invariant Lagrangian, then $\mathrm{E}(L)$ is a differential invariant, and hence $u_t = L$ is an invariant evolution equation; see **[190]**.

The evolution equation (7.50) provides the most general invariant evolution equation; however, choosing I constant does not necessarily yield the simplest one. For symmetry groups of importance in image processing, there is an alternative way of characterizing invariant evolution equations.

Theorem 7.48. *Let G be a subgroup of the projective group* $\mathrm{SL}(3)$. *Let $ds = P\,dx$ denote the contact-invariant one-form of lowest order (the G-invariant arc length element) and κ its fundamental differential invariant (the G-invariant curvature). Then every G-invariant evolution equation has the form*

$$u_t = \frac{u_{xx}}{P^2}\, I, \tag{7.56}$$

where $I = F\left(\kappa, \dfrac{d\kappa}{ds}, \dfrac{d^2\kappa}{ds^2}, \ldots, \dfrac{d^n\kappa}{ds^n}\right)$ *is an arbitrary differential invariant.*

Proof: The proof relies on the fact that the projective group $\mathrm{SL}(3)$ is the symmetry group of the differential equation $u_{xx} = 0$. The calculations in Example 6.6 imply that if $\mathbf{v} = \xi(x,u)\partial_x + \varphi(x,u)\partial_u$ is an infinitesimal generator of $G \subset \mathrm{SL}(3)$, then it satisfies

$$\mathbf{v}^{(2)}[u_{xx}] = (\varphi_u - 2\xi_x - 3\xi_u u_x)u_{xx} = (Q_u - 2D_x\xi)u_{xx},$$

i.e., the terms in $\mathbf{v}^{(2)}[u_{xx}]$ which do not depend on u_{xx} must add up to zero. Thus, in this case, u_{xx} is a relative invariant of weight $Q_u - 2D_x\xi$. Since $ds = P\,dx$ is a contact-invariant one-form, P is a relative invariant of weight $-D_x\xi$. Therefore the ratio u_{xx}/P^2 is a relative invariant of weight Q_u, as is u_t, and hence we have produced the desired invariant evolution equation. *Q.E.D.*

Note that, as a corollary of Theorems 7.45 and 7.48 we find that, for subgroups of the projective group, the Euler–Lagrange expression associated with any G-invariant Lagrangian, including the G-invariant arc-length functional, has the form $\mathrm{E}(L) = JL^3/u_{xx}$ for some differential invariant J. For the similarity (Euclidean plus scaling), special affine, affine, and full projective groups, (7.56) with I constant is distinguished as the *unique* G-invariant evolution equation of *lowest* order. For the Euclidean group, the simplest nontrivial invariant evolution equation is given by $u_t = c\sqrt{1+u_x^2}$ since, in this case, the curvature invariant κ has order 2, so we can take $J = c/\kappa$; see Exercise 7.47. In the Euclidean case, the evolution equation (7.56) defines the fundamental *curve shortening flow*, in which ones moves a curve defined by the graph of a function $u = f(x)$ in its normal direction by an amount proportional to the curvature. This flow, and its higher dimensional counterparts, are of immense interest in modern geometry; see particularly the foundational work of Gage and Hamilton, [**76**], and Grayson, [**101**]. The special affine version of curve shortening, in which one moves in proportion to the affine-invariant curvature, is also a second order diffusion equation: $u_t = u_{xx}^{1/3}$. Geometrically, up to reparametrization, the curve $u = f(x)$ moves in its Euclidean normal direction in proportion to $\kappa^{1/3}$, where κ is the Euclidean curvature. See [**201**], [**189**], [**190**] for applications of the affine-invariant flows to image processing.

Exercise 7.49. Generalize Theorem 7.48 to subgroups of the ten parameter contact group SO(5), which is Case 4.3 in the Tables. See [**161**] for the classification of contact symmetries of scalar evolution equations.

Chapter 8

Equivalence of Coframes

By definition, a coframe on a manifold is a "complete" collection of one-forms in the sense that, at each point, it provides a basis for the cotangent space. Two coframes are said to be equivalent if they are mapped to each other by a diffeomorphism. The equivalence problem for coframes is, in fact, the most important of the equivalence problems that we are to treat, because it ultimately includes all the others as special cases. Indeed, the remarkable and powerful Cartan equivalence method, [**39**], [**78**], [**230**], provides an *explicit*, *practical* algorithm for reducing most other equivalence problems to an equivalence problem for a suitable coframe. (The exceptions are those problems admitting an infinite dimensional symmetry group, which will be handled by similar, but more sophisticated methods to be discussed later on.) Therefore, it is crucial that we learn how to deal properly with this apparently special, but, in reality, quite general equivalence problem. In this chapter, we describe the basic constructions required to solve the fundamental equivalence problem for coframes. The solution will include a complete list of fundamental invariants associated with the coframe, and consequential necessary and sufficient conditions for equivalence.

Frames and Coframes

We begin by presenting the basic definitions. Let M be a smooth manifold of dimension m. If $x = (x^1, \ldots, x^m)$ are local coordinates on M, then the translational vector fields $\partial/\partial x^1, \ldots, \partial/\partial x^m$ provide a basis for the tangent space $TM|_x$ at each point x in the coordinate chart; these vector fields are known as the "coordinate frame" associated with the local coordinate system. Similarly, the differentials $dx^1, \ldots, dx^m$ of the coordinate functions provide the dual basis for the cotangent space $T^*M|_x$ at each point x, known as the corresponding "coordinate coframe". Every vector field on M can be uniquely written as a linear combination (with variable coefficients) of the coordinate frame vectors; similarly, every differential one-form on M can be uniquely written as a linear combination of the coordinate coframe elements. Thus the coordinate frame and coframe form the basic building blocks for computations with vector fields and one-forms. In many problems, though, these are not the most convenient bases to consider, and we introduce the general concepts of frame and coframe which will prove to be of such essential use in the subsequent development of the Cartan equivalence method.

Definition 8.1. Let M be a smooth m-dimensional manifold. A *frame* on M is an ordered set of vector fields $\mathcal{V} = \{\mathbf{v}_1, \ldots, \mathbf{v}_m\}$ having the property that they form a basis for the tangent space $TM|_x$ at each point $x \in M$. Dually, a *coframe* on M is an ordered set of one-forms $\boldsymbol{\theta} = \{\theta^1, \ldots, \theta^m\}$ which form a basis for the cotangent space $T^*M|_x$ at each point $x \in M$.

A set of one-forms $\boldsymbol{\theta} = \{\theta^1, \ldots, \theta^m\}$ defines a coframe on M if and only if their wedge product $\theta^1 \wedge \cdots \wedge \theta^m \neq 0$ does not vanish. Therefore, a coframe provides an orientation on the manifold M, and a trivialization of the cotangent bundle $T^*M \simeq M \times \mathbb{R}^m$, whence the alternative term *absolute parallelism*, [**41**]. Consequently, there *are* global obstructions to the existence of coframes (and frames) on a manifold, although, since all our considerations are local, this will not concern us.

A frame and the coframe are *dual* to each other if and only if they form dual bases for the tangent and cotangent spaces to M at each point. Given a coframe $\{\theta^1, \ldots, \theta^m\}$, we shall denote the dual frame vector fields by $\{\partial/\partial\theta^1, \ldots, \partial/\partial\theta^m\}$, so that

$$\left\langle \theta^i \,;\, \frac{\partial}{\partial\theta^j} \right\rangle = \delta^i_j, \qquad i, j = 1, \ldots, m. \tag{8.1}$$

If we introduce local coordinates $x = (x^1, \ldots, x^m)$ on M, then the coframe can be written in terms of the coordinate coframe, so that $\theta^i = \sum_j a^i_j(x) dx^j$, where $A(x) = \big(a^i_j(x)\big)$ is a nonsingular $m \times m$ matrix of functions. The dual frame is given by the vector fields $\partial/\partial\theta^j = \sum_i b^i_j(x)\partial_{x^i}$, where $B(x) = (b^i_j(x)) = A(x)^{-1}$. The frame vector fields play the role of "directional derivatives" in the "directions" specified by the coframe. Indeed, the differential of a function can be re-expressed in the coframe-adapted form

$$dF = \sum_{j=1}^{m} \frac{\partial F}{\partial \theta^j}\, \theta^j. \tag{8.2}$$

We will refer to the resulting coefficients $\partial F/\partial \theta^j = \sum_i b^i_j(x)\partial F/\partial x^i$ as the *coframe derivatives* of the function F. An important remark is that the coframe derivatives $\partial/\partial\theta^j$, being quite general vector fields, do not necessarily commute — see (8.10) below.

If $\boldsymbol{\theta} = \{\theta^1, \ldots, \theta^m\}$ is a coframe, then any other one-form can be written as a linear combination $\omega = h_1\theta^1 + \ldots + h_m\theta^m$, where the coefficients $h_i = h_i(x)$ are uniquely defined smooth functions. Even more, any differential k-form Ω can be written as a linear combination of k-fold exterior products of the elements of the coframe:

$$\Omega = \sum_{I} h_I(x)\, \theta^{i_1} \wedge \cdots \wedge \theta^{i_k}. \tag{8.3}$$

The coefficient functions $h_I(x)$ are unique provided we sum over strictly increasing multi-indices: $I = (i_1, \ldots, i_k)$, $1 \leq i_1 < i_2 < \ldots < i_k \leq m$.

Example 8.2. A particularly important example of a coframe is the Maurer–Cartan coframe on a Lie group G. Let $\{\mathbf{v}_1, \ldots, \mathbf{v}_r\}$ be a basis for the Lie algebra $\mathfrak{g}$ of G, so each $\mathbf{v}_i$ is a right-invariant† vector field on the group G. At each $g \in G$, the tangent vectors $\mathbf{v}_1|_g, \ldots, \mathbf{v}_r|_g$ are linearly independent, and hence $\{\mathbf{v}_1, \ldots, \mathbf{v}_r\}$ forms a (right-invariant) frame on G. The corresponding right-invariant Maurer–Cartan forms $\{\alpha^1, \ldots, \alpha^r\}$, which are just the dual basis elements for the dual space $\mathfrak{g}^*$, form the dual *Maurer–Cartan coframe* on G. Note that the Maurer–Cartan coframe does depend on a choice of basis $\mathbf{v}_i = \partial/\partial\alpha^i$ for $\mathfrak{g}$. The existence of the Lie algebra frame and Maurer–Cartan coframe implies that the tangent bundle, and the cotangent bundle, of any Lie group are topologically trivial, so we can naturally identify them with the Cartesian products $TG \simeq G \times \mathfrak{g}$ and $T^*G \simeq G \times \mathfrak{g}^*$. Since the Maurer–Cartan coframe plays a key role in our analysis, the reader should review its analytical construction presented in formula (2.35) and the subsequent discussion.

† One can, of course, also use the left-invariant vector fields and forms. However, all our applications will rely on right invariance, and so we shall restrict our attention to these from here on.

Example 8.3. Frames and coframes naturally arise in the study of quasi-linear systems of first order evolution equations, which are used to model a wide variety of physical systems, including gas dynamics, nonlinear elasticity, and shallow water theory, cf. [**200**]. Consider a system of partial differential equations of the form

$$u_t = A(u)\, u_x. \tag{8.4}$$

Here $u(x,t) = (u^1(x,t),\ldots,u^q(x,t))^T$, and $A(u) = \bigl(A^\alpha_\beta(u)\bigr)$ is a $q \times q$ matrix depending on u alone. The system is called *simple* if A has q distinct (complex) eigenvalues $\lambda_1(u),\ldots,\lambda_q(u)$, known as the *characteristic speeds* of the system, and *hyperbolic* if the eigenvalues (characteristic speeds) are all real. If A is diagonalizable, then the associated eigenvectors $\{\mathbf{v}_1,\ldots,\mathbf{v}_q\}$, which satisfy $A\cdot\mathbf{v}_\kappa = \lambda_\kappa \mathbf{v}_\kappa$, form a frame on the manifold coordinatized by the u's. The dual coframe $\boldsymbol{\omega} = \{\omega^1,\ldots,\omega^q\}$ consists of one-forms, which we can identify with the corresponding *left* eigenvectors of the matrix A, so $\omega^\kappa \cdot A = \lambda_\kappa \omega^\kappa$. Since we can multiply the eigenvectors by nonvanishing functions, the "eigen-frame" and "eigen-coframe" are not intrinsically associated with the system (8.4), although the individual directions indicated by the frame and coframe elements are intrinsic. The system (8.4) is called *diagonalizable* if there is a change of dependent variables $v = \Phi(u)$ placing it in diagonal form

$$\frac{\partial v^\alpha}{\partial t} = \widetilde{\lambda}_\alpha(v)\,\frac{\partial v^\alpha}{\partial x}, \qquad \alpha = 1,\ldots,q. \tag{8.5}$$

Note that, under a change of dependent variable, $\widetilde{\lambda}_\alpha(v) = \widetilde{\lambda}_\alpha(\Phi(u)) = \lambda_\alpha(u)$, i.e., the eigenvalues are invariant functions for the system. Diagonalizability of the system is stronger than mere (pointwise) diagonalizability of the matrix A, since the diagonalizing transformation must coincide with the Jacobian matrix of the change of variables Φ. The diagonalizing coordinates $v = (v^1,\ldots,v^q)$ are referred to as a complete system of *Riemann invariants* for the system; their existence or lack thereof is of great importance in the analysis as well as physical applications. In terms of the Riemann invariants, the basic frame and dual coframe are just given by their coordinate versions $\mathbf{v}_\alpha = \partial/\partial v^\alpha$, and $\omega^\alpha = dv^\alpha$. Since the pull-back of the coordinate change respects the coframe directions, we conclude that $\Phi^*(dv^\alpha) = d\Phi^\alpha(u)$ is a left eigenvector of the original matrix $A(u)$, and hence, in the simple case, equals a scalar multiple of the original coframe element: $\omega^\alpha = \kappa_\alpha(u) d\Phi^\alpha(u)$. Therefore, a complete system of Riemann invariants that diagonalizes (8.4) exists if and only if the coframe elements ω^α are one-forms having Darboux rank 1 or 2, as per Definition 1.41. We conclude that the conditions $\omega^\alpha \wedge d\omega^\alpha = 0$, $\alpha = 1, \ldots, q$, are necessary and sufficient for the existence of a complete system of Riemann invariants.

Exercise 8.4. Prove that a hyperbolic system involving one or two dependent variables is always diagonalizable. Find a three-dimensional system which cannot be diagonalized, i.e., has no Riemann invariants.

The eigen-coframe is not the only coframe of importance for the geometry of a hyperbolic system. The *symmetry coframe* of a simple system (8.4) is defined in terms of the eigen-frame and coframe by

$$\theta^\alpha = \sum_{\substack{\beta=1 \\ \beta\neq\alpha}}^{q} \frac{\mathbf{v}_\beta(\lambda_\alpha)}{\lambda_\alpha - \lambda_\beta}\, \omega^\beta, \qquad \alpha = 1, \ldots, q. \tag{8.6}$$

By definition, the one-forms $\theta^1, \ldots, \theta^q$ form a coframe if and only if the system is "genuinely nonlinear". It is not hard to prove that the symmetry coframe is intrinsically associated with the simple system (8.4) under a change of dependent variables. Moreover, it can be shown that two diagonalizable hyperbolic systems define symmetries of each other (meaning that the flows commute) if and only if they can be simultaneously diagonalized and their symmetry coframes are identical. The solution to the equivalence problem for coframes will therefore provide a complete classification of the symmetry classes of diagonalizable hyperbolic systems. See Doyle, [**63**], [**64**], for details and further applications to the Hamiltonian structure of such systems.

The Structure Functions

Suppose we are given two coframes, $\boldsymbol{\theta} = \{\theta^1, \ldots, \theta^m\}$ on a manifold M, and $\overline{\boldsymbol{\theta}} = \{\overline{\theta}^1, \ldots, \overline{\theta}^m\}$ on a manifold $\overline{M}$ of the *same* dimension. In the sequel, we shall consistently use plain (unbarred) letters to refer to objects on M and their barred counterparts to refer to objects on $\overline{M}$. In many instances $M = \overline{M}$ are the same manifold, but it will be useful to maintain the distinction in this case, too. The basic equivalence problem for coframes is to determine whether the two coframes can be mapped to each other by a diffeomorphism $\Phi: M \to \overline{M}$, so that

$$\Phi^* \,\overline{\theta}^i = \theta^i, \qquad i = 1, \ldots, m. \tag{8.7}$$

In our treatment, we shall only look at *local equivalence*, meaning that the diffeomorphism Φ is only required to be defined on a suitable open subset of M — see the end of Chapter 14 for some remarks on the global equivalence problem. Cartan made the fundamental observation that the invariance of the exterior derivative operator d under smooth maps, cf. Theorem 1.36, is the key to the solution of the coframe equivalence problem. Thus, if (8.7) holds, we also necessarily have

$$\Phi^* \, d\overline{\theta}^i = d\theta^i, \qquad i = 1, \ldots, m. \tag{8.8}$$

The solution to the equivalence problem for coframes lies in the detailed analysis of the differentiated conditions (8.8).

According to (8.3), since the θ's form a coframe, we can rewrite the two-forms $d\theta^i$ in terms of wedge products of the θ's. This produces the fundamental *structure equations*

$$d\theta^i = \sum_{\substack{j,k=1 \\ j<k}}^{m} T^i_{jk}\, \theta^j \wedge \theta^k, \qquad i = 1, \ldots, m, \tag{8.9}$$

associated with the given coframe. The $\frac{1}{2}m^2(m-1)$ *structure functions* $T^i_{jk} = T^i_{jk}(x)$ are uniquely defined (smooth) functions, many of which may be constant or even 0. We note that the structure functions measure the "degree of non-commutativity" of the corresponding coframe derivatives, since we have the dual Lie bracket formulae

$$\left[\frac{\partial}{\partial \theta^j}, \frac{\partial}{\partial \theta^k} \right] = -\sum_{i=1}^{m} T^i_{jk} \frac{\partial}{\partial \theta^i}\,. \tag{8.10}$$

(Note the change in sign!) Formulas (8.9) and (8.10) are equivalent owing to the formula (1.25) relating differentials and Lie brackets, coupled with the original definition (8.1) of dual frame and coframe. For example, if $\boldsymbol{\theta}$ is the Maurer–Cartan coframe for a Lie group G relative to some choice of basis for the Lie algebra $\mathfrak{g}$, then the structure equations (8.9) coincide with the Maurer–Cartan structure equations (2.34). Thus, in this case, the structure functions agree (up to sign) with the structure constants for $\mathfrak{g}$ relative to the chosen basis: $T^i_{jk} = -C^i_{jk}$. The commutation formula (8.10) therefore reduces to the standard bracket formula (2.13) for the Lie algebra of G.

Any equivalent coframe $\overline{\boldsymbol{\theta}}$ will have analogous structure equations

$$d\overline{\theta}^i = \sum_{\substack{j,k=1 \\ j<k}}^{m} \overline{T}^i_{jk}\,\overline{\theta}^j \wedge \overline{\theta}^k, \qquad i = 1, \ldots, m, \tag{8.11}$$

on $\overline{M}$, and so has its own set of structure functions $\overline{T}^i_{jk}(\bar{x})$. Now, substituting the two structure equations (8.9), (8.11) into the invariance condition (8.8), and recalling the invariance of the coframe elements themselves, (8.7), we find

$$\sum_{\substack{j,k=1 \\ j<k}}^{m} T^i_{jk}(x)\,\theta^j \wedge \theta^k = d\theta^i = \Phi^* d\overline{\theta}^i = \sum_{\substack{j,k=1 \\ j<k}}^{m} \overline{T}^i_{jk}(\Phi(x))\,\theta^j \wedge \theta^k.$$

Since the θ^i are linearly independent one-forms, the coefficients of each $\theta^j \wedge \theta^k$ in this equation must agree. This immediately implies the invariance of the structure functions:

$$\overline{T}^i_{jk}(\bar{x}) = T^i_{jk}(x), \quad \text{when} \quad \bar{x} = \Phi(x), \quad i,j,k = 1, \ldots, m, \ \ j < k. \tag{8.12}$$

Therefore, given a coframe, we readily deduce the existence of intrinsically associated scalar invariants provided by the structure functions T^i_{jk}. Moreover, the invariance conditions (8.12) provide an important set of necessary conditions for the two coframes. For instance, if a structure function T^i_{jk} for the coframe $\boldsymbol{\theta}$ is constant, the corresponding structure function $\overline{T}^i_{jk}$ (i.e., the one with the same indices) for any equivalent coframe $\overline{\boldsymbol{\theta}}$ must necessarily assume the same constant value. On the other hand, if a structure function is not constant, then it is of less direct use, since its local coordinate formula is *not* invariant. Just because an invariant $I = T^i_{jk}$ has the formula $I = x^2 + y$ in one coordinate system, and $\overline{I} = \tan(\bar{x} - \bar{y})$ in another, this says nothing one way or the other as to whether the two coframes are equivalent. As we discuss in detail below, the key is not the invariants themselves, but rather their functional interrelationships.

Example 8.5. It is instructive to look at the simplest nontrivial case in some detail. A coframe on $M \subset \mathbb{R}^2$ consists of a pair of one-forms

$$\theta^1 = A\,dx + B\,dy, \qquad \theta^2 = C\,dx + D\,dy, \tag{8.13}$$

where A, B, C, D are smooth functions of x, y, satisfying $AD - BC \neq 0$, so that $\theta^1 \wedge \theta^2 = (AD - BC)\,dx \wedge dy \neq 0$. The structure equations of the coframe (8.13) are readily computed:

$$d\theta^1 = J\,\theta^1 \wedge \theta^2, \qquad d\theta^2 = K\,\theta^1 \wedge \theta^2, \tag{8.14}$$

where the two structure functions are explicitly given by

$$T^1_{12} = J = \frac{B_x - A_y}{AD - BC}, \qquad T^2_{12} = K = \frac{D_x - C_y}{AD - BC}. \tag{8.15}$$

Any change of variables $\bar{x} = \xi(x,y)$, $\bar{y} = \eta(x,y)$ must preserve both invariants, meaning that $\bar{J}(\bar{x},\bar{y}) = J(x,y)$, $\bar{K}(\bar{x},\bar{y}) = K(x,y)$, a fact that can be verified by a tedious direct chain rule computation. In particular, the two invariants vanish, $J = K = 0$, if and only if both coframe forms are (locally) exact: $\theta^1 = df$, $\theta^2 = dg$ for functions f, g. Clearly, any equivalent coframe must also consist of a pair of locally exact forms, and hence have vanishing invariants $\bar{J} = \bar{K} = 0$. More generally, if J and K are constant, then so must be $\bar{J}$ and $\bar{K}$; later we will see that this necessary condition is also sufficient to conclude that the two coframes are equivalent.

Exercise 8.6. Determine the symmetry coframe (8.6) associated with a two-dimensional hyperbolic system (8.4), where $u \in \mathbb{R}^2$, and determine the associated invariants of the system.

Derived Invariants

The structure functions associated with a coframe thus provide us with quite a few invariant functions — $\frac{1}{2}m^2(m-1)$ to be precise — but that's not all! Suppose $I(x)$ is any scalar invariant, which is mapped to a corresponding invariant $\bar{I}(\bar{x})$ under the given change of variables: $\bar{I}(\bar{x}) = \bar{I}(\Phi(x)) = I(x)$. Then the differentials dI and $d\bar{I}$ must also agree: $\Phi^* d\bar{I} = dI$. On the other hand, we can re-express dI and $d\bar{I}$ in terms of the respective coframes; using the definition (8.2) of the coframe derivatives and the equivalence conditions (8.7), we find

$$\sum_{j=1}^{p} \frac{\partial I}{\partial \theta^j}(x)\,\theta^j = dI(x) = \Phi^* d\bar{I}(\bar{x}) = \sum_{j=1}^{p} \frac{\partial \bar{I}}{\partial \bar{\theta}^j}(\Phi(x))\,\theta^j. \tag{8.16}$$

Again, since the one-forms θ^j are pointwise linearly independent, we can conclude that the coframe derivatives of any invariant function are themselves invariant functions:

$$\frac{\partial \bar{I}}{\partial \bar{\theta}^j}(\bar{x}) = \frac{\partial I}{\partial \theta^j}(x), \qquad \text{when} \qquad \bar{x} = \Phi(x), \qquad j = 1, \ldots, m. \tag{8.17}$$

Clearly we can repeat this process, starting with any invariant I to produce an infinite collection of potentially different invariant functions by taking higher and higher order coframe derivatives: I, $\partial I/\partial \theta^j$, $\partial^2 I/\partial \theta^j \partial \theta^k$, etc., known as the *derived invariants* associated with the original invariant I. Note that, since the coframe derivatives do not, in general, commute, we need to fix our notational convention for the higher order derived invariants, which will be

$$\frac{\partial^2 I}{\partial \theta^j \partial \theta^k} = \frac{\partial}{\partial \theta^j}\left(\frac{\partial I}{\partial \theta^k}\right). \tag{8.18}$$

Example 8.7. Consider the general coframe (8.13) on $M \subset \mathbb{R}^2$. The corresponding coframe derivatives are

$$\frac{\partial F}{\partial \theta^1} = \frac{D}{\Delta}\frac{\partial F}{\partial x} - \frac{C}{\Delta}\frac{\partial F}{\partial y}, \qquad \frac{\partial F}{\partial \theta^2} = -\frac{B}{\Delta}\frac{\partial F}{\partial x} + \frac{A}{\Delta}\frac{\partial F}{\partial y}, \tag{8.19}$$

where $\Delta = AD - BC$. Thus, starting with the basic invariants J and K given by (8.15), we can construct an infinite series $\partial J/\partial \theta^1$, $\partial J/\partial \theta^2$, $\partial K/\partial \theta^1$, $\partial K/\partial \theta^2$, $\partial^2 J/(\partial \theta^1)^2$, $\partial^2 J/\partial \theta^1 \partial \theta^2$, etc., of progressively more and more complicated invariants, depending on higher and higher derivatives of the original coefficients A, B, C, D. As with J and K, each of these complicated differential functions is invariant under the general change of variables $\bar{x} = \xi(x,y)$, $\bar{y} = \eta(x,y)$. Thus, if one of the derived invariants happens to be constant in one coordinate system, then it must also necessarily achieve the same constant value in any other coordinate system — see Example 8.12 below.

Not all of the derived invariants associated with a given invariant are independent. First, the Lie bracket identities (8.10) allow us to permute coframe derivatives, at the expense of introducing lower order invariants:

$$\frac{\partial^2 I}{\partial \theta^k \partial \theta^j} - \frac{\partial^2 I}{\partial \theta^j \partial \theta^k} = \sum_{i=1}^{m} T^i_{jk} \frac{\partial I}{\partial \theta^i}. \tag{8.20}$$

Additional identities can be found by writing out the Jacobi identity (1.12) for the frame vector fields $\partial/\partial\theta^i$ using the commutation relations (8.10) (and keeping in mind that the coefficients T^i_{jk} are not necessarily constant). These reduce to the general *Jacobi identities* for the structure functions of a coframe

$$\frac{\partial T^i_{jk}}{\partial \theta^l} + \frac{\partial T^i_{kl}}{\partial \theta^j} + \frac{\partial T^i_{lj}}{\partial \theta^k} = \sum_{n=1}^{m} \left[T^i_{jn} T^n_{kl} + T^i_{kn} T^n_{lj} + T^i_{ln} T^n_{jk} \right], \qquad 1 \le i, j, k, l \le m. \tag{8.21}$$

Here $T^i_{kj} = -T^i_{jk}$, and $T^i_{jj} = 0$, by convention. For example, in the case of a Maurer–Cartan coframe, the structure functions are all constant, and so the general coframe Jacobi identities (8.21) reduce to the usual Lie algebra Jacobi identities (2.14) for the structure constants.

Exercise 8.8. Prove the Jacobi identities (8.21) directly by taking the exterior derivative of the structure equations (8.9).

Classifying Functions

Now we are confronted with an overabundance of riches — every coframe gives us an infinite collection of invariants (even taking the commutator and Jacobi identities into account), consisting of the structure functions and all their coframe derivatives (of arbitrarily high order). How do we make effective use of this plethora of invariants? As we remarked above, the invariants themselves, unless constant, are of very little direct use in solving the equivalence problem. Much more fundamental are the functional relationships among the invariants. Suppose, to take a simple example, that $I_1(x), I_2(x), I_3(x)$ are three invariants in the plain coordinates, and $\bar{I}_1(\bar{x}), \bar{I}_2(\bar{x}), \bar{I}_3(\bar{x})$ are the corresponding invariants in the barred coordinates. If the original invariants are functionally dependent, say $I_3 = H(I_1, I_2)$ for some function $H(z_1, z_2)$, then, by invariance, the barred invariants must satisfy the *same* functional relationship, $\bar{I}_3 = H(\bar{I}_1, \bar{I}_2)$ for the *same* function H; thus the function H is an invariant of the problem which *is completely independent of the coordinates*! We will call the collection of all such functional relationships (including the constant ones) the *classifying functions* associated with the given coframe. For instance, if $I_3 = I_1^2 + 3I_2$, then we also need $\bar{I}_3 = \bar{I}_1^2 + 3\bar{I}_2$, and, in this case, the associated classifying function is $H(z_1, z_2) = z_1^2 + 3z_2$. Clearly, if two coframes are to be equivalent, their classifying functions must all be identical.

Thus, starting with the coframe, we can construct a large collection of invariants, whose functional interrelationships provide a large number of necessary conditions for equivalence. The crucial question is whether these necessary conditions are also sufficient, and, on a more practical note, how many of these conditions must we in fact verify in order to be assured of the equivalence of the two coframes. The complete answer is based on Cartan's theory of exterior differential systems. Before stating the general theorem, though, we need to be a bit more precise on the nature of the classifying functions. In fact, in themselves, they do not provide the complete story, since they ignore common complications such as domains of definition and multiple-valuedness, both of which are important. In order to set up the theory on a firm and practical foundation, we describe a more efficient means of keeping track of these issues via the introduction of the "classifying manifold" associated with a coframe, an object which incorporates all the functional relationships among the structure invariants.

Let us continue by introducing a convenient notation

$$T_\sigma = \frac{\partial^s T^i_{jk}}{\partial \theta^{l_s} \partial \theta^{l_{s-1}} \cdots \partial \theta^{l_1}}, \qquad \text{where} \qquad \sigma = (i, j, k, l_1, \ldots, l_s), \tag{8.22}$$

for the most general *structure invariant* associated with the coframe, cf. (8.18). In (8.22), the indices i, j, k, l_κ all run from 1 to m, with $j < k$. The integer $s = \operatorname{order} \sigma$ is the *order* of the derived invariant (8.22). According to our calculations, if two coframes are equivalent, then the structure functions and all their coframe derivatives must agree, so that the required diffeomorphism $\bar{x} = \Phi(x)$ must satisfy the complete system of *invariance equations*

$$\bar{T}_\sigma(\bar{x}) = T_\sigma(x), \qquad \text{when} \qquad \bar{x} = \Phi(x), \qquad \operatorname{order} \sigma \geq 0. \tag{8.23}$$

As we shall see, the algebraic solvability of the system of invariance equations is, in the "regular" case, both necessary and sufficient for the equivalence of the two coframes. The structure invariants serve to define the components of the general structure map.

Definition 8.9. The sth order *classifying space* $\mathbb{K}^{(s)} = \mathbb{K}^{(s)}(m)$ associated with an m-dimensional manifold M is the Euclidean space of dimension $q_s(m) = \frac{1}{2}m^2(m-1)\binom{m+s}{m}$, which is coordinatized by $z^{(s)} = (\ldots, z_\sigma, \ldots)$. The entries of $z^{(s)}$ are labeled by nondecreasing multi-indices $\sigma = (i, j, k, l_1, \ldots, l_r)$, $j < k$, $1 \leq l_1 \leq l_2 \leq \cdots \leq l_r \leq m$, $0 \leq r \leq s$. The sth order *structure map* associated with a coframe $\boldsymbol{\theta}$ on M is the map $\mathbf{T}^{(s)}: M \to \mathbb{K}^{(s)}$ whose components are the structure invariants: $z_\sigma = T_\sigma(x)$, for order $\sigma \leq s$.

In formulating this definition,† we have noted that not all of the equations in (8.23) are independent. In particular, the commutation identities (8.10) allow us to permute coframe derivatives (at the expense of introducing lower order derived invariants), so we need only consider the $q_s(m)$ derived invariants corresponding to nondecreasing multi-indices. The Jacobi identities (8.21), which must also automatically hold, could be used to further reduce the collection of "basic" structure invariants; however, to avoid extra complications involved in keeping track of these in a consistent manner, we shall ignore them for the time being.

If $\boldsymbol{\theta}$ and $\overline{\boldsymbol{\theta}}$ are equivalent coframes, the invariance equations (8.23) must be satisfied, which implies that, for each $s = 0, 1, 2, \ldots$, the corresponding structure maps have the same image:

$$\overline{\mathbf{T}}^{(s)}(\bar{x}) = \mathbf{T}^{(s)}(x), \qquad \text{where} \qquad \bar{x} = \Phi(x). \tag{8.24}$$

Definition 8.10. The sth order *classifying set* $\mathcal{C}^{(s)} = \mathcal{C}^{(s)}(\boldsymbol{\theta}, U)$ associated with a coframe $\boldsymbol{\theta}$ on an open subset $U \subset M$ is defined as the image of the structure map $\mathbf{T}^{(s)}$:

$$\mathcal{C}^{(s)}(\boldsymbol{\theta}, U) = \{\mathbf{T}^{(s)}(x) \mid x \in U\} \subset \mathbb{K}^{(s)}. \tag{8.25}$$

† A more intrinsic definition of the classifying space $\mathbb{K}^{(s)}$ is that it forms the fiber of the sth order jet bundle $\mathrm{J}^s(TM \otimes \wedge^2 T^*M)$. The coframe provides a natural connection on the anti-symmetric tensor bundle $TM \otimes \wedge^2 T^*M$ of type $(1,2)$ whose fiber coordinates are given by the structure invariants. This allows us to project to $\mathbb{K}^{(s)}$ and thus define the structure map intrinsically. However, for practical applications, the local definition presented in the text suffices.

Proposition 8.11. *Suppose $\boldsymbol{\theta}$ and $\overline{\boldsymbol{\theta}}$ are equivalent coframes under $\Phi: M \to \overline{M}$. Then, for each $s \geq 0$, the sth order classifying sets are the same. Thus, $\mathcal{C}^{(s)}(\overline{\boldsymbol{\theta}}, \overline{U}) = \mathcal{C}^{(s)}(\boldsymbol{\theta}, U)$, where $U \subset M$ is the domain and $\overline{U} = \Phi(U) \subset \overline{M}$ the range of the local equivalence map Φ.*

Example 8.12. Consider a coframe on $M \subset \mathbb{R}^2$ as discussed earlier in Examples 8.5 and 8.7. The zeroth order classifying set $\mathcal{C}^{(0)}$ is the subset of the classifying space $\mathbb{K}^{(0)} \simeq \mathbb{R}^2$, with coordinates (z, w), parametrized by the two structure functions $z = J(x, y)$, $w = K(x, y)$. If J, K are functionally independent, then $\mathcal{C}^{(0)}$ will be an open subset of $\mathbb{K}^{(0)}$, and very little equivalence information can be extracted from it. On the other hand, if J and K are functionally dependent, then $\mathcal{C}^{(0)}$ will be either a curve or a point, and much more data is available. Similarly, the first order classifying set $\mathcal{C}^{(1)}$ is the subset of the classifying space $\mathbb{K}^{(1)} \simeq \mathbb{R}^6$ parametrized by the first (and zeroth) order structure invariants

$$z^{(1)} = (z, w, z_1, z_2, w_1, w_2) = (J, K, J_1, J_2, K_1, K_2), \tag{8.26}$$

where $J_1 = \partial J / \partial \theta^1$, etc. Since we are working over $M \subset \mathbb{R}^2$, there are (locally) at most two functionally independent invariants. Thus $\mathcal{C}^{(1)}$ will be either a two-dimensional surface, a one-dimensional curve, or a point in $\mathbb{K}^{(1)}$, corresponding to the cases where there are, respectively, two, one, and zero functionally independent first order structure invariants.

For example, consider the particular coframe

$$\theta^1 = \frac{\lambda\, dx}{x + y}, \qquad \theta^2 = \frac{\mu\, dy}{x + y}, \tag{8.27}$$

where λ, μ are nonzero constants, on $M = \{x+y > 0\} \subset \mathbb{R}^2$. Comparing with (8.13), we have $A = \lambda(x+y)^{-1}$, $D = \mu(x+y)^{-1}$, $B = C = 0$, and so the invariants (8.15) are both constant: $J = 1/\mu$, $K = -1/\lambda$. Therefore, the zeroth order classifying set is the single point $(\mu^{-1}, -\lambda^{-1})$, and any equivalent coframe must also have the same constant, nonzero invariants $\overline{J} = 1/\mu$, $\overline{K} = -1/\lambda$. In particular, this immediately implies that two coframes of the form (8.27) with *different* constants λ, μ are not equivalent — there is no change of variables $\bar{x} = \xi(x, y)$, $\bar{y} = \eta(x, y)$, mapping one to the other, as the reader can easily verify by direct computation. The first order classifying set adds no new information since in this case all the derived invariants are zero, so $\mathcal{C}^{(1)} = \{(\mu^{-1}, -\lambda^{-1}, 0, 0, 0, 0)\}$ is also a single point; however, any equivalent coframe already must have constant invariants, and so its derived invariants are also automatically zero.

Next, consider a slightly more general coframe

$$\theta^1 = \lambda(x+y)^n\,dx, \qquad \theta^2 = \mu(x+y)^n\,dy, \tag{8.28}$$

on $M = \{x+y>0\}$. The associated invariants are

$$J = -n\mu^{-1}(x+y)^{-n-1}, \qquad K = n\lambda^{-1}(x+y)^{-n-1}.$$

Therefore, for $n \neq 0, -1$, the zeroth order classifying set

$$\mathcal{C}^{(0)} = \{(-n\mu^{-1}t, n\lambda^{-1}t) \mid t>0\} \subset \mathbb{K}^{(0)}$$

is a half-line. Any equivalent coframe must have the same classifying set, and therefore its invariants must satisfy the linear relation $\mu J + \lambda K = 0$, and the inequality $n\mu J < 0$. Two coframes, both of the form (8.28), but with different ratios $\lambda/\mu \neq \overline{\lambda}/\overline{\mu}$ cannot be equivalent; even if the ratios are the same, if $n\mu$ and $\bar{n}\overline{\mu}$ have different signs, the coframes are not equivalent. To obtain even more equivalence information, we compute the first order derived invariants using (8.19),

$$\begin{aligned}
\frac{\partial J}{\partial \theta^1} &= \frac{n(n+1)}{\lambda\mu}(x+y)^{-2n-2}, & \frac{\partial K}{\partial \theta^1} &= -\frac{n(n+1)}{\lambda^2}(x+y)^{-2n-2},\\
\frac{\partial J}{\partial \theta^2} &= \frac{n(n+1)}{\mu^2}(x+y)^{-2n-2}, & \frac{\partial K}{\partial \theta^2} &= -\frac{n(n+1)}{\lambda\mu}(x+y)^{-2n-2}.
\end{aligned}$$

Therefore, the first order classifying set $\mathcal{C}^{(1)} \subset \mathbb{K}^{(1)} \simeq \mathbb{R}^6$ is the "positive half" of a quadratic curve:

$$\left\{ \left(-\frac{n}{\mu}t, \frac{n}{\lambda}t, \frac{n(n+1)}{\lambda\mu}t^2, \frac{n(n+1)}{\mu^2}t^2, \frac{n(n+1)}{\lambda^2}t^2, \frac{n(n+1)}{\lambda\mu}t^2 \right) \;\middle|\; t>0 \right\}.$$

As a consequence, two coframes (8.28) will have the same first order classifying sets if and only if the ratios $\lambda/\mu = \overline{\lambda}/\overline{\mu}$ are the same, the powers $n = \bar{n}$ agree, and the coefficients μ and $\overline{\mu}$ have the same sign. In particular, this implies that the coframes corresponding to different powers $n \neq \bar{n}$ are not equivalent. On the other hand, it is easy to see that any two coframes of the form (8.28) satisfying the three equivalence conditions are, in fact, equivalent under a simple rescaling of the variables $(\bar{x}, \bar{y}) = (\gamma x, \gamma y)$ for suitable $\gamma > 0$. The second order classifying set can be similarly determined by computing the second order derived invariants; for example,

$$\frac{\partial^2 J}{(\partial \theta^1)^2} = -\frac{\partial^2 K}{\partial \theta^1 \partial \theta^2} = -\frac{2n(n+1)^2}{\lambda^2 \mu}(x+y)^{-3n-3} = \frac{2(n+1)^2}{n^2}\left(\frac{\mu}{\lambda}\right)^2 J^3.$$

The resulting set is a cubic curve in $\mathbb{K}^{(2)} \simeq \mathbb{R}^{12}$. However, no new equivalence information can be gleaned from $\mathcal{C}^{(2)}$, a fact that will result from a more general argument to be presented shortly.

Exercise 8.13. Discuss the classifying set for the coframe having the same formula (8.28), but defined on the opposite half plane $\widetilde{M} = \{x + y < 0\}$. When is this coframe equivalent to a coframe of the same form on $M = \{x + y > 0\}$?

The Classifying Manifolds

Returning to the general framework, the goal now is to determine in what sense our necessary conditions for equivalence are also sufficient. In order to make progress, we need to impose some regularity conditions.

Definition 8.14. A coframe $\boldsymbol{\theta}$ is called *fully regular* if, for each $s \geq 0$, the sth order structure map $\mathbf{T}^{(s)}: M \to \mathbb{K}^{(s)}$ is regular.

In the regular case, we let $\rho_s = \operatorname{rank} \mathbf{T}^{(s)}$ be the (constant) rank of the structure map, which is computed as the rank of its Jacobian matrix. (See also (8.39) below.) According to Definition 1.12, the classifying set $\mathcal{C}^{(s)}$ is a ρ_s-dimensional submanifold (possibly with self-intersections) of the classifying space $\mathbb{K}^{(s)}$. From now on, to emphasize our regularity assumption, we will refer to $\mathcal{C}^{(s)}$ as the sth order *classifying manifold* associated with the coframe $\boldsymbol{\theta}$. Furthermore, according to Theorem 1.10, ρ_s equals the number of functionally independent structure invariants up to order s associated with the coframe. Let

$$\mathcal{F}^{(s)} = \mathcal{F}^{(s)}(\boldsymbol{\theta}) = \{T_\sigma \mid \operatorname{order} \sigma \leq s\}, \qquad s = 0, 1, \ldots, \tag{8.29}$$

denote the family of functions consisting of all the structure invariants up to order s, so that $\mathcal{F}^{(0)} \subset \mathcal{F}^{(1)} \subset \mathcal{F}^{(2)} \cdots$. Then, locally, we can choose

ρ_s functionally independent structure invariants $I_\nu = T_{\sigma_\nu}$, $\nu = 1, \ldots, \rho_s$, which *generate* the full set of sth order structure invariants $\mathcal{F}^{(s)}$, in the sense that every other invariant of order $\leq s$ can be expressed as function of the basic invariants:

$$T_\sigma = H_\sigma(I_1, \ldots, I_{\rho_s}), \qquad \text{order } \sigma \leq s. \tag{8.30}$$

In this way, the invariants $I_1, \ldots, I_{\rho_s}$, furnish local coordinates on the ρ_s–dimensional classifying manifold $\mathcal{C}^{(s)}$, which can thereby be locally identified with an open subset of the graph of the function $\mathbf{H}^{(s)} = (\ldots, H_\sigma, \ldots)$. The components H_σ, which express the functional relations among the structure invariants, are the *classifying functions* associated with the coframe, and, in most treatments of the equivalence problem, form the primary focus of interest; see, for instance, [**78**], [**127**]. However, the fact that they are only locally well defined, and often globally multiply valued, makes them rather tricky to work with — for example, one must pay careful attention to both the domains of definition (since the classifying functions do not include inequality constraints presented by the classifying manifold), and the particular branches when they are multiply valued. The technical complications stemming from reliance on the classifying functions indicates that it is advantageous to focus on the classifying manifold as our principal object of interest.

Since we are primarily interested in local equivalence of coframes, the classifying manifolds need only agree locally, i.e., on the domain of the equivalence map Φ. Therefore, in view of Proposition 8.11, necessary conditions for the local equivalence of coframes are that, for each $s \geq 0$, their s^{th} order classifying manifolds overlap, in the sense of Definition 1.19. Moreover, by (8.23), the domain of the equivalence map $\bar{x} = \Phi(x)$ will be contained in (but not necessarily equal) the inverse image, under $\mathbf{T}^{(s)}$, of the overlap $\mathcal{C}^{(s)}(\boldsymbol{\theta}) \cap \mathcal{C}^{(s)}(\overline{\boldsymbol{\theta}})$ of the two classifying manifolds, whereas its range is contained in the inverse image of the overlap under $\overline{\mathbf{T}}^{(s)}$. Further, note that overlapping of the sth order classifying manifolds automatically implies overlapping of all lower order classifying manifolds (but not, in general, conversely).

The fundamental theorem underlying the solution to the (local) equivalence problem for coframes states that (in the fully regular case) these necessary conditions are also sufficient. The proof of this result, which is a direct consequence of the Frobenius Theorem governing the existence of solutions to certain systems of partial differential equations, can be found in Chapter 14 — see Theorem 14.24.

Theorem 8.15. *Let $\boldsymbol{\theta}$ and $\overline{\boldsymbol{\theta}}$ be smooth, fully regular coframes defined, respectively, on m-dimensional manifolds M and $\overline{M}$. There exists a local diffeomorphism $\Phi: M \to \overline{M}$ mapping the coframes to each other, $\Phi^*\overline{\boldsymbol{\theta}} = \boldsymbol{\theta}$, if and only if for each $s \geq 0$, their s^{th} order classifying manifolds $\mathcal{C}^{(s)}(\boldsymbol{\theta})$ and $\mathcal{C}^{(s)}(\overline{\boldsymbol{\theta}})$ overlap.*

As a first application of our general equivalence theorem, consider the case of a coframe with constant structure functions. In this case, the zeroth order classifying manifold is a single point, defined by the constant values of the invariants $T^i_{jk} = -C^i_{jk}$, $i, j, k = 1, \ldots, m$, $j < k$. All derived invariants are automatically zero, and hence every higher order classifying manifold $\mathcal{C}^{(s)}$ is also a single point. Therefore $\rho_s = 0$ for $s \geq 0$, and we will refer to such a coframe as having *rank zero*. Any equivalent coframe must have the same constant structure functions, $\overline{T}^i_{jk} = -C^i_{jk}$, in order that its zeroth order classifying manifold be the same. The derived invariants are also all zero, and so the higher order classifying manifolds are automatically the same. Therefore, any two rank zero coframes with the same constant structure functions are locally equivalent. Note that, as a consequence of the Jacobi identities (8.21), the constants C^i_{jk} form the structure constants for a Lie algebra $\mathfrak{g}$, and so a particular example of such a coframe is furnished by the Maurer–Cartan coframe on the associated Lie group G. As a direct consequence of Theorem 8.15, we have established the following result, proving that all such rank zero coframes are locally equivalent.

Theorem 8.16. *Let $\boldsymbol{\theta}$ be a rank zero coframe on an m-dimensional manifold M, with constant structure functions $T^i_{jk} = -C^i_{jk}$, $i, j, k = 1, \ldots, m$. Let G be an m-dimensional Lie group whose Lie algebra $\mathfrak{g}$ has structure constants C^i_{jk} relative to a basis $\{\mathbf{v}_1, \ldots, \mathbf{v}_m\}$, and let $\{\alpha^1, \ldots, \alpha^m\}$ denote the dual basis of Maurer–Cartan forms. Then there exists a local diffeomorphism $\Phi: M \to G$ mapping the given coframe to the Maurer–Cartan coframe on G, i.e., $\Phi^*\alpha^i = \theta^i$, $i = 1, \ldots, m$.*

Example 8.17. According to Theorem 8.15, every coframe (8.13) on $\mathbb{R}^2$ with nonzero constant invariants $J = c \neq 0$, $K = d \neq 0$, can, by a suitable change of variables, be mapped to the canonical coframe (8.27) with $\lambda = -1/d$, $\mu = 1/c$. To prove this result directly, we note that Darboux' Theorem 1.42 implies that we can choose coordinates such that the coframe has the diagonal form

$$\theta^1 = A(x,y)\,dx, \qquad \theta^2 = D(x,y)\,dy, \tag{8.31}$$

for nonvanishing functions A and D, which can be chosen to be constant if and only if the two invariants J, K vanish. For the coframe in the reduced form (8.31), equation (8.15) implies that the invariants $J = c$ and $K = d$ are constant if and only if

$$A_y = -cAD, \qquad D_x = dAD. \tag{8.32}$$

The differential equations (8.32) imply that $dA_y = -cD_x$, and hence, for $c, d \neq 0$, we find $A = -(\log F)_x/d$, $D = (\log F)_y/c$ for some function $F(x, y)$. Substituting these expressions into (8.32) we deduce that F must be a solution of the linear wave equation $F_{xy} = 0$, and so $F(x, y) = f(x) + g(y)$. Therefore the general solution to (8.32) is

$$A(x, y) = -\frac{f'(x)}{d[f(x) + g(y)]}, \qquad D(x, y) = \frac{g'(y)}{c[f(x) + g(y)]}, \tag{8.33}$$

where $f(x)$ and $g(y)$ are arbitrary functions with nonzero derivatives (otherwise (8.31) would not form a coframe). Even in this very simple example, the power of the Cartan method over a direct approach becomes evident.

Moreover, according to Theorem 8.16, we can map the coframe (8.27) to a Maurer–Cartan coframe on the affine group A(1), consisting of matrices $A = \left(\begin{smallmatrix} a & b \\ 0 & 1 \end{smallmatrix}\right)$, $a \neq 0$. The standard Maurer–Cartan coframe, computed in Example 2.95, is given by $\alpha^1 = da/a$, $\alpha^2 = db - (b/a)\, da$, and any other Maurer–Cartan coframe is provided by a pair of linearly independent combinations of α^1 and α^2. The map $\Phi: M \to \mathrm{A}(1)$ given explicitly by $a = 1/(x+y)$, $b = (x-y)/(x+y)$ maps the coframe (8.27) to the Maurer–Cartan coframe $\widehat{\alpha}^1 = \frac{1}{2\lambda}(-\alpha^1 + \alpha^2)$, $\widehat{\alpha}^1 = -\frac{1}{2\mu}(\alpha^1 + \alpha^2)$ associated with the structure constants: $C^1_{12} = -1/\mu$, $C^2_{12} = 1/\lambda$.

The one remaining difficulty in applying Theorem 8.15 is that it requires us, in general, to verify an infinite number of conditions, one for each order $s \geq 0$. Fortunately, as in the rank zero case, we can reduce the problem to one of finite order, and hence to one that can be solved. Since the structure invariants of order s also appear as components of $\mathbf{T}^{(s+1)}$, the ranks are nondecreasing, $\rho_s \leq \rho_{s+1}$, and bounded by $\rho_s \leq m$, the dimension of M. Therefore, the ranks must eventually stabilize, and, in fact, once they stabilize they remain fixed.

Proposition 8.18. *Let $\boldsymbol{\theta}$ be a fully regular coframe, and let ρ_s denote the rank of the sth order structure map $\mathbf{T}^{(s)}$. The smallest s for which $\rho_s = \rho_{s+1}$ is called the* order *of the coframe, and we have*

$$0 \le \rho_0 < \rho_1 < \cdots < \rho_s = \rho_{s+1} = \rho_{s+2} = \cdots = r \le m. \tag{8.34}$$

The stabilizing rank r will be referred to as the rank *of the coframe.*

The proof of Proposition 8.18 will follow from a straightforward chain rule computation; see Exercise 8.24 below. In practice, to compute the order and the rank of a coframe, we begin with the original set of structure functions $\mathcal{F}^{(0)} = \{T^i_{jk}\}$, and let ρ_0 denote its rank or number of functionally independent structure functions T^i_{jk}. Next, compute the rank $\rho_1 \ge \rho_0$ of the set $\mathcal{F}^{(1)} \supset \mathcal{F}^{(0)}$ consisting of the structure functions and their first coframe derivatives. If $\rho_1 = \rho_0$, then every first order coframe derivative can be (locally) expressed as a function of the original structure functions, and the coframe has order $s = 0$ and rank $r = \rho_0$; moreover, according to Proposition 8.18, $\rho_t = r$ for all $t \ge 0$, meaning that there are no higher order derived invariants which are functionally independent of the original structure functions. On the other hand, if $\rho_0 < \rho_1$, then we compute the rank $\rho_2 \ge \rho_1$ of the set $\mathcal{F}^{(2)} \supset \mathcal{F}^{(1)}$ of second order structure invariants. If $\rho_2 = \rho_1$, then the coframe has rank $r = \rho_1$, and order $s = 1$; in this case all the second and higher order derived invariants are functions of the original structure functions and their first order coframe derivatives. The general situation is now clear. The order and rank of a coframe can therefore be practically computed by differentiating the structure functions until there are no more functionally independent invariants; the minimal s at which the ranks stabilize, $\rho_s = \rho_{s+1}$, determines both the order s and the rank $r = \rho_s$. In particular, since the rank is bounded by m, the dimension of M, the order s is bounded by $m - 1$, with equality only if the ranks steadily increase, with $\rho_k = k+1$ for $0 \le k \le m-1$, which happens only when there is just one independent nonconstant structure function, one of its first order coframe derivatives is functionally independent from it, one of its second order coframe derivatives is yet another independent function, and so on up to one of the $(m-1)^{\text{st}}$ order coframe derivatives, at which point we have the maximal number of independent functions that can exist on our m-dimensional manifold.

The rank ρ_s of the structure invariants of order at most s is the same as the dimension of the sth order classifying manifold $\mathcal{C}^{(s)}$. The fact that the ranks are nondecreasing is a reflection of the existence of a natural projection from $\mathcal{C}^{(s+1)}$ onto $\mathcal{C}^{(s)}$ stemming from the obvious projection $\mathbb{K}^{(s+1)} \to \mathbb{K}^{(s)}$ of classifying spaces. In particular, the order s of a coframe is the smallest integer for which $\dim \mathcal{C}^{(s)} = \dim \mathcal{C}^{(s+1)}$. As a consequence of these considerations, we discover that if $\boldsymbol{\theta}$ is a fully regular coframe of order s, then the projection $\mathcal{C}^{(s+1)} \to \mathcal{C}^{(s)}$ is a covering map. Once a branch of this covering is specified, then the higher order classifying manifolds are all uniquely determined, meaning that the higher order projections $\mathcal{C}^{(s+t)} \to \mathcal{C}^{(s+1)}$, $t \geq 2$, are all one-to-one! This is a consequence of the following fundamental theorem, which provides a complete solution to the equivalence problem for (regular) coframes.

Theorem 8.19. *Let $\boldsymbol{\theta}$ and $\overline{\boldsymbol{\theta}}$ be smooth, fully regular coframes defined, respectively, on m-dimensional manifolds M and $\overline{M}$. There exists a local diffeomorphism $\Phi: M \to \overline{M}$ mapping the coframes to each other, $\Phi^* \overline{\boldsymbol{\theta}} = \boldsymbol{\theta}$, if and only if they have the same order, $\bar{s} = s$, and their $(s+1)^{\mathrm{st}}$ order classifying manifolds $\mathcal{C}^{(s+1)}(\boldsymbol{\theta})$ and $\mathcal{C}^{(s+1)}(\overline{\boldsymbol{\theta}})$ overlap.*

Remark: Since, in the regular case, the order of a coframe is strictly less than the dimension of M, an alternative, order-independent version of Theorem 8.19 states that two fully regular m-dimensional coframes $\boldsymbol{\theta}$ and $\overline{\boldsymbol{\theta}}$ are locally equivalent if and only if their m^{th} order classifying manifolds $\mathcal{C}^{(m)}(\boldsymbol{\theta})$ and $\mathcal{C}^{(m)}(\overline{\boldsymbol{\theta}})$ overlap.

Remark: In the case of analytic coframes, the method of analytic continuation shows that the local overlapping of the two classifying manifolds implies strict overlapping, so that, on the common domains of definition, they coincide. Consequently, locally equivalent analytic coframes remain equivalent at all points where the analytic continuation of the local equivalence map remains well defined. This globalization result is clearly not true for more general smooth coframes.

Example 8.20. Return once again to the coframe on $\mathbb{R}^2$ discussed in Examples 8.5, 8.7 and 8.12. Since we are working on an (open subset of) $\mathbb{R}^2$, the coframe can have rank 0, 1, or 2 at each point, meaning, in the regular case, that either all invariants are constant, or there are at most 1 or 2 functionally independent invariants. If the rank is 0, then both J and K are constant, and the order is automatically zero. If the rank is 1, then there is a single functional relation connecting J and K, say $K = H(J)$, (which occurs provided J is not constant), and all the derived invariants can also be expressed locally as functions of the fundamental invariant J. In this case, the coframe is determined by its *classifying curve* $\mathcal{C}^{(1)} \subset \mathbb{K}^{(1)}$. If the coframe has rank 2, then there are two possible cases: either J and K are functionally independent, in which case the order is 0, and every derived invariant can be expressed as a function of J and K; the coframe is determined by its first order *classifying surface* $\mathcal{C}^{(1)} \subset \mathbb{K}^{(1)}$. Otherwise, the two original invariants J and K are functionally dependent, say $K = H(J)$, but one of the derived invariants, say $J_1 = \partial J/\partial \theta^1$, is independent of J, in which case the order is 1, and every other derived invariant can be expressed as a function of the two primary invariants J and J_1. In this case, $\mathcal{C}^{(0)}$ is a curve, $\mathcal{C}^{(1)}$ is a surface, and the equivalence class of the coframe is prescribed by its second order *classifying surface* $\mathcal{C}^{(2)} \subset \mathbb{K}^{(2)}$.

To make these considerations more precise, let us consider an interesting subclass of rank 1 coframes — the "constant curvature" coframes.† The Riemannian metric associated with a coframe $\{\theta^1, \theta^2\}$ is defined as $ds^2 = (\theta^1)^2 + (\theta^2)^2$. A computation (see Chapter 12 for details) proves that the Gaussian curvature of this metric is the first order structure invariant

$$\kappa = \frac{\partial J}{\partial \theta^2} - \frac{\partial K}{\partial \theta^1} - J^2 - K^2, \tag{8.35}$$

which we assume is constant. For simplicity, let us restrict our attention to the subcase in which *a*) $J = -K$, which implies that $\theta^1 + \theta^2$ is an exact one-form, and *b*) $\partial J/\partial \theta^1 = \partial J/\partial \theta^2 \equiv L$. In this case, owing to the imposed relation (8.35) with constant κ, the coframe necessarily has rank 0 or 1. In the rank zero case, J is constant, and hence $L = 0$. In the rank 1 case, J is not constant, and, owing to regularity, $dJ = L(\theta^1 + \theta^2)$ cannot vanish, hence $L \neq 0$ everywhere. Furthermore, equation (8.35) implies the classifying relations

$$L = \frac{\partial J}{\partial \theta^1} = \frac{\partial J}{\partial \theta^2} = H_\kappa(J), \qquad \text{where} \qquad H_\kappa(z) = z^2 + \tfrac{1}{2}\kappa.$$

Thus, in terms of the coordinates (8.26), the classifying curve is

$$\mathcal{C}^{(1)} = \left\{ (z, w, z_1, z_2, w_1, w_2) \;\middle|\; \begin{array}{l} w = -z, \\ z_1 = z_2 = -w_1 = -w_2 = z^2 + \tfrac{1}{2}\kappa \end{array} \right\}.$$

† Doyle, [**63**], [**64**], shows that the two-dimensional hyperbolic systems (8.4) whose symmetry coframes have constant curvature are the ones admitting multi-Hamiltonian structures.

Clearly, in this case we do not need to look at the full classifying curve, but, rather, its reduction to the two-dimensional reduced classifying space parametrized by $z = J(x,y)$, $u = L(x,y)$. The reduced classifying manifolds will be either points on the z-axis, or parabolic arcs which, by the above remarks, cannot cross the z-axis. With this information in hand, it is not hard to determine a complete list of canonical forms for these particular coframes.

Canonical Forms for Constant Curvature Coframes

1.	$\kappa = 0,$	$J = 0,$	$\{dx, dy\},$	
2.	$\kappa = 0,$	$J > 0,$	$\{e^{-x-y}\,dx, e^{-x-y}\,dy\},$	
3.	$\kappa = 0,$	$J < 0,$	$\{e^{x+y}\,dx, e^{x+y}\,dy\},$	
4.	$\kappa = -\dfrac{2}{\alpha^2},$	$J = \dfrac{1}{\alpha},$	$\left\{\dfrac{\alpha\,dx}{x+y}, \dfrac{\alpha\,dy}{x+y}\right\},$	$\alpha \neq 0,$ $x + y > 0,$
5.	$\kappa = -\dfrac{2}{\alpha^2},$	$L < 0,$	$\left\{\dfrac{\alpha\,dx}{\sin(x+y)}, \dfrac{\alpha\,dy}{\sin(x+y)}\right\},$	$\alpha > 0,$ $0 < x + y < \pi,$
6.	$\kappa = -\dfrac{2}{\alpha^2},$	$J > 0,$ $L > 0,$	$\left\{\dfrac{\alpha\,dx}{\sinh(x+y)}, \dfrac{\alpha\,dy}{\sinh(x+y)}\right\},$	$\alpha > 0,$ $x + y > 0,$
7.	$\kappa = -\dfrac{2}{\alpha^2},$	$J < 0,$ $L > 0,$	$\left\{\dfrac{\alpha\,dx}{\sinh(x+y)}, \dfrac{\alpha\,dy}{\sinh(x+y)}\right\},$	$\alpha < 0,$ $x + y > 0,$
8.	$\kappa = \dfrac{2}{\alpha^2},$	$L > 0,$	$\left\{\dfrac{\alpha\,dx}{\cosh(x+y)}, \dfrac{\alpha\,dy}{\cosh(x+y)}\right\},$	$\alpha > 0,$

where α is a constant.

Figure 3 illustrates the different branches. The points on the z–axis are zero-dimensional classifying manifolds ("classifying points"), and correspond to the coframes of rank zero, which are cases 1 and 4 in the preceding table. The genuine classifying curves are of three types, depending on whether the curvature κ is positive, zero, or negative: *a*) an entire parabola $u = z^2 + \frac{1}{2}\kappa$, $\kappa > 0$, corresponding to case 8; *b*) one of the two semi-parabolas $u = z^2$, with $z > 0$ or $z < 0$, corresponding to cases 2 and 3; or *c*) one of the three pieces left after slicing one of the parabolas $u = z^2 + \frac{1}{2}\kappa$, $\kappa < 0$, with the z-axis, corresponding to cases 5, 6, 7. Note that in the latter two types, even though the classifying function $H(z) = z^2 + \frac{1}{2}\kappa$ has the *same* functional form, the different coframes are *not* equivalent because the classifying curves are *disjoint*

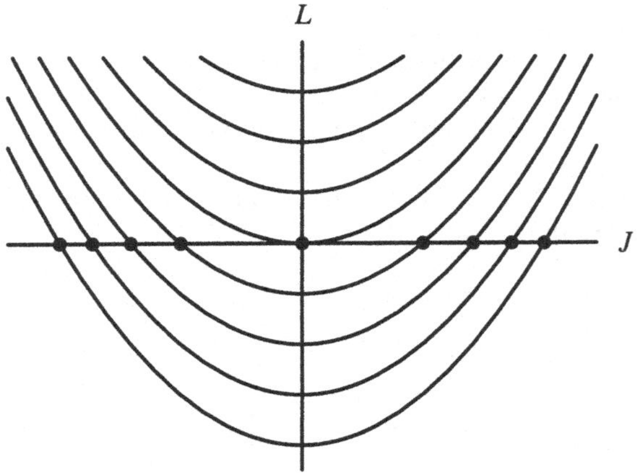

Figure 3. Classifying Curves for Constant Curvature Coframes

pieces of the graph of the classifying function. This means that the invariance equations $\bar{J}(\bar{x}, \bar{y}) = J(x, y)$, $\bar{L}(\bar{x}, \bar{y}) = L(x, y)$, which must be satisfied by any equivalence $(\bar{x}, \bar{y}) = \Phi(x, y)$, have *no* solution since the two pairs of structure invariants have disjoint images. This example illustrates the importance of emphasizing the classifying manifold, rather than the classifying functions, as the primary object of interest.

Symmetries of a Coframe

As we have remarked before, the solution to the equivalence problem for coframes given in Theorem 8.19 also determines the structure of the symmetry group of a given coframe.

Definition 8.21. Let $\boldsymbol{\theta} = \{\theta^1, \ldots, \theta^m\}$ be a coframe defined on a manifold M. The *symmetry group* of $\boldsymbol{\theta}$ is the group of self-equivalences, meaning local diffeomorphisms $\Phi: M \to M$ satisfying $\Phi^* \theta^i = \theta^i$ for $i = 1, \ldots, m$.

For any equivalence problem which can be reformulated as an equivalence problem for coframes, this definition of symmetry group coincides with the usual notion of symmetry group, the symmetries in question being, of course, restricted to lie in the prescribed class of equivalences.

Theorem 8.22. *If $\boldsymbol{\theta}$ is a coframe of rank r on an m-dimensional manifold M, then the symmetry group of $\boldsymbol{\theta}$ is a finite-dimensional Lie group of dimension $m - r$.*

A rigorous proof of Theorem 8.22 can be found in Chapter 14 — see Theorem 14.26. However, in order to motivate the result, consider the case when the coframe has rank zero, which means that all the structure functions are constant. According to Theorem 8.16, there is a local diffeomorphism $\Phi: M \to G$, mapping the coframe to the Maurer–Cartan coframe on the Lie group G whose structure constants coincide with the constant structure functions (up to sign). Since the Maurer–Cartan forms are right-invariant, the Lie group G itself, acting via right multiplication, provides the symmetry group of the coframe, proving Theorem 8.22 in this particular case.† Therefore, the structure of the symmetry group of such a coframe can be immediately determined from its structure equations.

At the other extreme, if the rank of a coframe equals the dimension of the underlying space, $r = m$, then the Inverse Function Theorem implies that the invariance equations (8.23) implicitly determine the relevant change of variables $\bar{x} = \Phi(x)$. In this case, the coframe admits at most a discrete symmetry group. In the intermediate cases, when the rank satisfies $0 < r < m$, we can introduce local coordinates $(y, z) = (y^1, \ldots, y^r, z^1, \ldots, z^{m-r})$ on M so that the y^i's form a complete system of fundamental invariants, meaning that, in the (y, z) coordinates, every other structure invariant can be written as a function of the y's alone. This is equivalent to the statement that each slice $N_c = \{y^1 = c^1, \ldots, y^r = c^r\}$, $c \in \mathbb{R}^r$, forms a level set for the structure map $\mathbf{T}^{(m)}$. Let us restrict the coframe $\boldsymbol{\theta}$ to a slice $N = N_c$, and then choose a set of $m - r$ independent one-forms $\theta^i \mid N$ which form a coframe on N. The torsion coefficients for this restricted coframe must be constant on N, since otherwise there would be invariants of the original coframe which depended essentially on the z's, contradicting our initial hypothesis. Therefore, on each slice N_c, we have reduced to the rank zero case, and thereby produced the requisite $(m - r)$-dimensional symmetry group. Remarkably, the resulting restricted symmetry groups are all isomorphic to the same $(m - r)$-dimensional Lie group G, and,

† One should, of course, check that there are no additional symmetries of the Maurer–Cartan coframe! See Proposition 14.19.

in fact, come from the restriction of an action of G on M itself which leaves the original coframe invariant.

Example 8.23. For the two-dimensional coframes discussed starting with Example 8.5, the rank is either 0, 1, or 2, and hence the symmetry group of such a coframe can have dimension 2, 1, or 0. Every coframe with a two-dimensional symmetry group has constant invariants, which determine the structure constants for the group: $J = -C^1_{12}$, $K = -C^2_{12}$. The invariants vanish, $J = K = 0$, if and only if the symmetry group is abelian. In this case, a canonical form for the coframe is provided by the two coordinate differentials $\{dx, dy\}$, and the symmetry group is just the group of translations in x and y. For the canonical coframe (8.27) with constant invariants, not both zero, the symmetry group is the two-parameter affine group $\mathrm{A}(1)$, acting on $\mathbb{R}^2$ via simultaneous translations and scalings: $(x, y) \mapsto (ax + b, ay - b)$. Any rank one coframe admits a one-parameter symmetry group. For the particular rank one coframes (8.28), with $n \neq -1$, the symmetry group is generated by the simultaneous translations $(x, y) \mapsto (x + b, y - b)$, and, if n is odd, the additional discrete reflectional symmetry $(x, y) \mapsto (-x, -y)$. Finally, in the rank two case, the coframe admits at most a discrete symmetry group.

Remarks and Extensions

Our basic Equivalence Theorem 8.19 admits a useful and readily established generalization which occasionally arises in examples. Besides the two coframes $\boldsymbol{\theta}$ and $\overline{\boldsymbol{\theta}}$, one might, in addition, require the invariance of one or more functions which do not appear among the structure functions or their coframe derivatives. In other words, we supplement the basic coframe invariance conditions (8.7) by an additional set of functional invariants, so that (omitting the pull-backs) the full equivalence conditions take the more general form

$$\overline{\theta}^i = \theta^i, \quad i = 1, \ldots, m, \qquad \overline{J}_\mu(\bar{x}) = J_\mu(x), \quad \mu = 1, \ldots, n. \tag{8.36}$$

The solution to this more general equivalence problem is straightforward. Let T_σ denote the usual structure invariants associated with the coframe, obtained by differentiating the structure functions, and, correspondingly, let J_τ, $\tau = (\mu, l_1, \ldots, l_s)$, denote the coframe derivatives of the additional functional invariants, cf. (8.22). In this case, one includes these additional invariants and their coframe derivatives in the definition of the *extended structure map* $\widehat{\mathbf{T}}^{(s)}$, whose components will now consist of all the structure invariants T_σ and J_τ up to order s. The *extended classifying manifolds* will now be parametrized by all these functions; the associated s^{th} order *extended classifying space* will be a Euclidean space of dimension $\left[\frac{1}{2}m^2(m-1)+n\right]\binom{m+s}{m}$. The rank and the order of such a coframe plus additional invariants are defined as before, and Theorem 8.19 holds without change. Examples will appear later.

In either case, as it is defined above, the classifying manifold of a coframe usually relies on far too much information. Indeed, if the underlying manifold M is three-dimensional, then $\mathcal{C}^{(3)}$ will be a parametrized manifold, of dimension at most three, in a $q_3(3) = 180$-dimensional Euclidean space! But (fortunately) we don't typically have to check all 180 components of the parametrization $\mathbf{T}^{(3)}$ in order to understand the equivalence problem. First of all, only in the extreme case when the order is maximal, i.e., 2, do we need to consider the third order classifying manifold; otherwise we can just use the first or second order one. Second, if we know the classifying function for any structure invariant T_σ in terms of some conveniently chosen fundamental set of invariants $I_1, \ldots, I_r$, then we automatically know the classifying functions for all derived invariants thereof, and so those coordinates are completely redundant. Indeed, given the functional dependency $T_\sigma = H_\sigma(I_1, \ldots, I_r)$ specified by the classifying function $H_\sigma(z^1, \ldots, z^r)$, we evaluate its differential using the chain rule and formula (8.2):

$$\sum_{i=1}^{m} \frac{\partial T_\sigma}{\partial \theta^j}\,\theta^j = dT_\sigma = \sum_{\nu=1}^{r} \frac{\partial H_\sigma}{\partial z^\nu}(I_1, \ldots, I_r)\, dI_\nu.$$

Since $dI_\nu = \sum_j \partial I_\nu/\partial\theta^j\,\theta^j$, the linear independence of the coframe elements θ^j implies that

$$\frac{\partial T_\sigma}{\partial \theta^j} = \sum_{\nu=1}^{r} \frac{\partial H_\sigma}{\partial z^\nu}(I_1, \ldots, I_r)\,\frac{\partial I_\nu}{\partial \theta^j}. \tag{8.37}$$

Therefore if we know the classifying functions for the derived invariants of our fundamental set,

$$\frac{\partial I_\nu}{\partial \theta^j} = H_{\nu,j}(I_1, \ldots, I_r), \qquad \nu = 1, \ldots, r, \quad j = 1, \ldots, m, \tag{8.38}$$

then we automatically know the classifying functions for all the derived invariants $\partial T_\sigma / \partial \theta^j$, and hence they play no role in the analysis of the classifying manifold. Additional automatic relations come from the Jacobi identities (8.21) (which will only play a role when dimension of M is three or more). Finally, many of the invariants may be automatically equal, e.g., they may be identically zero, and so we also do not need to consider them in our analysis of the classifying manifold. In practice, then, as we did in Example 8.12, we will only need to consider "reduced classifying manifolds", in which we utilize substantially fewer invariants to completely analyze the problem.

Exercise 8.24. Use the chain rule formula (8.37) to prove Proposition 8.18.

Interestingly, the ranks ρ_s of a coframe are completely determined by the values of the structure invariants. For example, ρ_0, being the rank of the family $\mathcal{F}^{(0)}$ of structure functions, is defined as the dimension of the space spanned by their differentials. However, we can write $dT^i_{jk} = \sum_l T^i_{jk,l}\theta^l$. Since the θ^i's form a coframe, ρ_0 equals the rank of the $\frac{1}{2}m^2(m-1)\times m$ matrix whose entries are the first order derived structure functions $T^i_{jk,l} = \partial T^i_{jk}/\partial\theta^l$. More generally, ρ_s equals the rank of the $\frac{1}{2}m^2(m-1) \times \left(\binom{m+s}{m} - 1\right)$ matrix $\widetilde{Z}^{(s)}$ whose entries are the derived invariants T_σ, $\sigma = (i, j, k, l_1, \ldots, l_r)$, of orders $1 \leq r \leq s+1$; here (i, j, k) labels the row and $(l_1, \ldots, l_r)$ the column where T_σ lies in $\widetilde{Z}^{(s)}$. This remark has the effect of restricting the classifying manifolds to lie within certain determinantal varieties of the full classifying space:

$$\mathcal{C}^{(s)} \subset \left\{ z^{(s)} \;\middle|\; \operatorname{rank} \widetilde{Z}^{(s)} = \rho_{s-1} \right\} \subset \mathbb{K}^{(s)}. \tag{8.39}$$

For example, in the case of a two-dimensional coframe discussed in Example 8.12, if the coframe has rank one, then its first order classifying curve is necessarily contained in the set of points $(z, w, z_1, z_2, w_1, w_2)$ satisfying $z_1 w_2 - z_2 w_1 = 0$.

Finally, note that Theorem 8.19 reduces the solution of the equivalence problem for coframes to the determination of whether two parametrized submanifolds of the classifying space overlap. Therefore, it is often possible to completely determine whether or not two given coframes are equivalent. However, in more complicated situations, we may be stuck with a rather difficult problem to resolve: When do two parametrized manifolds of Euclidean space overlap? (A more modest problem would be to ask when the two submanifolds are identical.) In the case of polynomial or rational parametrizations, Buchberger's Gröbner basis method, [**31**], can, at least in principle, solve this problem, although its practical implementation in the solution of equivalence problems remains undeveloped. I am not aware of any general algorithms for handling the overlapping or identification problems for more general smooth or analytically parametrized submanifolds!

Chapter 9

Formulation of Equivalence Problems

We now consider more general problems of equivalence, in which the equivalence conditions can be rewritten using differential forms, but a fully invariant coframe is not so immediately apparent. Nevertheless, in most cases, an invariant coframe which completely characterizes the equivalence problem does exist, and so the problem can actually be handled by the methods in the previous chapter. The difficulty is that the formulae for the invariant coframe are not obvious, and may require considerable ingenuity to construct *ab initio*. Fortunately, for those of us unendowed with the requisite insight, Cartan devised a general procedure that, starting with an equivalence problem suitably encoded using differential forms, eventually produces the explicit formulas for the desired invariant coframe. Thus, even though we are not able (or not clever enough) to immediately write down the invariant coframe, just being able to rewrite the problem in terms of differential forms allows us to apply the powerful algorithm developed by Cartan, and thereby deduce the solution to the original equivalence problem. As with our earlier analysis of the equivalence problem for coframes, Cartan's method reduces to an inspired exploitation of the invariance of the exterior derivative.

Equivalence Problems Using Differential Forms

In Cartan's approach, the conditions of equivalence of two objects (differential equations, polynomials, variational problems, differential operators, etc.) must be reformulated in terms of differential forms or, even more specifically, differential one-forms. We associate a collection of one-forms to an object under investigation in the original coordinates; the corresponding object in the new coordinates will have its own collection of one-forms. The fact that a certain class of transformations or changes of variables maps the original object to the new one will then be the same as writing the one-forms in the new coordinates as certain linear combinations of the one-forms in the original coordinates. Restrictions on the types of transformations allowed (e.g., linear, volume-preserving, contact, etc.) must also be encoded in the same language of one-forms. In this preliminary chapter we explain how a variety of equivalence problems can be reformulated in the language of differential forms. The invariant normalizations required to reduce the problem to an equivalence of coframes will be explained. The actual implementation of the Cartan algorithm, and the detailed solution of the various equivalence problems, will then be presented in the following chapters.

Example 9.1. *Equivalence of differential forms.* The simplest possible example is the problem of equivalence of one-forms, since this problem is already in the proper framework. (Of course, we already know that the solution to this problem, in the constant rank case, is given by Darboux' Theorem 1.42; nevertheless, it is of interest to see how the problem can be solved using the Cartan approach.) Two one-forms, ω defined on an m-dimensional manifold M, and $\overline{\omega}$ defined on a manifold $\overline{M}$ of the same dimension, are *equivalent* if and only if there is a smooth (locally defined) diffeomorphism $\Phi\colon M \to \overline{M}$ which pulls back $\overline{\omega}$ to ω, i.e., $\Phi^*\overline{\omega} = \omega$. Note the similarity with the equivalence problem for coframes, (8.7), except that now we do not have a complete coframe on the manifolds M and $\overline{M}$ to work with, and so the results of Chapter 8 do not directly apply.

The equivalence problem for two-forms is similar: a two-form Ω on M is equivalent to a two-form $\overline{\Omega}$ on $\overline{M}$ if and only if there is a smooth (local) diffeomorphism $\Phi\colon M \to \overline{M}$ such that $\Phi^*\overline{\Omega} = \Omega$. To place this problem into a form amenable to the Cartan approach, we are required to re-express the equivalence condition in terms of *one-forms*. According to formulae (1.21), (1.29), the *rank* r of a two-form is invariant under diffeomorphisms, hence equivalent two-forms must have the same rank. Assuming that each two-form has constant rank r, at each point we write the two-form in (algebraic) canonical form (2.6)

$$\Omega = \sum_{j=1}^{r} \omega^j \wedge \omega^{j+r}, \qquad \overline{\Omega} = \sum_{j=1}^{r} \overline{\omega}^j \wedge \overline{\omega}^{j+r}, \tag{9.1}$$

by introducing locally defined, independent one-forms $\{\omega^1, \ldots, \omega^{2r}\}$ on M, and $\{\overline{\omega}^1, \ldots, \overline{\omega}^{2r}\}$ on $\overline{M}$. According to Example 2.27, a linear transformation preserves the given algebraic canonical form if and only if it belongs to the symplectic group $\mathrm{Sp}(2r)$ — the isotropy group of a maximal rank two-form on $\mathbb{R}^{2r}$. Therefore, the two-forms (9.1) will be equivalent under the map Φ if and only if the corresponding sets of one-forms obey the transformation rule

$$\Phi^* \overline{\omega}^i = \sum_{j=1}^{2r} g^i_j(x)\, \omega^j, \qquad i = 1, \ldots, 2r, \tag{9.2}$$

where $g(x) = \big(g^i_j(x)\big)$ is a $(2r) \times (2r)$ matrix of functions on M taking values in $\mathrm{Sp}(2r)$.

Example 9.2. *Equivalence of Riemannian metrics.* The basic equivalence problem in Riemannian geometry asks when is there a local, metric-preserving diffeomorphism (*isometry* for short) mapping a given Riemannian manifold M, with metric tensor ds^2, to a second Riemannian manifold $\overline{M}$ of the same dimension, with metric $d\bar{s}^2$. In other words, the equivalence map $\Phi\colon M \to \overline{M}$ must satisfy $\Phi^*(d\bar{s}^2) = ds^2$ where defined. In order to recast this problem in terms of one-forms, we recall the algebraic canonical form for symmetric, positive definite forms (symmetric matrices) in Example 2.26. This allows us to introduce (local) coframes $\boldsymbol{\omega}$ on M, and $\overline{\boldsymbol{\omega}}$ on $\overline{M}$, which diagonalize the metrics at each point:

$$ds^2 = \sum_{i=1}^{m} (\omega^i)^2, \qquad d\bar{s}^2 = \sum_{i=1}^{m} (\omega^i)^2. \tag{9.3}$$

The two metrics will be equivalent if and only if their diagonalizing coframes satisfy

$$\Phi^* \overline{\omega}^i = \sum_{j=1}^{m} g^i_j(x)\, \omega^j, \qquad j = 1, \ldots, m, \tag{9.4}$$

where, in this case, the $m \times m$ matrix $g(x) = \big(g^i_j(x)\big)$ is required to take values in the orthogonal group $\mathrm{O}(m)$, the isotropy group of the algebraic canonical form. For pseudo-Riemannian manifolds, the metric tensor is no longer definite, but the equivalence problem can still be handled by the same approach. The metric is diagonalized $ds^2 = \sum_i \pm(\omega^i)^2$, the number of plus and minus signs being governed by its signature, cf. Example 2.26, which is an invariant. The structure group is the pseudo-orthogonal group $\mathrm{O}(p, m-p)$.

Example 9.3. *Equivalence of differential equations.* The general equivalence problem is to determine when two given differential equations can be mapped to each other by a suitable change of variables. Such problems were the focus of much research in the last century, and have assumed renewed importance in recent years. Here we show how to formulate the simplest such equivalence problem, that of a scalar second order ordinary differential equation, in the language of differential forms. (Actually, the simplest problem would be a scalar first order ordinary differential equation, but, as we know from Theorem 1.23, all first order equations, a.k.a. vector fields, are equivalent away from singularities, so that problem is essentially trivial. However, the reader may find it instructive to understand how this result could be deduced using the Cartan method.) Consider a general (smooth) scalar second order ordinary differential equation

$$\frac{d^2u}{dx^2} = Q\left(x, u, \frac{du}{dx}\right). \tag{9.5}$$

There are a number of possible classes of admissible changes of variables that come to mind, and each class will lead to a different equivalence problem with a different formulation in terms of differential forms. Here we explain how to set up the three most basic of these problems — equivalence under fiber-preserving, point, and contact transformations. The first step is to notice that, once the transformation of the independent and dependent variable is specified, the derivative variables $p = u_x$, $q = u_{xx}$, etc., must transform according to the associated prolonged transformations, cf. Example 4.11. The prolonged maps are contact transformations on the jet spaces J^n, $n \geq 1$, and hence map contact forms to contact forms. In particular, the first order contact forms must agree up to multiple,

$$d\bar{u} - \bar{p}\, d\bar{x} = \lambda(du - p\, dx), \tag{9.6}$$

where λ is a function of x, u, p, cf. (4.49). (For simplicity, we will omit the explicit mention of pull-backs in our formulae.) Similarly, the second order contact forms match up modulo first order ones:

$$d\bar{p} - \bar{q}\, d\bar{x} = \mu(du - p\, dx) + \nu(dp - q\, dx). \tag{9.7}$$

On the solutions to the differential equation (9.5), we can replace the derivative coordinate $q = u_{xx}$ by the right hand side $Q(x, u, p)$ of (9.5), and so (9.7) implies the additional one-form condition

$$d\bar{p} - \overline{Q}(\bar{x}, \bar{u}, \bar{p})\, d\bar{x} = \mu(du - p\, dx) + \nu(dp - Q(x, u, p)\, dx). \tag{9.8}$$

Conditions (9.8), (9.6), which are imposed on the first jet space $\mathrm{J}^1 \simeq \mathbb{R}^3$, constitute the required formulation of the equivalence problem for second order ordinary differential equations under contact transformations. It is a useful exercise for readers to convince themselves that any smooth transformation $(\bar{x}, \bar{u}, \bar{p}) = \Psi(x, u, p)$ which satisfies the two conditions (9.6) and (9.8) for functions λ, μ, ν is necessarily a contact transformation which maps the ordinary differential equation $u_{xx} = Q(x, u, p)$ to $\bar{u}_{\bar{x}\bar{x}} = \overline{Q}(\bar{x}, \bar{u}, \bar{p})$. Therefore, these two conditions are a completely equivalent reformulation of the original contact equivalence problem.

In the case of either point transformations or fiber-preserving transformations, we must impose additional restrictions, again prescribed by suitable one-form conditions. A fiber-preserving transformation requires that we not mix up independent and dependent variables, so $\bar{x} = \varphi(x)$ depends only on x, and hence its differential $d\bar{x}$ will pull back to a multiple of dx:

$$d\bar{x} = \alpha\, dx. \tag{9.9}$$

The reader can show that equations (9.6), (9.8) and (9.9) imply that the two ordinary differential equations are equivalent under a fiber-preserving transformation. (In particular, $\bar{u} = \psi(x, u)$ necessarily only depends on x and u.) Similarly, the condition

$$d\bar{x} = \alpha\, dx + \beta\, du, \tag{9.10}$$

will, in addition to (9.6) and (9.8) encode equivalence of the differential equations under point transformations $\bar{x} = \varphi(x, u)$, $\bar{u} = \psi(x, u)$. Note that if we are allowing general contact transformations, then $\bar{x}$ can depend on all three variables, x, u, p, and hence the (vacuous) condition

$$d\bar{x} = \alpha\, dx + \beta\, du + \gamma\, dp, \tag{9.11}$$

assumes the role of (9.9) or (9.10).

Example 9.4. The basic equivalence problem in the calculus of variations is to determine when two variational problems can be transformed into each other by a suitable change of variables. Here we formulate the simplest such problem, that of a first order variational problem

$$\mathcal{L}[u] = \int L\left(x, u, \frac{du}{dx}\right) dx, \tag{9.12}$$

in one independent variable, x, and one dependent variable, u. As in the equivalence problem for differential equations, we must precisely specify which change of variables, e.g., fiber-preserving, point, or contact transformations, are to be allowed. Moreover, there is an additional complication in specifying the type of equivalence we are to allow. The "standard equivalence problem" requires that the two functionals have the same values on functions: $\mathcal{L}[u] = \overline{\mathcal{L}}[\bar{u}]$. The more general "divergence equivalence problem" only requires that the Euler–Lagrange equations match up. Both equivalence problems were discussed in some detail in Chapter 7; let us see how to encode them in Cartan form. First, as in the preceding example, the prolonged transformation on J^1 must satisfy the basic contact condition (9.6). Second, according to (7.16), two first order variational problems are equivalent under a point transformation if and only if the associated Lagrangian forms are contact-equivalent under its first prolongation, so that

$$\overline{L}\, d\bar{x} = L\, dx + \mu(du - p\, dx) = L\, dx + \mu\, \theta, \tag{9.13}$$

for some function $\mu(x, u, p)$, where $\theta = du - p\, dx$ is the contact form. The reader should verify that the two one-form conditions (9.6), (9.13), completely encode the standard equivalence problem for variational problems under point transformations: If $\Psi\colon \mathrm{J}^1 \to \mathrm{J}^1$ satisfies (9.6) and (9.13) for some choice of functions λ, μ, then, necessarily, Ψ is *a*) a prolonged point transformation, which *b*) maps the variational problem with Lagrangian L to that with Lagrangian $\overline{L}$. In the more restrictive fiber-preserving case, we set $\mu = 0$ in (9.13), but retain (9.6) as it stands. On the other hand, a contact transformation does not map a first order Lagrangian directly to another first order Lagrangian, and so this equivalence problem plays no role here. (The equivalence of higher order Lagrangians under contact transformations is, however, of interest; see Chapter 10.)

As for the divergence equivalence problem, the contact forms transform as usual, (9.6). The Lagrangian transformation rule replacing

(9.13) was given in (7.27):

$$\overline{L}\, d\overline{x} = (L + D_x A)\, dx + \mu(du - p\, dx) = \nu L\, dx + \mu\, \theta, \tag{9.14}$$

where $\nu = 1 + (D_x A)/L$. There are two methods of encoding the transformation rules for the divergence term. The first, discussed in [**129**], is to extend the jet space J^1 by introducing an auxiliary real variable $w \in \mathbb{R}$. Any point transformation will induce a transformation of the extended space $\mathrm{J}^1 \times \mathbb{R}$, where the derivative coordinate p transforms via prolongation, and the auxiliary variable obeys the transformation rule $\overline{w} = w + A(x, u)$. Therefore, its differential satisfies

$$d\overline{w} = dw + dA = dw + D_x A\, dx + A_u \theta = dw + (\nu - 1) L\, dx + \rho\, \theta, \tag{9.15}$$

where ν is as in (9.14), and $\rho = A_u$. As above, the conditions (9.6), (9.14), (9.15) serve to completely encode the divergence equivalence problem under point transformations; furthermore, taking $\mu = 0$ will restrict to fiber-preserving transformations.

An alternative approach, advocated by Bryant, [**27**], is to use the invariance of the Cartan form

$$\eta_L = L\, dx + \frac{\partial L}{\partial p}\, \theta = L\, dx + \frac{\partial L}{\partial p}\, \theta \tag{9.16}$$

directly. According to Exercise 7.40, two first order Lagrangians are divergence equivalent if and only if their Cartan forms are equivalent modulo an exact form. Locally, by the Poincaré Lemma (Theorem 1.39), this condition can be replaced by the equivalence of their differentials, $d\overline{\eta}_{\overline{L}} = d\eta_L$. Using (9.16), we find $d\eta_L = \omega_L \wedge \theta$, where

$$\omega_L = -(L_u - L_{px} - pL_{pu})\, dx + L_{pp}\, dp = -\widetilde{\mathrm{E}}(L)\, dx + L_{pp}\, dp, \tag{9.17}$$

and $\widetilde{\mathrm{E}}(L)$ denotes the J^1 truncation of the Euler–Lagrange expression associated with L. As in (9.2), the invariance of the two-form $d\eta_L = \omega_L \wedge \theta$ can be replaced by the one-form conditions

$$\overline{\omega}_{\overline{L}} = \alpha\, \omega_L + \beta\, \theta, \qquad \overline{\theta} = \gamma\, \omega_L + \delta\, \theta, \qquad \text{where} \qquad \alpha\delta - \beta\gamma = 1.$$

However, since the prolonged transformation must preserve the contact form, $\gamma = 0$, and so the equivalence conditions are entirely encoded in the one-form identities

$$\overline{\theta} = \lambda\, \theta, \qquad \overline{\omega}_{\overline{L}} = \mu\, \theta + \frac{1}{\lambda}\, \omega_L, \tag{9.18}$$

together with either (9.9) or (9.10), depending on whether we are dealing with equivalence under fiber-preserving or point transformations. The reader should verify the claim that these one-form conditions are necessary and sufficient for the divergence equivalence of the two first order Lagrangians. The Bryant approach has the advantage of formulating the equivalence problem on the more natural space — one need not incorporate additional variables. However, as noted in Chapter 7, there is no suitable generalization of the Cartan form to higher order Lagrangians in several variables, so the latter approach is unavailable, whereas the extended space version does carry through to general Lagrangian equivalence problems; see [**132**] for more details.

Coframes and Structure Groups

As we saw in the preceding section, a rather wide class of equivalence problems can be completely reformulated as a system of equations involving one or more differential one-forms. However, unlike the coframe equivalence problem, these one-forms are either not fully invariant, or do not describe a complete coframe on the manifolds in question (or both). The goal of the Cartan equivalence method is to reduce the original one-form formulation of the problem to an equivalence problem involving a full invariant coframe, to which the results of Chapter 8 can be applied. The Cartan method will, in the cases where this is possible, accomplish this, but, in order to begin the algorithm, the problem must first be placed into a more amenable formulation based on a (not necessarily invariant) coframe on the underlying manifold. Some of the preceding problems, such as the equivalence problems for Riemannian metrics and for maximal rank two-forms, are already in the required form, in that the equivalence conditions are already based on a coframe. Others are "underdetermined" in that the equivalence conditions can be formulated using fewer one-forms than would be needed to form a coframe. However, provided the original one-forms are linearly independent at each point, it is easy to convert such underdetermined equivalence problems into the proper form merely by adjoining one or more additional one-forms so as to form a coframe; the transformation rules for the additional one-forms can (unless further information is readily available) be left completely arbitrary. A third possibility that occasionally arises is that the most natural formulation of the requisite equivalence conditions utilizes one-forms which are pointwise linearly dependent, and so cannot form (part of) a coframe. (Note that we are ignoring points where the dimension of the space spanned by the one-forms changes — such singularities are not readily handled by the Cartan method as it currently exists.) Here again, Cartan devised a straightforward algorithm which allows us to reduce such an "overdetermined equivalence problem" to one of the proper form.

Let us now describe exactly what we mean by the "proper form" (in the sense of Cartan) for an equivalence problem. Suppose we are given two coframes $\{\omega^1, \ldots, \omega^m\}$ and $\{\overline{\omega}^1, \ldots, \overline{\omega}^m\}$ defined on m-dimensional manifolds M and $\overline{M}$, respectively. We identify a coframe on M with a column vector $\boldsymbol{\omega} = (\omega^1, \ldots, \omega^m)^T$ whose entries are the one-forms in the coframe, and, similarly, set $\overline{\boldsymbol{\omega}} = (\overline{\omega}^1, \ldots, \overline{\omega}^m)^T$. Unlike the coframe equivalence problem, the one-forms are not mapped directly to each other; rather the pull-back $\Phi^* \overline{\omega}^{\,i}$ of each one-form on $\overline{M}$ is a linear combination of the ω^j's. The usual form of the Cartan method imposes certain additional constraints on what types of linear combinations are allowed. These restrictions are, in all the examples I know, automatically satisfied; however, as we shall see, they are not truly essential for the applicability of the general method; see the remarks at the end of Chapter 10. The constraints to be imposed require the introduction of a matrix Lie group $G \subset \mathrm{GL}(m)$, known as the *structure group* of the equivalence problem.

Definition 9.5. Let $G \subset \mathrm{GL}(m)$ be a Lie group. Let $\boldsymbol{\omega}$ and $\overline{\boldsymbol{\omega}}$ be coframes defined, respectively, on the m-dimensional manifolds M and $\overline{M}$. The *G-valued equivalence problem* for these coframes is to determine whether or not there exists a (local) diffeomorphism $\Phi: M \to \overline{M}$ and a G-valued function $g: M \to G$ with the property that

$$\Phi^* \, \overline{\boldsymbol{\omega}} = g(x) \, \boldsymbol{\omega}. \tag{9.19}$$

In full detail, the equivalence condition (9.19) has the form

$$\Phi^* \, \overline{\omega}^i = \sum_{j=1}^{m} g^i_j(x) \, \omega^j, \qquad \text{for} \qquad i = 1, \ldots, m, \tag{9.20}$$

where the functions $g^i_j(x)$ are the entries of the matrix $g(x)$, which is constrained to belong to the structure group G at each point $x \in M$. The simplest case is when $G = \{e\}$ is the trivial subgroup consisting of just the identity matrix. In this case, (9.20) takes the form $\Phi^* \overline{\omega}^i = \omega^i$,

$i = 1, \ldots, m$, and so the $\{e\}$-valued equivalence problem is exactly the same as the equivalence problem for coframes treated in Chapter 8. (Indeed, the equivalence problem for coframes often goes under the more exotic name "equivalence of $\{e\}$-structures", because the proper abstract framework for such equivalence problems is the geometric theory of G-structures, cf. [**207**].) At the other extreme, if $G = \mathrm{GL}(m)$, then *any* diffeomorphism Φ satisfies (9.19) for some invertible $g(x)$, and so trivially any two coframes are $\mathrm{GL}(m)$–equivalent. Thus, our principal interest will be in the intermediate cases $\{e\} \subsetneq G \subsetneq \mathrm{GL}(m)$. The goal of the equivalence method is, through a series of "invariant operations", to reduce the structure group G to the trivial structure group $\{e\}$, at which point the solution to the equivalence problem for coframes can be used to solve the problem completely. The reduction procedure provides an invariant specification of the entries of the unknown matrix-valued function $g(x)$, thereby determining an *invariant* coframe for the equivalence problem. This may not always be possible, and subsequent chapters will describe other possible outcomes of the equivalence procedure. However, before getting into the details of the algorithm, we first show how the previously described equivalence problems can be placed into the proper form of a G-valued equivalence problem for some matrix Lie group G.

Example 9.6. The equivalence problem for second order ordinary differential equations of Example 9.3 is already in the proper form. The coframe on $M = \mathrm{J}^1$ is

$$\omega^1 = du - p\,dx, \qquad \omega^2 = dp - Q(x,u,p)\,dx, \qquad \omega^3 = dx. \tag{9.21}$$

For the problem of equivalence under contact transformations, the equivalence conditions from (9.6), (9.8), and (9.11) can be rewritten (omitting the pull-back) as

$$\overline{\omega}^1 = a_1\,\omega^1, \qquad \overline{\omega}^2 = a_2\,\omega^1 + a_3\,\omega^2, \qquad \overline{\omega}^3 = a_4\,\omega^1 + a_5\,\omega^2 + a_6\,\omega^3, \tag{9.22}$$

for suitable functions $a_i(x,u,p)$, $i = 1, \ldots, 6$. The equations (9.22) have the required form (9.19), namely

$$\begin{pmatrix} \overline{\omega}^1 \\ \overline{\omega}^2 \\ \overline{\omega}^3 \end{pmatrix} = \begin{pmatrix} a_1 & 0 & 0 \\ a_2 & a_3 & 0 \\ a_4 & a_5 & a_6 \end{pmatrix} \begin{pmatrix} \omega^1 \\ \omega^2 \\ \omega^3 \end{pmatrix},$$

where the structure group is the six-dimensional lower triangular matrix group

$$G_c = \left\{ \begin{pmatrix} a_1 & 0 & 0 \\ a_2 & a_3 & 0 \\ a_4 & a_5 & a_6 \end{pmatrix} \;\middle|\; a_1 a_3 a_6 \neq 0 \right\}. \tag{9.23}$$

For equivalence of second order ordinary differential equations under point transformations, condition (9.10) replaces (9.11), and so $\overline{\omega}^3 = a_4\,\omega^1 + a_6\,\omega^3$. The structure group reduces to the subgroup $G_p \subset G_c$ obtained by setting $a_5 = 0$ in (9.23). For fiber-preserving equivalence, we use (9.9) instead, and so the structure group is the four-dimensional subgroup $G_f \subset G_c$ obtained by setting $a_4 = a_5 = 0$.

Example 9.7. For the (standard) Lagrangian equivalence problem, the conditions (9.6), (9.14) are not, by themselves, of the required form (9.20) since we need three independent one-forms to form a coframe on J^1, so the problem is "underdetermined". To obtain a coframe, besides the Lagrangian form and the contact form, we can include the one-form dp; the three one-forms

$$\omega^1 = du - p\,dx, \qquad \omega^2 = L(x,u,p)\,dx, \qquad \omega^3 = dp, \tag{9.24}$$

form a coframe as long as the Lagrangian L does not vanish. (Points at which $L = 0$ are singularities for our equivalence problem.) Constructing the analogous coframe for the variational problem with Lagrangian $\overline{L}(\bar{x},\bar{u},\bar{p})$, we see that, in the case of point transformations, the basic equivalence conditions take the form

$$\overline{\omega}^1 = a_1\omega^1, \qquad \overline{\omega}^2 = a_2\omega^1 + \omega^2, \qquad \overline{\omega}^3 = a_3\omega^1 + a_4\omega^2 + a_5\omega^3, \tag{9.25}$$

for functions $a_i(x,u,p)$, $i = 1,\ldots,5$. Therefore, the associated structure group will be the five-dimensional Lie group

$$\widetilde{G}_p = \left\{ \begin{pmatrix} a_1 & 0 & 0 \\ a_2 & 1 & 0 \\ a_3 & a_4 & a_5 \end{pmatrix} \;\middle|\; a_1 a_5 \neq 0 \right\}. \tag{9.26}$$

For the more restrictive problem of fiber-preserving equivalence, we restrict to the subgroup $\widetilde{G}_f$ obtained by setting $a_2 = 0$ in (9.26).

As for the divergence equivalence problem, in the extended set-up, we introduce the additional one-form $\omega^4 = dw$ to form a coframe on the extended space $\mathrm{J}^1 \times \mathbb{R}$, and, recalling (9.6), (9.14), (9.15), use the structure group

$$\widehat{G}_p = \left\{ \begin{pmatrix} a_1 & 0 & 0 & 0 \\ a_2 & a_3 & 0 & 0 \\ a_4 & a_5 & a_6 & 0 \\ a_7 & a_3 - 1 & 0 & 1 \end{pmatrix} \;\middle|\; a_1 \cdot a_3 \cdot a_6 \neq 0 \right\}$$

for equivalence under point transformations. The fiber-preserving problem is based on the subgroup $\widehat{G}_f$ obtained by setting $a_2 = 0$.

Exercise 9.8. Formulate the divergence equivalence problem on the space J^1 using the Cartan form approach discussed above. (See the proof of Theorem 11.12 below.)

The other equivalence problems discussed above are handled similarly. The problem for equivalence of Riemannian metrics is already of the required form (9.4), with orthogonal structure group $\mathrm{O}(m)$. Similarly, the equivalence of two-forms of maximal rank on a $2r$-dimensional manifold is the same as the general symplectic $\mathrm{Sp}(2r)$-valued equivalence problem. Finally, the cases of nonmaximal rank two-forms, and one-forms are underdetermined, but the equivalence conditions are readily completed to form a coframe — see Chapter 11.

Normalization

Once we have properly formulated an equivalence problem in terms of a coframe $\boldsymbol{\omega}$ and structure group G, we are in a position to apply Cartan's equivalence algorithm. The preliminary step is to modify the basic equivalence condition (9.19) so as to make the two sets of variables appear more symmetrically. In view of the group property of G, (9.19) will be satisfied if and only if we can find a pair of G-valued functions $\bar{g}(\bar{x})$ and $g(x)$ such that (omitting pull-backs for clarity)

$$\bar{g}(\bar{x})\,\overline{\boldsymbol{\omega}} = g(x)\,\boldsymbol{\omega}. \tag{9.27}$$

(Indeed, the $g(x)$ in (9.19) is replaced by the product $\bar{g}(\bar{x})^{-1} \cdot g(x)$ in (9.27).) This new form of the G-equivalence problem places the barred coordinates on the same footing as the original coordinates, making it more æsthetically pleasing, and, in fact, more useful.

Our goal is to reduce a given G-equivalence problem to a standard equivalence problem for coframes, and the way to do that is to specify the matrix entries of $g = g(x)$ and $\bar{g} = \bar{g}(\bar{x})$ as functions of their respective coordinates. Of course, this must be done in an invariant manner so as to be compatible with the equivalence condition (9.27). If this can be done, then the new coframes, defined by

$$\overline{\boldsymbol{\theta}} = \bar{g}(\bar{x}) \cdot \overline{\boldsymbol{\omega}}, \qquad \boldsymbol{\theta} = g(x) \cdot \boldsymbol{\omega}, \tag{9.28}$$

will obviously be invariant: $\Phi^* \overline{\boldsymbol{\theta}} = \boldsymbol{\theta}$, and thus the equivalence problem has been successfully reduced to an equivalence of coframes. The main complication is that it is by no means obvious how this can be effected: How indeed do we determine the functions $g(x)$ and $\bar{g}(\bar{x})$ in an algorithmic manner?

The key to answering this question is to find combinations of the group parameters g^i_j, and, possibly, the coordinates x^k, which are unchanged by the coordinate transformations, meaning that they retain the same formula in the transformed coordinates $\bar{g}^i_j$, $\bar{x}^k$. Each such combination will enable us to normalize one of the parameters and thereby reduce the dimension of the structure group by one. If we can find enough invariant combinations, we can effectively eliminate all the group parameters, reducing G to the trivial structure group $\widetilde{G} = \{e\}$, and hence our equivalence problem will reduce to the coframe equivalence problem, which we now know how to solve. More specifically, suppose that, for any coframe of the given form, we can find a scalar function $H(g, \omega)$ depending on the coordinates and the coframe elements† which is invariant under any diffeomorphism Φ, meaning that

$$H\big(\bar{g}(\bar{x}), \overline{\omega}|_{\bar{x}}\big) = H\big(g(x), \omega|_x\big), \quad \text{whenever} \quad \begin{aligned} \bar{x} &= \Phi(x), \\ \Phi^*\big(\bar{g}(\bar{x})\,\overline{\boldsymbol{\omega}}\big) &= g(x)\,\boldsymbol{\omega}. \end{aligned} \tag{9.29}$$

As Cartan observed, if H explicitly depends on the group parameters, we can "normalize" H to take on any convenient constant value, e.g., $H = 0$ or $H = 1$, by specifying one of the group parameters as a function of the coordinates and the other group parameters. Of course, our choice must be compatible with any restrictions which might be imposed by the group. The net effect is to eliminate one of the unspecified

† Actually, the notation is slightly misleading, since H may depend not only on the coefficient functions occurring in the coframe $\boldsymbol{\omega}$, but also their derivatives; thus we should really write $H(g, \omega^{(n)})$ as a function of the nth order jet of the coframe. However, this lapse in precision will not affect the discussion.

group parameters, and thus reduce the original equivalence problem to an equivalence problem whose structure group has dimension one less than the original. Before making this more precise, we consider a simple example.

Example 9.9. Consider the equivalence problem for second order ordinary differential equations under fiber-preserving transformations. The Cartan formulation is given by (9.21), (9.22), with $a_4 = 0$, or, in the symmetrical version (9.27),

$$\bar{a}_1 \,\overline{\omega}^1 = a_1 \,\omega^1, \qquad \bar{a}_2 \,\overline{\omega}^1 + \bar{a}_3 \,\overline{\omega}^2 = a_2 \,\omega^1 + a_3 \,\omega^2, \qquad \bar{a}_6 \,\omega^3 = a_6 \,\omega^3. \tag{9.30}$$

Consider a fiber-preserving transformation $\bar{x} = \varphi(x)$, $\bar{u} = \psi(x,u)$, with first prolongation $\bar{p} = (\psi_u p + \psi_x)/\varphi_x$, cf. (4.13). Substituting into (9.30), and recalling (9.21), (9.22), we find the explicit transformation rules

$$\bar{a}_1 \,\psi_u = a_1, \qquad \bar{a}_3 \,\frac{\psi_u}{\varphi_x} = a_3, \qquad \bar{a}_6 \,\varphi_x = a_6,$$

for the group parameters. By inspection, we discover that the ratio

$$\frac{\bar{a}_3 \bar{a}_6}{\bar{a}_1} = \frac{a_3 a_6}{a_1} \tag{9.31}$$

is an invariant combination. (This result also holds for equivalence under point and contact transformations.) We can equate both ratios to 1 (or, indeed, any other nonzero constant) without changing the invariant nature of the equivalence conditions — if the two coframes are equivalent for some choice of group parameters, they are also equivalent for a choice that makes the two ratios (9.31) equal to 1. Consequently, the normalizations $a_3 = a_1/a_6$, $\bar{a}_3 = \bar{a}_1/\bar{a}_6$, can be substituted into the original equivalence conditions (9.30) without affecting the equivalence of the two coframes. The net effect is a new equivalence problem

$$\bar{a}_1 \,\overline{\omega}^1 = a_1 \,\omega^1, \qquad \bar{a}_2 \,\overline{\omega}^1 + \frac{\bar{a}_1}{\bar{a}_6} \,\overline{\omega}^2 = a_2 \,\omega^1 + \frac{a_1}{a_6} \,\omega^2, \qquad \bar{a}_6 \,\omega^3 = a_6 \,\omega^3, \tag{9.32}$$

involving the same base coframe, but whose structure group has been reduced to the three-dimensional subgroup

$$\left\{ \begin{pmatrix} a_1 & 0 & 0 \\ a_2 & a_6^{-1} a_1 & 0 \\ 0 & 0 & a_6 \end{pmatrix} \;\middle|\; a_1 a_6 \neq 0 \right\},$$

thereby simplifying the problem under consideration. Note that the resulting normalizations are a direct consequence of the basic transformation rules for the prolonged action, and so, if we had been a bit smarter at the outset, could have been incorporated into our original formulation of the equivalence problem. A more detailed analysis of the transformation rules for the coframe can, with enough persistence and ingenuity, lead to additional invariant combinations of group parameters, thereby producing further reductions of the structure group. Of course, unless we are endowed with unusually perceptive insight, the appropriate invariant combinations of variables and group parameters may not be so readily apparent. The algorithmic calculation of the required combinations is the thrust of the powerful Cartan equivalence method, to be discussed in detail in the following three chapters.

The general situation is similar: One utilizes known invariant combinations, as in (9.29), in order to normalize group parameters. If the invariant combination explicitly depends on the coframe, the resulting normalization will also involve the coordinates x. In most situations, one can, by suitably modifying the original coframe, separate out the coordinate dependent parts of the normalization, and reconstitute the original equivalence problem as a reduced equivalence problem of the type discussed above, incorporating some lower dimensional, reduced structure group. Occasionally, however, this is not possible, the normalization being of "nonconstant type", and one must enlarge the class of equivalence problems which must be considered.

To be more specific, consider a G-valued equivalence problem, based on coframes ω on M and $\overline{\omega}$ on $\overline{M}$. Let $H(g,\omega)$ be an invariant combination (possibly vector-valued), satisfying (9.29). We let N denote the range of H. Note that, for a fixed equivalence map Φ, the equivalence conditions (9.27) are unique up to multiplication by a group-valued function. In other words, if $g^*(x) = \bar{g}^*(\Phi(x)) = \bar{g}^*(\bar{x})$ is any G-valued function, then clearly

$$\bar{g}^*(\bar{x})\bar{g}(\bar{x})\,\overline{\omega} = g^*(x)g(x)\,\omega. \tag{9.33}$$

The invariance of H, cf. (9.29), implies that

$$H(\bar{g}^*\bar{g},\overline{\omega}) = H(g^*g,\omega) \tag{9.34}$$

also. Since (9.34) holds for all such functions $g^*(x) = \bar{g}^*(\bar{x})$, we can conclude that the left action of G on itself preserves the level sets of

H, i.e., (9.33) implies that $H(g^*\bar{g},\bar{\omega}) = H(g^*g,\omega)$ for any $g^* \in G$. Consequently, there is an induced global action of G on the range of H, denoted by $z \mapsto h \cdot z$ for $z \in N$, $h \in G$, defined so that

$$g^* \cdot H(g,\omega) = H(g^* \cdot g,\omega), \qquad g^* \in G. \tag{9.35}$$

In particular, $H(g,\omega) = g \cdot H(e,\omega)$, so (9.33) implies that if $\bar{\omega}$ and ω are equivalent coframes under $\bar{x} = \Phi(x)$, then the points $H(e,\bar{\omega}|_{\bar{x}})$ and $H(e,\omega|_x)$ must lie in the *same* orbit of G on N.

Further analysis rests on the precise structure of group action of G on the space N. The simplest case, which is the one occurring most often and illustrated by Example 9.9, with $N = \mathbb{R} \setminus \{0\}$, happens when G acts transitively on N. More general, but equally easy to handle, are the normalizations of "constant type".

Definition 9.10. An invariant function $H(g,\omega)$ is said to define a normalization of *constant type* for the coframe ω if, for each $x \in M$, the values $H(e,\omega|_x)$ lie in a single orbit $\mathcal{O}_\omega \subset N$ of the induced action of G on the range of H.

Note that transitive normalizations are trivially of constant type. In the constant type case, we normalize the group parameters by fixing a convenient canonical form in the prescribed orbit, i.e., we choose a preferred point $z \in \mathcal{O}_\omega$. The invariance of H implies that there exists a group-dependent function $g_0(x) \in G$ such that $H(g_0(x),\omega|_x) = z$ on all of M; we shall assume H is sufficiently regular that $g_0(x)$ can be chosen to vary smoothly with $x \in M$. Performing a similar normalization of $\bar{\omega}$ then allows us to replace the original equivalence problem by one with a reduced structure group.

Proposition 9.11. *Assume ω and $\bar{\omega}$ are both of constant type for the invariant function $H(g,\omega)$. If ω and $\bar{\omega}$ are equivalent, then their associated orbits in N are necessarily the same $\mathcal{O}_{\bar{\omega}} = \mathcal{O}_\omega$. Moreover, fixing a point z in the orbit, we let $\widetilde{G} = G_z \subset G$ be the associated isotropy subgroup. We construct smooth G-valued functions $\bar{g}_0(\bar{x})$ and $g_0(x)$ so that $H(\bar{g}_0,\bar{\omega}) = z = H(g_0,\omega)$, which we use to define the modified coframes $\varpi = g_0\omega$, $\bar{\varpi} = \bar{g}_0\bar{\omega}$. The original two coframes are equivalent, $\bar{g}\,\bar{\omega} = g\,\omega$, for some specification of the group parameters $g,\bar{g} \in G$, if and only if the modified coframes are equivalent,*

$$\bar{h}(\bar{x})\,\bar{\boldsymbol{\varpi}} = h(x)\,\boldsymbol{\varpi} \qquad \text{when} \qquad \bar{x} = \Phi(x), \tag{9.36}$$

for some specification of the reduced group parameters $\bar{h}, h \in \widetilde{G}$.

Proof: Since $H(g,\omega) = H(\bar{g},\bar{\omega})$ are assumed to both lie in the same single orbit $\mathcal{O}$ of G, then, by regularity and invariance of H, we can choose a smooth group-valued function $g^*(x) = \bar{g}^*(\bar{x})$ so that $H(\bar{g}^*\bar{g},\bar{\omega}) = z = H(g^*g,\omega)$ for fixed $z \in \mathcal{O}$. Replacing the original equivalence condition (9.27) by the modified condition (9.33), we see that if the two coframes are equivalent, then they are equivalent for some specification of group parameters that fixes $H = z$. Define the G-valued function $h(x)$ so that $g^* \cdot g = h \cdot g_0$. Since $H(h \cdot g_0, \omega) = z = H(g_0, \omega)$, we deduce that $h(x)$ takes values in the isotropy group $\widetilde{G} = G_z$. Similarly define $\bar{h}(\bar{x}) \in \widetilde{G}$ so that $\bar{g}^* \cdot \bar{g} = \bar{h} \cdot \bar{g}_0$. Substituting these formulae into the modified equivalence condition (9.33), we deduce the equivalence (9.36) of the modified coframes. *Q.E.D.*

Thus the reduced structure group $\widetilde{G} \subset G$ can be identified with the isotropy subgroup of the chosen canonical form for the function H. In particular, if G acts freely on N, then the isotropy subgroup is trivial $G_z = \{e\}$. In this case, we have reduced the structure group to the identity, and reduced the equivalence problem to an equivalence of coframes, as we wanted.

Remark: If H is a scalar function, then, away from singular points, there are but two possibilities. Either the action of G on $N \subset \mathbb{R}$ is trivial, in which case $H = H(x)$ does not depend on g at all, and so defines an invariant function for our problem in the sense of Chapter 8, which will, ultimately, constitute one of the structure invariants for the reduced coframe equivalence problem. Such functions clearly cannot be used for normalization purposes. Otherwise, the action of G is nontrivial, in which case it is automatically transitive, and hence the normalization is of constant type. Therefore, the more complicated nonconstant type normalizations only occur when there are several invariant functions which must be simultaneously normalized.

The cases of nonconstant type are more difficult, but still can be handled by Cartan's method. The main complication is that the normalizations take us (slightly) outside the usual equivalence framework of structure group plus base coframe, which explains the reluctance of more recent authors, [**78**], to pursue these problems any further. A simple example will suffice to indicate what happens. Suppose the structure group $G = \mathbb{R}^2$ is the two-dimensional abelian group, parametrized by $g = (a, b) \in \mathbb{R}^2$. Suppose the invariant combination has two components $H(g, \omega) = (I(x), a + b + I(x))$, where $I(x) = \bar{I}(\bar{x})$ is an ordinary (non group-dependent) invariant. Note that the orbits of G on $N \simeq \mathbb{R}^2$ are the vertical lines, but the image of H is not contained in a single G-orbit. As Cartan explains, [**39**], we can still normalize the second component of H by setting $a = -b - I(x)$ and substituting back into the original cofame. The problem now is that we can no longer separate out the remaining group parameter b from the invariant function I, and so the reduced structure group will vary from point to point. Therefore, the normalized problem lies outside our original formulation; nevertheless, the reduction and normalization procedure of Cartan can still be practically continued. Since such problems do not appear in any of the relatively simple examples that we treat here (with the notable exception of the multi-dimensional Lagrangian equivalence problem discussed at the end of Chapter 10), we will concentrate on the Cartan framework discussed earlier and merely comment at appropriate places how one might generalize the constructions to include equivalence problems resulting from normalizations of nonconstant type.

Overdetermined Equivalence Problems

Before proceeding with the details of the Cartan equivalence method, let us complete our preliminary discussions with the formulation of the overdetermined equivalence problem. Most of the equivalence problems that one encounters are either in the proper Cartan form of a G-valued equivalence problem, or are underdetermined, and so can be readily put into the proper form by adjoining additional one-forms whose precise transformation properties can be left completely general. Occasionally, though, one encounters an equivalence problem which cannot be so readily encoded by a collection of linearly independent one-forms, and are not so immediately fitted into a Cartan framework. Equivalence problems which are expressed using a (pointwise) linearly dependent collection of one-forms are called *overdetermined*. As Cartan discovered, in this case invariant combinations of group parameters can be easily determined, and the problem straightforwardly reduces to one in the required form.

The overdetermined reduction algorithm proceeds along the following lines. One divides the defining one-forms into linearly independent subsets, each of which is either a coframe or can be readily completed to one. (As usual, we avoid singular points.) For simplicity, let us assume that there are just two such classes, and we begin by completing each to a coframe. Therefore, we have a pair of coframes $\boldsymbol{\omega}$ and $\boldsymbol{\varpi}$ on the manifold M, and corresponding coframes $\overline{\omega}$ and $\overline{\varpi}$ on $\overline{M}$. Also assume that we have two distinct matrix subgroups $G, H \subset \mathrm{GL}(m)$, so that the equivalence conditions are represented by the combined coframe conditions

$$\Phi^* \overline{\omega} = g\,\omega, \qquad \Phi^* \overline{\varpi} = h\,\varpi,$$

which must be satisfied by the (same) equivalence map $\Phi: M \to \overline{M}$ for particular choices $g = g(x)$ and $h = h(x)$ of G- and H-valued functions on M. More symmetrically, as in (9.27), (and, for simplicity, dropping the pull-backs),

$$\bar{g}\,\overline{\omega} = g\,\omega, \qquad \bar{h}\,\overline{\varpi} = h\,\varpi. \tag{9.37}$$

As in the case of a single coframe, we are allowed to normalize the group parameters $g, \bar{g}$ and $h, \bar{h}$ in any consistent, invariant manner. Now, since both ω and ϖ are coframes, we can express the ω's in terms of the ϖ's (and vice versa), so that

$$\overline{\omega} = \overline{A}\,\overline{\varpi}, \qquad \omega = A\,\varpi, \tag{9.38}$$

where $\overline{A}(\bar{x}) = \big(\bar{a}^i_j(\bar{x})\big)$ and $A(x) = \big(a^i_j(x)\big)$ are invertible $m \times m$ matrices of functions of the indicated variables. Combining (9.37) and (9.38), we find

$$\bar{g}\,\overline{A}\,\overline{\varpi} = gA\,\varpi, \qquad \text{hence} \qquad \bar{g}\,\overline{A}\,\bar{h}^{-1}(\bar{h}\,\overline{\varpi}) = gAh^{-1}(h\,\varpi). \tag{9.39}$$

Comparing the second equation in (9.39) with the second equivalence equation in (9.37), and using the fact that ϖ is a coframe, we deduce that

$$\bar{g}\,\overline{A}\,\bar{h}^{-1} = gAh^{-1} \qquad \text{whenever} \qquad \bar{x} = \Phi(x).$$

Thus, we have proved the following result.

Proposition 9.12. *Consider an overdetermined equivalence problem involving coframes ω and ϖ, and corresponding structure groups G and H, respectively, as in (9.37). Suppose the two coframes are related by $\omega = A(x)\,\varpi$. Then the entries of the matrix*

$$g \cdot A(x) \cdot h^{-1}, \qquad g \in G, \quad h \in H, \tag{9.40}$$

are invariant under changes of variables $\bar{x} = \Phi(x)$.

Any entries of the matrix (9.40) which explicitly depend on the group parameters g, h can be normalized as before. Non-constant entries which do not depend on any group parameters provide invariant functions for the equivalence problem, and must be incorporated into its solution. Under the requisite normalizations, the two equivalence conditions will become consistent, and the overdetermined problem will reduce to an ordinary equivalence problem involving a single structure group.

Example 9.13. An interesting example is the problem of equivalence of Lagrangians $L(x,u,p)$ under the (finite-dimensional) Euclidean group E(2) of planar isometries. In this case, the point transformation must preserve the Euclidean metric, $\Phi^*(d\bar{x}^2 + d\bar{u}^2) = dx^2 + du^2$. As remarked above, cf. (9.4), the isometry condition can be recast in terms of one-forms as

$$d\bar{x} = \cos\phi\, dx - \sin\phi\, du, \qquad d\bar{u} = \sin\phi\, dx + \cos\phi\, du.$$

On the other hand, the Lagrangian equivalence conditions are given by (9.6), (9.13), and there is no immediately obvious way (at least to me) to extract a coframe from these four conditions. In order to apply the overdetermined reduction method of Proposition 9.12, we complete the Lagrangian equivalence conditions to form a coframe, as in (9.24). Similarly, the isometry conditions can be placed into the required form by using the coordinate coframe

$$\varpi^1 = dx, \qquad \varpi^2 = du, \qquad \varpi^3 = dp, \tag{9.41}$$

and associated structure group

$$H = \left\{ \begin{pmatrix} \cos\phi & -\sin\phi & 0 \\ \sin\phi & \cos\phi & 0 \\ c_1 & c_2 & c_3 \end{pmatrix} \;\middle|\; c_3 \neq 0 \right\}.$$

The two coframes (9.24), (9.41) are connected by the matrix

$$\omega = A \cdot \varpi, \qquad \text{where} \qquad A = \begin{pmatrix} -p & 1 & 0 \\ L & 0 & 0 \\ 0 & 0 & 1 \end{pmatrix}.$$

Proposition 9.12 states that we can normalize the entries of the 3×3 matrix gAh^{-1}, which equals

$$\begin{pmatrix} -a_1(\sin\phi + p\cos\phi) & a_1(\cos\phi - p\sin\phi) & 0 \\ L\cos\phi - a_2(\sin\phi + p\cos\phi) & L\sin\phi + a_2(\cos\phi - p\sin\phi) & 0 \\ w_1 & w_2 & a_5/c_3 \end{pmatrix}$$

where $h \in H$, $g \in \widetilde{G}_p$, and

$$w_1 = -a_3(\sin\phi + p\cos\phi) + a_4 L\cos\phi + \frac{a_5}{c_3}(c_1\cos\phi + c_2\sin\phi),$$

$$w_2 = a_3(\cos\phi - p\sin\phi) + a_4 L\sin\phi - \frac{a_5}{c_3}(c_1\sin\phi + c_2\cos\phi).$$

(The latter formulae are not really important, since dp occurs in both coframes, so we can clearly normalize $c_1 = La_4 - pa_3$, $c_2 = a_3$, $c_3 = a_5$.) The following normalizations make the (1,1), (1,2), and (2,1) entries of gAh^{-1} equal to $-1, 0, 0$ respectively:

$$\cot\phi = p, \qquad a_1 = \sin\phi = \frac{1}{\sqrt{1+p^2}}, \qquad a_2 = L\sin\phi\cos\phi = \frac{pL}{1+p^2}. \tag{9.42}$$

The (2,2) entry then provides an invariant function for the problem,

$$I(x,u,p) = \frac{L(x,u,p)}{\sqrt{p^2+1}}, \tag{9.43}$$

which reflects the fact that both $L\,dx$ and $\sqrt{p^2+1}\,dx$ are contact-invariant one-forms, so their ratio is a differential invariant; see Chapter 7. The reader can check directly that if the Lagrangians L and $\overline{L}$ are mapped to each other via an isometry, then $\overline{I} = I$ are equal. The normalizations (9.42) have completely specified the group parameters a_1, a_2 in an invariant manner, and so provide two invariant one-forms, namely

$$\theta^1 = a_1\omega^1 = \frac{du - p\,dx}{\sqrt{p^2+1}}, \tag{9.44}$$

and

$$\widetilde{\theta}^{\,2} = a_2\omega^1 + \omega^2 = (1+p^2)^{-1} L\,\{\,dx + p\,du\}.$$

Since we can multiply any invariant coframe element by an invariant function, we make use of the invariant (9.43) to replace $\widetilde{\theta}^2$ by the slightly simpler one-form

$$\theta^2 = \frac{dx + p\,du}{\sqrt{p^2+1}}. \tag{9.45}$$

Moreover, the differential of our invariant function (9.43) provides a third invariant one-form

$$\theta^3 = dI = \frac{L_x\,dx + L_u\,du}{\sqrt{1+p^2}} + \frac{(1+p^2)L_p - pL}{(1+p^2)^{3/2}}\,dp. \tag{9.46}$$

The three one-forms (9.44), (9.45), (9.46) will provide an "adapted" invariant coframe, unless the coefficient I_p of dp in θ^3 is zero. Therefore, apart from "singular points", the only variational problems for which we cannot readily deduce an invariant coframe are the inhomogeneous arc length functionals $\mathcal{I}[u] = \int h(x,u)\,ds = \int h(x,u)\sqrt{1+u_x^2}\,dx$, which form a separate equivalence class. Such variational problems govern the paths traced by rays in an inhomogeneous optical medium, cf. [**85**].

Let us complete the solution to the equivalence problem, using the explicit form (9.44), (9.45), (9.46) for the invariant coframe. A straightforward computation proves that the structure equations are

$$\begin{aligned} d\theta^1 &= K_1\,\theta^1\wedge\theta^2 + J\,\theta^2\wedge\theta^3, \\ d\theta^2 &= K_2\,\theta^1\wedge\theta^2 - J\,\theta^1\wedge\theta^3, \end{aligned} \qquad d\theta^3 = 0, \tag{9.47}$$

where the structure functions have the explicit formulae

$$\begin{aligned} J &= \frac{1}{(1+p^2)I_p} = \frac{\sqrt{1+p^2}}{(1+p^2)L_p - pL}, \\ K_1 &= \frac{L_u - pL_x}{1+p^2}\,J, \qquad K_2 = \frac{L_x + pL_u}{1+p^2}\,J. \end{aligned} \tag{9.48}$$

It is interesting that the original invariant I, given in (9.43), does *not* appear among the structure functions of the adapted coframe. Nor does it appear among the resulting derived invariants, based on the adapted coframe derivatives

$$\begin{aligned}
\frac{\partial}{\partial\theta^1} &= \frac{1}{\sqrt{1+p^2}}\left(-p\,\frac{\partial}{\partial x}+\frac{\partial}{\partial u}\right)-(1+p^2)K_1\,\frac{\partial}{\partial p}\,,\\
\frac{\partial}{\partial\theta^2} &= \frac{1}{\sqrt{1+p^2}}\left(\frac{\partial}{\partial x}+p\,\frac{\partial}{\partial u}\right)-(1+p^2)K_2\,\frac{\partial}{\partial p}\,,\\
\frac{\partial}{\partial\theta^3} &= (1+p^2)J\,\frac{\partial}{\partial p}\,.
\end{aligned} \tag{9.49}$$

Indeed, we have been a little careless in our set-up of the equivalence problem. The coframe (9.44), (9.45), (9.46) does *not* in fact properly encode the isometric Lagrangian equivalence problem, but must be supplemented by the additional invariant I in order to effect the correct solution to the problem. The resulting equivalence problem has the nonstandard form indicated at the end of Chapter 8, in which the invariant coframe must be supplemented by an additional invariant function; see (8.36). In the present situation, there are three invariants provided by the structure functions (9.48) as well as one additional invariant I, all of whose derived invariants must be taken into account when discussing the solution to the problem. Thus, for example, the zeroth order extended classifying manifold is parametrized by I, J, K_1, K_2, the first order classifying manifold by them and their coframe derivatives, etc. Note, however, that since $dI = \theta_3$, the coframe derivatives of I are trivial: $\partial I/\partial\theta^1 = \partial I/\partial\theta^2 = 0$, $\partial I/\partial\theta^3 = 1$, and hence do not need to be taken into account in the analysis of the higher order classifying manifolds.

Let us consider the special case when the Lagrangian only depends on p, a condition which is clearly invariant under Euclidean isometries. (In particular, if $L(p) = \sqrt[n]{R(p)}$ where $R(p)$ is a polynomial of degree n, then the solution to the equivalence problem for the invariant coframe will provide a solution to the equivalence problem for binary forms under the orthogonal group $O(2)$.) Assuming I is not constant, we find $K_1 = K_2 = 0$, which is necessary and sufficient for L to depend only on p. In this case, the equivalence class of the variational problem is characterized by the first order (reduced) classifying curve, which is the curve parametrized by the two invariants $(I(p), J(p))$.

Note: Theorem 8.19 indicates that we should also consider the derived invariant $J_3(p) = \partial J/\partial\theta^3 = (1+p^2)JJ_p$. However, according to our chain rule argument, (8.37), once the classifying function $J = H(I)$ is known, then $\partial J/\partial\theta^3 = H'(I)\partial I/\partial\theta^3 = H'(I)$ is determined, and hence the derived invariant $\partial J/\partial\theta^3$ is not needed. Note also that we can always (locally) choose I as the fundamental invariant, since if $I = \widetilde{H}(J)$, then $1 = \partial I/\partial\theta^3 = \widetilde{H}'(J)\partial J/\partial\theta^3$, hence, away from singularities, $\widetilde{H}'(J) \neq 0$, so that we can always solve for $J = H(I)$. If J is constant, then the symmetry group is the full three-dimensional Euclidean group; otherwise it is reduced to the two-dimensional translational subgroup.

Exercise 9.14. Determine all Lagrangians admitting a three-dimensional group of isometric symmetries. Discuss their geometric significance (if any).

Exercise 9.15. Suppose one omits the invariant I from the preceding equivalence problem. What does the invariance of the three one-forms (9.44), (9.45), (9.46) say about the corresponding Lagrangians?

Chapter 10

Cartan's Equivalence Method

Once an equivalence problem has been reformulated in the proper Cartan form, in terms of a coframe $\boldsymbol{\omega}$ on the m-dimensional base manifold M, along with a structure group $G \subset \mathrm{GL}(m)$, we are ready to apply the Cartan equivalence method. As remarked in the previous chapter, the goal is to normalize the structure group valued coefficients in a suitably invariant manner, and this is accomplished through the determination of a sufficient number of invariant combinations thereof. The Cartan method provides an algorithmic approach for finding such invariant combinations. Each group-dependent invariant combination allows us to normalize one group parameter. In the favorable cases, we can determine enough different invariant combinations so as to normalize all the group parameters, thereby reducing the structure group to the trivial group $\{e\}$. In this case, the equivalence problem has been reduced to an equivalence problem for coframes, which we now know how to solve. In this chapter, we shall describe the basic steps in the Cartan equivalence method, and illustrate its application to a variety of problems, chosen so that a complete normalizing reduction can be successfully implemented. The more complicated cases where the method does not produce a full complement of normalizations for the group parameters will be dealt with in the following two chapters.

The Structure Equations

Consider a coframe $\boldsymbol{\omega} = \{\omega^1, \ldots, \omega^m\}$ on the m-dimensional manifold M, which, as in (9.19), we identify with a column vector of one-forms, and associated structure group $G \subset \mathrm{GL}(m)$. The corresponding G-valued equivalence problem has the form prescribed in Definition 9.5, or, in a more symmetrical form, equations (9.28), (9.36). This motivates the preliminary step in the Cartan solution to the equivalence problem, which is to introduce the *lifted coframe*

$$\boldsymbol{\theta} = g \cdot \boldsymbol{\omega}, \qquad \text{or, in full detail,} \qquad \theta^i = \sum_{j=1}^{m} g^i_j \, \omega^j. \tag{10.1}$$

In this formula, the group parameters (which we are trying to normalize) are left arbitrary, so that the one-forms θ^i in the lifted coframe are, in fact, defined on the Cartesian product space† $M \times G$. Technically speaking, the θ^i's do *not* form a coframe on $M \times G$, since they are only m in number, and only involve the differentials dx^k on M, so the term "lifted coframe" is slightly misleading. (We can, however, easily complete $\boldsymbol{\theta}$ to a full coframe by adjoining the Maurer–Cartan coframe on G itself — see below.) Any explicit normalization $g = g(x)$ of the group parameters, then, amounts to restricting the lifted coframe $\boldsymbol{\theta}$ to the graph of a function $g = g(x)$, i.e., a section of the space $M \times G$. Thus, one can reinterpret the final goal of the equivalence method as the determination of suitable sections of the two spaces on which the two lifted coframes agree under pull-back.

As in our solution to the equivalence problem for coframes, the invariance of the exterior derivative operator is the key! Let us begin by computing the differentials of the lifted coframe elements (10.1):

$$d\theta^i = d\left(\sum_{j=1}^{m} g^i_j \, \omega^j \right) = \sum_{j=1}^{m} \left\{ dg^i_j \wedge \omega^j + g^i_j \, d\omega^j \right\}.$$

Now, since the ω^i's form a coframe on M, the two-forms $d\omega^j$ can be re-expressed in terms of sums of wedge products of the ω^i's. Moreover, in view of (10.1), these in turn can be rewritten as wedge products of the θ^k's, so that

$$d\theta^i = \sum_{j=1}^{m} \gamma^i_j \wedge \theta^j + \sum_{\substack{j,k=1 \\ j<k}}^{m} T^i_{jk}(x,g) \, \theta^j \wedge \theta^k, \qquad i = 1, \ldots, m. \tag{10.2}$$

† A more intrinsic, global context for the construction is to replace the product space $M \times G$ by a principal G bundle over M, which serves to define the G–structure; see, for instance, [**207**]. As all our considerations are local, we shall not lose anything by sticking with the conceptually simpler product space here.

The *torsion coefficients* T^i_{jk} are functions, which may be constant, or may depend on the base variables x and/or the group parameters g; they are typically *not* invariants for the problem, although, as we shall see, some of them may be. The γ^i_j in (10.2) are the one-forms

$$\gamma^i_j = \sum_{k=1}^{m} dg^i_k \, (g^{-1})^k_j, \qquad \text{or, in matrix notation,} \qquad \gamma = dg \cdot g^{-1}. \tag{10.3}$$

In view of (2.35), γ forms the matrix of Maurer–Cartan forms on the structure group G and can therefore be readily determined. Indeed, if $\alpha^1, \ldots, \alpha^r$ are a basis for the space of Maurer–Cartan forms (usually corresponding to a local system of coordinates $a = (a^1, \ldots, a^r)$ in a neighborhood of the identity of G) then each γ^i_j is a certain linear combination of the Maurer–Cartan basis:

$$\gamma^i_j = \sum_{\kappa=1}^{r} A^i_{j\kappa} \, \alpha^\kappa, \qquad i, j = 1, \ldots, m. \tag{10.4}$$

The fact that the a^κ parametrize a group implies that the coefficients $A^i_{j\kappa}$ are constant,† their values determined solely by the matrix representation of $G \subset \mathrm{GL}(m)$. In terms of the Maurer–Cartan forms, then, the final *structure equations* for our lifted coframe have the general form

$$d\theta^i = \sum_{\kappa=1}^{r} \sum_{j=1}^{m} A^i_{j\kappa} \, \alpha^\kappa \wedge \theta^j + \sum_{\substack{j,k=1 \\ j<k}}^{m} T^i_{jk}(x, g) \, \theta^j \wedge \theta^k, \qquad i = 1, \ldots, m. \tag{10.5}$$

Performing an analogous computation for the barred coframe $\overline{\boldsymbol{\theta}} = \bar{g} \cdot \overline{\boldsymbol{\omega}}$ on $\overline{M}$ leads to the corresponding structure equations

$$d\overline{\theta}^i = \sum_{\kappa=1}^{r} \sum_{j=1}^{m} A^i_{j\kappa} \, \overline{\alpha}^\kappa \wedge \overline{\theta}^j + \sum_{\substack{j,k=1 \\ j<k}}^{m} \overline{T}^i_{jk}(\bar{x}, \bar{g}) \, \overline{\theta}^j \wedge \overline{\theta}^k, \qquad i = 1, \ldots, m, \tag{10.6}$$

† In the more general equivalence problem considered by Cartan, the equivalence conditions (9.27) no longer require that the functions $g(x)$ and $\bar{g}(\bar{x})$ belong to a subgroup of $\mathrm{GL}(m)$; in this case, the $A^i_{j\kappa}$ are no longer constant, but provide additional invariants for the problem. If they depend on the group parameters, they can be effectively normalized to reduce the problem as before. Such cases can arise in the implementation of the method when one confronts a normalization of "nonconstant type", as discussed in Chapter 9.

in which $\overline{T}^{\,i}_{jk}$ are the associated torsion coefficients, and $\overline{\alpha}^\kappa$ are the same Maurer–Cartan forms on the structure group $\overline{G}$, but written in terms of the barred group parameters $\overline{g}^i_j$. Note particularly that the coefficients $A^i_{j\kappa}$ are the *same* as in the unbarred case, since the group parameters, and hence the Maurer–Cartan forms, enter both lifted coframes in an identical manner, cf. (9.36). We now analyze the invariant consequences of the structure equations (10.5), (10.6) in detail.

Absorption and Normalization

Any consistent specification of the group parameters as functions $g = g(x)$ of $x \in M$ will reduce the Maurer–Cartan forms α^κ on G to one-forms $\widetilde{\alpha}^\kappa = \sum_i v^\kappa_i(x)\,dx^i$ on the base M, which we can therefore re-express in terms of the coframe $\boldsymbol{\omega}$, or, as before, in terms of the lifted coframe elements: $\widetilde{\alpha}^\kappa = \sum_k z^\kappa_j(x)\,\theta^j$. The problem is, of course, that we still don't know what functions the $g^i_j(x)$ should be, and hence have no a priori knowledge of what the coefficients $z^\kappa_j(x)$ are. Nevertheless, some data are available, and the manner in which the Maurer–Cartan forms enter into the structure equations (10.5) will allow us to determine which of the associated torsion coefficients T^i_{jk} are invariant. Let us reduce the Maurer–Cartan forms α^κ back to the base manifold M by replacing them by general linear combinations of coframe elements

$$\alpha^\kappa \longmapsto \sum_{k=1}^{r} z^\kappa_j\,\theta^j. \tag{10.7}$$

In (10.7), the z^κ_j's are as yet unspecified coefficients, whose explicit dependence on x will ultimately be determined once the $g^i_j(x)$'s have all been pinned down. Substituting (10.7) into the structure equations (10.5) leads to a system of two-forms

$$\Theta^i = \sum_{\substack{j,k=1\\ j<k}}^{m} \left\{ B^i_{jk}[\mathbf{z}] + T^i_{jk}(x,g) \right\} \theta^j \wedge \theta^k, \qquad i = 1,\ldots,m. \tag{10.8}$$

Here

$$B^i_{jk}[\mathbf{z}] = \sum_{\kappa=1}^{r} \left(A^i_{k\kappa}\, z^\kappa_j - A^i_{j\kappa}\, z^\kappa_k \right), \tag{10.9}$$

are certain linear functions of the coefficients $\mathbf{z} = (z_k^\kappa)$, whose (constant) coefficients are determined by the specific representation of the structure group G as a subgroup of $\mathrm{GL}(m)$, and so do not depend on the coordinate system. Performing a similar computation for the barred coframe elements $d\bar\theta^i$, cf. (10.6), and then substituting for the group parameters $\bar g(\bar x)$ leads to an analogous set of two-forms

$$\overline{\Theta}^i = \sum_{\substack{j,k=1 \\ j<k}}^{m} \left\{ B^i_{jk}[\bar{\mathbf{z}}] + \overline{T}^i_{jk}(\bar x, \bar g) \right\} \bar\theta^j \wedge \bar\theta^k, \qquad i = 1, \ldots, m, \tag{10.10}$$

involving the barred torsion coefficients, and the *same* linear functions (10.9) of the $\bar z_j^\kappa$'s. If the two original coframes are to be mapped to each other, then, for some choice of the as yet unspecified coefficients $\mathbf{z} = \mathbf{z}(x)$ and $\bar{\mathbf{z}} = \bar{\mathbf{z}}(\bar x)$, each two-form $\overline{\Theta}^i$ will map back to the two-form $\Theta^i = \Phi^*\overline{\Theta}^i$. Since the lifted coframe elements are also mapped directly to each other, $\Phi^*\bar\theta^j = \theta^j$, and are independent, we conclude that the coefficients of the two forms (10.8), (10.10) are themselves invariants of the problem:

$$B^i_{jk}[\bar{\mathbf{z}}] + \overline{T}^i_{jk}(\bar x) = B^i_{jk}[\mathbf{z}] + T^i_{jk}(x), \qquad \text{when} \qquad \bar x = \Phi(x). \tag{10.11}$$

Of course, the dependence of the $\mathbf{z}$'s on x and the $\bar{\mathbf{z}}$'s on $\bar x$ is, at present, unknown, so the equations in which they explicitly occur are not particularly helpful. However, if we find a coefficient of Θ^i which does *not* depend on $\mathbf{z}$, which means $B^i_{jk}[\mathbf{z}] = 0 = B^i_{jk}[\bar{\mathbf{z}}]$ for some j, k, then equation (10.11) implies that the corresponding torsion coefficient will be invariant: $\overline{T}^i_{jk}(\bar x, \bar g) = T^i_{jk}(x, g)$ for any specification of group parameters $\bar g(\bar x)$, $g(x)$! Such torsion coefficients are referred to as *essential torsion* for the structure equations (10.5). If an essential torsion coefficient doesn't explicitly depend on the group parameters, then it provides a true invariant for the problem, just as in the simple equivalence of coframes. On the other hand, if an essential torsion coefficient does explicitly depend on the group parameters, then it provides an invariant combination which, by our earlier remarks, can be normalized by eliminating one of the group parameters, and thereby reducing the structure group.

Remark: There is another, slightly more complicated mechanism for extracting essential torsion from the modified torsion coefficients

$K^i_{jk} = B^i_{jk}[\mathbf{z}] + T^i_{jk}$. A constant coefficient linear combination $\sum c^i_{jk} K^i_{jk}$ which does not explicitly depend on $\mathbf{z}$ provides us with an additional invariant. Suppose, for example, that we find that $K^1_{12} = z^1_1 + T^1_{12}$ and $K^2_{23} = 2z^1_1 + T^2_{23}$ are two of the coefficients in question. Then, in view of (10.11), the combination $2K^1_{12} - K^2_{23} = 2T^1_{12} - T^2_{23}$ is invariant, and so will also contribute to the essential torsion. This more general type of reduction will occur in later examples.

The general process of eliminating the unknown coefficients $\mathbf{z}$ from the full torsion coefficients is known as *absorption of torsion*, and is one of the two fundamental techniques in the implementation of the Cartan equivalence method, the other being *normalization* of the resulting invariant torsion coefficients, as described above. The inessential torsion is *absorbed* in the structure equation (10.5) by replacing each Maurer–Cartan form α^κ by the modified one-form

$$\pi^\kappa = \alpha^\kappa - \sum_{i=1}^{p} z^\kappa_i \, \theta^i, \qquad \kappa = 1, \ldots, r, \tag{10.12}$$

where the $z^\kappa_i = z^\kappa_i(x, g)$ are the solutions to the absorption equations. The net result is that the structure equations take on the simpler "absorbed form"

$$d\theta^i = \sum_{\kappa=1}^{r} \sum_{j=1}^{m} A^i_{j\kappa} \, \pi^\kappa \wedge \theta^j + \sum_{\substack{j,k=1 \\ j<k}}^{m} U^i_{jk} \, \theta^j \wedge \theta^k, \qquad i = 1, \ldots, m, \tag{10.13}$$

where the remaining nonzero coefficients U^i_{jk} consist only of essential torsion. Usually the absorption process is simply done by inspection, although it can readily be formalized into an algorithm, which can be straightforwardly programmed into a computer algebra system such as MATHEMATICA or MAPLE. Write out the linear system of *absorption equations* $\mathbf{B}[\mathbf{z}] = -\mathbf{T}$, or, more explicitly,

$$\sum_{\kappa=1}^{r} A^i_{j\kappa} \, z^\kappa_k - A^i_{k\kappa} \, z^\kappa_j = -T^i_{jk}, \tag{10.14}$$

and solve for the unknowns $\mathbf{z}$ using the standard Gaussian elimination method, as described, for instance, in [**208**]. The result will be a reduced system of the form $\mathbf{U}[\mathbf{z}] = \mathbf{W}$, where the matrix $\mathbf{U}$ is in row echelon form. This can be accomplished by only using the two elementary row operations of adding a multiple of one row to another and interchanging two rows. (In particular, we are not allowing the third row operation of multiplying a row by a nonzero scalar.) Under this restriction, a complete list of essential torsion coefficients is provided by the nonconstant entries of the right hand side $\mathbf{W}$ corresponding to the zero rows of $\mathbf{U}$.

In summary, then, the solution to the G-valued equivalence problem proceeds in a series of essentially algorithmic steps. First, the structure equations (10.2) are computed using the explicit form of the base coframe ω and the structure group G. Second, as many of the torsion coefficients as possible are absorbed by replacing the group part (10.5) of the structure equations by its expanded version (10.12) by suitably choosing the coefficients z_i^κ. Third, the remaining "essential torsion" coefficients which depend explicitly on the group parameters are normalized (if possible) by setting them to convenient constant values (usually ± 1 or 0) by solving for some of the group parameters. This constitutes one loop through the equivalence method. To continue, one substitutes the formulae deduced by the normalization part of the procedure back into the coframe, and then recomputes, going through the loop once again. The process continues until one of two things happens. In favorable cases, all the group parameters can eventually be normalized, and the process ends with an explicit invariant coframe, thereby reducing the equivalence problem to the equivalence problem for coframes, which we know how to solve. In more complicated cases, we are left with one or more unspecified group parameters, but no essential torsion coefficients which depend on group parameters, and so the current version of the method does not tell us how to normalize the remaining parameters. We shall defer the discussion of what to do then until the following chapters. We will now illustrate how the general equivalence method, as developed so far, is applied to particular examples. The examples here are chosen so that the method will reach a conclusion and provide a complete invariant coframe for the problem.

Equivalence Problems for Differential Operators

We begin with a relatively simple example that we already know how to solve — the problem of equivalence of second order differential operators. Although the results in Chapter 2 show how the problem can be solved directly, it nevertheless forms a useful exercise to verify that the

Cartan method does produce correct results. Also, the calculations are sufficiently simple that the problem serves as a good warm-up for the more substantial equivalence problems to be treated later. See [**130**] for additional details.

Consider a second order differential operator

$$\mathcal{D} = f(x)D_x^2 + g(x)D_x + h(x), \tag{10.15}$$

where f, g, h are smooth functions of the real variable $x \in \mathbb{R}$. (Here we concentrate on the real-valued equivalence problem, although the method can be readily adapted to complex-valued operators.) We discuss the problem of equivalence under general fiber-preserving transformations which are linear in the dependent variable

$$\bar{x} = \xi(x), \qquad \bar{u} = \eta(x)u, \tag{10.16}$$

where $\eta(x) \neq 0$. We first consider the *direct equivalence problem*, which identifies the two linear differential functions

$$\bar{f}(\bar{x})\bar{q} + \bar{g}(\bar{x})\bar{p} + \bar{h}(\bar{x})\bar{u} = \overline{\mathcal{D}}[\bar{u}] = \mathcal{D}[u] = f(x)q + g(x)p + h(x)u, \tag{10.17}$$

under the change of variables (10.16). This induces the transformation rule (6.81) on the differential operators themselves, and our task is to find explicit conditions on the coefficients of the two differential operators that guarantee that they satisfy (10.17) for some change of variables of the form (10.16).

In order to apply Cartan's equivalence method, we first need to recast the problem into a problem involving differential forms. The appropriate space to work in will be the second jet space J^2, which has coordinates $x, u, p = u_x, q = u_{xx}$, and the immediate goal is to construct a coframe on J^2 which will encode the relevant transformation rules for our problem. Note first that a point transformation will be in the desired linear form (10.16) if and only if, for some pair of functions α, β, the one-form equations

$$d\bar{x} = \alpha\, dx, \qquad \frac{d\bar{u}}{\bar{u}} = \frac{du}{u} + \beta\, dx, \tag{10.18}$$

hold on the subset of J^2 where $u \neq 0$. Indeed, the first equation implies that $\bar{x} = \xi(x)$, with $\alpha = \xi_x$, whereas the second necessarily requires

$\bar{u} = \eta(x)u$, with $\beta = \eta_x/\eta$. Second, in order that the derivative coordinates p and q transform correctly, we must prolong the transformation (10.16) to J^2, which requires that we impose the usual contact conditions (9.6), (9.7). Now, the combination of the first contact condition $d\bar{u} - \bar{p}\,d\bar{x} = \lambda(du - p\,dx)$ coupled with the linearity condition (10.18) already constitutes part of an overdetermined equivalence problem, and we can use Proposition 9.12 to reduce to an equivalence problem of standard form. Noting that

$$\begin{pmatrix} \alpha\,dx \\ u^{-1}\,du + \beta\,dx \end{pmatrix} = \begin{pmatrix} 1 & 0 \\ \alpha^{-1}(u^{-1}p + \beta) & \lambda^{-1}u \end{pmatrix} \begin{pmatrix} \alpha\,dx \\ \lambda(du - p\,dx) \end{pmatrix},$$

we conclude that the entries of the 2×2 matrix on the right hand side of this equation are invariants for the overdetermined problem (which will hold even after we complete both pairs of one-forms to a coframe on J^2), hence we should normalize $\beta = -p/u$, $\lambda = 1/u$. Therefore, the one-form

$$\frac{d\bar{u} - \bar{p}\,d\bar{x}}{\bar{u}} = \frac{du - p\,dx}{u}, \tag{10.19}$$

is invariant, and (10.19) can replace both the second equation in (10.18) and the first contact condition (9.6). (Of course, this observation is sufficiently simple that one could have deduced it by inspection, without recourse to the overdetermined equivalence algorithm.) Thus, we take as the first three elements of our eventual coframe the one-forms

$$\omega^1 = dx, \qquad \omega^2 = \frac{du - p\,dx}{u}, \qquad \omega^3 = dp - q\,dx. \tag{10.20}$$

Finally, in view of (10.17), the function $I(x, u, p, q) = \mathcal{D}[u]$ is an invariant for the problem, and hence its differential

$$\omega^4 = dI = f\,dq + g\,dp + h\,du + (f_x q + g_x p + h_x u)\,dx, \tag{10.21}$$

is an invariant one-form, which we take as our final coframe constituent. The four one-forms (10.20), (10.21) form a coframe on the subset $M = \{f(x) \neq 0 \neq u\} \subset \mathrm{J}^2$. It is not hard to prove that these one-form conditions do successfully encode our equivalence problem.

Proposition 10.1. *Two linear differential operators are directly equivalent if and only if their associated coframes satisfy*

$$\overline{\omega}^1 = \alpha\,\omega^1, \qquad \overline{\omega}^2 = \omega^2, \qquad \overline{\omega}^3 = \mu\,\omega^2 + \nu\,\omega^3, \qquad \overline{\omega}^4 = \omega^4. \tag{10.22}$$

Proof: Suppose conditions (10.22) hold. The first and second imply that the transformation has the desired linear form (10.16), with the transformation of the derivative coordinate p being given by the first prolongation $\bar{p} = (\eta p + \eta_x u)/\xi_x$. The third condition implies that the second derivative coordinate q transforms correctly. Finally, the fourth condition requires that $\bar{I} = I + c$ for some constant c. However, in view of the transformation rules for u, p, and q, the constant c must necessarily vanish, and so (10.17) holds, as required. *Q.E.D.*

In view of (10.22), the structure group associated with the equivalence problem is the three-dimensional matrix group

$$G = \left\{ \begin{pmatrix} a_1 & 0 & 0 & 0 \\ 0 & 1 & 0 & 0 \\ 0 & a_2 & a_3 & 0 \\ 0 & 0 & 0 & 1 \end{pmatrix} \;\middle|\; a_1, a_2, a_3 \in \mathbb{R}, \quad a_1 \cdot a_3 \neq 0 \right\}. \tag{10.23}$$

We are now ready to apply Cartan's reduction algorithm for this equivalence problem so as to prescribe invariant normalizations of the three group parameters a_1, a_2, a_3. We begin by defining the lifted coframe

$$\theta^1 = a_1 \omega^1, \qquad \theta^2 = \omega^2, \qquad \theta^3 = a_2 \omega^2 + a_3 \omega^3, \qquad \theta^4 = \omega^4. \tag{10.24}$$

The next step is to calculate the differentials of the lifted coframe elements (10.24). An explicit computation leads to the preliminary structure equations

$$\begin{aligned} d\theta^1 &= \alpha^1 \wedge \theta^1, \\ d\theta^2 &= \qquad\qquad\qquad\qquad\quad T^2_{12}\, \theta^1 \wedge \theta^2 + T^2_{13}\, \theta^1 \wedge \theta^3, \\ d\theta^3 &= \alpha^2 \wedge \theta^2 + \alpha^3 \wedge \theta^3 + T^3_{12}\, \theta^1 \wedge \theta^2 + T^3_{13}\, \theta^1 \wedge \theta^3 + T^3_{14}\, \theta^1 \wedge \theta^4, \\ d\theta^4 &= 0, \end{aligned} \tag{10.25}$$

with

$$\alpha^1 = \frac{da_1}{a_1}, \qquad \alpha^2 = \frac{a_3\, da_2 - a_2\, da_3}{a_3}, \qquad \alpha^3 = \frac{da_3}{a_3},$$

forming a basis for the right-invariant Maurer–Cartan forms on the Lie group G. The torsion coefficients in the structure equations (10.25) are explicitly given by

$$\begin{aligned} T^2_{12} &= -\frac{a_2 + a_3 p}{a_1 a_3 u}, \qquad T^2_{13} = \frac{1}{a_1 a_3 u}, \qquad T^3_{14} = \frac{a_3}{a_1 f}, \\ T^3_{12} &= \frac{a_2 g - a_3 h u}{a_1 f} - \frac{a_2^2 + a_2 a_3 p}{a_1 a_3 u}, \qquad T^3_{13} = \frac{a_2}{a_1 a_3 u} - \frac{g}{a_1 f}. \end{aligned} \tag{10.26}$$

(As we shall soon see, much of this computation was not really necessary.)

In the absorption part of Cartan's process, we replace each Maurer–Cartan form in the structure equations by general linear combination of coframe elements, so $\alpha^\kappa \mapsto z_1^\kappa\theta^1 + z_2^\kappa\theta^2 + z_3^\kappa\theta^3 + z_4^\kappa\theta^4$, where the coefficients z_j^κ are allowed to depend on both the base variables x, u, p, q and the group parameters a_1, a_2, a_3. The resulting two-forms are

$$\begin{aligned}
\Theta^1 &= -z_2^1\,\theta^1\wedge\theta^2 - z_3^1\,\theta^1\wedge\theta^3 - z_4^1\,\theta^1\wedge\theta^4,\\
\Theta^2 &= T_{12}^2\,\theta^1\wedge\theta^2 + T_{13}^2\,\theta^1\wedge\theta^3,\\
\Theta^3 &= (z_1^2 + T_{12}^3)\,\theta^1\wedge\theta^2 + (z_1^3 + T_{13}^3)\,\theta^1\wedge\theta^3 + T_{14}^3\,\theta^1\wedge\theta^4\\
&\quad + (-z_3^2 + z_2^3)\,\theta^2\wedge\theta^3 - z_4^2\,\theta^2\wedge\theta^4 - z_4^3\,\theta^3\wedge\theta^4,\\
\Theta^4 &= 0.
\end{aligned}$$

Coefficients of the basis two-forms $\theta^j\wedge\theta^k$ in each Θ^i which happen to be independent of the parameters z_j^κ are invariants of the problem, and can be normalized so as to reduce the structure group. In the present case, the essential torsion components are T_{12}^2, T_{13}^2, T_{14}^3, as given in (10.26). It is important to note that this fact could have been deduced by direct inspection of the structure equations (10.25): First, since there are no Maurer–Cartan forms in the equation for $d\theta^2$, any torsion components appearing there are essential; second, since the Maurer–Cartan forms in $d\theta^3$ multiply either θ^2 or θ^3, they can never produce a multiple of the two-form $\theta^1\wedge\theta^4$ upon replacement. In most cases, one can avoid the complete determination of the absorption equations, and use such simple observations to find the essential torsion. The essential torsion coefficients all depend on the group parameters, so the next step in the process is to normalize them to as simple a form as possible. We first normalize $T_{12}^2 = 0$ by setting $a_2 = -a_3 p$, thereby eliminating the group parameter a_2. Second, we can normalize $T_{14}^3 = 1$ by setting $a_3 = a_1 f$. Note that we cannot normalize T_{14}^3 to zero without violating the group constraints that $a_1 a_3 \neq 0$. (The decision to normalize T_{14}^3 before T_{13}^2 is merely for convenience, and does not affect the final outcome in any significant way.) With these two normalizations, the third essential torsion coefficient becomes $T_{13}^2 = 1/(a_1^2 f u)$. Since we are using real-valued functions, and $f(x)u \neq 0$ by assumption, there are two possible normalizations, namely $T_{13}^2 = \pm 1$, depending on $\kappa_1 = \operatorname{sign}[f(x)u]$. Note that κ_1 constitutes the first (discrete) invariant of our problem, since, under a real transformation, we can never change the sign of fu. (On the other hand, the complex transformation $x \mapsto \sqrt{-1}\,x$ does change this sign, and, indeed, in this case we can always normalize $T^2_{13} = +1$.) Therefore, we normalize $a_1 = \pm|fu|^{-1/2}$. Because we are required to use the most general solution to the normalization equation(s) — since otherwise we risk making a non-invariant specification of the group parameters — it is important that the sign of a_1 be left ambiguous; see below for further discussion of this point. In the real case, then, there are two different "branches" for the equivalence problem, depending on the sign κ_1. (Such branching phenomena, in which the problem divides into several subcases, are very common in applications of the method.)

The group parameter normalizations are

$$a_1 = \frac{\varepsilon_1}{\sqrt{|fu|}}, \qquad a_2 = -\varepsilon_2 p \sqrt{\left|\frac{f}{u}\right|}, \qquad a_3 = \varepsilon_2 \sqrt{\left|\frac{f}{u}\right|}, \tag{10.27}$$

where $\varepsilon_1 = \pm 1$ is an undetermined sign and $\varepsilon_2 = \varepsilon_1 \cdot \operatorname{sign} f$. The normalizations (10.27) have the effect of eliminating all three group parameters, and so, with just one loop through the equivalence method, we have found an invariant coframe:

$$\begin{aligned} \theta^1 &= \frac{\varepsilon_1\, dx}{\sqrt{|fu|}}, & \theta^3 &= \varepsilon_2 \sqrt{\left|\frac{f}{u}\right|} \left\{ (dp - q\, dx) - \frac{p}{u}(du - p\, dx) \right\}, \\ \theta^2 &= \frac{du - p\, dx}{u}, & \theta^4 &= f\, dq + g\, dp + h\, du + (f_x q + g_x p + h_x u)\, dx. \end{aligned} \tag{10.28}$$

Of course, if we had been endowed with unusually perceptive insight, we could have ascertained the formulae for the invariant coframe by inspection of the original transformation formulae. The great advantage of Cartan's method is that one does not need to be exceptionally brilliant in order to systematically discover the invariant coframe formulae. The skeptical reader is invited to check that the one-forms (10.28) do constitute a completely invariant coframe for the standard equivalence of differential operators, i.e., under any equivalence map Φ, taking $\mathcal{D}$ to $\overline{\mathcal{D}}$, the two invariant coframes satisfy $\Phi^* \overline{\theta}^i = \theta^i$. Note also that if we allow orientation-reversing transformations, the ambiguous sign in the one-forms θ^1 and θ^2 is essential; indeed, under the prolonged reflection $(\bar{x}, \bar{u}, \bar{p}, \bar{q}) = (-x, u, -p, q)$, the form $|\bar{f}\bar{u}|^{-1/2}\, d\bar{x}$ is pulled back to $-|fu|^{-1/2}\, dx$.

The structure equations for the invariant coframe (10.28) are

$$d\theta^1 = \tfrac{1}{2}\theta^1 \wedge \theta^2, \quad d\theta^3 = -I\,\theta^1 \wedge \theta^2 + \kappa_1 J\,\theta^1 \wedge \theta^3 + \theta^1 \wedge \theta^4 + \tfrac{1}{2}\theta^2 \wedge \theta^3,$$
$$d\theta^2 = \kappa_1 \theta^1 \wedge \theta^3, \quad d\theta^4 = 0. \tag{10.29}$$

The functions

$$I = fq + gp + hu, \qquad J = \varepsilon_2 \sqrt{\left|\frac{u}{f}\right|}\left(\frac{uf_x + 3pf}{2u} - g\right), \tag{10.30}$$

are the fundamental structure invariants of the problem. Note that we have recovered our original invariant (10.17) as one of the torsion components in the structure equations (10.30). Again, the ambiguous sign in front of J is essential for maintaining its invariance under orientation-reversing transformations. Note finally that the three previously normalized torsion coefficients have retained their normalized values in the final structure equations, a general fact that follows from the mechanics of the method.

We can now apply the solution to the equivalence problem for coframes developed in Chapter 8 to re-solve our differential operator equivalence problem. Comparing (8.2) and the formulae (10.28) for the invariant coframe, we find that coframe derivatives of a function F are given explicitly by

$$\frac{\partial F}{\partial \theta^1} = \varepsilon_1 \sqrt{|fu|}\,\widehat{D}_x F, \qquad \frac{\partial F}{\partial \theta^3} = \varepsilon_2 \sqrt{\left|\frac{f}{u}\right|}\left(\frac{\partial F}{\partial p} - \frac{g}{f}\frac{\partial F}{\partial q}\right),$$
$$\frac{\partial F}{\partial \theta^2} = u\frac{\partial F}{\partial u} + p\frac{\partial F}{\partial p} - \frac{pg + hu}{f}\frac{\partial F}{\partial q}, \quad \frac{\partial F}{\partial \theta^4} = \frac{1}{f}\frac{\partial F}{\partial q}, \tag{10.31}$$

where

$$\widehat{D}_x = \frac{\partial}{\partial x} + p\frac{\partial}{\partial u} + q\frac{\partial}{\partial p} + R\frac{\partial}{\partial q}, \quad R = -\frac{gq + hp + f_x q + g_x p + h_x u}{f}. \tag{10.32}$$

Note that if we differentiate the associated linear ordinary differential equation $I = \mathcal{D}[u] = 0$ with respect to x and solve for the third order derivative $r = u_{xxx}$, we recover the formula for R. Therefore $\widehat{D}_x$ can be identified with the the restriction of the total derivative operator D_x to the solution space to $\mathcal{D}[u] = 0$. Computing the differentials of the two structure functions, we find

$$dI = \theta^4, \qquad dJ = \kappa_1 K\,\theta^1 + \tfrac{1}{2}J\,\theta^2 - \tfrac{3}{2}\theta^3.$$

This means that all coframe derivatives of I are constant, and, in fact, the only independent derived invariant is

$$\begin{aligned} K &= \kappa_1 \frac{\partial J}{\partial \theta^1} = \kappa_1 \varepsilon_1 \sqrt{|fu|}\, \widehat{D}_x J \\ &= -\frac{3}{2} fq + \frac{3}{4}\frac{fp^2}{u} + \frac{1}{2} p(f_x + g) + u\, \frac{2ff_{xx} - f_x^2 + 2f_x g - 4fg_x}{4f}. \end{aligned} \tag{10.33}$$

We can continue to differentiate K to produce higher order derived invariants; for example

$$dK = L\,\theta^1 + \left(K - \tfrac{3}{2}I\right)\theta^2 - J\,\theta^3 + \tfrac{3}{2}\theta^4,$$

where $L = \partial K/\partial\theta^1 = \varepsilon_1 \sqrt{|fu|}\,\widehat{D}_x K$ is the only potentially new structure invariant.

To investigate the precise structure of the classifying manifolds, we proceed as follows. Note first that since $f \cdot u \neq 0$, the invariants I and J are always functionally independent. Solving (10.30) for p and q, and substituting into (10.33), we find that we can write

$$K = \tfrac{3}{2}a(x)u + \tfrac{1}{3}\kappa_1 J^2 - \tfrac{3}{2}I, \qquad a = \frac{5gf_x - 2f_x^2 - 2g^2}{9f} + h - \tfrac{2}{3}g_x + \tfrac{1}{3}f_{xx}. \tag{10.34}$$

If $a \equiv 0$, then K is a quadratic function of the functionally independent invariants I, J, and hence the coframe has rank 2 and order 0. Otherwise, we can take $\widetilde{K} = a(x)u$ as a new independent invariant, and compute the next nontrivial derived invariant

$$\begin{aligned} \frac{\partial \widetilde{K}}{\partial \theta^1} &= \varepsilon_1 \sqrt{|fu|}\,\widehat{D}_x \widetilde{K} \\ &= \kappa_1 b(x)|\widetilde{K}|^{3/2} - \tfrac{2}{3}J\widetilde{K}, \end{aligned} \qquad \text{where} \quad b(x) = \varepsilon_1 \frac{3a_x f + af_x - 2ag}{3\sqrt{|fa^3|}} \tag{10.35}$$

is an invariant which only depends on x. If b is constant, then I, J, $\widetilde{K}$ form a complete set of functionally independent invariants, and the coframe has rank 3 and order 1. Otherwise, we have a coframe of (maximal) rank 4 and order 2, with $I, J, \widetilde{K}, b$ comprising the four fundamental independent invariants. In the latter case, we complete the solution to the equivalence problem by computing one final derived invariant

$$\frac{\partial b}{\partial \theta^1} = \varepsilon_1 \sqrt{|fu|}\, b_x = c(x)\sqrt{|\widetilde{K}|}, \quad \text{so that} \quad c(x) = \varepsilon_1 \sqrt{\left|\frac{f(x)}{a(x)}\right|}\, b'(x) \tag{10.36}$$

is also an invariant. The invariants $b(x)$, $c(x)$ will serve to parametrize the (reduced) classifying curve $\mathcal{C} \subset \mathbb{R}^2$. Two differential operators will be equivalent if and only if they have the same classifying curve.

For example, in the case of the simple Schrödinger operator $\mathcal{D} = D_x^2 + h(x)$, the function $a(x)$ in (10.34) coincides with the potential $h(x)$. The two basic invariants are

$$b(x) = \varepsilon_1 \frac{h_x}{|h|^{3/2}}, \qquad \text{and} \qquad c(x) = \varepsilon_1 \frac{b_x}{\sqrt{|h|}} = \frac{2hh_{xx} - 3h_x^2}{2h^3}. \tag{10.37}$$

Thus, given any classifying curve $c = H(b)$, we can reconstruct an appropriate Schrödinger operator by solving the second order ordinary differential equation

$$2hh_{xx} - 3h_x^2 = 2h^3 H\left(\frac{h_x}{|h|^{3/2}}\right) \tag{10.38}$$

for the potential $h(x)$. The two integration constants reflect the translational and scaling invariance of the class of Schrödinger operators. Thus, we have reverified the fact that any second order operator is equivalent to a Schrödinger operator with potential $a(x)$. The coframe has rank two if and only if $\mathcal{D}$ is equivalent to the operator D_x^2; the associated two-parameter symmetry group consists of translations and scalings. The rank three cases are differential operators which can be mapped to a radial Laplace operator $D_x^2 + kx^{-2}$, with constant invariant $b = \pm\sqrt{6|k|}$, and one-dimensional scaling symmetry group. All other Schrödinger operators admit at most a discrete symmetry group.

Exercise 10.2. Given an explicit classifying function H, solve equation (10.38) for the corresponding family of Schrödinger potentials using Lie's method, as described in Chapter 6.

A similar computation provides the solution to the problem of gauge equivalence of differential operators. In this case, the operators are subject to the alternative transformation rule (6.79) under the linear change of variables (10.16). The Cartan formulation of this problem will use the same initial three one-forms (10.20), but now the final coframe element is the differential $\omega^4 = dI$ of the slightly more complicated gauge invariant

$$I(x, u, p, q) = \frac{\mathcal{D}[u]}{u} = \frac{f(x)q + g(x)p}{u} + h(x). \tag{10.39}$$

The structure group (10.23) is exactly the same, and so the equivalence method has the same intrinsic structure, although the explicit formulae are slightly different. The preliminary structure equations are (10.25), the unabsorbable torsion components being

$$T^2_{12} = -\frac{a_2 + a_3 p}{a_1 a_3 u}, \qquad T^2_{13} = \frac{1}{a_1 a_3 u}, \qquad T^3_{14} = \frac{a_3}{a_1 f u}.$$

Again, there are two branches, this time depending on $\kappa_1 = \operatorname{sign} f$, and we can normalize all the group parameters by setting

$$a_1 = \frac{\varepsilon_1}{u\sqrt{|f|}}, \qquad a_2 = -\varepsilon_2 p \sqrt{|f|}, \qquad a_3 = \varepsilon_2 \sqrt{|f|},$$

where $\varepsilon_1 = \pm 1$, $\varepsilon_2 = \varepsilon_1 \kappa_1$. The invariant coframe is now given by

$$\begin{aligned} \theta^1 &= \frac{\varepsilon_1\, dx}{u\sqrt{|f|}}, \qquad \theta^2 = \frac{du - p\, dx}{u}, \\ \theta^3 &= \varepsilon_2 \sqrt{|f|} \left\{ (dp - q\, dx) - \frac{p}{u}(du - p\, dx) \right\}, \\ \theta^4 &= \frac{f}{u}\, dq + \frac{g}{u}\, dp + \frac{fq + gp}{u^2}\, du + \left\{ \frac{f_x q + g_x p}{u} + h_x \right\} dx. \end{aligned} \tag{10.40}$$

The final structure equations take a slightly different form

$$d\theta^1 = d\theta^4 = 0, \quad d\theta^2 = \kappa_1 \theta^1 \wedge \theta^3, \quad d\theta^3 = -2J\, \theta^1 \wedge \theta^3 + \theta^1 \wedge \theta^4, \tag{10.41}$$

where

$$J = \frac{\varepsilon_1}{4\sqrt{|f|}} \left(2g - f_x + \frac{4pf}{u} \right) \tag{10.42}$$

is the only nontrivial structure function. Note that the original invariant I, given in (10.39), does *not* appear among the structure functions of the adapted coframe. Nor can it appear among the derived invariants, since the coframe derivatives are

$$\begin{aligned} \frac{\partial F}{\partial \theta^1} &= \varepsilon_1 \sqrt{|f|}\, \widehat{D}_x F, \qquad & \frac{\partial F}{\partial \theta^3} &= \frac{\varepsilon_2 u}{\sqrt{|f|}} \left(\frac{\partial F}{\partial p} - \frac{g}{f} \frac{\partial F}{\partial q} \right), \\ \frac{\partial F}{\partial \theta^2} &= \frac{\partial F}{\partial u} + p \frac{\partial F}{\partial p} - \frac{fq + pg(1-u)}{f} \frac{\partial F}{\partial q}, \qquad & \frac{\partial F}{\partial \theta^4} &= \frac{u}{f} \frac{\partial F}{\partial q}, \end{aligned} \tag{10.43}$$

where

$$\begin{aligned} \widehat{D}_x &= \frac{\partial}{\partial x} + p \frac{\partial}{\partial u} + q \frac{\partial}{\partial p} + R \frac{\partial}{\partial q}, \\ R &= \frac{fqp + gp^2}{u} - (gq + f_x q + g_x p + h_x u). \end{aligned} \tag{10.44}$$

Note that if we differentiate the eigenvalue equation $\mathcal{D}[u] = \lambda u$ (or, equivalently, the invariant equation $I = \lambda$) with respect to x and solve for the third order derivative $r = u_{xxx}$, we recover the formula for R. The reason that the original invariant I does not appear among the full set of structure invariants for the coframe (10.40) is that, in point of fact, we did *not* properly encode the gauge equivalence problem at the outset. Indeed, the invariance of the differential dI only implies that the two invariants agree up to an additive constant, $\bar{I} = I + \lambda$, or, in terms of the original differential operator, the operator $\mathcal{D}$ is gauge equivalent to $\overline{\mathcal{D}} + \lambda$ — thus the problem we have actually solved is the problem of gauge equivalence of the eigenvalue problems, *not* gauge equivalence of the two operators! However, this can easily be rectified — we must supplement the invariant coframe (10.40) by the additional scalar invariant I itself, leading to an equivalence problem similar to the isometric Lagrangian equivalence problem discussed at the end of Chapter 9.

Continuing with the solution to the equivalence problem, we see that the only independent derived invariants are

$$K = \frac{\partial J}{\partial \theta^1} = \varepsilon_1 \sqrt{|f|}\, \widehat{D}_x J, \qquad L = \frac{\partial K}{\partial \theta^1} = \varepsilon_1 \sqrt{|f|}\, \widehat{D}_x K.$$

In the case of gauge equivalence, there is always a one-parameter symmetry group of any differential operator, namely the scaling $u \mapsto \mu u$. Since the invariants must respect this symmetry, there can be at most three functionally independent invariants, so the rank can be at most three. Solving (10.39), (10.42) for p and q, we find

$$K = -a(x)+I-\kappa_1 J^2, \quad \text{where} \quad a = \frac{8gf_x - 3f_x^2 - 4g^2}{16f} + h - \tfrac{1}{2}g_x + \tfrac{1}{4}f_{xx}. \tag{10.45}$$

According to (6.88), the invariant $a(x)$ is the potential of the equivalent Schrödinger operator $\pm D_x^2 + a(x)$. The only degenerate case is when $a \equiv 0$, where the coframe has rank 2 and order 0, and admits a two-parameter symmetry group; in this case the operator is equivalent to $\pm D_x^2$. Otherwise, the coframe has rank 3, with only the possibility of discrete symmetries in addition to the universally admitted scaling symmetry. The invariants $a(x)$ and $\widehat{b}(x) = \partial a/\partial\theta^1 = \varepsilon_1\sqrt{|f|}\,a_x$ will parametrize the (reduced) classifying curve. Alternatively, replacing $\widehat{b}$ by $b(x) = \widehat{b}(x)^2$ will eliminate the ambiguous sign.

Fiber-preserving Equivalence of Scalar Lagrangians

A more substantial application of the Cartan equivalence method is to the equivalence problems arising in the calculus of variations. We begin with the simplest possible case — a first order variational problem in one independent and one dependent variable. Several basic equivalence problems for scalar Lagrangians were formulated in Examples 9.4 and 9.7. Here, for brevity, we shall only deal with the standard equivalence problems under fiber-preserving and point transformations, and refer the reader to [**129**] for the solution to the divergence equivalence problems. (See also Chapter 11.) After a thorough treatment of the scalar case, we present a more abbreviated treatment of second order scalar problems and first order problems in several independent and dependent variables.

We begin with the standard equivalence problem for first order scalar variational problems under fiber-preserving transformations. The Cartan formulation of Example 9.7 uses the base coframe

$$\omega^1 = du - p\,dx, \qquad \omega^2 = L(x,u,p)\,dx, \qquad \omega^3 = dp, \tag{10.46}$$

which is valid provided $L \neq 0$. The structure group $\widetilde{G}_f$ is given by (9.26) with $a_2 = 0$, so that the lifted coframe is

$$\theta^1 = a_1\omega^1, \qquad \theta^2 = \omega^2, \qquad \theta^3 = a_3\omega^1 + a_4\omega^2 + a_5\omega^3. \tag{10.47}$$

Note that, for the standard equivalence problem, the Lagrangian form $\omega^2 = L\,dx$ is already an invariant one-form since no group parameters occur in the definition of θ^2. To find the structure equations, we compute the differentials of the lifted coframe:

$$\begin{aligned} d\theta^1 &= \alpha^1 \wedge \theta^1 && + T^1_{12}\,\theta^1 \wedge \theta^2 + T^1_{23}\,\theta^2 \wedge \theta^3, \\ d\theta^2 &= && \quad T^2_{12}\,\theta^1 \wedge \theta^2 + T^2_{23}\,\theta^2 \wedge \theta^3, \\ d\theta^3 &= \alpha^3 \wedge \theta^1 + \alpha^4 \wedge \theta^2 + \alpha^5 \wedge \theta^3 && + T^3_{12}\,\theta^1 \wedge \theta^2 + T^3_{23}\,\theta^2 \wedge \theta^3, \end{aligned} \tag{10.48}$$

where $\alpha^1, \alpha^3, \alpha^4, \alpha^5$ are the Maurer–Cartan forms on the structure group $\widetilde{G}_f$. The essential torsion coefficients are

$$T^1_{23} = \frac{a_1}{a_5 L}, \qquad T^2_{12} = \frac{a_5 L_u - a_3 L_p}{a_1 a_5 L}, \qquad T^2_{23} = -\frac{L_p}{a_5 L}, \tag{10.49}$$

since the others can be absorbed by the Maurer–Cartan forms; for example, replacing α^1 by $\pi^1 = \alpha^1 - T^1_{12}\theta^2$ will absorb the coefficient T^1_{12} in the first structure equation. If the derivative L_p does not vanish, then all three essential torsion coefficients (10.49) depend explicitly on the group parameters, and we can normalize all of them. On the other hand, if $L_p \equiv 0$, then the Lagrangian $L = L(x, u)$ does not depend on the derivative coordinate $p = u_x$ at all; the variational problem is degenerate, and will be omitted from our analysis. Thus, in order to normalize the torsion coefficients consistently, we need only avoid "singularities" where $L_p = 0$, just as we avoided the singularities where $L = 0$ in our original construction of the coframe. Assuming $L_p \neq 0$, then, we normalize the essential torsion coefficients (10.49) to $1, 0, -1$, respectively, by setting

$$a_1 = L_p, \qquad a_3 = \frac{L_u}{L}, \qquad a_5 = \frac{L_p}{L}. \tag{10.50}$$

The normalizations (10.50) have the effect of reducing the original structure group $\widetilde{G}_f$ to a one-parameter subgroup, with a_4 the only remaining undetermined parameter.

In the second loop through the equivalence procedure, we substitute the expressions (10.50) into the formulae for the lifted coframe (10.47), and recompute the differentials. The revised structure equations are

$$\begin{aligned} d\theta^1 &= && T^1_{12}\,\theta^1 \wedge \theta^2 + T^1_{13}\,\theta^1 \wedge \theta^3 + \theta^2 \wedge \theta^3, \\ d\theta^2 &= && -\,\theta^2 \wedge \theta^3, \\ d\theta^3 &= \alpha^4 \wedge \theta^2 && + T^3_{12}\,\theta^1 \wedge \theta^2 + T^3_{23}\,\theta^2 \wedge \theta^3. \end{aligned} \tag{10.51}$$

Note that our previously normalized torsion coefficients have retained their normalized values (as they must). We can still absorb all the torsion in the expression for $d\theta^3$, hence there are just two essential torsion coefficients, both in the first structure equation. The first of these is

$$T^1_{12} = -\frac{L_p(L_u - L_{xp} - pL_{up}) + a_4 L^2 L_{pp}}{LL_p^2} = -\frac{L_p \widetilde{\mathrm{E}}(L) + a_4 L^2 L_{pp}}{LL_p^2},$$

where

$$\widetilde{\mathrm{E}}(L) = L_u - L_{xp} - pL_{up} \tag{10.52}$$

is the truncated Euler operator, obtained by omitting the second derivative term from the Euler–Lagrange equation

$$\mathrm{E}(L) = L_u - D_x L_p = L_u - L_{xp} - pL_{up} - qL_{pp} = 0. \tag{10.53}$$

If we assume that the Lagrangian satisfies the *nondegeneracy condition*

$$L_{pp} \neq 0, \tag{10.54}$$

then we can normalize $T^1_{12} = 0$ by setting

$$a_4 = -\frac{L_p \widetilde{\mathrm{E}}(L)}{L^2 L_{pp}} = -\frac{L_p Q}{L^2}, \qquad \text{where} \qquad Q(x,u,p) = \frac{\widetilde{\mathrm{E}}(L)}{L_{pp}}. \tag{10.55}$$

The function Q plays a particularly important role in the subsequent development. Solving the Euler–Lagrange equation (10.53) for the second derivative of u leads to the normal form $u_{xx} = Q(x, u, u_x)$. The invariance of the Euler–Lagrange equation, cf. Theorem 7.11, implies that Q obeys the same transformation rules as the second derivative of u, and, in the language of [**132**], is a "second order derivative covariant" for the fiber-preserving equivalence problem — see below for details.

The nondegeneracy condition (10.54) plays a crucial role in the analysis of the variational problem, and its appearance in the solution to the equivalence problem should come as no surprise. The convexity condition $L_{pp} > 0$ is of key importance for the existence theory of minimizers and the construction of the Legendre transformation, which maps a Lagrangian problem to the associated classical Hamiltonian system, cf. [**85**]. Points at which $L_{pp} = 0$ are singular points of the Euler–Lagrange equation (10.53). Thus (at least away from singularities) the only case not covered by the normalization (10.55) is when $L_{pp} \equiv 0$ which means that $L = a(x,u)p + b(x,u)$ is an affine function of the derivative. Such a Lagrangian is also essentially trivial; it is equivalent, modulo the addition of a suitable divergence, to a degenerate Lagrangian that does not depend on the derivative coordinate. Therefore, we also omit any further discussion of this degenerate case. Finally, the other unabsorbable torsion component in (10.51) is $T^1_{13} = -LL_p^{-2}L_{pp}$, which is independent of the group parameters, and thus forms the first fundamental invariant of the problem.

We have finally normalized all the group parameters. Inserting their prescribed values (10.50), (10.55) into (10.47) yields an invariant coframe for the standard, fiber-preserving Lagrangian equivalence problem:

$$\theta^1 = L_p\,(du - p\,dx), \qquad \theta^2 = L\,dx, \tag{10.56}$$
$$\theta^3 = \frac{L_u}{L}(du - p\,dx) + \frac{L_p}{L}(dp - Q\,dx) = d(\log L) - \widehat{D}_x(\log L)\,dx.$$

Here Q is given by (10.55), and

$$\widehat{D}_x = \frac{\partial}{\partial x} + p\,\frac{\partial}{\partial u} + Q(x,u,p)\,\frac{\partial}{\partial p}\,, \tag{10.57}$$

is the "adapted total derivative", which, in view of the discussion in the preceding paragraph, agrees with the total derivative D_x when applied to solutions of the Euler–Lagrange equation (10.53). The structure equations for the invariant coframe (10.56) are then found to be

$$\begin{aligned} d\theta^1 &= -I_1\,\theta^1 \wedge \theta^3 + \theta^2 \wedge \theta^3, \\ d\theta^2 &= -\,\theta^2 \wedge \theta^3, \\ d\theta^3 &= I_2\,\theta^1 \wedge \theta^2 + I_3\,\theta^2 \wedge \theta^3, \end{aligned} \tag{10.58}$$

where I_1, I_2, I_3 are the fundamental invariants of the problem, one of which we already know from the preceding formula for T^1_{13}. The explicit formulae for the other two invariants are rather complicated rational combinations of the partial derivatives of the Lagrangian L. Rather than calculate them directly, we can effectively utilize the coframe derivatives associated with (10.56), which are explicitly given as

$$\frac{\partial F}{\partial \theta^1} = \frac{L_p F_u - L_u F_p}{L_p^2}, \quad \frac{\partial F}{\partial \theta^2} = \frac{1}{L}\widehat{D}_x F, \quad \frac{\partial F}{\partial \theta^3} = \frac{L}{L_p}\,\frac{\partial F}{\partial p}\,. \tag{10.59}$$

The fundamental invariants can then be written in a simple explicit form:

$$I_1 = \frac{L L_{pp}}{L_p^2} = \frac{1}{L_p} \frac{\partial L_p}{\partial \theta^3}, \qquad I_2 = -\frac{1}{L} \frac{\partial^2 L}{\partial \theta^1 \partial \theta^2}, \qquad I_3 = \frac{1}{L} \frac{\partial^2 L}{\partial \theta^3 \partial \theta^2}. \tag{10.60}$$

These formulae can be derived directly from the structure equations (10.58) using the basic definition (8.2) of the coframe derivatives. Applying (10.59) when $F = L$, we first note that $\partial L / \partial \theta^1 = 0$, $\partial L / \partial \theta^3 = L$. Therefore $dL = (\partial L / \partial \theta^2)\, \theta^2 + L\, \theta^3$. Furthermore, using the structure equations (10.58) themselves,

$$\begin{aligned} 0 = d^2 L &= d\left(\frac{\partial L}{\partial \theta^2}\right) \wedge \theta^2 + dL \wedge \theta^3 + \frac{\partial L}{\partial \theta^2}\, d\theta^2 + L\, d\theta^3 \\ &= \left\{\frac{\partial^2 L}{\partial \theta^1 \partial \theta^2} + L\, I_2\right\} \theta^1 \wedge \theta^2 + \left\{-\frac{\partial^2 L}{\partial \theta^3 \partial \theta^2} + L\, I_3\right\} \theta^2 \wedge \theta^3, \end{aligned}$$

which will hold if and only if the invariants I_2, I_3 are given as in (10.60). The formula for I_1 follows from a similar analysis of $d\theta^1$.

The Jacobi identities for the coframe derivatives are found by reapplying the exterior derivative to the structure equations (10.58). An easy calculation shows that $d^2 \theta^2 = 0$ automatically, while the identities $d^2 \theta^1 = d^2 \theta^3 = 0$ imply the following relations among the invariants:

$$I_3 = -\frac{1}{I_1} \frac{\partial I_1}{\partial \theta^2}, \qquad \frac{\partial I_2}{\partial \theta^3} + \frac{\partial I_3}{\partial \theta^1} + (I_1 + 1) I_2 = 0, \tag{10.61}$$

which will prove to be of use later.

This completes the solution to the fiber-preserving equivalence problem. The necessary and sufficient conditions for equivalence are found by inspection of the functional relations among the invariants and the derived invariants. The rank of the coframe (10.56) is either 0, 1, 2, or 3. Therefore, Theorem 8.22 provides a second proof of the following fact, established in Theorem 7.30 for point transformational symmetries.

Proposition 10.3. *A first order Lagrangian $L(x, u, u_x)$ which is not an affine function of u_x admits a fiber-preserving variational symmetry group of dimension at most three.*

Maximally symmetric Lagrangians are those for which the coframe (10.56) has rank 0, which occurs if and only if all the invariants are constant. According to the Jacobi identities (10.61), if I_1 is constant, then I_3 is necessarily zero, and, unless $I_1 = -1$, then $I_2 = 0$ also. Also $I_1 \neq 0$, as otherwise L_{pp} would be identically zero, and we have explicitly excluded such Lagrangians from consideration. The associated three-parameter fiber-preserving variational symmetry group has commutation relations

$$[\mathbf{v}_1, \mathbf{v}_2] = \gamma\, \mathbf{v}_3, \qquad [\mathbf{v}_1, \mathbf{v}_3] = -\beta\, \mathbf{v}_1, \qquad [\mathbf{v}_2, \mathbf{v}_3] = \mathbf{v}_1 - \mathbf{v}_2, \tag{10.62}$$

where $I_1 = \beta$, $I_2 = \gamma$ are the constant values of the two nonzero invariants; these can be directly deduced from the structure equations (10.58) which, in the rank zero case, must be the Maurer–Cartan structure equations (2.34) for the symmetry group. Two rank zero Lagrangians are equivalent if and only if they possess the same constant invariants; therefore, to determine canonical forms for such Lagrangians we need only produce one simple example for each possible value of the invariants. Examples were found in Theorem 7.30, but they can be constructed directly from the explicit formulae for the invariants without difficulty.

Lagrangian	I_1	I_2	I_3	*Symmetry Group*
u_x^α	$1-\alpha^{-1}$	0	0	$\partial_u, \partial_x + \alpha\partial_u, -x\partial_x - (1-\alpha^{-1})u\partial_u,$
$\exp(u_x)$	1	0	0	$\partial_u, \partial_x, -x\partial_x + (x-u)\partial_u,$
$\sqrt{u_x + \sigma u^2}$	-1	4	0	$\partial_x, x\partial_x - u\partial_u, \sigma x^2\partial_x - (2\sigma x u + 1)\partial_u.$

For example, a simple corollary of this classification is that a variational problem is equivalent to a free particle Lagrangian $\int \frac{1}{2}p^2 dx$ with linear Euler–Lagrange equation $u_{xx} = 0$ if and only if its associated invariants take on the constant values $I_1 = \frac{1}{2}$, $I_2 = I_3 = 0$.

Next we consider the case of a Lagrangian whose associated invariant coframe has rank one, so that there is just one independent nonconstant invariant, and all the other invariants and derived invariants can be expressed as functions thereof. The most important of such Lagrangians are when $L = L(p)$ is a function of $p = u_x$ alone. In this case, $I_2 = I_3 = 0$, and, assuming L is not a pure power function, $I_1 = I(p)$ will be a nonconstant function of the single variable p. The only nonzero derived invariant is

$$J = \frac{\partial I_1}{\partial \theta^3} = \frac{L^2 L_p L_{ppp} - 2L^2 L_{pp}^2 + L L_p^2 L_{pp}}{L_p^4}, \tag{10.63}$$

which is again a function of p alone, and, as such, can be (locally) re-expressed as a function of $I(p)$. The invariants $(I(p), J(p))$ parametrize the (reduced) classifying curve $\mathcal{C} \subset \mathbb{R}^2$ associated with such Lagrangians.

Theorem 10.4. *Let $L(x,u,p)$ be a Lagrangian that is not an affine function of the derivative variable $p = u_x$. Then L is equivalent, under a fiber-preserving transformation, to a Lagrangian that depends only on p if and only if its invariants satisfy $I_2 = I_3 = 0 = \partial I_1/\partial\theta^1$. Two such Lagrangians are locally equivalent if and only if their reduced classifying curves overlap.*

Proof: First note that the condition $\partial I/\partial\theta^2 = 0$, where $I = I_1$, follows automatically from the first Jacobi identity (10.61). Therefore, $dI = J\,\theta^3$, where $J = \partial I/\partial\theta^3$. Moreover, using the structure equations (10.58), if $J \neq 0$, we have $0 = d^2I = dJ \wedge \theta^3 + J d\theta^3 = d\log J \wedge dI$, and hence I and J are functionally dependent. Therefore, L has rank 1 and order 0, and its classifying curve agrees with the classifying curve of an (x,u)-independent Lagrangian. The result follows directly from Theorem 8.19. *Q.E.D.*

Exercise 10.5. Suppose that the reduced classifying curve parametrized by $I(p)$ and $J(p)$ coincides with the graph of the classifying function $J = H(I)$. Prove that any associated Lagrangian $L = L(p)$ must be a solution to the third order ordinary differential equation obtained by substituting the explicit formulae (10.60), (10.63), into the equation for the classifying curve. Show how to compute the three-parameter family of equivalent Lagrangians by using Lie's reduction method to solve the ordinary differential equation, cf. [**183**].

Exercise 10.6. Consider the two Lagrangians $L = \exp(1/u_x^2)$, and $\bar{L} = \exp(-1/\bar{u}_{\bar{x}}^2)$. Prove that their invariants have the same classifying functions, but, nevertheless, they are *not* equivalent under a (real) fiber-preserving transformation. Discuss.

An Inductive Approach to Equivalence Problems

Before proceeding to the equivalence problem under point transformations, we make an elementary, but powerful observation on the use of subgroups to "induce" solutions to complicated equivalence problems. Suppose we are given two equivalence problems, both based on the same coframe ω on the base manifold M, but involving different structure groups H and G, the first of which is a subgroup of the second $H \subset G$.

For example, the standard fiber-preserving and point transformation Lagrangian equivalence problems are connected in this manner since they use the same basic coframe (10.46), and the structure group for the fiber-preserving problem is a subgroup of that for the point equivalence problem. We will see how the solution to the simpler equivalence problem, based on the subgroup H, can aid us in the solution to the more complicated equivalence problem with larger structure group G.

In this situation, the two lifted coframes are of the form $\boldsymbol{\theta} = h \cdot \boldsymbol{\omega}$, for $h \in H$, and $\boldsymbol{\zeta} = g \cdot \boldsymbol{\omega}$, for $g \in G$. Suppose that we have solved the H-equivalence problem and, hence, have determined the invariant coframe

$$\boldsymbol{\theta} = h_0 \cdot \boldsymbol{\omega}, \tag{10.64}$$

for some specification $h_0 = h_0(x)$ of the parameters in H that normalizes all the torsion. For the Lagrangian example, h_0 would denote the group element determined by the normalizations (10.50), (10.55), found in our solution to the fiber-preserving equivalence problem. The idea underlying the inductive method is that, instead of merely reverting to the original base coframe $\boldsymbol{\omega}$ to solve the G-equivalence problem from scratch, we instead use the H-adapted coframe $\boldsymbol{\theta}$ to reformulate the G-equivalence problem, and thereby effect an easier calculation for the more complicated problem. In other words, to solve the G-equivalence problem, rather than using the "direct" lifted coframe $\boldsymbol{\zeta} = \widehat{g} \cdot \boldsymbol{\omega}$, $\widehat{g} \in G$, we work with the *adapted coframe*

$$\boldsymbol{\zeta} = g \cdot \boldsymbol{\theta} = g \cdot h_0 \cdot \boldsymbol{\omega}, \qquad g \in G, \tag{10.65}$$

whereby $\widehat{g} = g \cdot h_0$. Since the structure equations for the H-equivalence problem express the differentials of $\boldsymbol{\theta}$ in terms of H-invariants, we can use these expressions in the absorption and normalization processes to obtain structure equations for $\boldsymbol{\zeta}$ depending on the adapted group parameters, and, ultimately, obtain expressions for the G-invariants written in terms of the H-invariants and their derived invariants. Thus, we will automatically derive expressions for the invariants of the more complicated problem in terms of the invariants of the simpler problem, and thereby explicitly expose the relationship between the two. The implementation of the method will become clear in the following example.

Lagrangian Equivalence under Point Transformations

We now solve the point transformation Lagrangian equivalence problem by using the inductive approach based on our previous solution to the fiber-preserving equivalence problem. The lifted coframe in this case will have the form given by (10.65), so

$$\zeta^1 = b_1\theta^1, \qquad \zeta^2 = b_2\theta^1 + \theta^2, \qquad \zeta^3 = b_3\theta^1 + b_4\theta^2 + b_5\theta^3, \qquad (10.66)$$

where the θ^i's are the invariant coframe (10.56) associated with the more restrictive fiber-preserving equivalence problem. Before we begin the implementation of the equivalence procedure, we remark that the construction of the fiber-preserving coframe (10.56) requires that the functions L, L_p, and L_{pp} do not vanish,† which implies that the invariant I_1, given in (10.60), also cannot vanish in the domain under consideration. As usual, we compute the differentials of the lifted coframe (10.66), but now use the structure equations (10.58) to evaluate the differentials $d\theta^i$ of the fiber-preserving invariant coframe. For example, we find

$$\begin{aligned} d\zeta^1 &= db_1 \wedge \theta^1 + b_1 d\theta^1 = \beta_1 \wedge \zeta^1 - b_1 I_1\, \theta^1 \wedge \theta^3 + b_1\, \theta^2 \wedge \theta^3 \\ &= \beta^1 \wedge \zeta^1 + \frac{b_4 I_1 + b_3}{b_5}\, \zeta^1 \wedge \zeta^2 + \frac{b_2 - 1}{b_5}\, \zeta^1 \wedge \zeta^3 + \frac{b_1}{b_5}\, \zeta^2 \wedge \zeta^3, \end{aligned}$$

where $\beta^1 = db^1/b^1$ is the first Maurer–Cartan form on G_p. The torsion coefficient $T^1_{23} = b_1/b_5$ is essential, and can be normalized to 1 by setting $b_1 = b_5$. Similarly, the coefficient $T^2_{23} = (b_2 - 1)/b_5 = 0$ of $\zeta^2 \wedge \zeta^3$ in $d\zeta^2$ is also essential torsion, and is normalized to 0 by setting $b_2 = 1$. Note that the second normalization yields a familiar *invariant* one-form:

$$\zeta^2 = \theta^1 + \theta^2 = (L - pL_p)\, dx + L_p\, du. \qquad (10.67)$$

According to Definition 7.38, the one-form ζ^2 is the *Cartan form* associated with the variational problem with Lagrangian L. Its importance in the geometric approach to the calculus of variations, including existence theory, symmetries, and conservation laws, was discussed at length in Chapter 7. In particular, Theorem 7.39 demonstrated the invariance of the Cartan form under general point transformations. It is noteworthy how the Cartan equivalence method leads immediately, and algorithmically, to the Cartan form; its invariance is an immediate corollary of the

† The condition $L_p \neq 0$ is not necessary for the solution to the point equivalence problem, but is required to apply the inductive method. The final formulae are readily extended to include points where $L_p = 0$.

general method, thereby obviating the more complicated explicit proof described in Chapter 7.

In the second phase of the equivalence method, the two unabsorbable torsion coefficients are

$$T^2_{12} = b_4 b_5^{-2} I_1 \qquad \text{and} \qquad T^2_{13} = -b_5^{-2} I_1. \tag{10.68}$$

We can normalize $T^2_{12} = 0$, by setting $b_4 = 0$. As for T^2_{13}, since it depends on the square of the group parameter b_5, the normalizations will depend on whether we are considering the real or the complex equivalence problem — see the preceding discussion of the equivalence problem for differential operators. For simplicity, we shall just treat the complex case in detail, and normalize $T^2_{13} = -1$ by setting $b_5 = \sqrt{I_1}$. Note that, even here, there is an ambiguity in the normalized group parameter b_5 since we have not (and, in fact, cannot) specify which branch of the complex square root is to be used.

In the third and final phase of the equivalence method, we complete the group reduction by normalizing the essential torsion coefficient $T^1_{12} = 0$ (which also happens to normalize $T^3_{23} = 0$ at the same time), whereby

$$b_3 = \frac{1}{2\sqrt{I_1}} \frac{\partial I_1}{\partial \theta^2} = -\frac{I_3 \sqrt{I_1}}{2}.$$

The final invariant coframe is

$$\begin{aligned} \zeta^1 &= \sqrt{I_1}\, \theta^1 = \sqrt{L L_{pp}}\, (du - p\, dx), \\ \zeta^2 &= \theta^1 + \theta^2 = (L - pL_p)\, dx + L_p\, du, \\ \zeta^3 &= -\frac{1}{2} I_3 \sqrt{I_1}\, \theta^1 + \sqrt{I_1}\, \theta^3, \end{aligned} \tag{10.69}$$

with structure equations

$$\begin{gathered} d\zeta^1 = J_1\, \zeta^1 \wedge \zeta^3 + \zeta^2 \wedge \zeta^3, \qquad d\zeta^2 = -\zeta^1 \wedge \zeta^3, \\ d\zeta^3 = J_2\, \zeta^1 \wedge \zeta^2 + J_3\, \zeta^1 \wedge \zeta^3. \end{gathered} \tag{10.70}$$

There are three fundamental structure invariants

$$\begin{aligned} J_1 &= \frac{1}{I_1^{3/2}} \left(\frac{1}{2} \frac{\partial I_1}{\partial \theta^3} + I_1^2 + I_1 \right) = \frac{L L_{ppp} + 3 L_p L_{pp}}{2\sqrt{L L_{pp}^3}}, \\ J_2 &= \frac{1}{2} \frac{\partial I_3}{\partial \theta^1} + I_2 - \frac{1}{4} I_3^2, \\ J_3 &= \frac{1}{2 I_1^{3/2}} \left(\frac{I_3}{2} \frac{\partial I_1}{\partial \theta^3} + \frac{\partial I_1}{\partial \theta^1} + I_1 \frac{\partial I_3}{\partial \theta^3} + I_1^2 I_3 \right). \end{aligned} \tag{10.71}$$

Note that the ambiguity in the square root function that appears in the formulae for the coframe and the invariants is essential, since the elementary orientation-reversing map $(x, u) \mapsto (x, -u)$ will change the sign of both J_1 and J_3. We can avoid the ambiguous sign by using

$$\widehat{J}(x,u,p) = J_1(x,u,p)^2 = \frac{(LL_{ppp} + 3L_pL_{pp})^2}{4LL_{pp}^3} \tag{10.72}$$

as our first fundamental invariant. Similar remarks hold for J_3, although one must be careful since the branches of the square roots appearing in J_1, J_3 must be consistent. The coframe derivatives associated with (10.69) are

$$\begin{aligned} \frac{\partial F}{\partial \zeta^1} &= \frac{1}{\sqrt{I_1}}\left(\frac{\partial F}{\partial \theta^2} - \frac{\partial F}{\partial \theta^1} + \frac{1}{2}I_3\frac{\partial F}{\partial \theta^3}\right), \\ \frac{\partial F}{\partial \zeta^2} &= \frac{\partial F}{\partial \theta^2}, \qquad \frac{\partial F}{\partial \zeta^3} = \frac{1}{\sqrt{I_1}}\frac{\partial F}{\partial \theta^2}. \end{aligned} \tag{10.73}$$

The formulae $d^2\zeta^1 = d^2\zeta^3 = 0$ lead to the Jacobi identities

$$J_3 = -\frac{\partial J_1}{\partial \zeta^2}, \qquad \frac{\partial J_2}{\partial \zeta^3} - \frac{\partial J_3}{\partial \zeta^2} + J_1J_2 = 0. \tag{10.74}$$

This completes our general solution to the Lagrangian equivalence problem under complex point transformations.

Exercise 10.7. Find the invariant coframe in the real-valued case.

Let us now apply this result to a few specific examples. As usual, we begin by looking at the rank zero case. According to the Jacobi identities (10.74), if J_1 is constant, then $J_3 = 0$, and, moreover, if $J_1 \neq 0$, then $J_2 = 0$ also. It is easy to prove, using the formulae (10.71), that the two canonical families of rank zero fiber-preserving Lagrangians are also canonical for the point transformation equivalence problem. For the (complex) power Lagrangians $L_\alpha = p^\alpha$, $\alpha \neq 0, 1$, the invariants (10.71), (10.72), are $\widehat{J} = 4 + \frac{1}{\alpha(\alpha-1)}$, $J_2 = J_3 = 0$. Thus, two different power Lagrangians L_α, $L_{\overline{\alpha}}$, $\overline{\alpha} \neq \alpha$ have the same constant invariants if and only if $\overline{\alpha} = 1 - \alpha$. Indeed, the hodograph map $(x,u) \mapsto (u,x)$ interchanging independent and dependent variable maps p^α to $p^{1-\alpha}$. A variational problem is equivalent to a free particle Lagrangian $\int \frac{1}{2}p^2 dx$ with linear Euler–Lagrange equation $u_{xx} = 0$ if and only if its invariants have the particular constant values $\widehat{J} = \frac{9}{2}$, $J_2 = J_3 = 0$. The exponential Lagrangian $L = e^p$ has the limiting value $\widehat{J} = 4$, $J_2 = J_3 = 0$. The square root Lagrangians $\sqrt{p + \sigma u^2}$ have invariants $\widehat{J} = 0$, $J_2 = 2\sigma$, $J_3 = 0$, and so remain inequivalent.

Next, consider the rank 1 case when $L = L(p)$ is a function of p alone. In this case, $J_2 = J_3 = 0$, and, assuming L is not a pure power function, $\widehat{J} = \widehat{J}(p)$ will be a nonconstant function of the single variable p. The only nonzero derived invariant is

$$\begin{aligned}\widehat{K} = \frac{\partial J_1}{\partial \zeta^3} &= \sqrt{\frac{L}{L_{pp}}}\,\frac{\partial J_1}{\partial p} \\ &= \frac{2L^2 L_{pp} L_{pppp} - 2LL_p L_{pp} L_{ppp} + 6LL_{pp}^3 - 3L_p^2 L_{pp}^2 - 3L^2 L_{ppp}^2}{2LL_{pp}^3}.\end{aligned} \tag{10.75}$$

The pair of invariants $(\widehat{J}(p), \widehat{K}(p))$ will serve to parametrize the reduced classifying curve $\mathcal{C}$ associated with such a Lagrangian. A result similar to Theorem 10.4 will characterize those Lagrangians which are equivalent to a Lagrangian that depends only on the derivative variable p under a point transformation.

Theorem 10.8. *Let $L(x, u, p)$ be a Lagrangian that is not an affine function of p. Then L is equivalent to a Lagrangian that depends only on the derivative variable p under a point transformation if and only if its invariants have the form $J_2 = J_3 = 0$, while $\partial J_1/\partial \zeta^1 = 0$. Two such Lagrangians are locally equivalent if and only if their reduced classifying curves overlap.*

Any Lagrangian which depends only on p always admits the two-parameter translation symmetry group $(x, u) \mapsto (x + \delta, u + \varepsilon)$. If L is an affine function of p, then L admits an infinite-dimensional Lie pseudogroup of symmetries depending on two arbitrary functions. If L is not an affine function of p, and the invariant I is constant, then L admits an additional one-parameter group of symmetries, which, in the canonical form $L = p^\alpha$ is a scaling group. If the invariant I is not constant, then, besides the translations, the symmetry group of L is generated by at most a discrete set of additional symmetries.

Exercise 10.9. Let $\mathcal{C} \subset \mathbb{C}^2$ be a given reduced classifying curve. Show how to recover the associated four-parameter family of equivalent (x, u)–independent Lagrangians. (See Exercise 10.5.)

Remark: An interesting geometrical application of these results to surfaces with Finsler metrics appears in the work of Chern, [**48**], [**49**]. Any first order Lagrangian form $L(x, u, u_x)\,dx$ can be viewed as the arc length element for a Finsler metric on a surface. In [**50**], Chern shows how the classical Gauss–Bonnet Theorem, which applies when the Lagrangian is the geodesic arc length functional for a Riemannian surface, and so given by the square root of a quadratic polynomial in p, can be generalized to "Landsberg surfaces", which are defined as Finsler surfaces whose metric functional has vanishing invariant J_2.

Remark: There is a remarkable connection between the Lagrangian equivalence problem and the problem of equivalence of control systems under feedback transformations. See Gardner, [**78**], for details and applications to control theory.

Applications to Classical Invariant Theory

According to Proposition 7.7, there is a fundamental connection between the equivalence problem for binary forms and the equivalence problem for (x, u)-independent Lagrangians under point transformations. Given a homogeneous polynomial (binary form) $Q(x, y)$ of degree n and weight 0, we define the associated Lagrangian $L(p) = \sqrt[n]{Q(p, 1)}$ as the nth root of its inhomogeneous representative $F(p) = Q(p, 1)$. Two forms are equivalent under a linear transformation if and only if their associated Lagrangians are equivalent under a point transformation. Therefore, Theorem 10.8 also provides a solution, first formulated in [**183**], to the equivalence problem for the original binary form Q. We can translate this result into the language of classical invariant theory by evaluating the invariants $\widehat{J}$ and $\widehat{K}$ directly in terms of known covariants of the binary form Q. Recalling Theorem 3.46 and, more specifically, Example 3.48, we first note that

$$L_{pp} = \frac{1}{n^2(n-1)}\, L^{1-2n} H,$$

where $H = \frac{1}{2}(F, F)^{(2)}$ is the Hessian covariant of the form. The nondegeneracy condition $L_{pp} \neq 0$ therefore is equivalent to the condition that the Hessian H does not vanish. The only forms which are not covered by Theorem 10.8 are those for which $H \equiv 0$, so $L = ap + b$ is an affine function of p, or, equivalently $F(p) = (ap + b)^n$. (See Theorem 3.49.) If

$H \neq 0$, then we find

$$\widehat{J} = \frac{(n-1)}{4}\,\frac{T^2}{H^3}, \qquad \widehat{K} = -\frac{n(n-1)}{4(n-2)}\,\frac{FU}{H^3}, \tag{10.76}$$

where $T = (F, H)^{(1)}$ and $U = (H, T)^{(1)}$ are particular higher order covariants of the form. Thus, except for inessential multiples, we can identify the Lagrangian invariants $\widehat{J}$ and $\widehat{K}$ with the absolute rational covariants

$$J^* = \frac{T^2}{H^3}, \qquad K^* = \frac{FU}{H^3}. \tag{10.77}$$

The (reduced) classifying manifold is the curve (or, in the rank zero case, point) $\mathcal{C}_F \subset \mathbb{C}^2$ parametrized by $(J^*(p), K^*(p))$. Theorem 10.8 immediately implies that two binary forms are locally equivalent if and only if their classifying curves overlap. Moreover, analyticity implies that two locally equivalent polynomials are in fact globally equivalent, and hence we deduce the following solution to the equivalence problem of complex-valued binary forms.

Theorem 10.10. *Let $F(p) = Q(p,1)$ and $\overline{F}(\bar{p}) = \overline{Q}(\bar{p},1)$ be two binary forms of degree n, which are not n^{th} powers of linear forms. Then F and $\overline{F}$ are equivalent under a linear fractional transformation if and only if their classifying curves are identical: $\mathcal{C}_F = \mathcal{C}_{\overline{F}}$.*

Therefore a complete solution to the complex equivalence problem for binary forms depends on merely two absolute rational covariants — J^* and K^*! It is easy to see that, except in the case when the Lagrangian is an affine function of p, the symmetry groups of a binary form F and the corresponding Lagrangian differ only by the translations $(x, u) \mapsto (x + c, u + d)$. Therefore, we deduce the following theorem on symmetries of binary forms.

Theorem 10.11. *Let $F(p)$ be a binary form of degree n. If $H \equiv 0$, then F admits a two-parameter group of symmetries. Otherwise, either J^* is constant, and F admits a one-parameter group of symmetries, or J^* is not constant, and F admits at most a discrete symmetry group.*

The first case is proved by direct computation, using the fact that F is the nth power of a linear form, and hence equivalent to $\pm p^n$. The second and third cases are direct consequences of Theorem 10.10. The rank zero case, when J^* is constant, provides an immediate test for determining whether a given form is equivalent to a monomial.

Theorem 10.12. *A binary form $F(p)$ is complex-equivalent to a monomial, i.e., to p^k, if and only if the covariant T^2 is a constant multiple of H^3.*

If the invariants J^* and $\overline{J}^*$ are not constants, and F and $\overline{F}$ have identical classifying curves, then one can explicitly determine all the transformations mapping F to $\overline{F}$ by solving the equations

$$J^*(p) = \overline{J}^*(\bar{p}), \qquad K^*(p) = \overline{K}^*(\bar{p}). \tag{10.78}$$

Of course, the second of these two equations merely serves to delineate the appropriate branch of the curve, and so rule out spurious solutions to the first equation which map between different branches. This is important, since the equation $\overline{J}^* = J^*$, or, equivalently,

$$T(p)^2\, \overline{H}^*(\bar{p})^3 = \overline{T}^*(\bar{p})^2\, H(p)^3,$$

is, in general, a polynomial equation of degree $6n - 12$ for p in terms of $\bar{p}$; as such, many of its roots will be spurious, leading to the wrong branch of the classifying curve. In particular, the discrete symmetry group of a form with nonconstant invariant J^* is given by the set of solutions to the pair of rational equations (10.78).

Let us apply these results to binary cubics. The Hilbert basis for the covariants of a cubic is provided by the form F itself, the covariants H and T, and the discriminant Δ; see Exercise 3.53. It can be shown, cf. [**183**], that the reducible covariant U can be written in terms of these as $U = (H, T)^{(1)} = -2^3 3^6 \Delta F$. Moreover, the irreducible covariants satisfy the basic syzygy

$$T^2 + H^3 - 2^3 3^6 \Delta F^2 = 0. \tag{10.79}$$

Consequently, $T^2 + H^3 + FU = 0$, hence, using (10.77),

$$K^* = -J^* - 1, \tag{10.80}$$

and the classifying curve of a nondegenerate cubic is contained in the graph of a linear function. In the complex case, the entire line is covered, and so, according to Theorem 10.10, there are only three possibilities: *a*) the Hessian vanishes identically, $H \equiv 0$, in which case the form is the cube of a linear form, *b*) T^2 is a constant multiple of H^3, in which case the cubic has a double root, or c) the generic case happens when the cubic has three simple roots. In the real case, the third possibility splits into two subcases depending on the sign of the Hessian, since this governs the sign of the invariant J^*, and hence which half of the line (10.80) is covered by the classifying curve.

Turning to a binary quartic, to find the classifying curve, we use the fundamental syzygy

$$T^2 = -\tfrac{32}{9}H^3 + 2^9 3^2 i Q^2 H - 2^{12}3^4 j Q^3,$$

cf. (3.47), and the identity

$$U = 2^9 3^2 (108 j Q^2 - i Q H).$$

where i, j are the fundamental invariants, cf. (3.35). (See [**99**], [**183**] for proofs, although the reader should be warned that the normalizations of our covariants are different.) We introduce the rational covariant $S = Q/H$ in terms of which the absolute covariants J^* and K^* assume the parametric form

$$J^* = 2^9 3^2(-72 j S^3 + i S^2) - \tfrac{32}{9}, \qquad K^* = 2^9 3^2 (108 j S^3 - i S^2).$$

These two equations give a simple parametrization of the classifying curve, which has the implicit form

$$9 j^2 \left(J^* + K^* + \tfrac{32}{9}\right)^3 = i^3 \left(J^* - \tfrac{3}{2}K^* + \tfrac{32}{9}\right)^2.$$

Thus, the absolute invariant j^2/i^3 plays the crucial role in the equivalence problem, distinguishing the different canonical forms for quartic polynomials. In particular, the classifying curve is the graph of a single-valued linear function if and only if either $i = 0$ or $j = 0$; see Chapter 3 for the geometric interpretation of these conditions. See [**183**] for further details and additional developments.

Exercise 10.13. Prove that a binary form Q of degree $n \geq 3$ is complex-equivalent to a sum of two n^{th} powers, i.e., to $x^n + y^n$, if and only if its covariants H, T, U are related by the equation

$$QU + T^2 + \frac{8(n-2)^2}{(n-1)^2} H^3 = 0. \tag{10.81}$$

Exercise 10.14. Formulate and solve the equivalence problem for first order scalar Lagrangians under volume-preserving transformations. Apply your result to solve the equivalence problem for binary forms under the special linear group SL(2).

Second Order Variational Problems

Higher order and multi-dimensional Lagrangian equivalence problems can also be treated by the Cartan method, although the computations rapidly become extremely complicated. In this section we discuss the case of a second order scalar variational problem

$$\mathcal{I}[u] = \int L\left(x, u, \frac{du}{dx}, \frac{d^2u}{dx^2}\right) dx. \tag{10.82}$$

The Lagrangian $L(x,u,p,q)$ is a smooth function on the second order jet space J^2, and is assumed to satisfy the nondegeneracy conditions

$$L(x,u,p,q) \neq 0, \qquad L_q(x,u,p,q) \neq 0, \qquad L_{qq}(x,u,p,q) \neq 0, \tag{10.83}$$

on some subdomain $M \subset \mathrm{J}^2$. The Euler–Lagrange equation,

$$\mathrm{E}(L) = L_u - D_x(L_p) + D_x^2(L_q) = 0, \tag{10.84}$$

is then a fourth order ordinary differential equation, invariantly associated to the variational problem (10.82). As in the first order case (9.13), two second order Lagrangians are equivalent under a contact transformation $\Psi \colon \mathrm{J}^2 \to \mathrm{J}^2$ if and only if

$$\Psi^*(\bar{L}\, d\bar{x}) = L\, dx + \mu(dp - q dx) + \nu(du - p\, dx), \tag{10.85}$$

where μ, ν are functions on J^2. For point transformations, $\mu = 0$, while for fiber-preserving transformations, $\mu = \nu = 0$, and the Lagrangian form $L\, dx$ is invariant. To formulate the equivalence problem in Cartan form, we introduce the coframe

$$\omega^1 = du - p\, dx, \qquad \omega^2 = \frac{dp - q\, dx}{L}, \qquad \omega^3 = L\, dx, \qquad \omega^4 = \frac{dq}{L^2}. \tag{10.86}$$

For the contact equivalence problem, the structure group is

$$G_c = \left\{ \begin{pmatrix} a_1 & 0 & 0 & 0 \\ a_2 & a_1 & 0 & 0 \\ a_3 & a_4 & 1 & 0 \\ a_5 & a_6 & a_7 & a_1 \end{pmatrix} \;\middle|\; a_1 \neq 0 \right\}, \tag{10.87}$$

where the diagonal factors are the same in view of our choice of scaling of the coframe elements (10.86). The point transformation subgroup requires $a_4 = 0$, while for fiber-preserving transformations $a_3 = a_4 = 0$.

For brevity, we shall only discuss the solution to the fiber-preserving problem here; the other problems are treated in [**42**] and [**133**]; see also [**51**] for an interesting alternative approach to the equivalence problem for higher order Lagrangians. The lifted coframe is

$$\begin{aligned} \theta^1 &= a_1\omega^1, & \theta^3 &= \omega^3, \\ \theta^2 &= a_2\omega^1 + a_1\omega^2, & \theta^4 &= a_5\omega^1 + a_6\omega^2 + a_7\omega^3 + a_1\omega^4. \end{aligned} \tag{10.88}$$

In the first loop through the equivalence procedure, the nonconstant essential torsion is

$$\begin{aligned} T^2_{24} = -T^3_{34} &= \frac{LL_q}{a_1}, \qquad T^3_{23} = \frac{a_1L_p - a_6LL_q}{a_1^2}, \\ T^3_{13} &= \frac{a_1^2L_u - a_1a_2LL_p + (a_2a_6 - a_1a_5)L^2L_q}{a_1^3L}, \\ T^2_{23} - T^1_{13} &= \frac{\widetilde{D}_xL}{L^2} + \frac{a_6 - 2a_2 - a_7LL_q}{a_1}, \end{aligned}$$

where $\widetilde{D}_x = \partial_x + p\partial_u + q\partial_p$ denotes the J^2 truncation of the total derivative operator. In view of our nondegeneracy assumption (10.83), we can normalize the essential torsion to take the constant values $1, 0, 0, 0$, respectively, by setting

$$a_1 = LL_q, \qquad a_5 = \frac{L_u}{L}, \qquad a_6 = L_p, \qquad a_7 = \frac{L_p - 2a_2}{LL_q} + \frac{\widetilde{D}_xL}{L^2}.$$

In the second loop through the equivalence procedure, we find the essential torsion $T^1_{13} = T^2_{23} = -U$, where

$$U = \frac{a_2\{2LL_{qq} + L_q^2\} + L_q^2\widetilde{D}_xL_q - L_qL_{qq}\widetilde{D}_xL - L_pL_q^2 - LL_pL_{qq}}{LL_q^3}. \tag{10.89}$$

At this point, the equivalence problem splits into two branches. If the additional "nondegeneracy" condition

$$2LL_{qq} + L_q^2 \neq 0 \tag{10.90}$$

holds, then we can normalize the torsion coefficients (10.89) to be zero by solving for a_2, thereby eliminating all the group parameters. The special Lagrangians for which the left hand side of (10.90) vanishes identically, i.e., those of the form

$$L = \left[A(x,u,u_x)u_{xx} + B(x,u,u_x)\right]^{2/3}, \tag{10.91}$$

(which include the maximally symmetric, affine-invariant arc length arc length functionals, cf. Theorem 7.31), cannot be handled by our current version of the equivalence method, but will require the development of the theory of prolongation to be discussed in Chapter 12. Omitting this particular class of Lagrangians, we are now able to write down the final invariant coframe:

$$\begin{aligned} &\theta^1 = LL_q\,(du - p\,dx), \qquad \theta^2 = P(\,du - p\,dx) + L_q\,(dp - q\,dx), \\ &\theta^3 = L\,dx, \qquad \theta^4 = \frac{L_u(du - p\,dx) + L_p(dp - q\,dx) + L_q(dq - R\,dx)}{L}, \end{aligned} \tag{10.92}$$

Here

$$P(x,u,p,q) = L_p - \widetilde{D}_x L_q - RL_{qq}, \tag{10.93}$$

and

$$R(x,u,p,q) = \frac{LL_pL_q - \widetilde{D}_x(LL_q^2)}{(LL_q^2)_q}. \tag{10.94}$$

The first two coframe elements are contact forms; moreover, if we replace the differential function R by the third order derivative $r = u_{xxx}$, then θ^4 also reduces to a contact form (on J^3). The invariance of the "pseudo-contact" form θ^4 implies that the function R is, in the language of [**133**], a third order *derivative covariant*, meaning that it transforms in exactly the same way as the third order derivative $r = u_{xxx}$ does. More precisely:

Proposition 10.15. *Let*

$$\begin{aligned} \bar{x} = \chi(x), \qquad \bar{u} = \psi(x,u), \qquad \bar{p} = \pi(x,u,p), \\ \bar{q} = \varpi(x,u,p,q), \qquad \bar{r} = \rho(x,u,p,q,r), \end{aligned}$$

be the second prolongation of a fiber-preserving transformation, mapping the second order Lagrangian $L(x,u,p,q)$ to $\overline{L}(\bar{x},\bar{u},\bar{p},\bar{q})$. Then the associated differential functions R and $\overline{R}$ are related by the same formula as the derivative coordinate r, namely $\overline{R} = \rho(x,u,p,q,R)$.

A remarkable consequence of this fact is that the *third order* differential equation $r = R(x, u, p, q)$ is *invariantly* associated with the second order variational problem (10.82).† Explicitly, the invariant third order equation is

$$D_x(LL_q^2) = LL_pL_q, \tag{10.95}$$

and we conclude that if L is mapped to $\overline{L}$ under a fiber-preserving transformation, then the third order equation (10.95) for L is mapped to the corresponding third order equation for $\overline{L}$! Classically, only the fourth order Euler–Lagrange equation (10.84) is known to be invariantly associated to a second order variational problem, so the appearance of the invariant third order equation (10.95) comes as a surprise. Indeed, a theorem due to I. Anderson, [**5**], states that the Euler–Lagrange equation is the only differential equation depending *linearly* on the Lagrangian which is invariantly associated with the variational problem. The precise geometric significance (if any) of the nonlinear invariant equation (10.95) remains unknown.

Since the transformation rules for R and the third order derivative r are the same, we can invariantly replace third order derivatives r in any expression by the derivative covariant R; the resulting expression, which only involves second order derivatives, obeys the same transformation rules as the original one. In particular, the third order truncation of the total derivative operator can be modified to yield an operator

$$D_x^* = \widetilde{D}_x + R\frac{\partial}{\partial q} = \frac{\partial}{\partial x} + p\frac{\partial}{\partial u} + q\frac{\partial}{\partial p} + R\frac{\partial}{\partial q}, \tag{10.96}$$

which is, in fact, a vector field on J^2. The transformation rules for D_x^* are the same as those of D_x, and hence applying D_x^* recursively to the derivative coordinates provides an entire hierarchy of higher order derivative covariants; the next one is $S = D_x^* R$, which transforms in the same fashion as the fourth order derivative coordinate $s = u_{xxxx}$.

† Proposition 10.15 implies the invariance of the equation for nondegenerate variational problems, but it is possible to demonstrate that this holds for general second order variational problems; see [**133**].

The structure equations for the coframe (10.92) are

$$\begin{aligned} d\theta^1 &= I_1\,\theta^1\wedge\theta^2 - (I_4+1)\,\theta^1\wedge\theta^4 - \theta^2\wedge\theta^3, \\ d\theta^2 &= I_3\,\theta^1\wedge\theta^2 + I_5\,\theta^1\wedge\theta^3 + I_2\,\theta^1\wedge\theta^4 - I_4\,\theta^2\wedge\theta^4 + \theta^3\wedge\theta^4, \\ d\theta^3 &= -\,\theta^3\wedge\theta^4, \\ d\theta^4 &= I_6\,\theta^1\wedge\theta^3 + I_7\,\theta^2\wedge\theta^3 + I_8\,\theta^3\wedge\theta^4. \end{aligned} \tag{10.97}$$

The coframe derivatives are written in terms of Jacobian determinants:

$$\begin{aligned} \frac{\partial F}{\partial\theta^1} &= \frac{1}{L_q^3}\left(-L_q\,\frac{\partial(L,\,F)}{\partial(u,\,q)} + P\,\frac{\partial(L,\,F)}{\partial(p,\,q)}\right), \qquad & \frac{\partial F}{\partial\theta^3} &= \frac{D_x^* F}{L}, \\ \frac{\partial F}{\partial\theta^2} &= -\frac{1}{L_q^2}\,\frac{\partial(L,\,F)}{\partial(p,\,q)}, & \frac{\partial F}{\partial\theta^4} &= \frac{L}{L_q}F_q. \end{aligned} \tag{10.98}$$

The eight (!) fundamental invariants are given by the following formulae:

$$\begin{aligned} I_1 &= -\frac{\partial(\log|LL_q|)}{\partial\theta^2} = \frac{L_pL_{qq} - L_qL_{pq}}{L_q^3}, \qquad I_2 = -\frac{1}{L}\,\frac{\partial}{\partial\theta^4}\left(\frac{P}{L_q}\right), \\ I_3 &= -\frac{1}{L}\,\frac{\partial}{\partial\theta^2}\left(\frac{P}{L_q}\right) + \frac{\partial(\log L_q)}{\partial\theta^1}, \quad I_4 = -\frac{\partial(\log|LL_q|)}{\partial\theta^4} - 1 = \frac{LL_{qq}}{L_q^2}, \\ I_5 &= -\frac{1}{L}\,\frac{\partial}{\partial\theta^3}\left(\frac{P}{L_q}\right) + \frac{P^2 - PL_q + L_pL_q}{L^2L_q^2} = \frac{\mathrm{E}^*(L)}{L^2L_q}, \\ I_6 &= -\frac{1}{L}\,\frac{\partial^2 L}{\partial\theta^1\partial\theta^3}, \qquad I_7 = -\frac{1}{L}\,\frac{\partial^2 L}{\partial\theta^2\partial\theta^3}, \qquad I_8 = -\frac{1}{L}\,\frac{\partial^2 L}{\partial\theta^4\partial\theta^3}. \end{aligned}$$

The most interesting invariant is I_5, which can be written in terms of the modified Euler–Lagrange expression

$$\mathrm{E}^*(L) = L_u - D_x^*(L_p) - (D_x^*)^2(L_q), \tag{10.99}$$

which is obtained from the usual Euler–Lagrange equation (10.84) by replacing the derivative coordinates r and s by the derivative covariants $R(x,u,p,q)$ and $S(x,u,p,q) = D_x^* R$, respectively, wherever they occur. Therefore, I_4 provides an "invariant second order version of the Euler–Lagrange equation", which is also invariantly associated with the variational problem.

Exercise 10.16. Solve the point and contact equivalence problems for second order Lagrangians using the inductive method. In particular, find the Cartan form for a second order Lagrangian. See [**42**] and [**133**] for details.

Multi-dimensional Lagrangians

As a final example, we consider the standard equivalence problem for a general first order variational problem

$$\mathcal{I}[u] = \int_\Omega L(x, u^{(1)})\, dx \tag{10.100}$$

in p independent variables and q dependent variables, under the class of point transformations. This equivalence problem has yet to be completely solved; however, following the results in [**187**], we show how even a partial implementation of the Cartan method yields interesting information, including an invariant candidate for the Cartan form and associated nondegeneracy conditions. We shall assume, without loss of generality, that the Lagrangian is positive, $L > 0$, on the domain of interest. To formulate the equivalence problem on the first jet space J^1, we introduce the basic contact forms

$$\omega^\alpha = du^\alpha - \sum_{i=1}^p u_i^\alpha\, dx^i, \qquad \alpha = 1, \ldots, q. \tag{10.101}$$

According to Proposition 7.9, two variational problems $\mathcal{I}[u]$ and $\overline{\mathcal{I}}[\bar{u}]$ are (standard) equivalent under a point transformation if and only if their Lagrangian forms are contact-equivalent, so

$$\Psi^*(\overline{L}\, d\bar{x}) \equiv L\, dx + \Theta, \tag{10.102}$$

where Θ is a contact p-form on J^1. We can restate this condition in terms of the p one-forms

$$\eta^i = L^{1/p}\, dx^i, \quad i = 1, \ldots, p, \qquad \text{so that} \qquad \eta^1 \wedge \cdots \wedge \eta^p = L\, dx. \tag{10.103}$$

Introducing a similar set of forms $\overline{\eta}^i = \overline{L}^{1/p}\, d\bar{x}^i$ in the barred variables, we find that the Lagrangian equivalence conditions (10.102) can be reformulated as

$$\Psi^*\overline{\eta}^i = \sum_{j=1}^p J_j^i\, \eta^j + \sum_{\beta=1}^q \widetilde{M}_\beta^i\, \omega^\beta, \qquad i = 1, \ldots, p, \tag{10.104}$$

where $\det J = \det(J^i_j) = 1$. The equivalence conditions (10.104), together with the contact conditions $\Psi^* \overline{\omega}^\alpha = \sum_\beta A^\alpha_\beta \omega^\beta$, constitute an underdetermined equivalence problem; as with the scalar case, we complete the one-forms (10.101), (10.103) to form a coframe on J^1 by appending the pq additional one-forms du^α_i, whose transformation rules can remain arbitrary. The structure group consists of the block matrices

$$G_p = \left\{ \begin{pmatrix} A & 0 & 0 \\ J \cdot M & J & 0 \\ C & D & E \end{pmatrix} \;\middle|\; A \in \mathrm{GL}(q), J \in \mathrm{SL}(p), E \in \mathrm{GL}(pq) \right\}.$$

The corresponding lifted coframe is given by the $pq + p + q$ one-forms

$$\begin{aligned} \theta^\alpha &= \sum_{\beta=1}^q A^\alpha_\beta \omega^\beta, \\ \xi^i &= \sum_{j=1}^p J^i_j \left(L^{1/p}\, dx^j + \sum_{\beta=1}^q M^j_\beta \omega^\beta \right), \qquad && i = 1, \ldots, p, \quad \alpha = 1, \ldots, q. \\ \pi^\alpha_i &= \sum_{\beta=1}^q C^\alpha_{\beta,i} \omega^\beta + \sum_{j=1}^p D^\alpha_{ij} L^{1/p}\, dx^j + \sum_{\beta=1}^q \sum_{j=1}^p E^{\alpha,j}_{i,\beta}\, du^\beta_j, \end{aligned} \tag{10.105}$$

In the first loop through Cartan's equivalence algorithm, the coefficients of $\xi^i \wedge \pi^\beta_j$ in $d\theta^\alpha$ provide essential torsion; they are

$$L^{-1/p} \sum_{\gamma=1}^q \sum_{k=1}^p A^\alpha_\gamma K^k_i F^{\gamma,j}_{k,\beta}, \qquad i, j = 1, \ldots, p, \quad \alpha, \beta = 1, \ldots, q,$$

where $K = J^{-1}$, $F = E^{-1}$. These torsion coefficients can be normalized to be a product of Kronecker deltas, namely $\delta^i_j \delta^\alpha_\beta$, by setting

$$E^{\alpha,j}_{i,\beta} = L^{-1/p} A^\alpha_\beta K^j_i, \qquad i, j = 1, \ldots, p, \quad \alpha, \beta = 1, \ldots, q. \tag{10.106}$$

Formulae (10.106) merely serve to incorporate the formulae for the first prolongation of our point transformation into the equivalence conditions, and, with a little more thought, could have been incorporated into our formulae for the lifted coframe at the outset. Second, the structure equations for $d\xi^i$ have the form

$$d\xi^i \equiv \sum_{j=1}^{p} \varphi^i_j \wedge \xi^j + \tau^i,$$

where φ^i_j are the Maurer–Cartan one-forms on the Lie group SL(p) corresponding to the group parameters $J = (J^i_j)$. Now, since J is restricted to have determinant 1, the matrix of Maurer–Cartan forms (φ^i_j) is trace-free, i.e., $\sum_i \varphi^i_i = 0$. The trace of the torsion coefficients

$$Q^{ik}_{\beta j} = \sum_{\alpha=1}^{q} \sum_{r,s,t=1}^{p} B^\alpha_\beta J^i_r J^k_s \bar{J}^t_j P^{rs}_{\alpha t}, \qquad P^{ik}_{\alpha j} = \bar{p} L^{(1-p)/p} \frac{\partial L}{\partial u^\alpha_j} \delta^i_j - M^i_\alpha \delta^k_j,$$

of $\pi^\beta_k \wedge \xi^j$ in $d\xi^i$ yields the essential torsion components

$$\sum_{i=1}^{p} Q^{ik}_{\beta i} = \sum_{\alpha=1}^{q} \sum_{l,i=1}^{p} B^\alpha_\beta J^k_l P^{li}_{\alpha i}. \tag{10.107}$$

We normalize the components of the trace (10.107) to zero by requiring that P itself have trace zero, $\sum_i P^{ik}_{\beta i} = 0$, which is achieved by normalizing the group parameters

$$M^i_\alpha = L^{(1-p)/p} \frac{\partial L}{\partial u^\alpha_i}. \tag{10.108}$$

With these normalizations,

$$\xi^i = \sum_j J^i_j \varpi^j, \qquad \text{where} \qquad \varpi^i = \sqrt[p]{L} \left\{ dx^i + \frac{1}{L} \frac{\partial L}{\partial u^\alpha_i} \theta^\alpha \right\}. \tag{10.109}$$

Since $\det J = 1$, the p-form

$$\Theta_C = \xi^1 \wedge \cdots \wedge \xi^p = \varpi^1 \wedge \cdots \wedge \varpi^p \tag{10.110}$$

is invariant under general point transformations. In analogy with the scalar case, cf. (10.67), the invariant p-form Θ_C is our candidate for the

Cartan form associated with the first order variational problem (10.100). It can be shown that it provides the connection between variational symmetries and conservation laws analogous to that described in the scalar case in Chapter 7. In fact, there are several competing Cartan forms for multiple integral problems in the calculus of variations, each of which leads to different multi-dimensional versions of the classical field theory and nondegeneracy conditions for scalar problems. The form Θ_C appears in the theory of Carathéodory, [**32**]; its most well-studied competitor was proposed by Weyl, [**222**], and is found by omitting all terms in Θ_C which are quadratic or higher degree in the contact forms. The Cartan equivalence method singles out the Carathéodory version as the essentially unique Cartan form which is invariant under general point transformations, a property not enjoyed by the Weyl and other versions. See [**98**], [**187**], for extensive discussions of these and other issues. The latter paper also analyzes the equivalence problem for general higher order variational problems, and, in the second order case, produces an invariant Cartan form. However, technical complications interfere in the third and higher order cases, and a fully invariant Cartan form is not as yet known.

Inserting the normalizations (10.107) back into the lifted coframe (10.105), we recompute the differentials $d\xi^i$, and look at the essential torsion terms $Y^{ik}_{\alpha\beta}\,\pi^\alpha_k \wedge \vartheta^\beta$, where

$$\begin{aligned} Y^{ik}_{\alpha\beta} &= \sum_{\gamma,\delta=1}^{q} \sum_{j,l=1}^{p} B^\gamma_\alpha B^\delta_\beta J^i_j J^k_l Z^{jl}_{\gamma\delta}, \\ Z^{ik}_{\alpha\beta} &= L^{(2-2p)/p} \left\{ L \frac{\partial^2 L}{\partial u^\alpha_i \partial u^\beta_k} + \frac{\partial L}{\partial u^\alpha_k} \frac{\partial L}{\partial u^\beta_i} - \frac{\partial L}{\partial u^\alpha_i} \frac{\partial L}{\partial u^\beta_k} \right\}. \end{aligned} \tag{10.111}$$

The first term in the *augmented Hessian tensor* $\mathbf{Z}$ is the $(pq) \times (pq)$ Hessian matrix $\mathbf{H}$ of L with respect to the first order derivatives of u; however, the additional first order derivative terms are unusual and unexpected. Note that if $i = k$, or if $\alpha = \beta$, then the first order derivative terms cancel out. Consequently, in the cases of mechanics $(p = 1)$ and scalar field theory $(q = 1)$ there *are* no first derivative terms in $\mathbf{Z}$, and hence the augmented Hessian tensor $\mathbf{Z}$ is a multiple of the standard Hessian $\mathbf{H}$. The fact that these terms appear as essential torsion in the equivalence method implies that the augmented Hessian tensor is invariant under general point transformations, in the sense that $\overline{\mathbf{Y}} = \mathbf{Y}$ for suitable values of the group elements $B(x, u^{(1)})$ and $J(x, u^{(1)})$. Surprisingly, the standard Hessian tensor $\mathbf{H}$ is *not* invariant if both p and q are greater than one!

The next move in the Cartan procedure would be to normalize the torsion tensor $\mathbf{Y}$ by suitable choice of the group parameters A^α_β, J^i_j, a process which depends on the algebraic character of the augmented Hessian tensor $\mathbf{Z}$. When $p = 1$ or $q = 1$, both $\mathbf{Z}$ and $\mathbf{Y}$ are symmetric matrices, and we are essentially normalizing a quadratic form on either $X = \mathbb{R}^p$, or $U = \mathbb{R}^q$, a problem which is well understood. In these cases, the invariants are the rank and signature of the usual Hessian matrix of L with respect to the derivative variables. The nondegeneracy conditions for a Lagrangian are just the nondegeneracy conditions that its Hessian matrix be nonsingular, and the analysis is reasonably straightforward, cf. [**78**], [**80**]. If both p and q are greater than 1, far less is known. The symmetric part of the augmented Hessian tensor can be identified with the "biquadratic biform"

$$S[\xi, \lambda] = S^{ik}_{\alpha\beta}\xi_i\xi_k\lambda^\alpha\lambda^\beta = L^{2(1-p)/p}\left\{\frac{\partial^2 L}{\partial u^\alpha_i \partial u^\beta_k}\,\xi_i\xi_k\lambda^\alpha\lambda^\beta\right\}, \qquad (10.112)$$

which is the *symbol* of the variational problem, and transforms naturally under the action of $\mathrm{GL}(p) \times \mathrm{GL}(q)$. The Legendre–Hadamard strong ellipticity condition, [**171**], requires that the symbol be *positive definite*, meaning that $S[\xi, \lambda] > 0$ for all $0 \neq \xi, \lambda$. The invariance of the symbol implies that the Legendre–Hadamard condition is invariant under arbitrary point transformations. The normalization step requires a complete determination of canonical forms for biquadratic biforms, which is equivalent to the problem of classifying quadratic Lagrangians. The canonical form problem has only been solved in the elementary cases $p = 1$, or $q = 1$, or $p = q = 2$, [**181**]. In the positive definite case, this classification is the same as a determination of canonical forms for linear elasticity, under general linear changes of variables; see [**182**] for the planar case, and [**185**] for the case $p = 2$, $q = 3$, governing planar displacements of a three-dimensional body. When $p, q > 1$, there are additional invariant moduli governing the canonical form of the symbol, so the required normalization is of "nonconstant type"; hence the further analysis of the equivalence problem is considerably more difficult than what we have encountered above.

Chapter 11

Involution

We now turn to the cases when the Cartan equivalence procedure, as presented so far, does not lead to a complete reduction of the structure group. Thus, we may assume that, after perhaps one or more loops through the absorption and reduction algorithm, we are left with a coframe that still involves one or more of the original group parameters, yet none of the remaining essential torsion coefficients depend explicitly on the group parameters. Therefore, the method fails to produce any further invariant combinations of variables and group parameters that will allow us to normalize the remaining group parameters in a consistent manner. The question is then: What do we do now?

Why might we expect such a situation to arise? Consider the possible symmetry groups of an equivalence problem. If the Cartan method leads to a complete reduction, resulting in an invariant coframe $\boldsymbol{\theta}$, then the associated symmetry group is, as we have seen, an $(m-r)$-dimensional Lie group, where m is the dimension of the underlying manifold (or number of coframe elements) and r is the rank of the coframe, or number of functionally independent structure functions — see Theorem 8.22. Consequently, the dimension of the symmetry group can *never exceed* the dimension of the manifold on which the problem is formulated as a Cartan equivalence problem. On the other hand, we have already encountered a number of equivalence problems wherein the symmetry group has a larger dimension than the space on which the problem is most naturally formulated. For example, the equivalence problem for scalar second order ordinary differential equations is most naturally formulated in Cartan form on the three-dimensional jet space J^1, as in Example 9.3, but the dimension of the point transformation symmetry group can be as large as eight, e.g., for the equation $u_{xx} = 0$ — see Example 6.6 and Theorem 6.25. Thus we cannot expect the equivalence problem for (at least some) second order ordinary differential equations to reduce to an equivalence of coframes on any manifold having dimension less than eight, and so we will be unable to obtain an invariant coframe on J^1. Even more dramatically, according to Example 6.6, every scalar second order ordinary differential equation admits an infinite-dimensional group of contact symmetries, and hence we can *never* realize the equivalence problem for second order ordinary differential equations under contact transformations as an equivalence of coframes on *any* finite-dimensional manifold!

Fortunately, Cartan figured out how to deal with both these types of problems as well. There is an essential dichotomy between the two, and they require different tools to complete the solution. The first, in which the symmetry group is finite-dimensional, but of larger dimension than the underlying space, is solved by the process of "prolongation" (not to be confused with the process of prolongation of transformation groups and vector fields discussed in Chapter 4). We will deal with these types of equivalence problems in Chapter 12. The second type of problem, in which the symmetry group is infinite-dimensional, cannot be handled by the relatively simple Frobenius Theorem, but requires the Cartan–Kähler Existence Theorem to complete its solution. Since this powerful theorem is a consequence of the Cauchy–Kovalevskaya Existence Theorem for analytic systems of partial differential equations, it is only applicable to analytic equivalence problems, and we must therefore accordingly restrict the problems in this case. The existence of an infinite-dimensional symmetry group is a consequence of the "involutivity" of a certain differential system prescribed by the structure equations for the lifted coframe. The decision as to whether we need to prolong an equivalence problem (finite-dimensional symmetry group) or not (infinite-dimensional symmetry group) is handled by an arithmetic test for involutivity due to Cartan. In this chapter, we will discuss how to implement Cartan's test practically, and, in the case of an involutive system, how the infinite-dimensional symmetry group and canonical forms are determined. The theory and proofs underlying the method will be deferred until Chapter 15.

Example 11.1. To illustrate the distinction between the two possible outcomes, we consider the three basic equivalence problems for a second order ordinary differential equation $u_{xx} = Q(x, u, p)$, $p = u_x$, under, respectively, fiber-preserving transformations, point transformations and, most generally, contact transformations. As discussed in Example 9.6, we introduce the base coframe

$$\omega^1 = du - p\,dx, \qquad \omega^2 = dp - Q(x,u,p)\,dx, \qquad \omega^3 = dx, \tag{11.1}$$

on the first jet space J^1. For the contact equivalence problem, the lifted coframe is

$$\theta^1 = a_1\omega^1, \qquad \theta^2 = a_2\omega^1 + a_3\omega^2, \qquad \theta^3 = a_4\omega^1 + a_5\omega^2 + a_6\omega^3, \tag{11.2}$$

where the a_i parametrize the group G_c given in (9.23). For point transformation equivalence, we use the subgroup $G_p \subset G_c$ obtained by setting $a_5 = 0$, whereas the fiber-preserving equivalence problem uses the subgroup $G_f \subset G_c$ with $a_4 = a_5 = 0$.

We first see how far the Cartan equivalence method from Chapter 10 will go toward normalizing group parameters. We write the structure equations associated with the lifted coframe (11.2) in the absorbed form (10.13), so

$$\begin{aligned} d\theta^1 &= \pi^1 \wedge \theta^1 + T^1_{23}\,\theta^2 \wedge \theta^3, \\ d\theta^2 &= \pi^2 \wedge \theta^1 + \pi^3 \wedge \theta^2, \\ d\theta^3 &= \pi^4 \wedge \theta^1 + \pi^5 \wedge \theta^2 + \pi^6 \wedge \theta^3. \end{aligned} \tag{11.3}$$

Each one-form $\pi^\kappa = \alpha^\kappa + \sum_i z_i^\kappa \theta^i$, $\kappa = 1, \ldots, 6$, is obtained from the corresponding Maurer–Cartan form α^κ on the Lie group G_c by adding in the solution to the absorption equations, as given in (10.12). For example, $\pi^1 = \alpha^1 - T^1_{12}\,\theta^1 \wedge \theta^2 - T^1_{13}\,\theta^1 \wedge \theta^3$, etc. In the point transformation case $\pi^5 = 0$, while for fiber-preserving equivalence, $\pi^4 = \pi^5 = 0$; otherwise the structure equations have the same form, although the explicit formulae are different. In all cases, the only essential torsion coefficient is

$$T^1_{23} = -\frac{a_1}{a_3 a_6}, \tag{11.4}$$

which is normalized to -1 by setting $a_6 = a_1/a_3$. (As we noted in Example 9.9, this normalization actually follows trivially from the transformation formula for the derivative coordinate, cf. (4.13), and so could have been stipulated in advance. It is always reassuring to see how the Cartan method inevitably comes through even if we neglect to include relatively simple consequences of the basic transformation rules.) This normalization effectively reduces the structure group to a codimension 1 subgroup.

In the second loop through the equivalence procedure, we first recompute the structure equations, and then absorb all the inessential torsion. The resulting structure equations take the form

$$\begin{aligned} d\theta^1 &= \pi^1 \wedge \theta^1 - \theta^2 \wedge \theta^3, \\ d\theta^2 &= \pi^2 \wedge \theta^1 + \pi^3 \wedge \theta^2, \\ d\theta^3 &= \pi^4 \wedge \theta^1 + \pi^5 \wedge \theta^2 + (\pi^1 - \pi^3) \wedge \theta^3. \end{aligned} \tag{11.5}$$

The cases of point transformations and fiber-preserving transformations lead to similar structure equations, with $\pi^5 = 0$ in the former, and $\pi^4 = \pi^5 = 0$ in the latter. In the structure equations (11.5), there is no nonconstant torsion to normalize, and, as we had predicted above, the reduction procedure of Chapter 10 has hit a dead end. Now we must figure out whether the symmetry group of the equivalence problem is infinite-dimensional, which is indicated by the system (11.5) being in involution, or finite-dimensional, so that the system (11.5) is not in involution, and we must prolong.

Cartan's Test

Cartan devised a straightforward arithmetic test for distinguishing between the two possibilities of involution versus prolongation. The complete story is told in Chapter 15; here we merely explain how to practically apply Cartan's involutivity test to particular examples. Assume that we are given an equivalence problem based on the lifted coframe $\boldsymbol{\theta} = \{\theta^1, \ldots, \theta^m\}$, which are one-forms on the $(m+r)$-dimensional product space $M \times G$, since their coefficients also depend on the group parameters. Here r denotes the dimension of the (reduced) structure group, which we denote by G for simplicity. The structure equations for the coframe have the form

$$d\theta^i = \sum_{j=1}^{m} \sum_{\kappa=1}^{r} A^i_{j\kappa}\, \alpha^\kappa \wedge \theta^j + \sum_{j,k=1}^{m} T^i_{jk}\, \theta^j \wedge \theta^k, \qquad i = 1, \ldots, m, \tag{11.6}$$

where the α^κ are a basis for the right-invariant Maurer–Cartan forms on G, and the structure coefficients $A^i_{j\kappa}$ are constants, cf. (10.5). (More generally, in cases resulting from normalizations of nonconstant type, the $A^i_{j\kappa}$ can be nonconstant invariants; this does not significantly alter the basic analysis.) Moreover, we are assuming that none of the essential (unabsorbable) torsion coefficients depend explicitly on the group parameters, since we are assuming that there is no more torsion left to normalize. Replacing each Maurer–Cartan form α^κ by an arbitrary linear combination of the coframe elements $\sum z^\kappa_j\,\theta^j$ results in the linear absorption equations

$$\sum_{\kappa=1}^{r} \left(A^i_{j\kappa} z^\kappa_k - A^i_{k\kappa} z^\kappa_j\right) = T^i_{jk}, \qquad i,j,k = 1,\ldots,m, \quad j < k, \qquad (11.7)$$

for the unknown coefficients z^κ_j, cf. (10.14). In solving the linear system (11.7), some of the variables z^κ_j will be specified (these are sometimes referred to in linear algebra as "basic variables") while others can be given arbitrary values (referred to as "free variables").

Definition 11.2. The *degree of indeterminancy* $r^{(1)}$ of a lifted coframe is the number of free variables in the solution to the associated linear absorption system.

In other words, $r^{(1)}$ is the dimension of the solution space to the corresponding homogeneous system of linear equations

$$\sum_{\kappa=1}^{r} \left(A^i_{j\kappa} z^\kappa_k - A^i_{k\kappa} z^\kappa_j\right) = 0, \qquad i,j,k = 1,\ldots,m, \quad j < k. \qquad (11.8)$$

The degree of indeterminancy $r^{(1)}$ of the lifted coframe $\boldsymbol{\theta}$ plays an extremely important role in the subsequent development, both in the prolongation and involution cases. Note that $r^{(1)}$ only depends on how the (reduced) structure group G is realized as a subgroup of $\mathrm{GL}(m)$, and not on the particular formulae for the base coframe elements.

Example 11.3. Consider the structure equations (11.5) for the equivalence of a second order ordinary differential equation under contact transformations. To compute the degree of indeterminancy $r^{(1)}$, we neglect the torsion terms, and replace each Maurer–Cartan form α^κ (or, equivalently, each π^κ) by the combination $z^\kappa_1\theta^1 + z^\kappa_2\theta^2 + z^\kappa_3\theta^3$ of lifted coframe elements. Equating the resulting coefficients of the basis two-forms $\theta^j \wedge \theta^k$ to zero leads to the homogeneous linear system

$$z^1_2 = z^1_3 = 0, \qquad z^3_1 - z^2_2 = z^2_3 = z^3_3 = 0,$$
$$z^5_1 - z^4_2 = z^1_1 - z^3_1 - z^4_3 = z^1_2 - z^3_2 - z^5_3 = 0,$$

which is the same as the left hand side of the full absorption equations. The general solution is

$$\begin{gathered} z_2^1 = z_3^1 = z_3^2 = z_3^3 = 0, \\ z_1^1 = z_1^3 + z_3^4, \qquad z_2^2 = z_1^3, \qquad z_2^3 = -z_3^5, \qquad z_2^4 = z_1^5. \end{gathered} \tag{11.9}$$

The seven parameters $z_1^2, z_1^3, z_1^4, z_3^4, z_1^5, z_2^5, z_3^5$ can be chosen arbitrarily — they are the free variables for the absorption equations — and the remaining eight parameters are prescribed in terms of these. Therefore, for the contact equivalence problem, the degree of indeterminancy is $r^{(1)} = 7$. For point transformations, $\pi^5 = 0$, so the absorption equations have the same form, but parameters z_j^5 do not appear; therefore $r^{(1)} = 4$ in this case. For fiber-preserving transformations, $\pi^4 = \pi^5 = 0$, so z_j^4, z_j^5 do not appear, and the degree of indeterminancy is only $r^{(1)} = 2$. (The final equation in (11.9) is vacuous in this case.)

To continue to describe Cartan's test for involutivity, we next introduce the reduced Cartan characters associated with the structure equations (11.6). First, given a vector $\mathbf{v} = (v^1, \ldots, v^m) \in \mathbb{R}^m$, define $L[\mathbf{v}]$ to be the $m \times r$ matrix with entries

$$L_\kappa^i[\mathbf{v}] = \sum_{j=1}^m A_{j\kappa}^i v^j, \qquad i = 1, \ldots, m, \quad \kappa = 1, \ldots, r. \tag{11.10}$$

Thus, $L[\mathbf{v}]$ is just the coefficient matrix for the Maurer–Cartan forms α^κ in the structure equations (11.6) upon replacing each coframe element θ^i by the corresponding entry v^i of the vector $\mathbf{v}$ (and deleting the wedge products). The rank of $L[\mathbf{v}]$ will, of course, depend on the particular vector $\mathbf{v}$. (For instance, if $\mathbf{v} = 0$, then the rank is obviously zero.) The important numerical quantity is the *maximal* rank of the matrix $L[\mathbf{v}]$ over all possible vectors $\mathbf{v} \in \mathbb{R}^m$; this is called the first reduced character for the lifted coframe, denoted by

$$s_1' = \max \{\operatorname{rank} L[\mathbf{v}] \mid \mathbf{v} \in \mathbb{R}^m\}.$$

The second reduced character s_2' is obtained by computing the maximal rank of a $2m \times r$ matrix which is obtained by stacking two copies of the previous matrix, corresponding to two different vectors, on top of each other. Specifically, we define s_2' by the equation

$$s_1' + s_2' = \max \left\{ \operatorname{rank} \begin{pmatrix} L[\mathbf{v}_1] \\ L[\mathbf{v}_2] \end{pmatrix} \;\middle|\; \mathbf{v}_1, \mathbf{v}_2 \in \mathbb{R}^m \right\}.$$

The higher order reduced characters — except for the last one s'_m — are defined inductively by a similar construction.

Definition 11.4. Let $\boldsymbol{\theta}$ be a lifted coframe on the m-dimensional manifold M with r-dimensional structure group G. The first $m-1$ *reduced characters* $s'_1, \ldots, s'_{m-1}$ of the coframe are defined as

$$s'_1 + s'_2 + \cdots + s'_k = \max \left\{ \operatorname{rank} \begin{pmatrix} L[\mathbf{v}_1] \\ L[\mathbf{v}_2] \\ \vdots \\ L[\mathbf{v}_k] \end{pmatrix} \;\middle|\; \mathbf{v}_1, \ldots, \mathbf{v}_k \in \mathbb{R}^m \right\}, \tag{11.11}$$

for $k = 1, \ldots, m-1$, where $L[\mathbf{v}]$ is the $m \times r$ matrix in (11.10). The final *reduced character* s'_m is, however, defined by the equation

$$s'_1 + \cdots + s'_{m-1} + s'_m = r. \tag{11.12}$$

Note that each $s'_i \geq 0$ is nonnegative, since the rank of the matrix in (11.11) is bounded by r — the number of columns.

The motivation for this definition of the reduced characters is a consequence of Cartan's theory of exterior differential systems, and is discussed in full detail in Chapter 15. In the present chapter, we shall be content to familiarize ourselves with the definition and to learn how the reduced characters aid us to distinguish between lifted coframes in involution (for which there is an infinite-dimensional symmetry group) and those which are not (which we must prolong).

Example 11.5. Consider the structure equations for a second order ordinary differential equation under contact transformations (11.5). According to (11.10), the corresponding 3×5 matrix used to define the reduced characters is

$$L[\mathbf{v}] = \begin{pmatrix} v^1 & 0 & 0 & 0 & 0 \\ 0 & v^1 & v^2 & 0 & 0 \\ v^3 & 0 & -v^3 & v^1 & v^2 \end{pmatrix}, \qquad \mathbf{v} = (v^1, v^2, v^3) \in \mathbb{R}^3.$$

The rows of L correspond to the three structure equations, and the columns correspond to the five Maurer–Cartan forms $\alpha^1, \ldots, \alpha^5$. The first reduced character s'_1 is just the maximal rank of $L[\mathbf{v}]$ for all possible vectors $\mathbf{v} \in \mathbb{R}^3$. Clearly $s'_1 = 3$; for example, this is achieved by taking $\mathbf{v} = (1,0,0)$, say. The second Cartan character is then defined by

$$s'_1 + s'_2 = \max \operatorname{rank} \begin{pmatrix} v^1 & 0 & 0 & 0 & 0 \\ 0 & v^1 & v^2 & 0 & 0 \\ v^3 & 0 & -v^3 & v^1 & v^2 \\ \widehat{v}^1 & 0 & 0 & 0 & 0 \\ 0 & \widehat{v}^1 & \widehat{v}^2 & 0 & 0 \\ \widehat{v}^3 & 0 & -\widehat{v}^3 & \widehat{v}^1 & \widehat{v}^2 \end{pmatrix} = 5,$$

the maximum being taken over all pairs of vectors $\mathbf{v} = (v^1, v^2, v^3), \widehat{\mathbf{v}} = (\widehat{v}^1, \widehat{v}^2, \widehat{v}^3) \in \mathbb{R}^3$. The fourth row of this matrix is always a multiple of the first, and so the maximum rank is 5, achieved for $\mathbf{v} = (1,0,0)$, $\widehat{\mathbf{v}} = (0,1,0)$. Therefore, the second reduced character is $s'_2 = 5 - s'_1 = 2$. The final reduced character $s'_3 = 5 - s'_1 - s'_2 = 0$ is defined by (11.12).

In the case of equivalence under point transformations, the only difference in the structure equations (11.5) is that the form α^5 does not appear. Therefore, the corresponding matrices defining the reduced characters can be obtained from the contact versions merely by omitting the last column. The reduced characters are readily found to be $s'_1 = 3$, $s'_2 = 1$, $s'_3 = 0$. Similarly, for the fiber-preserving equivalence problem, we set $\alpha^4 = \alpha^5 = 0$, so the matrices defining the reduced characters can be obtained from the contact versions by omitting the last two columns. The reduced characters are $s'_1 = 3$, $s'_2 = 0$, $s'_3 = 0$.

Exercise 11.6. Consider the structure equations

$$\begin{aligned} d\theta^1 &= \pi^1 \wedge \theta^1 + (\pi^2 + \pi^3) \wedge \theta^2 + \pi^4 \wedge \theta^3, \\ d\theta^2 &= \pi^2 \wedge \theta^1 + \pi^1 \wedge \theta^2, \end{aligned} \qquad d\theta^3 = 0.$$

Find the matrix $L[\mathbf{v}]$. Prove that the reduced characters are $s'_1 = s'_2 = 2$, $s'_3 = 0$. Note that if $\mathbf{v}_1 = (1,0,0)$, then $\operatorname{rank} L[\mathbf{v}_1] = 2$ is maximal, but there is no vector $\mathbf{v}_2$ such that $\operatorname{rank} \begin{pmatrix} L[\mathbf{v}_1] \\ L[\mathbf{v}_2] \end{pmatrix} = 4$ is maximal. This shows that one must, occasionally, use some care in calculating the reduced characters.

Loosely speaking, a system of partial differential equations is said to be "involutive" if there are no integrability conditions. In our context, involutivity is equivalent to the existence of an infinite-dimensional symmetry group for the coframe under consideration. The key tool is Cartan's algebraic test for involutivity, which we state here as a definition. (In Chapter 15 we will see how the test is motivated and proved.)

Definition 11.7. Let $\boldsymbol{\theta}$ be a lifted coframe with reduced characters $s'_1, \ldots, s'_m$, and degree of indeterminancy $r^{(1)}$. Then $\boldsymbol{\theta}$ is *involutive* if and only if it satisfies the *Cartan test*

$$s'_1 + 2s'_2 + \cdots + ms'_m = r^{(1)}. \tag{11.13}$$

We remark that, in all cases, the left hand side of (11.13) provides an upper bound for the degree of indeterminancy of the coframe, so $r^{(1)} \leq s'_1 + 2s'_2 + \cdots + ms'_m$, with equality if and only if the coframe is involutive.

The Transitive Case

The equivalence problem for involutive analytic coframes has a complete solution. We begin by stating the simpler (and much more common) case when the essential torsion coefficients in the structure equations are all constant. The basic result here is that *all* involutive coframes having the same structure equations are equivalent, and all have a transitive, infinite-dimensional symmetry group. Thus, such cases are commonly referred to as *transitive* structure equations, the cases with nonconstant essential torsion being called *intransitive*.

Theorem 11.8. *Let $\boldsymbol{\theta}$ and $\overline{\boldsymbol{\theta}}$ be analytic lifted coframes on m-dimensional manifolds M and $\overline{M}$ having the same structure group $G \subset$ GL(m). Assume that both $\boldsymbol{\theta}$ and $\overline{\boldsymbol{\theta}}$ are involutive, meaning that the Cartan test (11.13) is satisfied, and that all the essential torsion coefficients in the two sets of structure equations are constant. Then $\boldsymbol{\theta}$ is locally equivalent to $\overline{\boldsymbol{\theta}}$ if and only if they have the same constant essential torsion. Moreover, if $s'_k > 0$ is the last nonzero character, so $s'_l = 0$ for $l > k$, then the set of (analytic) equivalences depends on s'_k arbitrary analytic functions, each depending on k variables.*

Corollary 11.9. *An involutive analytic coframe has an infinite-dimensional symmetry group depending on s'_k arbitrary analytic functions of k variables.*

Remark: The infinite-dimensional symmetry groups arising from involutive analytic coframes are commonly referred to as infinite *pseudogroups*. Beyond the usual (local) group axioms, their key defining property is that they satisfy a system of partial differential equations. These objects, which are the infinite-dimensional counterparts of Lie groups of transformations, were originally introduced and studied by Lie, [**155**], who obtained a complete classification in two dimensions, [**151**], and, subsequently, by Cartan, [**37**], who classified the primitive, transitive cases. Strangely, there is no proper intrinsic formulation of infinite pseudo-groups that does not rely on their realization as groups of transformations. See [**146**], [**147**], [**204**] for more recent treatments.

Example 11.10. According to Examples 11.3 and 11.5, the lifted coframe for the contact equivalence problem for a second order ordinary differential equation has degree of indeterminancy $r^{(1)} = 7$. The reduced characters are $s_1' = 3$, $s_2' = 2$, $s_3' = 0$. The Cartan involutivity test (11.13) is satisfied in this case, since $s_1' + 2s_2' + 3s_3' = 3 + 2 \cdot 2 + 3 \cdot 0 = 7$. Therefore, the structure equations for the contact equivalence of second order ordinary differential equations are involutive. Since the essential torsion in (11.5) is constant, we can apply Theorem 11.8. Since $s_2' = 2$, while $s_3' = 0$, the general equivalence map will depend on two arbitrary analytic functions, each depending on two variables.

Theorem 11.11. *All nonsingular, analytic, second order ordinary differential equations are (locally) equivalent under contact transformations. Moreover, every analytic second order ordinary differential equation admits an infinite-dimensional symmetry group of contact transformations depending on two arbitrary analytic functions of two variables, and hence can be mapped to any other one in infinitely many ways.*

Remark: In the case of the simple equation $u_{xx} = 0$, the structure of the contact symmetry group was explicitly found in (6.14) — $A(z,p)$ and $B(z,p)$ are the two analytic functions; the fact that any other second order equation is contact-equivalent to $u_{xx} = 0$ proves its validity in general.

On the other hand, for the equivalence problem under point transformations, the degree of indeterminancy is $r^{(1)} = 4$, whereas the reduced characters are $s_1' = 3$, $s_2' = 1$, $s_3' = 0$. Therefore $s_1' + 2s_2' + 3s_3' = 5 > r^{(1)} = 4$, and the Cartan involutivity test (11.13) does not hold. Thus, the structure equations are *not* in involution, and we must prolong. Similarly, for the fiber-preserving equivalence problem, the degree of indeterminancy is $r^{(1)} = 2$, the reduced characters are $s_1' = 3$, $s_2' = 0$, $s_3' = 0$, and $s_1' + 2s_2' + 3s_3' = 3 > r^{(1)} = 2$. Again (not surprisingly) the Cartan involutivity test (11.13) fails.

Divergence Equivalence of First Order Lagrangians

As a second application, we consider the divergence equivalence problem for first order Lagrangians under contact transformations. Our aim is to prove the following theorem.

Theorem 11.12. *All nondegenerate analytic first order Lagrangians in one independent variable and one dependent variable are (locally) equivalent under a contact transformation. Moreover, the infinite-dimensional contact divergence symmetry group of a nondegenerate first order Lagrangian depends on a single arbitrary analytic function of two variables.*

Proof: We shall employ the approach based on the Cartan form; see (9.18). Given a Lagrangian $L(x,u,p)$, we take

$$\omega^1 = du - p\,dx, \qquad \omega^2 = -\widetilde{\mathrm{E}}(L)\,dx + L_{pp}\,dp, \qquad \omega^3 = dx, \tag{11.14}$$

where $\widetilde{\mathrm{E}}(L) = L_u - L_{px} - pL_{pu}$, as the base coframe. Note that this does form a coframe as long as L satisfies the nondegeneracy condition $L_{pp} \neq 0$. (Indeed, it is not hard to prove that even contact transformations preserve the singularities where $L_{pp} = 0$, and so we can never map a degenerate Lagrangian to a nondegenerate one.) For the contact equivalence problem, the lifted coframe is

$$\theta^1 = a_1\omega^1, \qquad \theta^2 = a_2\omega^1 + \frac{1}{a_1}\omega^2, \qquad \theta^3 = a_3\omega^1 + a_4\omega^2 + a_5\omega^3, \tag{11.15}$$

where the a_i parametrize the five parameter structure group. The structure equations associated with the lifted coframe (11.15) have the form

$$\begin{aligned} d\theta^1 &= \alpha^1 \wedge \theta^1 && + \Theta^1, \\ d\theta^2 &= \alpha^2 \wedge \theta^1 - \alpha^1 \wedge \theta^2 && + \Theta^2, \\ d\theta^3 &= \alpha^3 \wedge \theta^1 + \alpha^4 \wedge \theta^2 + \alpha^5 \wedge \theta^3 && + \Theta^3, \end{aligned} \tag{11.16}$$

where the α^κ's are the Maurer–Cartan forms, and where the torsion terms have the form $\Theta^i = T^i_{12}\,\theta^1 \wedge \theta^2 + T^i_{13}\,\theta^1 \wedge \theta^3 + T^i_{23}\,\theta^2 \wedge \theta^3$. There are two essential torsion terms. The first is $T^1_{23} = -a_1^2/(a_5 L_{pp})$, which is normalized to $T^1_{23} \mapsto -1$ by setting $a_5 = a_1^2/L_{pp}$. The second is $T^1_{13}+T^2_{23}$, which, in view of the formula for $\widetilde{\mathrm{E}}(L)$, happens to be identically zero;

thus only one normalization is possible. (Actually, this fact is not an accident: the form ω^2 was defined in terms of the Cartan form by $d\eta_L = \omega^2 \wedge \omega^1$, so $0 = d\omega^2 \wedge \omega^1 - \omega^2 \wedge d\omega^1 = d\omega^2 \wedge \omega^1$. Therefore $d\omega^2$ cannot contain any terms involving $\omega^2 \wedge \omega^3$.)

Substituting our normalization for a_5 and recomputing, we find that there is no longer any essential torsion. After absorption, the structure equations take the form

$$\begin{aligned} d\theta^1 &= \pi^1 \wedge \theta^1 - \theta^2 \wedge \theta^3, \\ d\theta^2 &= \pi^2 \wedge \theta^1 - \pi^1 \wedge \theta^2, \\ d\theta^3 &= \pi^3 \wedge \theta^1 + \pi^4 \wedge \theta^2 - 2\pi^1 \wedge \theta^3. \end{aligned} \tag{11.17}$$

(The modified Maurer–Cartan forms in (11.17) involve slightly different linear combinations of the θ^i's than those in (11.16); using the same notation, though, will not cause any confusion.) We now check for involutivity. The degree of indeterminancy of (11.17) is $r^{(1)} = 5$, since we can replace

$$\begin{aligned} &\pi^1 \mapsto z_1\theta^1, && \pi^3 \mapsto z_3\theta^1 - z_4\theta^2 - 2z_1\theta^3, \\ &\pi^2 \mapsto z_2\theta^1 - z_1\theta^2, && \pi^4 \mapsto z_4\theta^1 + z_5\theta^2, \end{aligned}$$

without changing the torsion coefficients. On the other hand, the Cartan characters are readily computed: $s'_1 = 3$, $s'_2 = 1$, $s'_3 = 0$. The Cartan test (11.13) is satisfied: $s'_1 + 2s'_2 + 3s'_3 = 3 + 2 \cdot 1 + 3 \cdot 0 = 5 = r^{(1)}$. Therefore, the structure equations (11.17) are involutive, which suffices to prove the theorem. *Q.E.D.*

Exercise 11.13. Re-prove Theorem 11.12 using the alternative formulation on the extended space $\mathrm{J}^1 \times \mathbb{R}$ discussed in Chapter 9. In particular, show how the Cartan form naturally arises in this case.

Exercise 11.14. Solve the divergence equivalence problems for first order Lagrangians under fiber-preserving and point transformations. (*Warning*: The calculations are quite complicated; see [**129**].)

The Intrinsic Method

As the reader has noticed, many of the explicit formulae for torsion coefficients, group parameters, etc., that arise in any application of the Cartan equivalence method are not really required in order to find the final solution of the equivalence problem. This is particularly true for equivalence problems leading to involutive structure equations, since (at least in the transitive case) there are no invariants, and so, during the absorption and normalization procedure (not to mention the verification of the involutivity test), all of the explicit formulae are essentially irrelevant.

An alternative approach, popularized by Gardner, [**77**], is to work, not explicitly as we have been doing, but rather *intrinsically*, without any need for the explicit formulae beyond the form of the structure equations. As we shall see, the action of the group on the essential torsion coefficients can be determined by differentiating the structure equations, thereby obviating those formulae for the normalization part of the process too. The resulting method is known as the *intrinsic* approach, and is distinguished from the so-called *parametric* approach we have employed so far. The intrinsic method can also be successfully employed in equivalence problems reducing to equivalence of coframes; the calculations produce the general form of the structure equations, so that one can read off the number of structure invariants, and the possible dimensions and structure of symmetry groups. The advantages are that the computations can often be carried through by hand, and the general form of the final structure equations can be deduced. Disadvantages are that the explicit expressions for the invariants are not forthcoming, and moreover, one might be led down spurious branches of the equivalence procedure owing to unexpected normalizations or cancellations due to the explicit forms of the coframe under consideration. Nevertheless, the intrinsic method can often be put to good use as a guide to expedite navigation through the far more computationally detailed parametric method that we have emphasized here.

Furthermore, although the equivalence method can be applied, in principle, to any reasonable equivalence problem, in practice the calculations can get rapidly out of hand, and one must resort to a computer algebra system, such as MATHEMATICA or MAPLE, to make progress with the explicit computations. Even so, one is often confronted with computer output consisting of several pages of intricate formulae for invariants, which prove quite useless for practical application. Even worse, the calculations can explode beyond even the power of these systems. (As we have seen, equivalence calculations often lead to rational algebraic functions, which are particularly difficult for current computer algebra packages to handle.) In such cases, the intrinsic method may still be practical, and provide important information on the number of structure invariants, the possible dimensions of symmetry groups, and so on.

To illustrate the intrinsic approach, let us return to the proof of Theorem 11.12. Consider the structure equations (11.16). Note that the form of these equations can be inferred directly from that of the lifted coframe, without any explicit (parametric) computations. To find out how the group acts on the essential torsion coefficient $T = T^1_{23}$ we differentiate the first structure equation:

$$\begin{aligned} 0 = d^2\theta^1 &= d\pi^1 \wedge \theta^1 - \pi^1 \wedge d\theta^1 + dT \wedge \theta^2 \wedge \theta^3 + T d\theta^2 \wedge \theta^3 - T\,\theta^2 \wedge d\theta^3 \\ &= \left[dT - 2T\pi^1 + T\pi^5\right] \wedge \theta^2 \wedge \theta^3 + \left[d\pi^1 - T\pi^2 \wedge \theta^3 + T\pi^3 \wedge \theta^2\right] \wedge \theta^1. \end{aligned}$$

Since $d^2\theta^1 = 0$, the coefficient of $\theta^2 \wedge \theta^3$ must be a multiple of θ^1, so

$$dT = T(2\pi^1 - \pi^5) + A\,\theta^1 + B\,\theta^2 + C\,\theta^3 = T(2\alpha^1 - \alpha^5) + \widetilde{A}\,\theta^1 + \widetilde{B}\,\theta^2 + \widetilde{C}\,\theta^3,$$

for certain coefficients $A, B, \ldots, \widetilde{C}$, whose precise form will not matter here. Thus, the infinitesimal variation of T as a function of the group parameters is given by the first term:

$$dT \equiv T(2\alpha^1 - \alpha^5), \tag{11.18}$$

where $\equiv$ means equality modulo linear combinations of the θ's. This congruence means that T is being scaled by the group action, and hence (provided it does not vanish) can be normalized to $T = 1$. Indeed, taking the differential of our earlier parametric formula (11.4), we find (omitting the terms involving the θ's coming from the differential of L_{pp})

$$\frac{dT}{T} = \frac{a_5 L_{pp}}{a_1^2}\, d\left(\frac{a_1^2}{a_5 L_{pp}}\right) \equiv 2\frac{da_1}{a_1} - \frac{da_5}{a_5} = 2\alpha^1 - \alpha^5,$$

which reconfirms the intrinsic computation. Once T has been normalized to be constant, its differential is 0, and hence (11.18) implies that the corresponding Maurer–Cartan forms must satisfy $\alpha^5 \equiv 2\alpha^1$, or, equivalently, $\pi^5 \equiv 2\pi^1$ modulo combinations of the θ's. We can substitute this "infinitesimal normalization" back in the structure equations, leading directly to the involutive system (11.17). Note particularly that the intrinsic approach required *no* explicit computations. Actually, this is not entirely correct, since we did need to verify that $T \neq 0$, and, further, the other essential torsion coefficient vanishes: $T^1_{13} + T^2_{23} = 0$; these facts follow from the particular form of the base coframe. Often, such details are not so apparent from purely intrinsic calculations, which points up one of the weaknesses of the approach — a "normalizable torsion coefficient" may, because of the particular formulae for the coframe, be identically zero and so an apparent intrinsic normalization may not, in point of fact, be there. Thus, although the intrinsic method can go far to a painless resolution of an equivalence problem, one must always watch out for particular parametric forms of the essential torsion that can influence the continuation of the method.

Contact Transformations

As an application of the intrinsic method, let us look at the equivalence problem for contact forms. Since every symmetry of the space of contact forms is, by definition, a contact transformation, our calculations will include, as a by-product, a proof of Bäcklund's Theorem 4.32 that every (analytic) contact transformation is the prolongation of a first order contact transformation or a point transformation. In order to keep the presentation as simple as possible, we shall just treat the scalar, second order case, leaving the generalization to higher order and multi-dimensional contact transformations to the interested reader. Thus, we begin with the underdetermined equivalence problem

$$\overline{\omega}^0 = a_1\,\omega^0 + a_2\,\omega^1, \qquad \overline{\omega}^1 = a_3\,\omega^0 + a_4\,\omega^1, \tag{11.19}$$

based on the contact forms $\omega^0 = du - p\,dx$ and $\omega^1 = dp - q\,dx$ on the second jet space J^2. We complete these to a coframe by adding in the basis forms dx and dq. The resulting lifted coframe is

$$\begin{aligned} \theta^1 &= a_1\,\omega^0 + a_2\,\omega^1, & \theta^3 &= b_1\,\omega^0 + b_2\,\omega^1 + c_1\,dx + c_2\,dq, \\ \theta^2 &= a_3\,\omega^0 + a_4\,\omega^1, & \theta^4 &= b_3\,\omega^0 + b_4\,\omega^1 + c_3\,dx + c_4\,dq. \end{aligned}$$

After absorption, the structure equations take the form

$$\begin{aligned} d\theta^1 &= \alpha^1 \wedge \theta^1 + \alpha^2 \wedge \theta^2 && + T_1\,\theta^3 \wedge \theta^4, \\ d\theta^2 &= \alpha^3 \wedge \theta^1 + \alpha^4 \wedge \theta^2 && + T_2\,\theta^3 \wedge \theta^4, \\ d\theta^3 &= \beta^1 \wedge \theta^1 + \beta^2 \wedge \theta^2 + \gamma^1 \wedge \theta^3 + \gamma^2 \wedge \theta^4, \\ d\theta^3 &= \beta^3 \wedge \theta^1 + \beta^4 \wedge \theta^2 + \gamma^3 \wedge \theta^3 + \gamma^4 \wedge \theta^4, \end{aligned} \tag{11.20}$$

where the α's, β's, and γ's are equivalent, modulo combinations of the θ's, to the Maurer–Cartan forms on the twelve-parameter structure group. The two essential torsion coefficients T_1, T_2 are not so difficult to compute directly; however, we shall use the intrinsic approach to determine how the group acts on them, by computing the differentials of the first two structure equations. Using the basic properties of the differential, combined with the structure equations themselves, we find

$$0 = d^2\theta^1 = \left[dT_1 + T_1(\gamma^1 + \gamma^4 - \alpha^1) - T_2\alpha^2\right] \wedge \theta^3 \wedge \theta^4 + \cdots,$$
$$0 = d^2\theta^2 = \left[dT_2 - T_1\alpha^3 + T_2(\gamma^1 + \gamma^4 - \alpha^4)\right] \wedge \theta^3 \wedge \theta^4 + \cdots,$$

where the omitted terms all contain either θ^1 or θ^2 (or both). Therefore, modulo linear combinations of the θ's, we find

$$\begin{aligned} dT_1 &\equiv T_1(\alpha^1 - \gamma^1 - \gamma^4) + T_2\alpha^2, \\ dT_2 &\equiv T_1\alpha^3 + T_2(\alpha^4 - \gamma^1 - \gamma^4). \end{aligned} \tag{11.21}$$

Thus, as in the simple example above, equations (11.21) provide the infinitesimal version of the group action on the essential torsion coefficients — it means that the vector (T_1, T_2) is being acted on by matrix multiplication by the 2×2 matrix A with entries a_i, and scaled by the reciprocal of the determinant of the matrix C with entries c_i. Thus, assuming $(T_1, T_2) \neq 0$, we can normalize $T_1 = 0$, $T_2 = 1$, by suitably specifying two of the group parameters. Substituting these values into (11.21), we find that the resulting group normalizations imply that the Maurer–Cartan forms are related by $\alpha^2 \equiv 0$, and $\alpha^4 \equiv \gamma^1 + \gamma^4$. Note that the first normalization implies that we have reduced the matrix A to a lower triangular matrix, and hence the first lifted coframe element is $\theta^1 = a_1\omega^0$. This observation has the consequence that the zeroth order contact forms are mapped to multiples of each other, which is one of the requirements that the transformation must satisfy if it is to reduce to a first order contact transformation.

To proceed further, substitute the normalizations for the Maurer–Cartan forms into the structure equations (11.20). The third and fourth are unchanged (although the explicit formulae will be different owing to the normalizations); the first two become

$$\begin{aligned} d\theta^1 &= \alpha^1 \wedge \theta^1 + S_1\,\theta^2 \wedge \theta^3 + S_2\,\theta^2 \wedge \theta^4, \\ d\theta^2 &= \alpha^3 \wedge \theta^1 + (\gamma^1 + \gamma^4) \wedge \theta^2 + \theta^3 \wedge \theta^4. \end{aligned}$$

There are two new essential torsion coefficients, which can no longer be absorbed by the Maurer–Cartan forms. (There is no $\theta^3 \wedge \theta^4$ term in $d\theta^1$ since it's already been normalized to 0.) To find how the reduced structure group acts on them, we compute the terms in $d^2\theta^1$ that do not involve θ^1, leading to the equations

$$\begin{aligned} dS_1 &\equiv S_1(\alpha^1 - 2\gamma^1 - \gamma^4) - S_2\gamma_3, \\ dS_2 &\equiv -S_1\gamma^2 + S_2(\alpha^1 - \gamma^1 - 2\gamma^4). \end{aligned} \tag{11.22}$$

Again, as long as $(S_1, S_2) \neq 0$, we can normalize $S_1 = 1$, $S_2 = 0$, which implies the infinitesimal group normalizations $\gamma^2 \equiv 0$, $\alpha^1 \equiv 2\gamma^1 + \gamma^4$. In particular, the first one implies that the full structure group is lower triangular. Therefore, under a contact transformation, $d\bar{x}$, $d\bar{u}$, and $d\bar{p}$, which are linear combinations of the first three lifted coframe elements, must be linear combinations of dx, du, and dp. This implies that the transformation reduces to a first order contact transformation on J^1, which completes the equivalence proof of Bäcklund's Theorem 4.32 in this particular situation. The structure equations now take the form

$$\begin{aligned} d\theta^1 &= (2\gamma^1 + \gamma^4) \wedge \theta^1 + \theta^2 \wedge \theta^3, \\ d\theta^2 &= \alpha^3 \wedge \theta^1 + (\gamma^1 + \gamma^4) \wedge \theta^2 + \theta^3 \wedge \theta^4, \\ d\theta^3 &= \beta^1 \wedge \theta^1 + \beta^2 \wedge \theta^1 + \gamma^1 \wedge \theta^3, \\ d\theta^4 &= \beta^3 \wedge \theta^1 + \beta^4 \wedge \theta^2 + \gamma^3 \wedge \theta^3 + \gamma^4 \wedge \theta^4, \end{aligned} \tag{11.23}$$

and we now determine whether or not they are in involution. A straightforward calculation shows that the degree of indeterminancy is $r^{(1)} = 13$. (Actually, in establishing this fact, the reader will discover that there is an essential torsion coefficient, namely $T^1_{14} - T^2_{24} - T^3_{34}$. However, this turns out to be identically 0, and so does not lead to any further group normalization.) On the other hand, the reduced characters are $s'_1 = 4$, $s'_2 = 3$, and $s'_3 = 1$, so the Cartan involutivity test (11.13) is satisfied. We conclude that the set of second order contact transformations, i.e., symmetries of the contact coframe, depends on a single arbitrary function of three variables, which we can identify with the solution of the implicit equation $H(x, u, \bar{x}, \bar{u}) = 0$ prescribed by the generating function H — see Theorem 4.28.

The equivalence problems for higher order and multi-dimensional contact transformations are solved similarly. The only complications are that the absorption and normalization procedure must be iterated more, and the involutivity criterion must be carefully checked. The details are left to the ambitious reader; see also Chapter 13.

Exercise 11.15. Set up and solve the equivalence problem for contact transformations on the space J^1E, where $E \simeq \mathbb{R}^1 \times \mathbb{R}^2$ has one independent and two dependent variables. Use your calculations to prove that every contact transformation is the first prolongation of a point transformation.

Darboux' Theorem

As a second application, we consider the problem of equivalence of a single differential one-form. Our computations will lead to a proof of Darboux' Theorem 1.42 (in the analytic category) based on the Cartan method. Again, for simplicity, we restrict our attention to a simple case, although the arguments readily generalize. Let ω, $\overline{\omega}$ be one-forms defined on open subsets of $\mathbb{R}^3$, and consider the problem of finding a local diffeomorphism mapping ω to $\overline{\omega}$. This problem can be readily encoded in Cartan form. Assuming $\omega \neq 0$, we can complete ω to a coframe by including two additional one-forms ζ^1, ζ^2, subject only to the nondegeneracy condition $\zeta^1 \wedge \zeta^2 \wedge \omega \neq 0$. Since the ζ's transform more or less arbitrarily, the structure group is the planar affine group, realized as a subgroup of $\mathrm{GL}(3)$ in the standard fashion:

$$\mathrm{A}(2) = \left\{ \begin{pmatrix} A & b \\ 0 & 1 \end{pmatrix} \;\middle|\; A \in \mathrm{GL}(2), \quad b \in \mathbb{R}^2 \right\},$$

cf. Example 2.11. The lifted coframe is then

$$\theta^1 = a_1\,\zeta^1 + a_2\,\zeta^2 + b_1\,\omega, \qquad \theta^2 = a_3\,\zeta^1 + a_4\,\zeta^2 + b_2\,\omega, \qquad \theta^3 = \omega, \tag{11.24}$$

where $a_1, a_2, a_3, a_4, b_1, b_2$ are the group parameters, with $a_1a_4 - a_2a_3 \neq 0$. In the initial set of structure equations, only the last will contain essential torsion, and we find, after absorption,

$$\begin{aligned} d\theta^1 &= \alpha^1 \wedge \theta^1 + \alpha^2 \wedge \theta^2 + \beta^1 \wedge \omega, \\ d\theta^2 &= \alpha^3 \wedge \theta^1 + \alpha^4 \wedge \theta^2 + \beta^2 \wedge \omega, \\ d\omega &= S\,\theta^1 \wedge \theta^2 + T_1\,\omega \wedge \theta^1 + T_2\,\omega \wedge \theta^2. \end{aligned} \tag{11.25}$$

The group action on the torsion coefficients is found by differentiation:

$$\begin{aligned} 0 = d^2\omega = {} & \left[dS + S(\alpha^1 + \alpha^4)\right] \wedge \theta^1 \wedge \theta^2 + \\ & + \left[dT_1 - S\,\beta^2 + T_1\,\alpha^1 + T_2\,\alpha^3\right] \wedge \omega \wedge \theta^1 \\ & + \left[dT_2 + S\,\beta^1 + T_1\,\alpha^2 + T_2\,\alpha^4\right] \wedge \omega \wedge \theta^2. \end{aligned}$$

Thus the group acts by scaling S, and also scaling and translating, by a multiple of S, the vector $T = (T_1, T_2)$. Consequently, the problem naturally splits into three branches, depending on whether S and/or T vanishes.

Case 1: If $S \neq 0$,† we can normalize $S = 1$ by scaling. Furthermore, we can then normalize $T = 0$ by translating, leading to a net reduction of three in the group parameters. The reduced structure group is obtained by the Maurer–Cartan form normalizations $\alpha^1 + \alpha^4 \equiv \beta^1 \equiv \beta^2 \equiv 0$. Continuing with the second loop through the Cartan procedure, after absorption, the resulting structure equations have the form

$$d\theta^1 = \alpha^1 \wedge \theta^1 + \alpha^2 \wedge \theta^2, \qquad d\theta^2 = \alpha^3 \wedge \theta^1 - \alpha^1 \wedge \theta^2, \qquad d\omega = \theta^1 \wedge \theta^2. \tag{11.26}$$

As there are no further essential torsion coefficients to normalize, we must investigate the involutivity of the structure equations (11.26). The degree of indeterminancy is $r^{(1)} = 4$, whereas the reduced characters are $s'_1 = 2$, $s'_2 = 1$, $s'_3 = 0$. Therefore, the system satisfies Cartan's test and is in involution. Since there is no nonconstant torsion, Theorem 11.8 implies that all such coframes are equivalent, and all possess an infinite-dimensional symmetry group depending on a single arbitrary function of two variables. The canonical form for such a coframe can be taken to be $\omega = e^x\, dy + dz$, $\zeta^1 = e^x dx$, $\zeta^2 = dy$, providing the canonical form for a one-form ω of Darboux rank two.

Case 2: If $S = 0$ but $T \neq 0$, we can then normalize $T_1 = 1$, $T_2 = 0$, by a suitable normalization of the matrix A, which implies that $\alpha^1 \equiv \alpha^2 \equiv 0$. After absorption, the resulting structure equations have the form

$$d\theta^1 = \beta^1 \wedge \omega, \qquad d\theta^2 = \alpha^3 \wedge \theta^1 + \alpha^4 \wedge \theta^2 + \beta^2 \wedge \omega, \qquad d\omega = \omega \wedge \theta^1.$$

The degree of indeterminancy is $r^{(1)} = 7$. The reduced characters are $s'_1 = 2$, $s'_2 = 1$, $s'_3 = 1$, so the system is in involution. The canonical form for such a coframe is $\omega = e^{-x}\, dy$, $\zeta^1 = dx$, $\zeta^2 = dz$, giving a one-form ω of Darboux rank one. The associated infinite-dimensional symmetry group depends on a single arbitrary function of three variables.

† All our assumptions will be generic, holding in an open subset; thus we avoid singularities such as isolated points where $S = 0$.

Case 3: If S and T are both 0, then there are no normalizations. The degree of indeterminancy of (11.25) is $r^{(1)} = 12$. The reduced characters are $s'_1 = 2$, $s'_2 = 2$, $s'_3 = 2$, and again the system is in involution. The canonical form for such a coframe is $\omega = dx$, $\zeta^1 = dy$, $\zeta^2 = dz$, giving a one-form of Darboux rank zero. The associated infinite-dimensional symmetry group depends on two arbitrary functions of three variables.

This calculation can be generalized to the multi-dimensional situation, but is a bit complicated, and is clearly not the simplest available proof of the Darboux Theorem. Nevertheless, it does indicate how one might proceed with generalizations to the problem of equivalence of a pair (or pencil) of one-forms. Indeed, despite important applications to the canonical forms of biHamiltonian systems, [**162**], [**184**], very little is known beyond some results of Debever, [**60**], for pencils of a particular algebraic type in $\mathbb{C}^4$.

The Intransitive Case

Let us now return to our general discussion. The intransitive case, which arises when there are nonconstant essential torsion coefficients (not depending on any remaining group parameters), is more complicated, but still eminently tractable. The essential torsion coefficients will define invariants of the problem, which must be included into the equivalence conditions. As in the equivalence of coframes, we must also compute their coframe derivatives, now with respect to the lifted coframe, since these also define invariants. Now, it might well happen, because there are still unnormalized group parameters occurring in the lifted coframe elements, that one or more of these derived invariants depends on the remaining group parameters. If this occurs, then each such derived invariant will provide a further group-dependent invariant combination which, just like the essential torsion coefficients themselves, can be normalized to any convenient constant value by specifying one or more of the remaining group parameters. There will then be a further group reduction corresponding to such a normalization, and one can continue to turn the Cartan equivalence crank, possibly yet achieving a final reduction to an invariant coframe. Consequently, in order to analyze the intransitive case, we need to assume that, not only are all the essential torsion coefficients independent of the remaining group parameters, but so are all their coframe derivatives. (As with our chain rule calculations of the classifying functions for a coframe, cf. (8.37), this can be practically checked by just computing coframe derivatives until no further functionally independent invariants appear.) As before, the collection of all such derived invariants up to some order s provides the components of the structure map $\mathbf{T}^{(s)}$, which, in the regular case, parametrizes the classifying manifold $\mathcal{C}^{(s)}(\boldsymbol{\theta})$. The definition of order and rank of such a lifted coframe is the same as in Proposition 8.18. Moreover, the involutivity condition for such coframes stated in Definition 11.7 is also exactly the same as above. We can now state the general result.

Theorem 11.16. *Let $\boldsymbol{\theta}$ and $\overline{\boldsymbol{\theta}}$ be regular, analytic lifted coframes on m-dimensional manifolds M having the same structure group G. Assume that both $\boldsymbol{\theta}$ and $\overline{\boldsymbol{\theta}}$ are involutive, meaning that the Cartan test* (11.13) *is satisfied. Assume further that all the essential torsion coefficients and their coframe derivatives are independent of the group parameters, forming a regular family of functions, and hence parametrizing the classifying manifolds for each coframe. The two coframes $\boldsymbol{\theta}$ and $\overline{\boldsymbol{\theta}}$ are locally equivalent if and only if they have the same order, $\bar{s} = s$, and their* $(s+1)^{\mathrm{st}}$ *order classifying manifolds $\mathcal{C}^{(s+1)}(\boldsymbol{\theta})$ and $\mathcal{C}^{(s+1)}(\overline{\boldsymbol{\theta}})$ overlap. Moreover, the symmetry group of such a coframe (and therefore the set of equivalences) is infinite-dimensional, depending on s'_k arbitrary functions of k variables, where s'_k denotes the last nonzero reduced character of the structure equations.*

Warning: When checking for involutivity, it is very important to check that none of the essential torsion coefficients, or their coframe derivatives, depends on the group parameters. Failure to do this can lead to incorrect conclusions.

Equivalence of Nonclosed Two-Forms

The solution to the Darboux problem of equivalence of one-forms ω induces a solution to the equivalence problem for closed two-forms $\Omega = d\omega$. A complete set of canonical forms was described in Theorem 1.44. Although not relevant to symplectic geometry, the problem of equivalence of more general two-forms is of interest. The four-dimensional case was investigated by Debever, [**60**]. More recently, motivated by work of David and Holm, [**59**], on multi-Hamiltonian systems of ordinary differential equations in optics, Gardner and Shadwick, [**81**], investigated a more general version of the three-dimensional problem. This problem appears to be the simplest naturally occurring equivalence problem in which an intransitive involutive structure appears.†

Consider a nonvanishing two-form Ω defined on an open subset $M \subset \mathbb{R}^3$. Since $\Omega \neq 0$, it must have rank one at all points of M, and hence we can introduce a pair of linearly independent one-forms ω^1, ω^2 such that (locally) $\Omega = \omega^1 \wedge \omega^2$. If Ω is closed, $d\Omega = 0$, then we already know there is just one canonical form, namely the rank one Darboux form $\Omega = dx \wedge dy$, cf. (1.30). Let us therefore assume that Ω is *not* closed, which means that we can find a third independent one-form ω^3, completing the coframe, such that

$$d\Omega = \omega^1 \wedge \omega^2 \wedge \omega^3 \neq 0. \tag{11.27}$$

Two such two-forms Ω and $\overline{\Omega}$ will be equivalent if and only if their associated coframes satisfy

$$\begin{pmatrix} \overline{\omega}^1 \\ \overline{\omega}^2 \\ \overline{\omega}^3 \end{pmatrix} = \begin{pmatrix} a_1 & a_2 & 0 \\ a_3 & a_4 & 0 \\ b_1 & b_2 & 1 \end{pmatrix} \begin{pmatrix} \omega^1 \\ \omega^2 \\ \omega^3 \end{pmatrix}, \qquad \text{where} \qquad a_1 a_4 - a_2 a_3 = 1,$$

leading to a five parameter structure group. The lifted coframe is then

$$\theta^1 = a_1\,\omega^1 + a_2\,\omega^2, \qquad \theta^2 = a_3\,\omega^1 + a_4\,\omega^2, \qquad \theta^3 = b_1\,\omega^1 + b_2\,\omega^2 + \omega^3. \tag{11.28}$$

After absorption, the initial set of structure equations has the form

$$\begin{aligned} d\theta^1 &= \alpha^1 \wedge \theta^1 + \alpha^2 \wedge \theta^2, \\ d\theta^2 &= \alpha^3 \wedge \theta^1 - \alpha^1 \wedge \theta^2 - T\,\theta^2 \wedge \theta^3, \\ d\theta^3 &= \beta^1 \wedge \theta^1 + \beta^2 \wedge \theta^2, \end{aligned} \tag{11.29}$$

where the essential torsion coefficient $T = -T^1_{13} - T^2_{23}$ is minus the sum of the coefficients of $\theta^1 \wedge \theta^3$ in $d\theta^1$ and $\theta^2 \wedge \theta^3$ in $d\theta^2$. Our choice of sign stems from noting that $\Omega = \theta^1 \wedge \theta^2 = \omega^1 \wedge \omega^2$, hence

$$d\Omega = d\theta^1 \wedge \theta^2 - \theta^1 \wedge d\theta^2 = T\,\theta^1 \wedge \theta^2 \wedge \theta^3 = T\,\omega^1 \wedge \omega^2 \wedge \omega^3. \tag{11.30}$$

† Actually, the only other example I know of in the literature is the work of Kamran and Shadwick, [**134**], on second order hyperbolic partial differential equations in two variables. The intransitive cases arise for linear equations, in accordance with Theorem 6.46.

Equation (11.27) implies that, in this case, $T = 1$ is constant, and there are no normalizations possible. The structure equations (11.29) are involutive: The degree of indeterminancy is $r^{(1)} = 7$, whereas the reduced characters are $s_1' = 3$, $s_2' = 2$, $s_3' = 0$. Therefore, we have a transitive, involutive structure, so Theorem 11.8 implies the following result. For brevity, we call a nonclosed form Ω *nonsingular* if $\Omega \neq 0 \neq d\Omega$ everywhere.

Theorem 11.17. *All nonsingular, analytic, nonclosed two-forms on a three-dimensional manifold are locally equivalent. The particular example* $\Omega = e^z dx \wedge dy$ *provides a canonical form for such two-forms. The symmetry group depends on two arbitrary functions of two variables.*

Exercise 11.18. Prove that all nonclosed two-forms of rank two on $\mathbb{R}^4$ are locally equivalent, and find a canonical form; see Debever, [**60**]. *A research project*: Investigate the general equivalence problem for nonclosed two-forms on $\mathbb{R}^n$.

Gardner and Shadwick, [**81**], generalize the preceding example by considering the simultaneous equivalence of a two-form Ω and a vector field $\mathbf{v}$ on $\mathbb{R}^3$. The equivalence map Φ must satisfy $\Phi^* \overline{\Omega} = \Omega$ and $d\Phi^{-1}(\overline{\mathbf{v}}) = \mathbf{v}$. The problem naturally splits into two branches; the one treated in detail is when $\mathbf{v} \lrcorner \Omega = 0$. In this case, we factor the two-form $\Omega = \omega^1 \wedge \omega^2$ as before. The third coframe element ω^3 is then fixed, modulo linear combinations of ω^1 and ω^2, by the requirement that $\langle \omega^3 ; \mathbf{v} \rangle = 1$. Therefore, in place of (11.27), we have

$$d\Omega = T\,\omega^1 \wedge \omega^2 \wedge \omega^3, \tag{11.31}$$

for some function T. The lifted coframe (11.28) is the same, so the structure equations have exactly the same form (11.29), where T is given in (11.31); in particular, T does not depend on any of the group parameters. (The reader can verify this fact directly by computing $d^2\theta^2$.) If T is constant, then the structure equations are involutive. There is a one-parameter family of canonical forms for such a coframe, with $\omega^1 = e^z\,dx$, $\omega^2 = dy$, $\omega^3 = dz/c$. The corresponding two-form is $\Omega = e^z\,dx \wedge dy$, the vector field is $\mathbf{v} = c\,\partial_z$, and the invariant is $T = c$.

If T is not constant, then its coframe derivatives $T_i = \partial T/\partial \theta^i$ do depend on the group parameters, and hence serve to prescribe some group normalizations. Since $dT = T_1\,\theta^1 + T_2\,\theta^2 + T_3\,\theta^3$, we find

$$0 = d^2T = dT_1 \wedge \theta^1 + dT_2 \wedge \theta^2 + dT_3 \wedge \theta^3 + T_1\,d\theta^1 + T_2\,d\theta^2 + T_3\,d\theta^3,$$

which, in view of (11.29), implies the congruences

$$\begin{aligned} dT_1 &\equiv -T_1\,\alpha^1 - T_2\,\alpha^3 - T_3\,\beta^1, \\ dT_2 &\equiv -T_1\,\alpha^2 + T_2\,\alpha^1 - T_3\,\beta^2, \end{aligned} \qquad dT_3 \equiv 0, \tag{11.32}$$

modulo the base coframe $\boldsymbol{\theta}$. In particular, the third coframe derivative T_3 does not depend on the group parameters and forms a second invariant of the problem. The equivalence problem now splits into two branches, depending on whether T_3 vanishes or not.

Case 1: If $T_3 = 0$, then $(T_1, T_2) \neq (0,0)$, as otherwise T would be constant. We can use the group action (11.32) to normalize $T_1 = 1$, $T_2 = 0$, and so $\alpha^1 \equiv \alpha^2 \equiv 0$ modulo the θ's. With this normalization, $dT = \theta^1$, hence θ^1 is closed, and the structure equations (11.29) reduce to

$$d\theta^1 = 0, \qquad d\theta^2 = \alpha^3 \wedge \theta^1 - T\,\theta^2 \wedge \theta^3, \qquad d\theta^3 = \beta^1 \wedge \theta^1 + \beta^2 \wedge \theta^2.$$

These are involutive, since the degree of indeterminancy is $r^{(1)} = 4$, while the reduced characters are $s_1' = 2$, $s_2' = 1$, $s_3' = 0$. All such coframes are locally equivalent, and all possess an infinite-dimensional symmetry group depending on a single arbitrary function of two variables. A canonical form for such a coframe is $\omega^1 = e^x\,dx$, $\omega^2 = e^z\,dy$, $\omega^3 = e^x\,dz$, so $\Omega = e^{x+z}\,dx \wedge dy$.

Case 2: If $T_3 \neq 0$, we can then normalize $T_1 = T_2 = 0$, hence $\beta^1 \equiv \beta^2 \equiv 0$ and $dT = T_3\,\theta^3$. Thus $0 = d^2T = dT_3 \wedge \theta^3 + T_3\,d\theta^3$, so the structure equation

$$d\theta^3 = S_1\,\theta^1 \wedge \theta^3 + S_2\,\theta^2 \wedge \theta^3,$$

contains two essential torsion coefficients: $S_i = \partial T_3/\partial\theta^i$, $i = 1,2$. The infinitesimal group action on S_1, S_2, is found by computing $d^2\theta^3$, which implies

$$dS_1 \equiv -S_1\alpha^1 - S_2\alpha^3, \qquad dS_2 \equiv -S_1\alpha^2 + S_2\alpha^1.$$

The problem now splits into two sub-branches.

Case 2a: If $S_1 = S_2 = 0$, then there is no further torsion to normalize. In this case, dT_3 is a multiple of θ^3, hence $dT \wedge dT_3 = 0$, and so T and $T_3 = H(T)$ are functionally dependent. The reduced structure equations

$$\begin{aligned} d\theta^1 &= \alpha^1 \wedge \theta^1 + \alpha^2 \wedge \theta^2, \\ d\theta^2 &= \alpha^3 \wedge \theta^1 - \alpha^1 \wedge \theta^2 - T\,\theta^2 \wedge \theta^3, \end{aligned} \qquad d\theta^3 = 0,$$

are in involution, with $r^{(1)} = 4$, and $s_1' = 2$, $s_2' = 1$, $s_3' = 0$. Therefore, the coframe has rank 1, and forms an *intransitive*, involutive structure. The equivalence classes are distinguished by the single classifying function H relating T and T_3 (or, more properly, the reduced classifying curve parametrized by the pair T, T_3). Setting $T = z$ to be one of the coordinate functions, we find a complete set of canonical forms for the coframes lying in this branch to be given by

$$\omega^1 = dx, \qquad \omega^2 = \exp\left[-\int \frac{z\,dz}{H(z)}\right] dy, \qquad \omega^3 = \frac{dz}{H(z)}, \tag{11.33}$$

valid for $z > 0$. The symmetry group of the coframe (11.33) depends on a single arbitrary analytic function of two variables.

Case 2b: If $(S_1, S_2) \neq (0,0)$, we normalize $S_1 = 1$, $S_2 = 0$, which requires $\alpha^1 \equiv \alpha^2 \equiv 0$. Then $dT_3 = T_3\,\theta^1 + U\,\theta^3$, where U is yet another invariant. Computing $0 = d^2 T_3$ proves that U does not depend on the remaining group parameter a_3. There are two further subbranches, depending on whether U is zero or not. If $U = 0$, the structure equations are involutive, intransitive, of rank two, and governed by the two independent invariants T and T_3. If $U \neq 0$, then its coframe derivatives can be used to normalize the final group parameter, reducing the problem to an equivalence problem for coframes! The detailed analysis of these last subcases is left to the reader as an exercise; see also [**81**].

Exercise 11.19. Set up and solve the second simultaneous equivalence problem of a two-form and a vector field on $\mathbb{R}^3$, when $\mathbf{v} \lrcorner \Omega \neq 0$.

Chapter 12

Prolongation of Equivalence Problems

Finally, we must tackle the problems in which the Cartan equivalence procedure does not lead to a complete reduction of the structure group, and, moreover, the structure equations are not involutive, so that (at least at the moment) we cannot deduce the existence of an infinite dimensional symmetry group. In practice, such problems, which include equivalence problems for differential equations and for Riemannian metrics, occur when the symmetry group has maximal dimension strictly larger than the dimension of the underlying manifold M upon which the original coframe was erected so as to encode the equivalence problem. (In principle, the structure equations could fail to be involutive even though an infinite-dimensional symmetry group is still present. However, I do not know any naturally occurring examples exhibiting this phenomenon, and, moreover, the prolongation procedure to be discussed below will handle this (remote) possibility as well.) In essence, then, the reason that the equivalence method has failed to produce the desired solution is because we have formulated the problem on a space whose dimension is too small to incorporate all possible symmetries of the problem.

Clearly, the resolution of the difficulty is to reformulate the problem on a suitably larger dimensional manifold and then, if necessary, reapply the Cartan reduction algorithm. The only difficulty, though, is how to construct the required higher dimensional equivalence problem. The prolongation method† of Cartan resolves this problem, giving a natural, readily implementable method for appending new coordinates and new one-forms to our original coframe which continue to properly encode the problem at hand.

† The term "prolongation" used here is *not* the same as the prolongation of transformations, vector fields, etc., to jet bundles which was extensively discussed in previous chapters. This conflict in terminology is unfortunate, but the meaning will always be clear from context.

Thus, we assume that, after perhaps one or more loops through the absorption and reduction algorithm, we have produced a coframe that still involves one or more of the original group parameters, yet none of the remaining essential torsion coefficients, nor any of their coframe derivatives, depends explicitly on the group parameters. For simplicity, let G denote the reduced structure group at this point, and let $r > 0$ be its dimension. Moreover, we are assuming that our coframe fails the Cartan involutivity test (11.13), so the relevant combination of reduced characters is strictly greater than the degree of indeterminancy: $r^{(1)} < s_1' + 2s_2' + \cdots + ms_m'$. The Cartan–Kähler Theorem gives us no indication as to whether an appropriate equivalence map exists, basically because there are additional constraints, stemming from as yet unknown invariants, that the map must satisfy. We are therefore forced to prolong the system by appending additional variables. Cartan noted that a natural choice of new variables to be appended are the group parameters or coordinates on the remaining structure group G itself! In other words, the prolonged equivalence problem will be naturally formulated in terms of a coframe on the $(m+r)$-dimensional manifold $M^{(1)} = M \times G$. The required coframe will contain the original coframe elements θ^i (lifted to $M \times G$) supplemented by suitably modified Maurer–Cartan forms on G. The prolonged coframe will contain a certain number of unspecified parameters — namely the solutions of the homogeneous absorption equations — which number $r^{(1)}$ in all, and serve to parametrize the new structure group. The goal is then, using the same basic absorption algorithm, to normalize the new parameters and, we hope, reduce the equivalence problem to one that we know how to solve. In many cases, the prolonged problem will also lead to additional invariant combinations of the original group parameters, which will enable us to further normalize the original equivalence problem, perhaps even reducing to an invariant coframe on the original manifold. On the other hand, we may not have increased the dimension of the space sufficiently, and so may be forced to prolong yet again.

The Determinate Case

Before getting into the general details, let us begin with the simplest special case. Suppose that, after we've performed all the normalizations and reductions that we can, we are left with a system of structure equations of the usual form (11.6) having degree of indeterminancy $r^{(1)} = 0$. We shall call such a system *determinate* so as to distinguish it from the more general cases when $r^{(1)} > 0$. Note that, in the determinate case, the structure equations cannot possibly be involutive, since the reduced characters cannot all be zero, and so Cartan's test (11.13) cannot hold. (Compare (11.12), or note that the matrix (11.10) is not the zero matrix since there is, by assumption, at least one unnormalized group parameter remaining in the structure group G.) At this stage, we complete the absorption procedure by absorbing all of the inessential torsion coefficients; the result is a system of equations of the absorbed form (10.13), so

$$d\theta^i = \sum_{j=1}^{m}\sum_{\kappa=1}^{r} A^i_{j\kappa}\,\pi^\kappa \wedge \theta^j + \sum_{j,k=1}^{m} U^i_{jk}\,\theta^j \wedge \theta^k, \qquad i = 1, \ldots, m. \tag{12.1}$$

Here U^i_{jk} denote the remaining unabsorbable torsion coefficients, which, by assumption, do not depend explicitly on any of the remaining group parameters. As before, the essential torsion coefficients are invariant, and so must agree, $\overline{U}^i_{jk} = U^i_{jk}$, if the two coframes are to be equivalent. The one-forms $\{\pi^1, \ldots, \pi^r\}$ are equal, modulo the θ's, to the Maurer–Cartan one-forms $\{\alpha^1, \ldots, \alpha^r\}$ on the group G. More explicitly:

$$\pi^\kappa = \alpha^\kappa - \sum_{j=1}^{m} z^\kappa_j\,\theta^j, \qquad \kappa = 1, \ldots, r, \tag{12.2}$$

where each absorption coefficient z^κ_j is replaced by its solution from the absorption equations (11.7). In general, there is some ambiguity in the determination of the absorption coefficients z^κ_j since we can add in any solution of the associated homogeneous linear system (11.8) without affecting the answer. However, in the present determinate case, we are assuming that $r^{(1)}$, which is defined as the dimension of the solution space to the homogeneous system, is 0. This means that the coefficients $z^\kappa_j = S^\kappa_j(x, g)$ are all uniquely specified, and the only solution to the homogeneous system of two-form equations

$$\sum_{j=1}^{m}\sum_{\kappa=1}^{r} A^i_{j\kappa}\,\pi^\kappa \wedge \theta^j = 0, \qquad i = 1, \ldots, m, \tag{12.3}$$

is the trivial one: $\pi^\kappa = 0$, $\kappa = 1, \ldots, r$.

Now, suppose that we have an equivalence $\Phi: M \to \overline{M}$, so that $\Phi^*\overline{\theta}^i = \theta^i$ preserves the lifted coframes, for some choice of the group parameters $g = g(x)$, $\overline{g} = \overline{g}(\overline{x})$. This implies that their differentials are also invariant, so $\Phi^*(d\overline{\theta}^i) = d\theta^i$. Substituting the structure equations (12.3) and their barred counterparts into these latter conditions, we find (omitting pull-backs as usual)

$$\sum_{j=1}^{m}\sum_{\kappa=1}^{r} A^i_{j\kappa}\,\overline{\pi}^\kappa \wedge \overline{\theta}^j + \sum_{j,k=1}^{m} \overline{U}^i_{jk}\,\overline{\theta}^j \wedge \overline{\theta}^k =$$
$$= \sum_{j=1}^{m}\sum_{\kappa=1}^{r} A^i_{j\kappa}\,\pi^\kappa \wedge \theta^j + \sum_{j,k=1}^{m} U^i_{jk}\,\theta^j \wedge \theta^k.$$

Since the lifted coframe elements and essential torsion coefficients are invariant, the second summations on each side of this system cancel out, and we are left with the invariance conditions

$$\sum_{j=1}^{m}\sum_{\kappa=1}^{r} A^i_{j\kappa}\left(\overline{\pi}^\kappa - \pi^\kappa\right) \wedge \overline{\theta}^j = 0, \qquad i = 1, \ldots, m.$$

Our assumption that the problem is determinate implies, cf. (12.3), that the modified Maurer–Cartan forms π^κ are also invariant: $\Phi^*\overline{\pi}^\kappa = \pi^\kappa$.

We now prolong the problem by increasing the number of variables. The natural candidates are the coordinates on the structure group G, which means that we should reformulate the equivalence problem on the prolonged space $M^{(1)} = M \times G$. We already know an invariant coframe on $M^{(1)}$; it consists of the original lifted coframe $\boldsymbol{\theta} = g \cdot \boldsymbol{\omega}$, in which the group parameters are regarded as new independent variables, and the modified Maurer–Cartan forms $\boldsymbol{\pi}$. Similarly, we construct a prolonged coframe $\overline{\boldsymbol{\theta}}$, $\overline{\boldsymbol{\pi}}$, corresponding to the lifted coframe on $\overline{M}$, which will be defined on the manifold $\overline{M}^{(1)} = \overline{M} \times \overline{G}$. Here $\overline{G} = G$ is the *same* structure group, but we will employ the barred notation so as to distinguish between the two sets of group parameters. The key point is that every equivalence between the original coframes extends to an equivalence between the prolonged coframes and conversely.

Proposition 12.1. *Let $\boldsymbol{\theta} = g \cdot \boldsymbol{\omega}$ and $\overline{\boldsymbol{\theta}} = \overline{g} \cdot \overline{\boldsymbol{\omega}}$ be lifted coframes having the same structure group G, no group dependent essential torsion coefficients, and degree of indeterminancy $r^{(1)} = 0$. Let $\boldsymbol{\pi}$ and $\overline{\boldsymbol{\pi}}$ be* *the modified Maurer–Cartan forms obtained by solving the absorption equations. Then there exists a diffeomorphism* $\Phi: M \to \overline{M}$ *mapping* $\overline{\boldsymbol{\theta}}$ *to* $\boldsymbol{\theta}$ *for some choice of group parameters* $g = g(x)$, $\bar{g} = \bar{g}(\bar{x})$ *if and only if there is a diffeomorphism* $\Psi: M \times G \to \overline{M} \times \overline{G}$ *mapping* $(\overline{\boldsymbol{\theta}}, \overline{\boldsymbol{\pi}})$ *to* $(\boldsymbol{\theta}, \boldsymbol{\pi})$.

Proof: First assume that the original equivalence problem has a solution, so there exists a map $\Phi: M \to \overline{M}$ and a G-valued matrix of functions $g_0(x)$ such that $\Phi^* \overline{\boldsymbol{\omega}} = g_0(x) \cdot \boldsymbol{\omega}$. The prolonged coframes $\boldsymbol{\theta} = g \cdot \boldsymbol{\omega}$ and $\overline{\boldsymbol{\theta}} = \bar{g} \cdot \overline{\boldsymbol{\omega}}$, where g and $\bar{g}$ now denote coordinates on the group G, will then be equivalent if and only if $\bar{g} = g \cdot g_0(x)^{-1}$. This serves to define the prolonged map $\Psi: M \times G \to \overline{M} \times \overline{G}$, which is given by

$$(\bar{x}, \bar{g}) = \Psi(x, g) = (\Phi(x), g \cdot g_0(x)^{-1}). \tag{12.4}$$

Our earlier analysis of the structure equations implies that Ψ preserves both parts of the prolonged coframe:

$$\Psi^* \overline{\boldsymbol{\theta}} = \boldsymbol{\theta}, \qquad \Psi^* \overline{\boldsymbol{\pi}} = \boldsymbol{\pi}. \tag{12.5}$$

Conversely, consider a map $\Psi: M \times G \to \overline{M} \times \overline{G}$ which preserves the prolonged coframe, i.e., satisfies (12.5). We must show that Ψ is necessarily of the form (12.4). The first set of conditions in (12.5) implies that each $\Psi^* d\bar{x}^i$ is a linear combination of the dx^j's. Therefore Ψ takes the projectable form $\Psi(x, g) = (\Phi(x), \Lambda(x, g))$. Write $\Lambda(x, g) = g \cdot h(x, g)$ for some G-valued function h. Then

$$\Psi^*(d\bar{g} \cdot \bar{g}^{-1}) = dg \cdot g^{-1} + g \cdot dh \cdot (gh)^{-1}. \tag{12.6}$$

But, according to (12.5), the Maurer–Cartan forms must match up to combinations of the dx's, so $\Psi^*(\overline{\alpha}^\kappa) = \alpha^\kappa + \sum_i w_i^\kappa \, dx^i$. Since the Maurer–Cartan forms are the entries of $dg \cdot g^{-1}$ and $d\bar{g} \cdot \bar{g}^{-1}$, (12.6) implies that the entries of dh can only depend on the dx's, and therefore $h = h(x)$ depends only on x. This means that the equivalence map must be of the previous form (12.4), with $g_0(x) = h(x)^{-1}$. It is now a simple matter to prove that, for this choice of $g_0(x)$, the projection Φ of Ψ to M satisfies $\Phi^* \overline{\boldsymbol{\omega}} = g_0(x) \cdot \boldsymbol{\omega}$, and hence we obtain the desired equivalence. *Q.E.D.*

Thus, the solution to a determinate equivalence problem can be completely effected by prolonging to the space $M^{(1)} = M \times G$, and using the coframe consisting of the original lifted coframe $\boldsymbol{\theta}$ and the modified Maurer–Cartan form $\boldsymbol{\pi}$. We illustrate how to handle these kinds of problems with a solution of the fundamental equivalence problem from Riemannian geometry.

Equivalence of Surfaces

Let us consider a basic equivalence problem from classical differential geometry: Determine whether two surfaces having prescribed metrics (first fundamental forms) can be mapped to each other by a metric-preserving diffeomorphism. This is a special case of the more general problem of isometric equivalence of Riemannian manifolds, which we shall discuss subsequently. It includes, as a special case, the problem of classifying the isometries of a Riemannian manifold, as these are just self-equivalences of the manifold that preserve the metric. As usual, we will concentrate on the local aspects of the problem in this book, and ignore equally interesting global, topological considerations.

In order to clarify the basic idea, we begin with a trivial version of this problem. Consider the problem of classifying all (local) isometries of the Euclidean plane. These will be (smooth) maps $(\bar{x}, \bar{y}) = \Phi(x, y)$ that preserve the Euclidean metric

$$\Phi^*\{d\bar{x}^2 + d\bar{y}^2\} = dx^2 + dy^2. \tag{12.7}$$

Since we are only dealing with local issues, we will also assume that the isometry is orientation preserving. (To find orientation reversing isometries, we can just compose any orientation preserving one with the reflection $(x, y) \mapsto (-x, y)$.) If we introduce the base coframe $\omega^1 = dx$, $\omega^2 = dy$, then the isometry condition (12.7) is the same as the condition

$$\Phi^* \begin{pmatrix} \overline{\omega}^1 \\ \overline{\omega}^2 \end{pmatrix} = \begin{pmatrix} \cos t & -\sin t \\ \sin t & \cos t \end{pmatrix} \begin{pmatrix} \omega^1 \\ \omega^2 \end{pmatrix}. \tag{12.8}$$

Thus, the structure group is the rotation group $G = \mathrm{SO}(2)$. In accordance with our usual procedure, we introduce the lifted coframe

$$\theta^1 = (\cos t)\,\omega^1 - (\sin t)\,\omega^2, \qquad \theta^2 = (\sin t)\,\omega^1 + (\cos t)\,\omega^2. \tag{12.9}$$

The associated structure equations are easily found:

$$d\theta^1 = -\alpha \wedge \theta^2, \qquad d\theta^2 = \alpha \wedge \theta^1, \tag{12.10}$$

where $\alpha = dt$ is the Maurer–Cartan form on SO(2). In this case, there is no torsion to absorb, and so the forms π^κ in (12.1) just consist of the Maurer–Cartan form $\pi = \alpha$ alone. Moreover, the degree of indeterminancy is clearly $r^{(1)} = 0$, since the linear absorption system (11.6), which is just $z_1 = z_2 = 0$, has no nonzero solutions, and so the problem must be prolonged. (In this case, the Cartan characters are $s_1' = 1, s_2' = 0$.)

Under the transformation $(\bar{x}, \bar{y}) = \Phi(x, y)$, there will be an induced map $\bar{t} = \Psi(x, y, t)$ which maintains the equivalence of the lifted coframes: $\bar{\theta}^i = \theta^i$, $i = 1, 2$. Comparing the structure equations with the corresponding structure equations in the barred variables, we conclude that $\bar{\alpha} = \alpha$ must agree as well (which implies $\bar{t} = t + k$ for some constant k). Thus, we can prolong the problem by adjoining the additional one-form α to our original lifted coframe. The prolonged structure equations have the form

$$d\theta^1 = -\alpha \wedge \theta^2, \qquad d\theta^2 = \alpha \wedge \theta^1, \qquad d\alpha = 0. \tag{12.11}$$

Since the structure coefficients are all constant, the prolonged coframe has rank zero. Equations (12.11) are simply the Maurer–Cartan structure equations for the Euclidean group $\mathrm{SE}(2) = \mathrm{SO}(2) \ltimes \mathbb{R}^2$. Therefore, we have proved that the connected group of isometries of the Euclidean plane is SE(2), i.e., isometries are combinations of translations and rotations. (Since the Cartan method is local, the existence of additional discrete isometries, such as reflections, is certainly also possible.)

Let us now be a bit more ambitious, and consider the general problem of isometric equivalence of surfaces. Let S be a surface with metric (first fundamental form)

$$ds^2 = E(x, y)\, dx^2 + 2F(x, y)\, dx\, dy + G(x, y)\, dy^2. \tag{12.12}$$

As in Example 9.2, we introduce a diagonalizing coframe

$$\omega^1 = A(x, y)\, dx + B(x, y)\, dy, \qquad \omega^2 = C(x, y)\, dx + D(x, y)\, dy, \tag{12.13}$$

which places the (positive definite) quadratic form ds^2 into canonical form $ds^2 = (\omega^1)^2 + (\omega^2)^2$. This is the same as the requirement that the coefficient matrix $H = \left(\begin{smallmatrix} A & B \\ C & D \end{smallmatrix}\right)$ of the diagonalizing coframe is a "square root" of the symmetric, positive definite matrix $L = \left(\begin{smallmatrix} E & F \\ F & G \end{smallmatrix}\right)$, meaning that $H^T H = L$. There are various possible choices for H. The more natural is, perhaps, to choose it symmetric, so $B = C$; however, choosing H to be upper triangular, so

$$A = \sqrt{E}, \qquad B = \frac{F}{\sqrt{E}}, \qquad C = 0, \qquad D = \sqrt{\frac{EG - F^2}{E}} = \sqrt{\frac{\Delta}{E}}, \tag{12.14}$$

with

$$\Delta = EG - F^2 = \det L = (\det H)^2 = (AD - BC)^2, \tag{12.15}$$

leads to slightly simpler parametric calculations.

With the preceding choice of coframe, the equivalence problem has the same reformulation (12.8) as in the Euclidean case, with the same structure group $G = \mathrm{SO}(2)$. The lifted coframe is given by equation (12.9), but the structure equations now include some torsion:

$$d\theta^1 = -\alpha \wedge \theta^2 + P\,\theta^1 \wedge \theta^2, \quad d\theta^2 = \alpha \wedge \theta^1 + Q\,\theta^1 \wedge \theta^2. \tag{12.16}$$

Here $\alpha = dt$ is the Maurer–Cartan form on SO(2), and the torsion coefficients

$$P = J \cos t - K \sin t, \qquad Q = J \sin t + K \cos t \tag{12.17}$$

are obtained by rotating the fundamental structure invariants

$$J = \frac{B_x - A_y}{\sqrt{\Delta}} = \frac{B_x - A_y}{AD - BC}, \qquad K = \frac{D_x - C_y}{\sqrt{\Delta}} = \frac{D_x - C_y}{AD - BC}, \tag{12.18}$$

associated with the base coframe (12.13). These are defined by the structure equations

$$d\omega^1 = J\,\omega^1 \wedge \omega^2, \qquad d\omega^2 = K\,\omega^1 \wedge \omega^2; \tag{12.19}$$

see (8.15). We can clearly absorb both torsion terms by setting

$$\pi = \alpha - P\,\theta^1 - Q\,\theta^2 = \alpha - J\,\omega^1 - K\,\omega^2, \tag{12.20}$$

leading to structure equations of the same form as (12.10):

$$d\theta^1 = -\pi \wedge \theta^2, \qquad d\theta^2 = \pi \wedge \theta^1. \tag{12.21}$$

As before, the degree of indeterminancy is $r^{(1)} = 0$, so the modified Maurer–Cartan form π is uniquely determined. We must prolong to the three-dimensional manifold $S^{(1)} = S \times \mathrm{SO}(2)$ with coframe provided by θ^1, θ^2, and π. The prolonged structure equations can, of course, be computed directly. However, note that (12.21) implies that

$$0 = d^2\theta^1 = -d\pi \wedge \theta^1 + \pi \wedge d\theta^2 = -d\pi \wedge \theta^1,$$

and, similarly, $0 = d^2\theta^2 = d\pi \wedge \theta^2$, hence the two-form $d\pi$ can only involve the first two coframe elements θ^1 and θ^2. Therefore, the structure equations must have the form

$$d\theta^1 = -\pi \wedge \theta^2, \qquad d\theta^2 = \pi \wedge \theta^1, \qquad d\pi = \kappa\, \theta^1 \wedge \theta^2, \tag{12.22}$$

where κ is a scalar invariant. Moreover, since $0 = d^2\pi = d\kappa \wedge \theta^1 \wedge \theta^2$, the invariant $\kappa = \kappa(x, y)$ cannot depend on the group parameter t, as otherwise there would be a term involving π in $d\kappa$. The equations (12.22) are the fundamental structure equations for intrinsic surface geometry; see [**106**; Chapter 10] for applications and connections with the extrinsic geometry.

What is the scalar invariant κ? The fact that it is an intrinsic quantity invariantly associated with the surface metric will convince anyone familiar with the geometry of surfaces that it must be the Gaussian curvature! This can be verified through a direct computation using the second formula in equation (12.20) for π, and the fact that $\theta^1 \wedge \theta^2 = \omega^1 \wedge \omega^2$. We find

$$\begin{aligned} d\pi &= -dJ \wedge \omega^1 - dK \wedge \omega^2 - J\, d\omega^1 - K\, d\omega^2 \\ &= \left\{ \frac{\partial J}{\partial \omega^2} - \frac{\partial K}{\partial \omega^1} - J^2 - K^2 \right\} \theta^1 \wedge \theta^2, \end{aligned}$$

where $\partial / \partial \omega^i$ denote the coframe derivatives associated with the base coframe (12.13); see (8.19). Therefore, we deduce the well-known explicit formula

$$\kappa = \frac{\partial J}{\partial \omega^2} - \frac{\partial K}{\partial \omega^1} - J^2 - K^2 = \frac{A J_y - B J_x + C K_y - D K_x}{AD - BC} - J^2 - K^2 \tag{12.23}$$

for the Gaussian curvature in terms of the first fundamental form. This computation provides an intrinsic formulation of Gauss' Theorema Egregium — that the Gaussian curvature of a surface is an isometric invariant, depending only on its first fundamental form, not the particular embedding into $\mathbb{R}^3$. See, for instance, [**100**], [**210**] for the classical version, and [**106**] for more details on the differential form approach to surface theory.

Exercise 12.2. Prove that the Gaussian curvature can be expressed in terms of the metric coefficients as $\kappa = (EG - F^2)^{-2}\left[D_1 - D_2\right]$, where D_1 and D_2 are the following two determinants:

$$\begin{vmatrix} -\frac{1}{2}E_{yy} + F_{xy} - \frac{1}{2}G_{xx} & \frac{1}{2}E_x & F_x - \frac{1}{2}E_y \\ F_y - \frac{1}{2}G_x & E & F \\ \frac{1}{2}G_y & F & G \end{vmatrix}, \qquad \begin{vmatrix} 0 & \frac{1}{2}E_y & \frac{1}{2}G_x \\ \frac{1}{2}E_y & E & F \\ \frac{1}{2}G_x & F & G \end{vmatrix};$$

see [**210**; p. 112].

According to Theorem 8.22, the surfaces with maximal isometry group are those whose coframes have rank zero, meaning that the fundamental invariant κ is a constant. Thus, we immediately deduce the following classical result.

Proposition 12.3. *Let S be a surface with first fundamental form ds^2 and Gaussian curvature κ. Then the group of isometries of S is a Lie group of dimension at most three. Moreover, the isometry group is three-dimensional if and only if the surface has constant Gaussian curvature, in which case the connected component of the symmetry group has the structure of* $\mathrm{SL}(2,\mathbb{R})$ *if* $\kappa < 0$ *(hyperbolic geometry),* $\mathrm{SE}(2)$ *if* $\kappa = 0$ *(Euclidean geometry), or* $\mathrm{SO}(3)$ *if* $\kappa > 0$ *(spherical geometry).*

The surfaces with constant curvature, then, define coframes of rank zero. Higher rank coframes arise when $\kappa(x,y)$ is not constant. It is interesting that there are no coframes of rank one. Indeed, the first two nonzero derived invariants,

$$\begin{aligned} \widehat{\kappa}_1 &= \frac{\partial \kappa}{\partial \theta^1} = \cos t \, \frac{\partial \kappa}{\partial \omega^1} - \sin t \, \frac{\partial \kappa}{\partial \omega^2}, \\ \widehat{\kappa}_2 &= \frac{\partial \kappa}{\partial \theta^2} = \sin t \, \frac{\partial \kappa}{\partial \omega^1} + \cos t \, \frac{\partial \kappa}{\partial \omega^2}, \end{aligned} \tag{12.24}$$

do depend on the group parameter t, and hence cannot be functionally dependent upon κ itself! Here, we are using the formulae

$$\begin{aligned}
\frac{\partial}{\partial \theta^1} &= P\,\frac{\partial}{\partial t} + \cos t\,\frac{\partial}{\partial \omega^1} - \sin t\,\frac{\partial}{\partial \omega^2} \\
&= \cos t\left\{\frac{\partial}{\partial \omega^1} + J\,\frac{\partial}{\partial t}\right\} - \sin t\left\{\frac{\partial}{\partial \omega^2} + K\,\frac{\partial}{\partial t}\right\}, \\
\frac{\partial}{\partial \theta^2} &= Q\,\frac{\partial}{\partial t} + \sin t\,\frac{\partial}{\partial \omega^1} + \cos t\,\frac{\partial}{\partial \omega^2} \\
&= \sin t\left\{\frac{\partial}{\partial \omega^1} + J\,\frac{\partial}{\partial t}\right\} + \cos t\left\{\frac{\partial}{\partial \omega^2} + K\,\frac{\partial}{\partial t}\right\}, \\
\frac{\partial}{\partial \pi} &= \frac{\partial}{\partial t},
\end{aligned} \tag{12.25}$$

for the coframe derivatives associated with the invariant coframe (12.9), (12.20) on $S \times \mathrm{SO}(2)$. Note that, although the individual coframe derivatives $\kappa_i = \partial\kappa/\partial\omega^i$ are *not* invariant, the "norm"

$$\rho = \sqrt{(\widehat{\kappa}_1)^2 + (\widehat{\kappa}_2)^2} = \sqrt{(\kappa_1)^2 + (\kappa_2)^2} \tag{12.26}$$

does define an isometric invariant. In view of (12.25), the second order derived invariants are

$$\begin{aligned}
\widehat{\kappa}_{11} &= \frac{\partial^2\kappa}{(\partial\theta^1)^2} = (\cos t)^2\,\widetilde{\kappa}_{11} - 2\sin t\cos t\,\widetilde{\kappa}_{12} + (\sin t)^2\,\widetilde{\kappa}_{22}, \\
\widehat{\kappa}_{12} &= \frac{\partial^2\kappa}{\partial\theta^1\partial\theta^2} = \sin t\cos t\,\widetilde{\kappa}_{11} - 2\sin t\cos t\,\widetilde{\kappa}_{12} + \sin t\cos t\,\widetilde{\kappa}_{22}, \\
\widehat{\kappa}_{22} &= \frac{\partial^2\kappa}{(\partial\theta^2)^2} = (\sin t)^2\,\widetilde{\kappa}_{11} + 2\sin t\cos t\,\widetilde{\kappa}_{12} + (\cos t)^2\,\widetilde{\kappa}_{22},
\end{aligned} \tag{12.27}$$

where

$$\begin{aligned}
\widetilde{\kappa}_{11} &= \frac{\partial^2\kappa}{(\partial\omega^1)^2} - J\,\frac{\partial\kappa}{\partial\omega^2}, \qquad \widetilde{\kappa}_{22} = \frac{\partial^2\kappa}{(\partial\omega^2)^2} + K\,\frac{\partial\kappa}{\partial\omega^1} \\
\widetilde{\kappa}_{12} &= \frac{\partial^2\kappa}{\partial\omega^1\partial\omega^2} + J\,\frac{\partial\kappa}{\partial\omega^1} = \frac{\partial^2\kappa}{\partial\omega^2\partial\omega^1} - K\,\frac{\partial\kappa}{\partial\omega^2}.
\end{aligned}$$

Exercise 12.4. Explain why $\partial^2\kappa/\partial\theta^1\partial\theta^2 = \partial^2\kappa/\partial\theta^2\partial\theta^1$.

If we eliminate t between the first and second order structure invariants, we find three new surface invariants:

$$\begin{aligned} \sigma_1 &= \kappa_1\kappa_2(\widetilde{\kappa}_{11} - \widetilde{\kappa}_{22}) + 2(\kappa_1^2 - \kappa_2^2)\widetilde{\kappa}_{12}, \\ \sigma_2 &= \kappa_2^2\widetilde{\kappa}_{11} - 2\kappa_1\kappa_2\widetilde{\kappa}_{12} + \kappa_1^2\widetilde{\kappa}_{22}, \\ \sigma_3 &= \kappa_1^2\widetilde{\kappa}_{11} + 2\kappa_1\kappa_2\widetilde{\kappa}_{12} + \kappa_2^2\widetilde{\kappa}_{22}. \end{aligned} \tag{12.28}$$

Exercise 12.5. Show that $\Delta\kappa = \widetilde{\kappa}_{11} + \widetilde{\kappa}_{22}$ and $\widetilde{\kappa}_{11}\widetilde{\kappa}_{22} - \widetilde{\kappa}_{12}^2$ are also surface invariants, and show how to express them in terms of the invariants κ, ρ, and (12.28). Here Δ denotes the Laplace–Beltrami operator associated with the metric on S, cf. [**221**; Chapter 6].

If the coframe has rank two, then the structure functions κ and ρ are functionally dependent, so that, locally, $\rho = H(\kappa)$ for some classifying function H. However, this in itself is not sufficient to guarantee that the coframe has rank two since the second order structure functions σ_i should also depend on κ, but this does not follow automatically from the structure equations. According to Theorem 8.22, any rank two surface has a one-parameter isometry group. A classical result, cf. [**100**; p. 182], says that these are the surfaces of revolution, which are the subsets of $\mathbb{R}^3$ obtained by rotating a curve about a fixed axis.

Theorem 12.6. *A surface admits a one-parameter group of isometries if and only if it is isometric to (part of) a surface of revolution.*

Proof: Given a curve $z = f(x)$, $y = 0$, say, and taking the axis of revolution to be the z-axis, we see that such a surface is parametrized by

$$x = r\cos v, \qquad y = r\sin v, \qquad z = f(r).$$

The associated metric tensor is readily computed:

$$ds^2 = r^2\,dr^2 + \{1 + f'(r)^2\}\,dv^2.$$

This can be further simplified by reparametrizing the surface using $u = \int\sqrt{1 + f'(r)^2}\,dr$, with inverse $r = h(u)$, in terms of which

$$x = h(u)\cos v, \qquad y = h(u)\sin v, \qquad z = \int\sqrt{1 - h'(u)^2}\,du, \tag{12.29}$$

and the metric is $ds^2 = du^2 + h(u)^2\,dv^2$. The associated diagonalizing coframe for such a metric is $\omega^1 = du$, $\omega^2 = h(u)\,dv$. An easy calculation shows that the additional prolonged coframe element (12.20) is $\pi = \alpha - h'(u)\,dv$. Substituting into the final structure equation (12.22), we find the simple formula $\kappa = -h_{uu}/h$ for the Gaussian curvature of a surface of revolution parametrized as in (12.29). The associated coframe derivatives are $\partial\kappa/\partial\theta^1 = \kappa_u \cos t$, $\partial\kappa/\partial\theta^2 = \kappa_u \sin t$, so, assuming κ is not constant (i.e., the surface is not conical or cylindrical), the key classifying function is that relating $\rho = \kappa_u$ to κ. Conversely, given a classifying curve $\kappa_u = H(\kappa)$, there exists a three-parameter family of parametrized, locally isometric surfaces of revolution (12.29). Two of the parameters indicate the trivial translation reparamatrization $u \mapsto u + c$, and the rescaling $h \mapsto \lambda h$; the third free parameter (which does not exist in the case of constant curvature surfaces of revolution) is more interesting. For instance, if the classifying curve is $\kappa_u = \sqrt{-2\kappa^3}$, then $\kappa(u) = -2(u+c)^{-2}$, and hence $h(u) = a(u+c)^{-1} + b(u+c)^2$ describes the associated family of isometric surfaces of revolution. *Q.E.D.*

Exercise 12.7. Describe all constant curvature surfaces of revolution.

A surface whose coframe has rank three, and hence has at most a discrete isometry group, is uniquely determined by the functional interrelationships between, or, more precisely, the reduced classifying manifold parametrized by, the structure invariants (12.24), (12.28). An interesting question, to which I do not know the answer, is to characterize those surfaces, if any, which have rank 3 and order 1, so that $\rho = H(\kappa)$, but at least one of the σ_i's is independent of κ. More details on the geometric applications of the differential invariants of surfaces can be found in [**72**].

Once the prolonged coframe has been fully normalized, the resulting invariant structure functions will often involve the original group parameters. As such, they provide the invariant combinations that we were originally seeking before we were forced to introduce the higher-dimensional prolonged problem. Therefore, we can normalize any such group-dependent invariant, and thereby reduce the original structure group. This will allow us to proceed further with the reduction procedure back down on M, and, in favorable cases, allow us to construct an invariant coframe on M itself. In such cases, then, we can entirely dispense with the prolongation version, which merely served as a device to enable us to completely reduce the problem in an invariant manner.

For example, in the isometric equivalence of surfaces of nonconstant curvature, the two derived structure invariants (12.24) depend on the group parameter t. As before, we are allowed to normalize one of these invariants and thereby specify the rotation parameter t. The result is an invariant coframe back on the original surface S, which completes the solution to the equivalence problem. For example, we can normalize $\kappa_2 = 0$ by choosing $\tan t = -\kappa_2/\kappa_1$. Substituting into formula (12.9), we produce an "adapted" diagonalizing coframe

$$\theta^1 = \frac{\kappa_1\,\omega^1 + \kappa_2\,\omega^2}{\rho}, \qquad \theta^2 = \frac{-\kappa_2\,\omega^1 + \kappa_1\,\omega^2}{\rho}, \tag{12.30}$$

which is *invariantly* associated with any surface of nonconstant curvature. The structure equations for the reduced equivalence problem take the usual form (8.2) for a coframe on a two-dimensional manifold; the fundamental structure invariants are

$$J^* = -\frac{\sigma_1}{\rho^3}, \qquad K^* = \frac{\sigma_2}{\rho^3}, \tag{12.31}$$

cf. (12.28). The equivalence problem now is resolved using these two invariants and their first or second order coframe derivatives. Although the solution must be the same as that deduced using the prolonged coframe, the required structure invariants are different. In this approach, the Gaussian curvature does *not* appear as one of the fundamental invariants, although it will, of course, reappear as a first order derived invariant via the general formula (8.35), with J^*, K^* replacing J, K.

Conformal Equivalence of Surfaces

It is interesting to contrast the isometric surface equivalence problem with the problem of determining when two surfaces are equivalent up to a conformal factor. Here we only require that the diffeomorphism Φ map the metrics to scalar multiples of each other: $\Phi^*\, d\bar{s}^2 = \lambda^2\, ds^2$, where $\lambda(x,y)$ is a positive scalar function. Using the same coframe (12.8), (12.9), the condition implies that

$$\Phi^* \begin{pmatrix} \bar\omega_1 \\ \bar\omega_2 \end{pmatrix} = \begin{pmatrix} \lambda\cos t & -\lambda\sin t \\ \lambda\sin t & \lambda\cos t \end{pmatrix} \begin{pmatrix} \omega_1 \\ \omega_2 \end{pmatrix}, \tag{12.32}$$

so that the two-dimensional conformal structure group $G = \mathrm{SO}(2)\times\mathbb{R}^+$ consists of rotations and scalings. The lifted coframe is given by

$$\theta^1 = \lambda\big[(\cos t)\,\omega_1 - (\sin t)\,\omega_2\big], \quad \theta^2 = \lambda\big[(\sin t)\,\omega_1 + (\cos t)\,\omega_2\big],$$

with associated structure equations

$$\begin{aligned} d\theta^1 &= \rho \wedge \theta^1 - \alpha \wedge \theta^2 + P\,\theta^1 \wedge \theta^2, \\ d\theta^2 &= \alpha \wedge \theta^1 + \rho \wedge \theta^2 + Q\,\theta^1 \wedge \theta^2. \end{aligned} \tag{12.33}$$

Here $\alpha = dt$ and $\rho = d\log\lambda$ are the two Maurer–Cartan forms, and the torsion coefficients P, Q have similar (but not identical) formulae to those in (12.17). (The precise formulae will, as we shall momentarily see, not be required.) If we replace ρ by $z_1\,\theta^1 + z_2\,\theta^2$, and α by $w_1\,\theta^1 + w_2\,\theta^2$, then the linear system (11.7) governing the absorption reads $z_2 + w_1 = P$, $-z_1 + w_2 = Q$, so the degree of indeterminancy is $r^{(1)} = 2$. On the other hand, the Cartan characters associated with (12.33) are computed using (11.11), and we find $s_1' = 2$, while $s_2' = 0$. Therefore, $s_1' + 2s_2' = r^{(1)} = 2$, and Cartan's involutivity test (11.13) is satisfied. Theorem 11.8 then implies the following result.

Theorem 12.8. *Any two analytic surfaces are (locally) conformally equivalent. Moreover, the conformal symmetry group of a surface S with metric ds^2 is infinite-dimensional, depending on two arbitrary functions of a single variable.*

Thus, by Theorem 12.8, every surface can (locally) be conformally mapped to the Euclidean plane. Moreover, if we identify the Euclidean plane with the complex line $\mathbb{C}$, having coordinate $z = x + iy$, the (orientation preserving) conformal symmetries are defined by an arbitrary analytic function $\bar{z} = F(z)$. The two arbitrary functions mentioned in the theorem are the real and imaginary parts of the complex function $F(z)$.

Exercise 12.9. Set up, and solve, the equivalence problem for when two one-parameter families of plane curves can be mapped to each other via a conformal transformation. This problem forms the foundation of the theory of "web geometry", cf. [**39**].

Equivalence of Riemannian Manifolds

The higher dimensional counterpart of the isometric equivalence problem for surfaces is the general problem of equivalence for Riemannian manifolds. This problem was originally studied by Riemann, [**197**], who was motivated by a problem on heat flow in a solid, and also by Christoffel, [**52**]. (Riemann's paper was written before Christoffel's, but appeared later, posthumously; see [**143**] for further interesting historical details.) Remarkably, the explicit formulae for many of the fundamental objects of Riemannian geometry, such as connection and curvature, trace their origins back to Riemann and Christoffel's direct analysis of the Riemannian equivalence problem. Here we shall solve the equivalence problem using Cartan's method — the interested reader can also refer to [**229**] for a description of a variety of other approaches. As we shall see, the general discussion requires a more intrinsic approach to the resulting formulae, which is useful even in the two-dimensional case discussed above.

Let M be an m-dimensional Riemannian manifold with positive definite metric tensor. We introduce a coframe $\boldsymbol{\omega}$ which (locally) diagonalizes the metric

$$ds^2 = \sum_{i,j=1}^{m} g_{ij}(x)\, dx^i\, dx^j = \sum_{i=1}^{m} (\omega^i)^2.$$

Thus $\omega^i = \sum_j h^i_j\, dx^j$, where $H = (h^i_j(x))$ is any convenient (e.g., symmetric or upper triangular) square root of the positive definite metric tensor, meaning that $H^T H = G = (g_{ij}(x))$. (Pseudo-Riemannian manifolds can be treated similarly; we diagonalize the "metric" using the appropriate canonical form for symmetric matrices as in Example 2.26. For simplicity, though, we shall only discuss the positive definite case.)

The structure group is the orthogonal group $\mathrm{O}(m)$, and the lifted coframe is $\boldsymbol{\theta} = R\boldsymbol{\omega}$, where $R \in \mathrm{O}(m)$. The initial structure equations have the form

$$d\theta^i = \sum_{j=1}^{m} \rho^i_j \wedge \theta^j + \sum_{j<k} T^i_{jk}\, \theta^j \wedge \theta^k, \tag{12.34}$$

where $\boldsymbol{\rho} = (\rho^i_j) = dR \cdot R^{-1} = dR \cdot R^T$ is the skew-symmetric matrix of Maurer–Cartan forms on $\mathrm{O}(m)$, with components $\rho^i_j = -\rho^j_i = \sum_k r^j_k\, dr^i_k$. I claim that, just as in the simpler two-dimensional case, we can absorb all the torsion, so the equivalence problem is determinate: $r^{(1)} = 0$. To prove this, we replace each ρ^i_j by a linear combination $\sum_k z^i_{jk}\theta^k$, where the coefficients must respect the skew-symmetry condition $z^i_{jk} = -z^j_{ik}$. The linear absorption system (11.7) is

$$z^i_{kj} - z^i_{jk} = T^i_{jk}, \tag{12.35}$$

where T^i_{jk} is the coefficient of $\theta^j \wedge \theta^k$ in the ith structure equation for $d\theta^i$. Since z^i_{jk} is skew-symmetric in i and j, whereas T^i_{jk} is skew-symmetric in j and k, if we add (12.35) to the same equation with k, j, i, and also with j, k, i, replacing i, j, k, we find

$$-2z^i_{jk} = T^i_{jk} + T^j_{ik} + T^k_{ji}. \tag{12.36}$$

This shows that all the absorption equations (12.35) can be solved, proving the claim.

In order to compute the explicit formulae for the torsion coefficients, we introduce the structure coefficients associated with the diagonalizing coframe $\boldsymbol{\omega}$, which are defined by

$$d\omega^i = \sum_{j<k} K^i_{jk}\, \omega^j \wedge \omega^k, \qquad i = 1, \ldots, m. \tag{12.37}$$

Explicitly,

$$K^i_{jk} = \sum_{l,n=1}^{m} \widetilde{h}^l_j\, \widetilde{h}^n_k \left(\frac{\partial h^i_n}{\partial x^l} - \frac{\partial h^i_l}{\partial x^n} \right), \tag{12.38}$$

where the $\widetilde{h}^l_j$ are the entries of the inverse matrix $\widetilde{H} = H^{-1}$. (Note that by (8.10), the same structure coefficients appear in the Lie brackets

$$[\mathbf{v}_j, \mathbf{v}_k] = -\sum_{i=1}^{m} K^i_{jk} \mathbf{v}_i, \tag{12.39}$$

between the dual frame vector fields $\mathbf{v}_i = \partial/\partial\omega^i$.) A simple computation using (12.37) and the formula for the lifted coframe demonstrates that the torsion coefficients are just the rotated structure coefficients:

$$T^i_{jk} = \sum_{l,p,q=1}^{m} r^i_l r^j_p r^k_q K^l_{pq}. \tag{12.40}$$

(Note the positions of the indices.) Therefore, according to (12.36),

$$z^i_{jk} = -\sum_{l,p,q=1}^{m} r^i_l r^j_p r^k_q\, \widehat{\Gamma}^l_{qp}, \qquad \text{where} \qquad \widehat{\Gamma}^l_{pq} = \frac{1}{2}\left(K^l_{pq} + K^p_{lq} + K^q_{lp} \right). \tag{12.41}$$

The $\widehat{\Gamma}$'s are just the components of the symmetric Levi–Civita connection associated with the Riemannian metric, but expressed with respect to the diagonalizing frame $\{\mathbf{v}_1, \ldots, \mathbf{v}_m\}$. (The book [**57**] is a good reference for details on these and other basic formulae in Riemannian geometry.) If $\widehat{\nabla}_i$ denotes the covariant derivative with respect to the vector field $\mathbf{v}_i$, then

$$\widehat{\nabla}_k(\mathbf{v}_j) = \sum_{i=1}^{m} \widehat{\Gamma}^i_{jk} \mathbf{v}_i. \tag{12.42}$$

Indeed, substituting (12.39) and (12.42) into the symmetric connection condition

$$[\mathbf{v}_j, \mathbf{v}_k] = \widehat{\nabla}_j(\mathbf{v}_k) - \widehat{\nabla}_k(\mathbf{v}_j), \tag{12.43}$$

we find $K^i_{jk} = \widehat{\Gamma}^i_{jk} - \widehat{\Gamma}^i_{kj}$, which, upon comparison with (12.41), demonstrates the claim.

In view of (12.41), we define the modified Maurer–Cartan forms

$$\pi^i_j = \rho^i_j - \sum_{k=1}^{m} z^i_{jk}\, \theta^k, \tag{12.44}$$

which, along with the θ^i's, form a coframe on the prolonged space $M^{(1)} = M \times \mathrm{O}(m)$. Since

$$0 = d^2\theta^i = \sum_{j=1}^{m} \left[d\pi^i_j \wedge \theta^j - \pi^i_j \wedge d\theta^j \right] = \sum_{k=1}^{m} \left[d\pi^i_k - \sum_{j=1}^{m} \pi^i_j \wedge \pi^k_j \right] \wedge \theta^k,$$

the final structure equations must have the form

$$d\theta^i = \sum_{j=1}^{m} \pi^i_j \wedge \theta^j, \qquad d\pi^i_j = \sum_{k=1}^{m} \pi^i_k \wedge \pi^k_j - \sum_{k=1}^{m} S^i_{jkl}\, \theta^k \wedge \theta^l. \tag{12.45}$$

Note, in particular, that the first set of summands in the formula for $d\pi^i_j$ reproduces for the structure equations of the orthogonal group found in Exercise 2.97. Equations (12.45) are known as the *Cartan structure equations* for a Riemannian manifold, and generalize the structure equations (12.22) for surfaces.

The structure invariants S^i_{jkl} turn out to be the components of the Riemann curvature tensor $\mathbf{R}$, expressed in terms of the lifted coframe

$\boldsymbol{\theta}$. Before proving this, we determine the coframe derivatives for the prolonged coframe. Suppose $F(x,R)$ is any function on $M^{(1)}$. We find, using (12.44), (12.41),

$$\begin{aligned} dF &= \sum_{i=1}^{m} \mathbf{v}_i(F)\,\omega^i + \sum_{j,k,l=1}^{m} r^j_l \frac{\partial F}{\partial r^k_l}\,\rho^k_j \\ &= \sum_{i=1}^{m} \left(\sum_{j=1}^{m} r^i_j\,\mathbf{v}_j(F) + \sum_{j,k,l,n=1}^{m} \widehat{\Gamma}^k_{nj} \frac{\partial F}{\partial r^l_n}\, r^l_k r^i_j \right) \theta^i + \\ &\quad + \sum_{j<k} \sum_{l=1}^{m} \left(r^j_l \frac{\partial F}{\partial r^k_l} - r^k_l \frac{\partial F}{\partial r^j_l} \right) \pi^k_j . \end{aligned}$$

Therefore, the coframe derivatives take the explicit form

$$\begin{aligned} \frac{\partial F}{\partial \theta^i} &= \sum_{j=1}^{m} r^i_j \left(\mathbf{v}_j(F) + \sum_{k,l,n=1}^{m} \widehat{\Gamma}^k_{nj} \frac{\partial F}{\partial r^l_n}\, r^l_k \right), \\ \frac{\partial F}{\partial \pi^k_j} &= \sum_{l=1}^{m} \left(r^j_l \frac{\partial F}{\partial r^k_l} - r^k_l \frac{\partial F}{\partial r^j_l} \right). \end{aligned} \tag{12.46}$$

Let us recall the definition of a general *tensor field* $\mathbf{T}$ of *type* (p,q), i.e., a smooth section of the tensor product bundle $\bigotimes^p TM \otimes \bigotimes^q T^*M$. If we introduce a coframe $\{\omega^1,\ldots,\omega^m\}$ on M, with associated dual frame $\{\mathbf{v}_1,\ldots,\mathbf{v}_m\}$, where $\mathbf{v}_i = \partial/\partial\omega^i$, then the components of $\mathbf{T}$ are defined by evaluating it on the indicated basis vectors of the tangent and cotangent spaces, so $T^{ij\cdots}_{rs\ldots} = \mathbf{T}(\omega^i,\omega^j,\ldots,\mathbf{v}_r,\mathbf{v}_s,\ldots)$. For example, a differential form of degree k can be identified with a fully anti-symmetric tensor field of type $(0,k)$; its components with respect to the coframe $\boldsymbol{\omega}$ are merely its coefficients when writing it as a linear combination of k-fold wedge products of the coframe elements. Now, if $\mathbf{T}$ is a tensor field, then its components with respect to the lifted coframe $\boldsymbol{\theta}$ are found by suitably rotating its components with respect to the diagonalizing coframe $\boldsymbol{\omega}$; we will call these components the *invariant components* of $\mathbf{T}$. For example, if $\mathbf{T}$ is a tensor field of type $(1,1)$, its invariant components are $\widehat{T}^i_j = \mathbf{T}(\theta^k, \partial_{\theta^l}) = \sum_{k,l} r^i_k r^j_l T^k_l$, where $T^k_l = \mathbf{T}(\omega^k, \mathbf{v}_l)$. Note that the upper, covariant index is rotated as expected, whereas the lower, contravariant index is rotated by the inverse (or transposed) rotation $R^{-1} = R^T$; compare equation (12.40). The *covariant differential* of T, denoted $\nabla \mathbf{T}$, is the tensor of type $(p, q+1)$ whose components are the covariant derivatives of the components of $\mathbf{T}$. For example, if $\mathbf{T}$ is of type $(1,1)$, then the components of $\mathbf{W} = \nabla \mathbf{T}$ with respect to the diagonalizing coframe are

$$W^i_{jk} = \widehat{\nabla}_k(T^i_j) = \mathbf{v}_k(T^i_j) + \sum_{l=1}^{m} (\widehat{\Gamma}^i_{lk} T^l_j - \widehat{\Gamma}^l_{jk} T^i_l);$$

its invariant components are found by rotation: $\widehat{W}^i_{jk} = \sum_{l,p,q} r^i_l r^j_p r^k_q W^l_{pq}$. Finally, we let $\mathbf{T} \otimes \mathbb{1}$ denote the tensor of type $(p+1, q+1)$ obtained by tensoring $\mathbf{T}$ with the identity tensor $\mathbb{1}$, which has Kronecker components δ^i_j (with respect to any coframe). For instance, if $\mathbf{T}$ is of type $(1,1)$, then the components of $\mathbf{U} = \mathbf{T} \otimes \mathbb{1}$ with respect to the diagonalizing coframe are $U^{il}_{jk} = T^i_j \delta^l_k$. A straightforward computation using the formulae (12.46) for the coframe derivatives proves the following key result.

Proposition 12.10. *Let $\mathbf{T}$ be any tensor field on M. Then the coframe derivatives of the invariant components of $\mathbf{T}$, with respect to the lifted coframe $\boldsymbol{\theta}$ and $\boldsymbol{\pi}$, are the invariant components of the covariant differential $\nabla \mathbf{T}$, and the tensor product $\mathbf{T} \otimes \mathbb{1}$, respectively.*

In particular, evaluating the differentials (12.44) and comparing with the structure equations (12.45), we find that

$$S^i_{jkl} = -\frac{\partial z^i_{jl}}{\partial \theta^k} + \frac{\partial z^i_{jk}}{\partial \theta^l} - \sum_{n=1}^{m} \left[z^i_{nk} z^n_{jl} - z^i_{nl} z^n_{jk} \right] = \sum_{i',j',k',l'=1}^{m} r^i_{i'} r^j_{j'} r^k_{k'} r^l_{l'} \widehat{R}^{i'}_{j'k'l'}, \tag{12.47}$$

where

$$\widehat{R}^i_{jkl} = \mathbf{v}_k(\widehat{\Gamma}^i_{jl}) - \mathbf{v}_l(\widehat{\Gamma}^i_{jk}) + \sum_{n=1}^{m} \left[\widehat{\Gamma}^i_{nk} \widehat{\Gamma}^n_{jl} - \widehat{\Gamma}^i_{jn} \widehat{\Gamma}^n_{lk} - \widehat{\Gamma}^i_{nl} \widehat{\Gamma}^n_{jk} + \widehat{\Gamma}^i_{jn} \widehat{\Gamma}^n_{kl} \right]. \tag{12.48}$$

The latter formula reproduces the standard definition of the curvature tensor (evaluated with respect to a diagonalizing coframe), cf. [**57**; p. 273]. Therefore, we have demonstrated that *the structure functions for the prolonged coframe are the individual invariant components of the Riemann curvature tensor in terms of the lifted coframe $\boldsymbol{\theta}$ on M.* Note that, in general, the structure invariants S^i_{jkl} depend on the orthogonal group parameters r^i_j and hence do not immediately provide us with invariants on M itself. Indeed, the components (12.48) of the curvature tensor in the base (diagonalizing) coframe are *not* invariant, but, according to (12.47), are "equivariant" under the action of the orthogonal group $\mathrm{O}(m)$. In view of Proposition 12.10, the derived invariants associated with the prolonged coframe $\boldsymbol{\theta}$, $\boldsymbol{\pi}$ are the invariant components of the covariant differentials $\nabla^k\mathbf{R}$ of the Riemann curvature tensor, along with its various tensor products with the identity tensor; the latter, however, do not provide any new information. Therefore, we have reproduced the classical theorem of Christoffel, [**52**], giving necessary and sufficient conditions for the equivalence of Riemannian metrics.

Theorem 12.11. *A complete set of structure invariants for a Riemannian manifold are provided by the invariant components of the* higher order curvature tensors $\nabla^k\mathbf{R}$, $k = 0, 1, 2, \ldots$. *In the regular case, two Riemannian manifolds are locally isometric if and only if all their curvature tensors parametrize overlapping classifying manifolds.*

The *order* of a Riemannian metric is, by definition, the order of the coframe, which is just the highest order curvature tensor containing additional functionally independent invariant components. According to Theorem 8.19, we need only verify the overlapping condition for curvature tensors of order at most one more than the order of the metric. Unfortunately, very little appears to be known about the structure of the various classifying manifolds.

The basic Theorem 8.22 on symmetry groups of coframes immediately implies the classical result bounding the isometry group of an m-dimensional Riemannian manifold to have dimension at most $\frac{1}{2}m(m+1)$, which is the dimension of the bundle $M \times \mathrm{O}(m)$. The metric will define a coframe of rank zero and hence have an isometry group of maximal dimension if and only if the invariant components S^i_{jkl}, cf. (12.47), of the curvature tensor are constant, meaning that they do not depend on either the x's or the rotation parameters r^i_j. This means that *a*) the curvature components $\widehat{R}^i_{jkl}$ are constant, and, moreover *b*) the r^i_j's in (12.46) must cancel out. This will occur if and only if the curvature components are linear combinations of various products of Kronecker deltas, e.g., $\delta^i_j\delta^l_k$. The symmetry restrictions

$$\widehat{R}^i_{jkl} = -\widehat{R}^i_{jlk}, \qquad \widehat{R}^i_{jkl} + \widehat{R}^i_{ljk} + \widehat{R}^i_{klj} = 0, \qquad (12.49)$$

on the Riemann curvature tensor imply that $\widehat{R}^i_{jkl} = a(\delta^i_k \delta^l_j - \delta^i_l \delta^k_j)$ for some constant a, which means that the manifold has constant scalar curvature, [**67**; p. 83], [**138**; Theorem 3.1].

Theorem 12.12. *The group of isometries of an m-dimensional Riemannian manifold has dimension at most $\frac{1}{2}m(m+1)$. Moreover, the isometry group has maximal dimension $\frac{1}{2}m(m+1)$ if and only if M has constant scalar curvature, and hence is locally isometric to either $\mathbb{R}^m$, with isometry group $\mathrm{E}(m)$, to the sphere S^m, with isometry group $\mathrm{O}(m+1)$, or to hyperbolic space H^m with isometry group $\mathrm{O}(m,1)$.*

It is also known, [**138**; Theorem 3.2], that if $m \neq 4$, the dimension of the isometry group of a Riemannian manifold with nonconstant scalar curvature has dimension $\leq \frac{1}{2}m(m-1)+1$, a fact that also follows from a more detailed analysis of the dependency of the curvature tensor on the rotation parameters r^i_j.

Note that the kth order curvature tensor $\nabla^k \mathbf{R}$ has type $(1, k+3)$. The symmetry class of such tensors is governed by the consequences of the symmetry conditions (12.49) and the Jacobi identities for the structure functions, which, in this case, reduce to the classical *Bianchi identities*

$$\widehat{\nabla}_n \widehat{R}^i_{jkl} + \widehat{\nabla}_k \widehat{R}^i_{jln} + \widehat{\nabla}_l \widehat{R}^i_{jnk} = 0. \tag{12.50}$$

The coefficients of $\nabla^k \mathbf{R}$ with respect to the lifted coframe $\boldsymbol{\theta} = R\boldsymbol{\omega}$ are obtained from its coefficients with respect to the diagonalizing coframe $\boldsymbol{\omega}$ by rotating them in the appropriate manner using the orthogonal matrix R. The structure invariants which do not depend on the group parameters R, then, will be provided by the joint $\mathrm{SO}(m)$ invariants, in the sense of classical invariant theory, of the higher order curvature tensors. See [**74**] for a detailed analysis of the symmetry classes, invariants, and canonical forms for the curvature tensors of orders $k \leq 6$.

As in the case of surfaces, if there are $\frac{1}{2}m(m-1)$ curvature invariants which are functionally independent as functions of the rotation parameters r^i_j, then we can use these invariants to normalize the rotation matrix, and reduce to an equivalence problem of invariant diagonalizing coframes on the original Riemannian manifold. The one case where such a reduction procedure has been carried out is that of a Minkowski metric on a four-dimensional space-time manifold. Classification of the canonical forms of the Weyl tensor, which is one of the irreducible components of the Riemann curvature tensor (the other two being the Ricci tensor and the Ricci scalar) leads to the Petrov classification of metrics in general relativity, thereby specifying the group reduction and invariant diagonalizing coframe on the space-time manifold. See [**25**], [**136**], [**137**], and [**141**] for additional details in the relativistic context.

The Indeterminate Case

Let us now return to the general Cartan equivalence problem. The only case yet to be treated occurs when the problem reduces to a non-involutive set of structure equations with positive degree of indeterminancy: $r^{(1)} > 0$. In this, the *indeterminate case*, there are unresolved ambiguities in the absorption procedure. The prolongation process is essentially the same as before, but now every free parameter in the solution of the absorption equations provides a new group parameter which must eventually be normalized if we are to solve the prolonged problem.

The modified Maurer–Cartan forms are defined by the same formula, (12.2), where the z_j^κ's define the *general* solution to the absorption equations (11.6). In the determinate case, this completely specified the z_j^κ's in terms of the base variables and group parameters. Here, though, some of the z_j^κ's (more specifically, $r^{(1)}$ of them) are free variables, and cannot be invariantly prescribed by the absorption equations alone. Let us denote the free variables by w^μ, $\mu = 1, \ldots, r^{(1)}$, so that the general solution to the linear absorption system (11.6) takes the form $z = K[w] + S$, where the linear map K and the vector S (which is a "particular" solution) depend, in general, on x and g. Thus, the most general collection of one-forms (12.2) which solve the absorption equations is of the form

$$\boldsymbol{\pi} = \boldsymbol{\alpha} - (K[w] + S)\,\boldsymbol{\theta}. \tag{12.51}$$

The equivalence of the lifted coframes, and hence their differentials, implies that, for some choice of the group parameters $g = g(x)$ *and* the free variables $w = w(x)$, the new forms $\boldsymbol{\pi}$ are also invariant under an equivalence map.

Consequently, our prolongation procedure leads us to introduce a lifted coframe on the "prolonged" space, which, as in the determinate space, we take to be $M^{(1)} = M \times G$. The base coframe on $M^{(1)}$ will contain, first, the m one-forms in the original lifted coframe $\boldsymbol{\theta} = g \cdot \boldsymbol{\omega}$, now considered as one-forms on the space $M^{(1)}$, and, second, the r

modified Maurer–Cartan forms

$$\boldsymbol{\varpi} = \boldsymbol{\alpha} + S\,\boldsymbol{\theta}, \qquad \text{or, explicitly,} \qquad \varpi^i = \alpha^i + \sum_{j=1}^{m} S^i_j\,\theta^j, \tag{12.52}$$

which are obtained from the general solution (12.51) to the absorption equations by setting all the free variables to zero: $w = 0$. Note that $\boldsymbol{\theta}$ and $\boldsymbol{\varpi}$ do form a coframe on $M^{(1)}$. The free variables w will parametrize a new structure group $G^{(1)}$, which is the subgroup of $\mathrm{GL}(m+r)$ consisting of block lower triangular matrices of the form

$$G^{(1)} = \left\{ \begin{pmatrix} \mathrm{I} & 0 \\ K[w] & \mathrm{I} \end{pmatrix} \;\middle|\; w \in \mathbb{R}^{r^{(1)}} \right\}. \tag{12.53}$$

Thus the structure group $G^{(1)}$ for the prolonged problem has dimension $r^{(1)}$, the degree of indeterminancy of the coframe, and, interestingly, is always abelian! The lifted coframe will live on the prolonged space $M^{(1)} \times G^{(1)} = M \times G \times G^{(1)}$, and consist of the original lifted coframe $\boldsymbol{\theta}$ along with the modified Maurer–Cartan forms $\boldsymbol{\pi} = \boldsymbol{\varpi} + K[w]\boldsymbol{\theta}$, as defined in (12.51). Similarly, we construct the prolonged coframe corresponding to the coframe $\overline{\boldsymbol{\theta}}$ on $\overline{M}$, which will be defined on the manifold $\overline{M}^{(1)} = \overline{M} \times \overline{G}$, where, in accordance with our usual convention, $\overline{G} = G$. As with Proposition 12.1, we can prove the isomorphism between the original equivalence problem on M with structure group G and the new equivalence problem on $M^{(1)}$ with structure group $G^{(1)}$.

Proposition 12.13. *Let $(\boldsymbol{\theta}, \boldsymbol{\pi})$ and $(\overline{\boldsymbol{\theta}}, \overline{\boldsymbol{\pi}})$ be lifted prolonged coframes constructed as above. Then there exists a diffeomorphism $\Phi: M \to \overline{M}$ mapping $\overline{\boldsymbol{\theta}}$ to $\boldsymbol{\theta}$ for some choice of group parameters $g = g(x)$, $\bar{g} = \bar{g}(\bar{x})$ if and only if there is a diffeomorphism $\Psi: M \times G \to \overline{M} \times \overline{G}$ mapping $(\overline{\boldsymbol{\theta}}, \overline{\boldsymbol{\pi}})$ to $(\boldsymbol{\theta}, \boldsymbol{\pi})$ for some choice of prolonged group parameters $w = w(x)$, $\overline{w} = \overline{w}(\bar{x})$.*

Therefore, the original equivalence problem and its prolonged counterpart have (essentially) identical solutions, and we can concentrate on solving the latter. The goal now is to use the absorption and normalization procedure to eliminate the new group parameters $w \in G^{(1)}$. The procedure is exactly the same as before. Again, after one or more loops through the equivalence algorithm, several possibilities present themselves. First, during the course of the equivalence procedure for the prolonged coframe, we may discover a nonconstant essential torsion coefficient which depends, not on the new group parameters w, but only on the original group parameters g and the base variables x. This torsion coefficient is an invariant of the problem, and thereby determines an invariant combination of x's and g's. But this is what we were really after when we became stuck in the original reduction process! Therefore, we can safely normalize this new torsion coefficient to a convenient constant value, so as to eliminate one of the original group parameters, and thereby permit us to continue with the reduction process back on the original manifold M; the prolongation has served its purpose and can be safely forgotten. However, if there are yet further group parameters g to be normalized, it may well be advantageous to carry the reduction of the prolonged problem through to its completion in order possibly to find yet more useful essential torsion coefficients of this type, and thereby eliminate even more group parameters, which could avoid a second prolongation necessitated by a premature reversion to the base space M. Moreover, as we saw with the Riemannian equivalence problem, the reversion to the base M is not really necessary, as the equivalence problem on the prolonged space is entirely equivalent to the original one.

Second, after completing the equivalence procedure for the prolonged problem, there are, as with the original problem, three possible outcomes. In the most favorable case, we are able to reduce the prolonged problem to an invariant coframe on $M^{(1)} = M \times G$ by normalizing all of the new group parameters w. If the resulting structure functions are nonconstant, and involve g's, then the preceding reduction back to M is available. On the other hand, if all the structure functions (and their coframe derivatives) depend only on the base variables x, then the problem has been reduced as much as it can be, and we have solved the original equivalence problem by constructing an invariant coframe, not on the original manifold M, but rather on a higher dimension space $M^{(1)}$. The construction of the classifying manifold via coframe differentiation, the implementation of the equivalence criterion, and the dimensionality of the symmetry group, being $m+r-s$, where s is the number of functionally independent structure invariants, proceeds just as before. The other possibility is that we cannot eliminate all of the new group parameters w by normalizing essential torsion for the prolonged coframe. As with the original problem, the question then is whether the resulting prolonged coframe is involutive or not. If Cartan's involutivity test, as described in Definition 11.7, is passed, the prolonged coframe is involutive, and it (along with the original coframe) possesses an infinite-dimensional symmetry group. Otherwise, we must prolong yet again, leading in general to further reductions and yet further possible outcomes.

Remark: As always, when checking involutivity, one needs to be careful that there are no group-dependent invariants among the remaining unabsorbable torsion coefficients and their coframe derivatives, since these must all be normalized first. Here we must test for both the original group parameters and the new prolonged group parameters.

Does the process ever end? In part relying on earlier work of (among others), Tresse, [**214**], Cartan believed the answer to be affirmative — after a finite number of prolongations, every coframe reduces either to an invariant coframe, with finite-dimensional symmetry group, or to an involutive coframe, with infinite-dimensional symmetry group, cf. [**36**; p. 116]. Moreover, in every known practical example, this is indeed the case. (In fact, almost all examples arising in applications require at most two prolongations.) However, Cartan left unstated the precise regularity conditions required to reach such a conclusion, because, as he says, they are not easy to state. A rigorous version of the desired result with explicit, albeit complicated, regularity hypotheses was proved by Kuranishi, [**145**], and is known as the Cartan–Kuranishi Prolongation Theorem. However, in view of its rather complicated formulation and proof, we shall refer the interested reader to the discussion in [**28**] for details.

Remark: Even in the cases when the structure equations are involutive, one could still prolong them as described above. The resulting lifted coframe would, possibly after some additional absorption and normalization, again give an involutive equivalence problem, and the procedure could be continued. In this case, the process would never end — one can never reach a fully determined equivalence problem on any finite-dimensional space, as this would violate the fact that an involutive coframe has an infinite-dimensional symmetry group!

Second Order Ordinary Differential Equations

The general prolongation process will now be illustrated by a relatively simple example, which, nevertheless, demonstrates the degree of complication that can ensue from even basic equivalence problems. We shall apply the procedure to find a complete solution to the equivalence problem for a scalar second order ordinary differential equation

$$\frac{d^2u}{dx^2} = Q\left(x, u, \frac{du}{dx}\right), \tag{12.54}$$

under both fiber-preserving and point transformations. The latter analysis first appears in the work of Cartan, [**40**], in the context of projective connections. We begin with the fiber-preserving case. According to Example 11.1, after one elementary reduction, we are left with the base coframe

$$\omega^1 = du - p\,dx, \qquad \omega^2 = dp - Q(x,u,p)\,dx, \qquad \omega^3 = dx, \tag{12.55}$$

on the first jet space J^1, and structure group

$$\widehat{G}_f = \left\{ \begin{pmatrix} a_1 & 0 & 0 \\ a_2 & a_1 a_6^{-1} & 0 \\ 0 & 0 & a_6 \end{pmatrix} \;\middle|\; a_1 a_6 \neq 0 \right\}. \tag{12.56}$$

Thus, the lifted coframe is

$$\theta^1 = a_1\omega^1, \qquad \theta^2 = a_2\omega^1 + \frac{a_1}{a_6}\omega^2, \qquad \theta^3 = a_6\omega^3, \tag{12.57}$$

cf. (9.32). After absorption, the structure equations for (12.57) are

$$\begin{aligned} d\theta^1 &= \pi^1 \wedge \theta^1 - \theta^2 \wedge \theta^3, \\ d\theta^2 &= \pi^2 \wedge \theta^1 + (\pi^1 - \pi^6) \wedge \theta^2, \\ d\theta^3 &= \pi^6 \wedge \theta^3. \end{aligned} \tag{12.58}$$

Here

$$\pi^1 = \varpi^1 + r\theta^1, \qquad \pi^2 = \varpi^2 + s\theta^1 + r\theta^2, \qquad \pi^6 = \varpi^6, \tag{12.59}$$

are the modified Maurer–Cartan forms, as in (12.51), arising from the general solution to the linear absorption system. They depend on the two free variables $z_1^1 = z_2^2 = r$, $z_1^2 = s$, reflecting the degree of indeterminancy $r^{(1)} = 2$ of the coframe (12.57), and the additional one-forms

$$\begin{aligned} \varpi^1 &= \alpha^1 - \frac{a_2}{a_1}\theta^3, \qquad \varpi^6 = \alpha^6 - \frac{a_1 Q_p + 2a_2 a_6}{a_1 a_6}\,\theta^3, \\ \varpi^2 &= \alpha^2 + \frac{a_1^2 Q_u - a_1 a_2 a_6 Q_p - a_2^2 a_6^2}{a_1^2 a_6^2}\,\theta^3, \end{aligned} \tag{12.60}$$

where $\alpha^1, \alpha^2, \alpha^3$ are the Maurer–Cartan forms for the group $\widehat{G}_f$. The prolonged coframe thus consists of the original lifted coframe (12.57), now viewed as a collection of one-forms on the six-dimensional space $M^{(1)} = \mathrm{J}^1 \times \widehat{G}_f$ with coordinates x, u, p, a_1, a_2, a_6, together with the modified Maurer–Cartan forms (12.60). According to (12.53), the prolonged structure group is a two-dimensional abelian group having the 6×6 matrix representation

$$\widehat{G}^{(1)} = \left\{ \begin{pmatrix} \mathrm{I} & 0 \\ R & \mathrm{I} \end{pmatrix} \;\middle|\; R = \begin{pmatrix} r & 0 & 0 \\ s & r & 0 \\ 0 & 0 & 0 \end{pmatrix} \right\}.$$

The next step in our general procedure is to try to solve the new equivalence problem for the lifted coframe (12.57) and (12.59) on the six-dimensional space $M^{(1)}$. We compute the structure equations, which, after absorption, are

$$\begin{aligned} d\theta^1 &= \pi^1 \wedge \theta^1 - \theta^2 \wedge \theta^3, \\ d\theta^2 &= \pi^2 \wedge \theta^1 + (\pi^1 - \pi^6) \wedge \theta^2, \\ d\theta^3 &= \pi^6 \wedge \theta^3, \\ d\pi^1 &= \rho \wedge \theta^1 - \pi^2 \wedge \theta^3, \\ d\pi^2 &= \sigma \wedge \theta^1 + \rho \wedge \theta^2 - A\,\theta^2 \wedge \theta^3 + \pi^2 \wedge \pi^6, \\ d\pi^6 &= A\,\theta^1 \wedge \theta^3 + B\,\theta^2 \wedge \theta^3 - 2\,\pi^2 \wedge \theta^3. \end{aligned}$$

Note that the first three structure equations coincide with (12.58); indeed, this will automatically hold in general, and we do not, in fact, ever need to recompute this part of the structure equations. Consequently, essential torsion will only appear in the structure equations involving the new coframe elements π^α. The explicit formulae for the essential torsion coefficients are

$$A = 2s + \frac{a_2 a_6 Q_{pp} - a_1 Q_{pu}}{a_1^2 a_6}, \qquad B = 2r - \frac{Q_{pp}}{a_1}.$$

Both A and B can be translated to 0 by normalizing the two prolonged group parameters:

$$r = \frac{Q_{pp}}{2a_1}, \qquad s = \frac{a_1 Q_{pu} - a_2 a_6 Q_{pp}}{2a_1^2 a_6}.$$

Therefore, we immediately reduce the prolonged equivalence problem to an equivalence problem for the coframe (12.57), (12.59), and the structure equations are

$$\begin{aligned} d\theta^1 &= \pi^1 \wedge \theta^1 - \theta^2 \wedge \theta^3, \\ d\theta^2 &= \pi^2 \wedge \theta^1 + (\pi^1 - \pi^6) \wedge \theta^2, \\ d\theta^3 &= \pi^6 \wedge \theta^3, \\ d\pi^1 &= -J_1\, \theta^1 \wedge \theta^2 + J_2\, \theta^1 \wedge \theta^3 - \pi^2 \wedge \theta^3, \\ d\pi^2 &= J_3\, \theta^1 \wedge \theta^3 + J_2\, \theta^2 \wedge \theta^3 + \pi^2 \wedge \pi^6, \\ d\pi^6 &= -2\, \pi^2 \wedge \theta^3. \end{aligned} \tag{12.61}$$

The invariant structure functions for the prolonged coframe are

$$\begin{aligned} J_1 &= \frac{a_6}{2a_1^2}\, Q_{ppp}, \\ J_2 &= \frac{1}{2a_1 a_6}\, \big(Q_{pu} - \widehat{D}_x Q_{pp}\big), \\ J_3 &= \frac{1}{2a_1 a_6^2}\, \big(2Q_{uu} + Q_p Q_{pu} - \widehat{D}_x Q_{pu} - Q_u Q_{pp}\big) - \frac{a_2}{a_1} J_2, \end{aligned} \tag{12.62}$$

where

$$\widehat{D}_x = \frac{\partial}{\partial x} + p\,\frac{\partial}{\partial u} + Q(x,u,p)\,\frac{\partial}{\partial p} \tag{12.63}$$

is the "adapted total derivative", which agrees with the usual total derivative when restricted to solutions $u = f(x)$ of the differential equation (12.54).

Further reduction of the original group parameters depends on the precise values of the invariants J_1, J_2, J_3. If they all vanish, then there are no further reductions available. In this case, the prolonged coframe has rank zero, forming the Maurer–Cartan structure equations for the six-parameter fiber-preserving symmetry group of any associated differential equation. Moreover, according to Theorem 8.16, all such coframes are (locally) equivalent. Therefore, we have proved the following linearization theorem.

Theorem 12.14. *Let $u_{xx} = Q(x, u, u_x)$ be a second order ordinary differential equation. The following three conditions are equivalent:*

(*i*) *The equation admits a six-dimensional fiber-preserving symmetry group.*

(*ii*) *The equation can be mapped to* $u_{xx} = 0$ *by a fiber-preserving transformation.*

(*iii*) $Q(x, u, u_x) = \frac{1}{2} M_u u_x^2 + M_x u_x + N$, *where* M, N *are functions of* x, u, *and satisfy*

$$M_{xxu} + (N M_u)_u - M_x M_{xu} - 2 N_{uu} = 0. \tag{12.64}$$

The proof is immediate; condition (12.64) follows from the solution to the equations

$$Q_{ppp} = 0, \qquad \widehat{D}_x Q_{pp} = Q_{pu}, \qquad \widehat{D}_x Q_{pu} + Q_u Q_{pp} = Q_p Q_{pu} + 2 Q_{uu}, \tag{12.65}$$

implied by the vanishing of the invariants. Thus, only particular equations which are quadratic in the derivative variable are linearizable via fiber-preserving transformations

If the invariants are not all zero, then there are various possible reductions, but, in all cases, the equivalence problem reduces back to an invariant coframe on the base manifold $M \subset \mathrm{J}^1$. This implies that a second order ordinary differential equation admits a fiber-preserving symmetry group of dimension either 0, 1, 2, 3, or 6. These are exhaustively classified by Hsu and Kamran, [**117**], leading to a variety of interesting physical applications. The actual reductions depend on the values of the three lifted invariants J_1, J_2, J_3. In the generic case, $J_1 J_2 \neq 0$, so $Q_{ppp}(\widehat{D}_x Q_{pp} - Q_{pu}) \neq 0$, and we can eliminate the three group parameters a_1, a_2, a_6 by normalizing $J_1 = J_2 = 1$, $J_3 = 0$. Substituting these values into the original lifted coframe (12.57) results in an invariant coframe on J^1, with structure equations of the form

$$\begin{aligned} d\theta^1 &= (I_1 + I_2 - I_3)\,\theta^1 \wedge \theta^2 - I_1\,\theta^1 \wedge \theta^3 - \theta^2 \wedge \theta^3, \\ d\theta^2 &= -2 I_4\,\theta^1 \wedge \theta^2 + I_5\,\theta^1 \wedge \theta^3 + I_3\,\theta^2 \wedge \theta^3, \\ d\theta^3 &= I_4\,\theta^1 \wedge \theta^3 + I_2\,\theta^2 \wedge \theta^3, \end{aligned} \tag{12.66}$$

where the structure invariants $I_1, \ldots, I_5$ are complicated expressions in Q and its derivatives. In particular, the equation admits a three-dimensional symmetry group if and only if all five structure functions are constant. As a particular example, the equation $u_{xx} = \exp(u_x)$ lies in this branch, and has constant structure invariants $-I_1 = I_2 = I_3 = \sqrt[3]{2}$, $I_4 = I_5 = 0$; its three-dimensional symmetry group can be deduced using Lie's approach — see Table 7.

There are four additional branches to the equivalence problem — the full details can be found in [**117**]. As an example, let us consider the case when $J_1 = J_2 = 0$, which, in view of (12.62) implies that the right hand side of (12.54) is a quadratic function of the derivative variable of the particular form $Q = \frac{1}{2} R_u u_x^2 + R_x u_x + S$, where $R(x,u)$ and $S(x,u)$ depend only on the independent and dependent variable. In this case we can normalize the third invariant $J_3 \mapsto -1$ by the group normalization

$$a_1 = \frac{F}{2a_6^2} \qquad \text{where} \qquad F = \widehat{D}_x Q_{pu} + Q_u Q_{pp} - Q_p Q_{pu} - 2Q_{uu}.$$

The reduced structure equations are then of the form

$$\begin{aligned} d\theta^1 &= -2\pi^6 \wedge \theta^1 + K_1\, \theta^1 \wedge \theta^3 - \theta^2 \wedge \theta^3, \\ d\theta^2 &= \pi^2 \wedge \theta^1 - 3\pi^6 \wedge \theta^2 + K_2\, \theta^1 \wedge \theta^2 + K_1\, \theta^2 \wedge \theta^3, \\ d\theta^3 &= \pi^6 \wedge \theta^3, \\ d\pi^2 &= \pi^2 \wedge \pi^6 - \theta^1 \wedge \theta^3, \\ d\pi^6 &= -2\,\pi^2 \wedge \theta^3, \end{aligned}$$

involving two further essential torsion coefficients K_1, K_2. Let us restrict our attention to the branch where $K_2 = 0$, which occurs when $F(x,u) = f(x) \exp \frac{1}{2} R(x,u)$. We normalize $K_1 \mapsto 0$ using the parameter a_2, leading to

$$\begin{aligned} d\theta^1 &= -2\pi^6 \wedge \theta^1 - \theta^2 \wedge \theta^3, & d\theta^3 &= \pi^6 \wedge \theta^3, \\ d\theta^2 &= -3\pi^6 \wedge \theta^2 + K_3\, \theta^1 \wedge \theta^3, & d\pi^6 &= 0. \end{aligned}$$

Finally, since $K_3 \neq 0$ (Verify!) we can normalize $K_3 \mapsto \varepsilon = \pm 1$ up to sign. The final structure equations have the form

$$\begin{aligned} d\theta^1 &= 2I\, \theta^1 \wedge \theta^3 - \theta^2 \wedge \theta^3, \\ d\theta^2 &= -\tfrac{3}{2}\varepsilon\, \theta^1 \wedge \theta^2 + \varepsilon\, \theta^1 \wedge \theta^3 + 3I\, \theta^2 \wedge \theta^3, \end{aligned} \qquad d\theta^3 = \tfrac{1}{2}\varepsilon\, \theta^1 \wedge \theta^3,$$

so that, for this particular branch of the equivalence problem, there is a single basic structure invariant I.

Exercise 12.15. Prove that, in the generic case of this branch, the derived invariants I, $J = \partial I/\partial \theta^3$, and $K = \partial J/\partial \theta^3$, are functionally independent, and, moreover, the classifying manifold is completely determined by the classifying function $L = H(I, J, K)$ for the invariant $L = \partial K/\partial \theta^3$. Prove that the classifying function H is a polynomial if and only if the original differential equation is equivalent, under a fiber-preserving map, to the first Painlevé transcendent $u_{xx} = 6u^2 + x$, cf. [**116**]. Thus we have found an interesting intrinsic geometric characterization of the first Painlevé transcendent; see [**128**], [**135**] for details.

Exercise 12.16. Characterize all second order equations which are equivalent, via a fiber-preserving map, to one whose right hand side $Q = Q(u_x)$ depends only on the derivative coordinate. Discuss the possible fiber-preserving symmetry groups; in particular determine those equations whose symmetry group is three-dimensional.

Exercise 12.17. Use the Cartan method to demonstrate that the Duffing equation $u_{xx} + au_x - bu + u^3 = 0$ admits a one-parameter fiber-preserving symmetry group unless $b = -\frac{2}{9}a^2$, in which case the equation admits a second independent symmetry generator, [**117**]. Can you deduce this result using the Lie symmetry method? Compare the two methods.

Exercise 12.18. Set up and solve the equivalence problem for a second order ordinary differential equation under transformations $\bar{x} = x$, $\bar{u} = \psi(u)$ that only affect the dependent variable, cf. [**39**].

Turning to the point transformation equivalence problem, an application of the inductive procedure discussed in Chapter 10 might be of interest, but the complications and proliferation of branches already introduced by the prolongation procedure make this, at least to me, seem an unattractive route. Instead, we begin again from scratch. The base coframe is given in (12.55). The structure group has been reduced to

$$\widehat{G}_p = \left\{ \begin{pmatrix} a_1 & 0 & 0 \\ a_2 & a_1 a_6^{-1} & 0 \\ a_4 & 0 & a_6 \end{pmatrix} \middle| \quad a_1 a_6 \neq 0 \right\}, \tag{12.67}$$

with lifted one-forms

$$\theta^1 = a_1 \omega^1, \qquad \theta^2 = a_2 \omega^1 + \frac{a_1}{a_6}\omega^2, \qquad \theta^3 = a_4 \omega^1 + a_6 \omega^3. \tag{12.68}$$

After absorption, the structure equations are

$$\begin{aligned} d\theta^1 &= \pi^1 \wedge \theta^1 - \theta^2 \wedge \theta^3, \\ d\theta^2 &= \pi^2 \wedge \theta^1 + (\pi^1 - \pi^6) \wedge \theta^2, \\ d\theta^3 &= \pi^4 \wedge \theta^1 + \pi^6 \wedge \theta^3. \end{aligned} \tag{12.69}$$

The base coframe on $M^{(1)} = M \times \widehat{G}_p$ will consist of the forms (12.68) and the modified Maurer–Cartan forms

$$\begin{aligned}
\varpi^1 &= \alpha^1 + \frac{a_4}{a_1}\,\theta^2 - \frac{a_2}{a_1}\,\theta^3, \\
\varpi^2 &= \alpha^2 - \frac{a_1 a_4 Q_p + a_2 a_4 a_6}{a_1^2 a_6}\,\theta^2 + \frac{a_1^2 Q_u - a_1 a_2 a_6 Q_p - a_2^2 a_6^2}{a_1^2 a_6^2}\,\theta^3, \\
\varpi^4 &= \alpha^4 + \frac{a_4^2}{a_1^2}\,\theta^2 - \frac{a_2 a_4}{a_1^2}\,\theta^3, \\
\varpi^6 &= \alpha^6 - \frac{a_4}{a_1}\,\theta^2 - \frac{2a_2 a_6 + a_1 Q_p}{a_1 a_6}\,\theta^3,
\end{aligned} \tag{12.70}$$

found by solving the absorption equations and setting the free variables $z_1^1 = r_1$, $z_1^2 = r_2$, $z_1^6 = z_3^4 = r_3$, $z_1^4 = r_4$, to zero. The complete solution to the absorption system is given by

$$\begin{aligned}
\pi^1 &= \varpi^1 + r_1\theta^1, & \pi^2 &= \varpi^2 + r_2\theta^1 + (r_1 - r_3)\theta^2, \\
\pi^4 &= \varpi^4 + r_4\theta^1 + r_3\theta^3, & \pi^6 &= \varpi^6 + r_3\theta^1.
\end{aligned} \tag{12.71}$$

The prolonged structure group is a four-dimensional abelian group with 7×7 matrix representation

$$\widehat{G}^{(1)} = \left\{ \begin{pmatrix} \mathrm{I} & 0 \\ R & \mathrm{I} \end{pmatrix} \;\middle|\; R = \begin{pmatrix} r_1 & 0 & 0 \\ r_2 & r_1 - r_3 & 0 \\ r_4 & 0 & r_3 \\ r_3 & 0 & 0 \end{pmatrix} \right\}.$$

The structure equations for the lifted coframe (12.68), (12.71) consist of the original structure equations (12.69), and, after absorption, the equations

$$\begin{aligned}
d\pi^1 &= \rho^1 \wedge \theta^1 + \pi^4 \wedge \theta^2 - \pi^2 \wedge \theta^3, \\
d\pi^2 &= \rho^2 \wedge \theta^1 + (\rho^1 - \rho^3) \wedge \theta^2 + \pi^2 \wedge \pi^6, \\
d\pi^4 &= \rho^4 \wedge \theta^1 + \rho^3 \wedge \theta^3 + \pi^4 \wedge (\pi^1 - \pi^6), \\
d\pi^6 &= \rho^3 \wedge \theta^1 - \pi^4 \wedge \theta^2 - 2\,\pi^2 \wedge \theta^3 + T\,\theta^2 \wedge \theta^3.
\end{aligned}$$

The only essential torsion coefficient is

$$T = \frac{2r_1 - 4r_3 + (2a_4 Q_p - a_6 Q_{pp})}{a_1 a_6},$$

appearing in the final equation, and we normalize $T = 0$ by setting

$$r_3 = \frac{1}{2} r_1 + \frac{2a_4 Q_p - a_6 Q_{pp}}{4a_1 a_6}.$$

In the second loop through the reduction procedure, the last four structure equations become, after absorption,

$$\begin{aligned}
d\pi^1 &= \rho^1 \wedge \theta^1 + \pi^4 \wedge \theta^2 - \pi^2 \wedge \theta^3, \\
d\pi^2 &= \rho^2 \wedge \theta^1 + \tfrac{1}{2}\,\rho^1 \wedge \theta^2 + \pi^2 \wedge \pi^6 - U\,\theta^2 \wedge \theta^3, \\
d\pi^4 &= \rho^4 \wedge \theta^1 + \tfrac{1}{2}\,\rho^1 \wedge \theta^3 + \pi^4 \wedge (\pi^1 - \pi^6) - V\,\theta^2 \wedge \theta^3, \\
d\pi^6 &= \tfrac{1}{2}\,\rho^1 \wedge \theta^1 - \pi^4 \wedge \theta^2 - 2\pi^2 \wedge \theta^3 + V\,\theta^1 \wedge \theta^2 + U\,\theta^1 \wedge \theta^3.
\end{aligned}$$

The essential torsion coefficients are U, V, which are, respectively,

$$\begin{aligned}
&\tfrac{3}{2} r_2 + \frac{12 a_2^2 a_4 a_6^2 + 6 a_1^2 a_4 Q_u + 3 a_1 a_2 a_6^2 Q_{pp} + a_1^2 a_6 \widehat{D}_x Q_{pp} - 4 a_1^2 a_6 Q_{pu}}{4 a_1^3 a_6^2}, \\
&\tfrac{3}{2} r_4 + \frac{6 a_2 a_4^2 a_6 + 6 a_1 a_4^2 Q_p - 3 a_1 a_4 a_6 Q_{pp} + a_1 a_6^2 Q_{ppp}}{4 a_1^3 a_6^2}.
\end{aligned}$$

We translate both of these to 0 in the obvious manner. There is no more unabsorbable torsion left, so the final remaining group variable $r_1 = r$ cannot be normalized, and we must once again decide whether or not to prolong. However, here the system has degree of indeterminancy $r^{(2)} = 0$, and there is no need to compute the reduced characters (thank goodness!) — the problem is determinate, and we make a second prolongation merely by including the equation for $d\rho$, where $\rho = \rho_1$ is the one remaining Maurer–Cartan form for the group $G^{(1)}$. We deduce the final structure equations

$$\begin{aligned}
d\theta^1 &= \pi^1 \wedge \theta^1 - \theta^2 \wedge \theta^3, \\
d\theta^2 &= \pi^2 \wedge \theta^1 + (\pi^1 - \pi^6) \wedge \theta^2, \\
d\theta^3 &= \pi^4 \wedge \theta^1 + \pi^6 \wedge \theta^3. \\
d\pi^1 &= \rho \wedge \theta^1 + \pi^4 \wedge \theta^2 - \pi^2 \wedge \theta^3, \\
d\pi^2 &= \tfrac{1}{2}\,\rho \wedge \theta^2 + \pi^2 \wedge \pi^6 + K_2\,\theta^1 \wedge \theta^2, \\
d\pi^4 &= \tfrac{1}{2}\,\rho \wedge \theta^3 + \pi^4 \wedge (\pi^1 - \pi^6) + K_1\,\theta^1 \wedge \theta^3, \\
d\pi^6 &= \tfrac{1}{2}\,\rho \wedge \theta^1 - \pi^4 \wedge \theta^2 - 2\,\pi^2 \wedge \theta^3, \\
d\rho &= \rho \wedge \pi^1 - 2\pi^2 \wedge \pi^4 - K_3\,\theta^1 \wedge \theta^2 - K_4\,\theta^1 \wedge \theta^3,
\end{aligned} \tag{12.72}$$

which form an invariant coframe on the eight-dimensional space $M^{(2)} = \mathrm{J}^1 \times \widehat{G}_p \times \widehat{G}^{(1)}$, where $\widehat{G}^{(1)} \subset G^{(1)}$ is the subgroup given by setting $r_1 = 2r_3 = r$, $r_2 = r_4 = 0$. (The fact that this final space is eight-dimensional comes as no surprise, owing to our familiarity with the possible symmetry groups of a second order ordinary differential equation!) The structure functions for the prolonged coframe are

$$\begin{aligned}
K_1 &= \frac{a_6^2}{6a_1^3} Q_{pppp}, \\
K_2 &= \frac{\widehat{D}_x^2 Q_{pp} - 4\widehat{D}_x Q_{up} - Q_p \widehat{D}_x Q_{pp} 6Q_{uu} - 3Q_u Q_{pp} + 4Q_p Q_{up}}{6a_1 a_6^2}, \\
K_3 &= \frac{a_6}{6a_1^4}\left[a_1 \widehat{D}_x Q_{pppp} - (a_2 a_6 + 2a_1 Q_p) Q_{pppp}\right], \qquad (12.73) \\
K_4 &= \left(\frac{\partial}{\partial p} - \frac{a_4}{a_6}\right) K_2,
\end{aligned}$$

where the modified total derivative operator $\widehat{D}_x$ is as in (12.63).

Further reduction of the original group parameters depends on the values of these invariants. If K_1 and K_2 vanish, then K_3 and K_4 also automatically vanish, hence there is no further reduction, and we have produced an invariant coframe on the eight-dimensional space coordinatized by $x, u, p, a_1, a_2, a_4, a_6, r$. Any such differential equation admits an eight-dimensional symmetry group of point transformations, and, moreover, all such equations are equivalent.

Theorem 12.19. *The following three conditions for a second order ordinary differential equation $u_{xx} = Q(x, u, u_x)$ are equivalent:*

(*i*) *The equation admits an eight dimensional symmetry group of point transformations, which is necessarily isomorphic to* SL(3).

(*ii*) *The equation is equivalent to $u_{xx} = 0$ under a point transformation.*

(*iii*) $Q(x, u, u_x) = Ku_x^3 + 3Lu_x^2 + 3Mu_x + N$, *where K, L, M, N are functions of x, u and satisfy*

$$\begin{aligned}
K_{xx} - 2L_{xu} + M_{uu} - NK_u - 2KN_u + 3(KM)_x + 3LM_u - 6LL_x &= 0, \\
L_{xx} - 2M_{xu} + N_{uu} + KN_x + 2NK_x - 3(LN)_u - 3ML_x + 6MM_u &= 0.
\end{aligned} \qquad (12.74)$$

The proof is immediate; the vanishing of K_1 implies that Q is cubic in p, and the vanishing of K_2 provides the final two conditions (12.74). This particular theorem predates the work of Cartan, and is originally due to Lie, [**150**; part III, §1]; see also Tresse, [**215**], and the more recent references [**13**; §1.6], [**23**], [**92**], [**104**], and [**121**].

Example 12.20. As a specific example, consider the equation $u_{xx} = 3uu_x - u^3$. This equation satisfies the hypotheses of Theorem 12.19 and is hence equivalent to the equation $u_{xx} = 0$ under a point transformation. In fact, motivated by the explicit form of the symmetry algebra (see Proposition 6.2), one finds that the point transformation $x = y + 1/v$, $u = -(y/v) - \frac{1}{2}y^2$ maps the equation $u_{xx} = 0$ to $v_{yy} = 3vv_y - v^3$. On the other hand, the fiber-preserving invariants (12.62) for this equation are not all zero, and hence the equation cannot be linearized by any fiber-preserving transformation.

If the structure invariants (12.73) are not all zero, then, as with the fiber-preserving equivalence problem, there are various possible reductions, leading to different branches of the equivalence problem. The complete classification of equivalence reductions has not, so far as I know, been carried through in full detail.

Exercise 12.21. Prove that the equation $u_{xx} = 3uu_x - u^3$ admits a three-dimensional symmetry group of fiber-preserving transformations by completing the reduction of the fiber-preserving invariant coframe on $\mathrm{J}^1 \times G$ back to J^1. (See [**117**].)

Exercise 12.22. Prove that any linear second order ordinary differential equation is equivalent to the equation $u_{xx} = 0$. Discuss how to find the equivalence map.

Exercise 12.23. Prove that a second order equation of the form $u_{xx} = Q(x, u)$ is linearizable if and only if it is already linear.

Exercise 12.24. Discuss the equivalence problem for the first Painlevé transcendent $u_{xx} = 6u^2 + x$ under point transformations; see [**127**] for details.

In view of the complications inherent in the Cartan equivalence method, only the case of a scalar second order ordinary differential equation has been relatively thoroughly analyzed. However, there are a number of partial results, including some complete intrinsic solutions, for higher order equations, partial differential equations, and systems of differential equations. See Chern, [**46**], [**47**], and Fels, [**69**], [**70**], for results on the equivalence problem for systems of second and third order ordinary differential equations. Applications to second order partial differential equations appear in [**38**], [**134**], [**127**], [**69**], and [**173**]; see also [**8**], [**29**], [**30**] for recent applications to their conservation laws.

Finally, I should remark on other approaches to the solution of equivalence problems that can be viewed as competitors to the Cartan approach developed here. We have already remarked on Lie's approach, which relies on symmetry classification, and, although usually more effective when available, has limited use in higher dimensional problems. The reader might profitably compare the amount of effort needed to implement the Cartan equivalence method with the analysis of second order ordinary differential equations based on Lie's classification of transformation groups in the plane as was discussed in Chapter 6. The Cartan approach has the advantage of providing completely general equivalence results, and has a much wider range of applicability. On the other hand, the computations required are considerably more complicated than those based on a more direct analysis using Lie's methods (although our presentation of the Lie classification did omit the considerable work needed to establish it). Direct methods, beginning with the work of Riemann, [**197**], Christoffel, [**52**], and Tresse, [**215**], have already been referred to in the text; a recent preprint [**23**] shows how they can be effectively used to circumvent many of the complications in the Cartan approach. Their primary disadvantage is that they are not nearly as systematic, and are only practical when the particular form of the equation or geometric object is rather stringently specified. Vessiot, [**219**], proposes a "dual" method based on the theory of vector field systems instead of differential forms; Vessiot's approach was recently used by Vassiliou, [**218**], to analyze equivalence of wave equations. A symbolic method, originated by Maschke, [**167**], and based on the symbolic approach to classical invariant theory, is considerably less well known, but worth updating. Unfortunately, space limitations preclude any more extensive discussions and comparisons, which must be left to the motivated reader to pursue. In summary, the complexity of equivalence problems indicates that no one tool is a cure-all, but rather a variety of interconnected and complementary techniques should be utilized in order to make real progress.

Chapter 13

Differential Systems

The results that form the theoretical foundation of the Cartan equivalence method all ultimately rest on the existence of solutions to certain systems of partial differential equations which are defined by the vanishing of a collection of differential forms. The final part of the book is devoted to a fairly detailed exposition of the necessary existence theory for such systems of partial differential equations. The two fundamental results are the well-known Frobenius Theorem, to be covered in Chapter 14, and the more complicated Cartan–Kähler Theorem, which is discussed in Chapter 15. To properly formulate these two existence theorems, we first need to present a basic summary of the theory of differential systems, and this forms the topic of the present preliminary chapter. More extensive treatments of these matters can be found, for instance, in [**28**], [**36**], and [**230**]. We begin our discussion by formalizing the basic terminology.

Differential Systems and Ideals

In general, by a *differential system* we will mean a collection of differential forms $\{\omega^1, \omega^2, \ldots\}$ defined on an m-dimensional manifold M. A submanifold $N \subset M$ is called an *integral submanifold* if it annihilates all the forms in the differential system, meaning that each form ω^i vanishes when restricted (pulled back) to the submanifold: $\omega^i | N = 0$. Given a differential system, our principal goal is to find integral submanifolds $N \subset M$ of a prescribed dimension n. Now, if N is an integral submanifold, and η is any differential form, not necessarily in the system, then its wedge product $\eta \wedge \omega^i$ with a form in the system also vanishes on N, so that we may as well include all such differential forms in our differential system. This implies that we should work with, not just the generating forms ω^i, but rather the entire "ideal" which they generate in the space of all differential forms.

Definition 13.1. An *exterior ideal* is a collection $\mathcal{I}$ of differential forms defined on a manifold M having the properties *a*) if ω and ϖ are k-forms in $\mathcal{I}$, so is $\omega + \varpi \in \mathcal{I}$, and *b*) if $\omega \in \mathcal{I}$ and η is *any* differential form, then $\eta \wedge \omega \in \mathcal{I}$. In particular, multiplying a k-form $\omega \in \mathcal{I}$ by any smooth function f gives a k-form $f\omega \in \mathcal{I}$.

Definition 13.2. A submanifold $N \subset M$ is an *integral submanifold* of the differential system determined by an exterior ideal $\mathcal{I}$ if and only if $\mathcal{I}$ vanishes on N, i.e., for any $\omega \in \mathcal{I}$, we have $\omega | N = 0$.

Proposition 13.3. *If $N \subset M$ is an integral submanifold of a differential system, and $\widetilde{N} \subset N$ is any submanifold of N, then $\widetilde{N}$ is also an integral submanifold.*

Given an exterior ideal $\mathcal{I}$ on an m-dimensional manifold M, we let $\mathcal{I}^{(k)} = \{\omega \in \mathcal{I} \mid \deg \omega = k\}$, $0 \leq k \leq m$, denote its homogeneous component consisting of all k-forms in $\mathcal{I}$. Thus $\mathcal{I}^{(0)}$ will denote the space of all functions (0-forms) in $\mathcal{I}$, while $\mathcal{I}^{(1)}$ will denote the space of all one-forms, etc. Since differential forms are always of homogeneous degree, the ideal is the sum of its homogeneous components: $\mathcal{I} = \bigoplus_{k=0}^{m} \mathcal{I}^{(k)}$.

Lemma 13.4. *An n-dimensional submanifold N is an integral submanifold of the differential system determined by an exterior ideal $\mathcal{I}$ if and only if the n-forms in $\mathcal{I}$ vanish on N, i.e., $\Omega | N = 0$ for each $\Omega \in \mathcal{I}^{(n)}$.*

Proof: Let $\omega \in \mathcal{I}^{(k)}$ be any k-form in the ideal. If $k < n$, then the wedge product $\eta \wedge \omega$ of ω with *any* $(n-k)$-form η will produce an n-form $\eta \wedge \omega$ in $\mathcal{I}^{(n)}$ which must vanish on N. Since this holds for all $(n-k)$-forms η, we conclude that ω itself necessarily vanishes on N. On the other hand, if $k > n$, then ω automatically vanishes on N. *Q.E.D.*

A set of differential forms $\{\omega^1, \omega^2, \ldots\}$ is said to *generate* the ideal $\mathcal{I}$ if every form $\theta \in \mathcal{I}$ can be written as a finite "linear combination" $\theta = \sum_j \eta^j \wedge \omega^j$, where the η^j are arbitrary differential forms satisfying $\deg \theta = \deg \eta^j + \deg \omega^j$. Clearly, a submanifold N is an integral submanifold for $\mathcal{I}$ if and only if each generator vanishes on N. In most applications the ideal is *finitely generated*; although the assumption is not necessary for the validity of the main theorems, it does serve to simplify their proofs and practical implementations.

Example 13.5. Consider the differential system on $M = \mathbb{R}^3$ generated by a single one-form $\omega = a(x,y,z)\,dx + b(x,y,z)\,dy + c(x,y,z)\,dz$.

The associated exterior ideal consists of all multiples $\xi \wedge \omega$, where ξ is an arbitrary differential form (or function). A one-dimensional submanifold (curve), parametrized by $x = \phi(t)$, $y = \psi(t)$, $z = \chi(t)$, will be an integral submanifold if and only if its tangent vector is annihilated by ω, and so satisfies the single underdetermined ordinary differential equation $a\phi_t + b\psi_t + c\chi_t = 0$. Thus, assuming $\omega \neq 0$, one can find many integral curves passing through a given point.

Consider next a surface, which, for simplicity, we assume is given by the graph Γ_φ of a function $z = \varphi(x, y)$. The graph will be an integral submanifold if and only if φ satisfies the overdetermined, quasi-linear system of first order partial differential equations

$$c(x, y, \varphi(x, y)) \frac{\partial \varphi}{\partial x} = a(x, y, \varphi(x, y)),$$
$$c(x, y, \varphi(x, y)) \frac{\partial \varphi}{\partial y} = b(x, y, \varphi(x, y)).$$

If $c \neq 0$, there is a nontrivial integrability condition between this pair of differential equations, namely $(a/c)_y = (b/c)_x$, which, if not satisfied, precludes the existence of a solution $\varphi(x, y)$. This means that not every one-form on $\mathbb{R}^3$ possesses an integral surface. Note that if the one-form does have an integral surface S, then any curve $C \subset S$ is an integral curve, in accordance with Proposition 13.3.

Exercise 13.6. Let $\omega = x\,dx + y\,dy + z\,dz$. Given a point $(x, y, z) \neq (0, 0, 0)$, find all integral curves and integral surfaces passing through it. Prove that ω has no integral curves or surfaces passing through the origin.

Exercise 13.7. Let ω be a nonvanishing one-form on $M \subset \mathbb{R}^3$. Prove the equivalence of the following three conditions:

(*i*) ω has an integral surface passing through each point $(x, y, z) \in M$.

(*ii*) ω has Darboux rank at most two. (See Definition 1.41.)

(*iii*) ω has a (local) integrating factor, so $f\omega = dh$ for functions $f \neq 0$ and h.

Exercise 13.8. Suppose a surface S is the union of integral curves of a differential system $\mathcal{I}$. Is S an integral surface? Explain.

Roughly speaking, a differential system will be called *integrable* if we can find integral submanifolds of a prescribed dimension passing through each point. The preceding example shows that there are important restrictions on the differential system in order that it be integrable. First, if the ideal contains functions, i.e., $\mathcal{I}^{(0)} \neq \varnothing$, then they are required to vanish on integral submanifolds. Therefore, any integral submanifold must be a subset of the variety

$$\mathcal{S}_{\mathcal{I}} = \left\{ x \;\middle|\; F(x) = 0 \quad \text{for all} \quad F \in \mathcal{I}^{(0)} \right\}. \tag{13.1}$$

In all our applications, the variety $\mathcal{S}_{\mathcal{I}}$ is, in fact, a submanifold of M, which, in particular, holds if $\mathcal{I}^{(0)}$ is a regular family of functions, as per Definition 1.8. Thus, if $\mathcal{I}$ has an n-dimensional integral submanifold, the variety $\mathcal{S}_{\mathcal{I}}$ must have (local) dimension at least n, which, in the regular case, requires that $\mathcal{I}^{(0)}$ have rank at most $m-n$. In this case, an integral submanifold N of the original differential system must be a submanifold of $\mathcal{S}_{\mathcal{I}}$, and hence an integral submanifold of the differential system $\widehat{\mathcal{I}} = \mathcal{I} \,|\, \mathcal{S}_{\mathcal{I}}$ obtained by restricting the forms in $\mathcal{I}$ to the submanifold. In view of (13.1), the restricted differential ideal $\widehat{\mathcal{I}}$ contains no (non-zero) functions in it, so $\widehat{\mathcal{I}}^{(0)} = \{0\}$. Therefore, in all the cases we will be concerned with, we can always assume, without loss of generality, that the ideal does not contain any functions, $\mathcal{I}^{(0)} = \{0\}$, and so the variety $\mathcal{S}_{\mathcal{I}}$ is the entire manifold M.

A second set of necessary conditions for the existence of n-dimensional integral submanifolds is found by working infinitesimally. Suppose N is an n-dimensional integral submanifold of the ideal $\mathcal{I}$. Given $x \in N$, the tangent space $TN|_x$ is an n-dimensional subspace of the tangent space $TM|_x$. The differential forms in $\mathcal{I}$ vanish on N if and only if, when regarded as multi-linear maps on $TM|_x$, they annihilate the subspace $TN|_x$, i.e., $\langle \Omega\,; \mathbf{v}_1, \ldots, \mathbf{v}_k \rangle = 0$ whenever $\Omega \in \mathcal{I}^{(k)}$, $\mathbf{v}_i \in TN|_x$. This yields an immediate necessary condition that a subspace of the tangent space $TM|_x$ must satisfy in order than it be a possible candidate for the tangent space to an integral submanifold of the differential system.

Definition 13.9. An n-dimensional subspace $E \subset TM|_x$ is called an *integral element* of the differential system $\mathcal{I}$ if all the differential forms in $\mathcal{I}$ annihilate E, i.e., for each $k \geq 1$,

$$\langle \Omega\,; \mathbf{v}_1, \ldots, \mathbf{v}_k \rangle = 0, \quad \text{for every} \quad \Omega \in \mathcal{I}^{(k)}, \quad \mathbf{v}_1, \ldots, \mathbf{v}_k \in E. \tag{13.2}$$

Proposition 13.10. *A submanifold N is an integral submanifold of the ideal $\mathcal{I}$ if and only if its tangent space $TN|_x$ at each point $x \in N$ is an integral element of $\mathcal{I}$.*

As with integral submanifolds, for an n-dimensional subspace $E \subset TM|_x$ to be an integral element of an ideal $\mathcal{I}$, we need only check that the n-forms in $\mathcal{I}^{(n)}$ vanish on it, i.e., (13.2) holds for $k = n$; see Lemma 13.4. Alternatively, if we know a set of generators for $\mathcal{I}$, we need only check that E is annihilated by the generators of degree at most n.

The integration problem for a differential system defined by an ideal of differential forms $\mathcal{I}$ is to determine whether, given an n-dimensional integral element $E \subset TM|_x$, there exists an n-dimensional integral submanifold N passing through x whose tangent space coincides with the given integral element: $TN|_x = E$. If this is true for every n-dimensional integral element, then we will (provisionally) call the differential system *n-integrable.* In practice, though, only sufficiently regular integral elements will produce integral submanifolds, and so this condition should, and will, be appropriately relaxed. The required integral submanifold satisfies a system of partial differential equations governed by the differential system, and so the integral element can be viewed as "initial conditions" that the solution must satisfy. However, note that these initial conditions are *not* the usual Cauchy conditions for systems of partial differential equations, but are in the form of local solvability conditions, as described in Definition 6.3.

Dimensional considerations come into play to provide elementary necessary conditions for the existence of integral submanifolds. Since each one-form in the ideal $\mathcal{I}$ must vanish on the tangent space $TN|_x$ to an integral submanifold, the subspace of the cotangent space $T^*M|_x$ spanned by the one-forms in $\mathcal{I}$ can have dimension at most $m - n$. This dimension will be called the "rank" of the ideal.

Definition 13.11. Let $\mathcal{I}$ be an exterior ideal. The *rank* $r = r(x)$ of $\mathcal{I}$ at a point x is defined as the dimension of the subspace of $T^*M|_x$ spanned by the one-forms in $\mathcal{I}^{(1)}$.

Proposition 13.12. *If the ideal $\mathcal{I}$ has an n-dimensional integral submanifold passing through x, then its rank r at x must satisfy the inequality $r \leq m - n$.*

So far, we have just considered elementary algebraic conditions on our differential system that are necessary for the existence of integral submanifolds. Since we are working with differential forms, additional necessary conditions for integrability of a differential system arise by differentiation of the forms. Recall that one of the basic properties of the differential d is that if a differential form ω vanishes on a submanifold N, so does its exterior derivative $d\omega$; see Corollary 1.37. Thus, if N is an integral submanifold of the differential system defined by the ideal $\mathcal{I}$, and ω is any form in $\mathcal{I}$, then $d\omega$ also vanishes on N, and so should also be included in the ideal $\mathcal{I}$. In other words, the ideal should be "closed" under the exterior derivative operation.

Definition 13.13. An exterior ideal is called *closed* if whenever $\omega \in \mathcal{I}$, then $d\omega \in \mathcal{I}$ also. A *differential ideal* is, by definition, a closed exterior ideal.

Thus, if N is an integral submanifold of an exterior ideal $\mathcal{I}$, then it is also an integral submanifold of its *closure* $\widehat{\mathcal{I}}$, which is the ideal generated by all the forms in $\mathcal{I}$ and their exterior derivatives. As far as integral submanifolds are concerned, then, there is no loss in generality in assuming that we are dealing with a (closed) differential ideal.

Proposition 13.14. *If $\mathcal{I}$ is finitely generated by $\omega^1, \ldots, \omega^r$, its closure $\widehat{\mathcal{I}}$ is also finitely generated, by $\omega^1, \ldots, \omega^r$ and $d\omega^1, \ldots, d\omega^r$. In particular, $\mathcal{I}$ is closed if and only if it contains the exterior derivative, $d\omega^i \in \mathcal{I}$, of each generator ω^i, $i = 1, \ldots, r$.*

In all the applications that are considered in this book, we are interested in the integral submanifolds of a differential system generated by a collection of one-forms $\omega^1, \ldots, \omega^r$. The analysis of such differential systems was pioneered by Pfaff, [**194**], who studied the integrability of a differential system generated by a single non-exact one-form; see [**111**] for a detailed account of the history, and [**58**] for the application of the Darboux canonical form (1.28) to the solution to Pfaff's problem. Differential systems generated by a system of one-forms are known as *Pfaffian systems*; the associated exterior ideals will be distinguished by the following definition.

Definition 13.15. An exterior ideal is called *simply generated* if it is generated by a collection of one-forms.

If an exterior ideal $\mathcal{I}$ is simply generated by a finite collection of one-forms $\omega^1, \ldots, \omega^r$, then every differential form $\Omega \in \mathcal{I}^{(n)}$ can be expressed as a linear combination of the generators,

$$\Omega = \sum_{j=1}^{r} \eta^j \wedge \omega^j, \tag{13.3}$$

for (arbitrary) $(n-1)$-forms $\eta^1, \ldots, \eta^r$. The integral submanifolds of $\mathcal{I}$ must satisfy the Pfaffian system $\omega^i \mid N = 0$ prescribed by the vanishing of the generating one-forms (or Pfaffian forms). The ideal $\mathcal{I}$ is closed if and only if the differentials of the generating one-forms belong to the ideal, and hence can be written as

$$d\omega^i = \sum_{j=1}^{r} \zeta^i_j \wedge \omega^j, \qquad i = 1, \ldots, r, \tag{13.4}$$

for suitable one-forms ζ^i_j, $i, j = 1, \ldots, r$. In general, the integrability of exterior ideals depends crucially on whether or not they are closed. In the regular case, if an ideal is simply generated and closed, its integrability is guaranteed by the Frobenius Theorem. The case when the ideal $\mathcal{I}$ is not closed, meaning that some of the two-forms $d\omega^i$ are *not* in the exterior ideal generated by the one-forms is considerably more complicated, and will require the full force of the Cartan–Kähler Theorem.

Equivalence of Differential Systems

The basic equivalence problems for differential forms can all be subsumed under the general equivalence problem for differential systems. Given two exterior ideals $\mathcal{I}$ on M and $\overline{\mathcal{I}}$ on $\overline{M}$, both manifolds of the same dimension, when does there exist a local diffeomorphism $\Phi: M \to \overline{M}$ such that $\Phi^* \overline{\mathcal{I}} = \mathcal{I}$? A basic set of invariants is provided by the ranks of the various homogeneous components of the two ideals. Moreover, since the differential is unaffected by smooth maps, Φ must also preserve the closures of the two ideals. A slightly more subtle invariant is the following:

Definition 13.16. Given an exterior ideal $\mathcal{I}$, its *derived ideal* is defined as

$$\delta\mathcal{I} = \{\Omega \in \mathcal{I} \mid d\Omega \in \mathcal{I}\}. \tag{13.5}$$

In particular, $\mathcal{I}$ is closed if and only if $\delta\mathcal{I} = \mathcal{I}$. Higher order derived ideals are defined in the obvious way: $\delta^k\mathcal{I} = \delta(\delta^{k-1}\mathcal{I})$.

Proposition 13.17. *If $\Phi^* \overline{\mathcal{I}} = \mathcal{I}$ is an equivalence between exterior ideals, then it also defines an equivalence between all their derived ideals, $\Phi^*[\delta^k\overline{\mathcal{I}}] = \delta^k\mathcal{I}$, $k = 0, 1, 2, \ldots$.*

Proposition 13.17 can be used to further simplify our equivalence proof of Bäcklund's Theorem 4.32, that every contact transformation is the prolongation of a first order contact transformation; see Chapter 11. Let $\mathcal{I}_n$ denote the nth order *contact ideal*, consisting of all contact forms on the nth jet space J^n. For simplicity, let us restrict our attention to the scalar case, so that $\mathcal{I}_n$ is spanned by the basic contact forms $\theta_i = du_i - u_{i+1}\,dx$, $i = 0, \ldots, n-1$, where $u_i = D_x^i u$. By definition, a contact transformation $\Psi\colon \mathrm{J}^n \to \mathrm{J}^n$ defines a symmetry (self-equivalence) of the contact ideal, meaning $\Psi^* \mathcal{I}_n = \mathcal{I}_n$. The key observation is that the derived contact ideal can be identified with the $(n-1)^{\text{st}}$ order contact ideal: $\delta \mathcal{I}_n = \mathcal{I}_{n-1}$. This is because $d\theta_i = \theta_{i+1} \wedge dx$, hence $d\theta_i \in \mathcal{I}_n$ if and only if $i \leq n-2$. Therefore, $\delta^k \mathcal{I}_n = \mathcal{I}_{n-k}$, and so, according to Proposition 13.17, each lower order contact ideal must also be preserved under the contact transformation Ψ. In particular, Ψ must map the zeroth order contact form $du - u_x\,dx$ to a multiple of itself, and hence determine a first order contact transformation. The remaining details of the proof are similar to those in Chapter 11. Extensions of this argument to several dimensions are left to the reader.

Remark: Suppose M has odd dimension $m = 2p + 1$, and let $\mathcal{I}$ be a differential system generated by a single one-form ω. If the one-forms have maximal rank m, then Darboux' Theorem 1.42 shows that we can locally identify M with an open subset of the first jet space $\mathrm{J}^1 E$, where $E \simeq \mathbb{R}^p \times \mathbb{R}$, so that ω is identified with the standard contact form. In this way, the symmetries of the differential system are locally isomorphic to contact transformations.

Vector Field Systems

The dual objects to differential forms are vector fields. The integration of a differential system therefore translates into the "dual problem" of integrating a system of vector fields. In the case when the differential ideal is simply generated by one-forms, this duality is precise, and the theorem of Frobenius will imply the integrability of both the differential system and its dual vector field system.

In general, by a *vector field system* we mean a set $\mathcal{V}$ of vector fields on a manifold M which forms a linear space under the operations of addition and multiplication by smooth functions. Therefore, we require that if $\mathbf{v}, \mathbf{w} \in \mathcal{V}$, and $f, h \in C^\infty(M)$, then $f\mathbf{v} + h\mathbf{w} \in \mathcal{V}$. In most applications, $\mathcal{V}$ is finitely generated by vector fields $\mathbf{v}_1, \ldots, \mathbf{v}_r$, and so consists of all linear combinations $\sum_i f_i \mathbf{v}_i$ with arbitrary smooth functions f_i for coefficients. We define $\mathcal{V}|_x$ to be the subspace of $TM|_x$ spanned by the tangent vectors $\mathbf{v}|_x$ for all $\mathbf{v} \in \mathcal{V}$; in the finitely generated case $\mathcal{V}|_x = \operatorname{Span}\{\mathbf{v}_1|_x, \ldots, \mathbf{v}_r|_x\}$.

Definition 13.18. A submanifold $N \subset M$ is called an *integral submanifold* of the vector field system $\mathcal{V}$ if and only if its tangent space $TN|_x$ is contained in the subspace $\mathcal{V}|_x$ for each $x \in N$.

The *rank* of the vector field system at a point $x \in M$ is, by definition, the dimension of the subspace $\mathcal{V}|_x$. The dimension of any integral submanifold passing through x, then, is bounded by the rank of the system at x. In the standard approach, one looks exclusively for integral submanifolds of maximal dimension, meaning ones whose dimension equals the rank of a vector field system at each of its points. However, we have chosen to keep the more general definition so as to correspond more closely to the differential form case. Often the rank of the vector field system is assumed to be constant, and so all the maximal integral submanifolds have the same dimension, but the general vector field version of Frobenius' Theorem does not require this extra hypothesis.

Definition 13.19. A system of vector fields $\mathcal{V}$ is called *integrable* if through every point $x \in M$ there passes an integral submanifold of dimension $n = \dim \mathcal{V}|_x$.

Note that if a vector field system $\mathcal{V}$ is integrable, and $\mathbf{v}$ is a vector field having the property that it is tangent to every n-dimensional integral submanifold, then $\mathbf{v}$ necessarily belongs to $\mathcal{V}$. Consequently, an immediate necessary condition for the integrability of a vector field system is provided by the fact that the Lie bracket of any two vector fields tangent to a submanifold is also tangent to the submanifold, cf. Proposition 1.29. Thus, if $\mathbf{v}, \mathbf{w}$ are any two vector fields in the system, their Lie bracket $[\mathbf{v}, \mathbf{w}]$ must be tangent to each integral submanifold, and hence belong to the system. With this in mind, we make the following definition.

Definition 13.20. A system of vector fields $\mathcal{V}$ is *involutive* if, whenever $\mathbf{v}, \mathbf{w} \in \mathcal{V}$ are any two vector fields in $\mathcal{V}$, their Lie bracket $[\mathbf{v}, \mathbf{w}]$ also belongs to $\mathcal{V}$.

In the case that $\mathcal{V}$ is finitely generated, the basic properties (1.9) of the Lie bracket imply that we need only check the involutivity condition on a basis for the system. Therefore, a vector field system generated by $\mathbf{v}_1, \ldots, \mathbf{v}_r$ is involutive if and only if there exist smooth functions $a_{ij}^k(x)$, $i, j, k = 1, \ldots r$, such that

$$[\mathbf{v}_i, \mathbf{v}_j] = \sum_{k=1}^{r} a_{ij}^k \, \mathbf{v}_k, \qquad i, j = 1, \ldots, r. \tag{13.6}$$

Note that the vector fields $\mathbf{v}_i$ generating the system need *not* span a Lie algebra of vector fields since we are not necessarily requiring that the coefficients a_{ij}^k in (13.6) be constant.

Frobenius' Theorem for vector field systems states that a regular system is integrable if and only if it is involutive. The precise statement and proof will be given at the beginning of the following chapter. A simple example, though, is provided by a vector field system generated by a single vector field. Such a system is automatically involutive. On the other hand, the basic existence theorem for systems of ordinary differential equations provides the required integral curves (and equilibrium points) of the vector field. Therefore the system is also automatically integrable. Thus, for a vector field system generated by a single vector field, Frobenius' Theorem reduces to the usual existence theorem for systems of ordinary differential equations.

There is a natural duality between vector field systems and simply generated ideals. Given a system of vector fields $\mathcal{V}$, we let $\mathcal{I}^{(1)}$ denote the dual space of differential one-forms which vanish on the system, i.e., $\omega \in \mathcal{I}^{(1)}$ if and only if $\langle \omega \, ; \mathbf{v} \rangle = 0$ for every vector field $\mathbf{v} \in \mathcal{V}$. Let $\mathcal{I}$ denote the exterior ideal which is (simply) generated by these one-forms; note that $\mathcal{I}$ is the space of differential forms which vanish on the system $\mathcal{V}$, i.e., $\Omega \in \mathcal{I}^{(k)}$ if and only if $\langle \Omega \, ; \mathbf{v}_1, \ldots, \mathbf{v}_k \rangle = 0$ for any $\mathbf{v}_1, \ldots, \mathbf{v}_k \in \mathcal{V}$. If $\mathcal{V}$ is spanned by n vector fields $\mathbf{v}_1, \ldots, \mathbf{v}_n$ which are linearly independent at each point (so that the vector field system has constant rank n) then $\mathcal{I}^{(1)}$ will be generated by $m - n$ one-forms $\omega^1, \ldots, \omega^{m-n}$, which are also linearly independent at each point. In particular, the rank of the ideal $\mathcal{I}$ is $m - n$, the "co-rank" of the dual vector field system. Conversely, if $\mathcal{I}$ is a simply generated exterior ideal, then the dual vector field system $\mathcal{V}$ is defined as the set of all vector fields annihilated by the one-forms in $\mathcal{I}$, i.e., $\mathbf{v} \in \mathcal{V}$ if and only if $\langle \omega \, ; \mathbf{v} \rangle = 0$ for all $\omega \in \mathcal{I}^{(1)}$. In the constant rank case, then, these two concepts are naturally dual, and the notions of integral submanifold for the vector field system $\mathcal{V}$ and the dual exterior ideal $\mathcal{I}$ coincide.

Proposition 13.21. *Let $\mathcal{V}$ be a vector field system of constant rank n and $\mathcal{I}$ its simply generated dual exterior ideal of rank $m-n$. A submanifold $N \subset M$ is an integral submanifold of the ideal $\mathcal{I}$ if and only if it is an integral submanifold of the vector field system $\mathcal{V}$.*

The condition analogous to the involution criterion for vector field systems is the closure of the dual ideal $\mathcal{I}$ under exterior differentiation.

Proposition 13.22. *A constant rank vector field system $\mathcal{V}$ is involutive if and only if the dual exterior ideal $\mathcal{I}$ is closed, hence a differential ideal.*

Proof: Note first that, since $\mathcal{I}$ is generated by one-forms, we need only check that $d\omega \in \mathcal{I}$ for every one-form $\omega \in \mathcal{I}^{(1)}$. The proof rests on the important identity

$$\langle d\omega\,;\mathbf{v},\mathbf{w}\rangle = \mathbf{v}\,\langle \omega\,;\mathbf{w}\rangle - \mathbf{w}\,\langle \omega\,;\mathbf{v}\rangle - \langle \omega\,;[\mathbf{v},\mathbf{w}]\rangle, \tag{13.7}$$

that we already encountered in our intrinsic definition, (1.25), of the differential of a one-form. Thus, if the vector field system is involutive, and $\omega \in \mathcal{I}^{(1)}$ vanishes on it, then all three terms on the right hand side of (13.7) vanish, and so $d\omega$ also vanishes on the system. Conversely, if the ideal is closed, then (13.7) shows that if the vector fields $\mathbf{v}$ and $\mathbf{w}$ are annihilated by the ideal, so is their Lie bracket $[\mathbf{v},\mathbf{w}]$, which proves that the dual vector field system is involutive. *Q.E.D.*

Example 13.23. Let $M=\mathbb{R}^3$, and consider the vector field system generated by the single vector field $\mathbf{v} = -y\partial_x + x\partial_y + (1+z^2)\partial_z$; i.e., $\mathcal{V}$ consists of all multiples $f(x,y,z)\mathbf{v}$. The integral curves of $\mathbf{v}$ were explicitly computed in Example 2.75. The dual differential system $\mathcal{I}$ is generated by the pair of one-forms $\omega^1 = (1+z^2)\,dx + y\,dz$, $\omega^2 = (1+z^2)\,dy - x\,dz$. To check the closure of $\mathcal{I}$, we compute

$$\begin{aligned} d\omega^1 &= (dy - 2z\,dx)\wedge dz = \frac{dz}{1+z^2}\wedge\left(\omega^2 - 2z\,\omega^1\right),\\ d\omega^2 &= (dx + 2z\,dy)\wedge dz = \frac{dz}{1+z^2}\wedge\left(\omega^1 + 2z\,\omega^2\right), \end{aligned}$$

so both $d\omega^1$ and $d\omega^2$ belong to the ideal $\mathcal{I}$. It is not difficult to generalize this example to prove that any Pfaffian system of rank $m-1$ on an m-dimensional manifold is automatically closed, the dual vector field system being generated by a single vector field. (This explains why the equivalence problems for vector fields and for one-forms are so different.)

It must be emphasized that the duality between vector field systems and simply generated differential systems only works in the constant rank case. Indeed, as we shall see for variable rank systems, there is a crucial difference between vector fields and differential forms. Frobenius' Theorem will imply that finitely generated involutive vector field systems are always integrable, whereas this is certainly not the case for differential systems of variable rank.

Chapter 14

Frobenius' Theorem

The basic existence result in our subject (except in those situations where we must appeal to the more complicated Cartan–Kähler Theorem) is the Theorem of Frobenius characterizing solutions to involutive systems of first order partial differential equations. Frobenius' original version of this theorem, [**73**], was stated directly in the language of partial differential equations. Later, in view of its important applications in differential geometry, an equivalent formulation in terms of integral submanifolds of systems of vector fields on a manifold was popularized. Alternatively, as emphasized by Cartan, the theorem can be formulated in the language of differential forms, providing an existence theorem for integral submanifolds of closed differential ideals generated by one-forms. In this chapter, we begin with the statement and proof of this fundamental theorem; applications to the Cartan equivalence method will follow. The equivalence applications are based on Cartan's "technique of the graph" in which the required equivalence transformation is reconstructed from its graph, which forms an integral submanifold of an appropriate differential system on the Cartesian product of the domain and range manifolds.

Vector Field Systems

We shall begin our presentation with the vector field version of the theorem of Frobenius. Let M be a smooth manifold of dimension m and consider a regular vector field system $\mathcal{V}$ of constant rank n. In this case, any integral submanifold N of the system can have dimension at most n. Frobenius' Theorem says that, for constant rank systems in involution, this dimension is actually attained, and the vector field system is integrable.

Theorem 14.1. *Let $\mathcal{V}$ be a system of smooth vector fields on a manifold M of constant rank n. Then $\mathcal{V}$ is integrable if and only if it is involutive.*

Remark: If the vector field system is finitely generated, or consists of analytic vector fields, then the theorem remains true even if the rank varies: If $\mathcal{V}$ has rank $n = n(x)$ at x then there exists an n-dimensional integral submanifold N passing through x, and, moreover, at every point of N the rank of the vector field system is the same, namely n; see [**113**]. An interesting problem is whether allowing integral submanifolds of nonmaximal dimension enables us to extend this result to nonfinitely generated vector field systems; see Exercise 14.5 below.

Proof: Let $x_0 \in M$. We introduce local coordinates $x = (y, z) = (y^1, \ldots, y^n, z^1, \ldots, z^{m-n})$ by first translating x_0 to the origin, and then applying a linear transformation so that the subspace $\mathcal{V}|_0 \subset TM|_0$ corresponding to the vector field system at $x_0 = 0$ is spanned by the first n coordinate tangent vectors $\partial/\partial y^1, \ldots, \partial/\partial y^n$. By continuity, for $x = (y, z)$ in a neighborhood of $x_0 = 0$, the subspace $\mathcal{V}|_x \subset TM|_x$ is spanned by vector fields of the form

$$\widehat{\mathbf{v}}_i = \frac{\partial}{\partial y^i} + \sum_{l=1}^{m-n} \xi_i^l(y, z) \frac{\partial}{\partial z^l}, \qquad i = 1, \ldots, n, \tag{14.1}$$

where, at $x_0 = (0, 0)$, the coefficients satisfy $\xi_i^l(0, 0) = 0$ for all i, l. The Lie bracket of any two of these vector fields has the form

$$[\,\widehat{\mathbf{v}}_i, \widehat{\mathbf{v}}_j\,] = \sum_{l=1}^{m-n} \eta_{ij}^l(y, z) \frac{\partial}{\partial z^l}, \qquad i, j = 1, \ldots, n. \tag{14.2}$$

In order that the system be involutive, each vector field (14.2) must be a linear combination of the vector fields (14.1). However, owing to the form of the vector fields $\widehat{\mathbf{v}}_i$, this can only happen if all the coefficients are 0, so $\eta_{ij}^l(y, z) \equiv 0$ for all i, j, l. Therefore the vector fields pairwise commute: $[\widehat{\mathbf{v}}_i, \widehat{\mathbf{v}}_j] = 0$.

Let $\exp(t\,\widehat{\mathbf{v}}_i)$ denote the flow of the ith vector field. According to Theorem 1.28, commutativity of the vector fields implies that the flows commute, so that, where defined,

$$\exp(t\,\widehat{\mathbf{v}}_i) \circ \exp(s\,\widehat{\mathbf{v}}_j) = \exp(s\,\widehat{\mathbf{v}}_j) \circ \exp(t\,\widehat{\mathbf{v}}_i). \tag{14.3}$$

Our desired integral submanifolds will be found by starting at a given point and successively flowing in all directions prescribed by the vector fields (14.1). More explicitly, for $(t,s) = (t_1, \ldots, t_n, s_1, \ldots, s_{m-n})$ in a neighborhood U of the origin in $\mathbb{R}^m$, we define the map $\Phi: U \to M$ by

$$x = \Phi(t,w) = \exp(t_1 \widehat{\mathbf{v}}_1) \circ \exp(t_2 \widehat{\mathbf{v}}_2) \circ \cdots \circ \exp(t_n \widehat{\mathbf{v}}_n)(0,s), \qquad (14.4)$$

i.e., we start at the point $x = (0,s)$ and flow an amount t_j by the vector field $\widehat{\mathbf{v}}_j$. Note that, by commutativity, the order in which we perform the flows is immaterial. The integral submanifolds will be the images, under the map Φ, of the slices $H_s = \{(y,s) \mid y \in \mathbb{R}^n\} \subset \mathbb{R}^m$, for $s \in \mathbb{R}^{m-n}$ sufficiently near 0, which implies that (t,s) are the desired rectifying coordinates. More specifically, I claim that *a*) the map $\Phi: V \to M$ is a diffeomorphism in a neighborhood $0 \in V \subset U$, and *b*) for each $s \in \mathbb{R}^{m-n}$, the submanifold $N_s = \Phi[V \cap H_s]$ is an n-dimensional integral submanifold of $\mathcal{V}$. The two claims will suffice to prove the Theorem of Frobenius.

According to the Inverse Function Theorem, the first claim will follow if we show that the differential (Jacobian matrix) of the map Φ is nonsingular at the origin. To compute the t-derivatives of (14.4), we use the commutativity of the flows:

$$\begin{aligned} d\Phi\left(\left.\frac{\partial}{\partial t_i}\right|_{(t,s)}\right) &= \frac{\partial}{\partial t_i}\left[\exp(t_1 \widehat{\mathbf{v}}_1) \circ \exp(t_2 \widehat{\mathbf{v}}_2) \circ \cdots \circ \exp(t_n \widehat{\mathbf{v}}_n)(0,s)\right] \\ &= \frac{\partial}{\partial t_i} \exp(t_i \widehat{\mathbf{v}}_i)\Big[\exp(t_1 \widehat{\mathbf{v}}_1) \circ \cdots \circ \exp(t_{i-1} \widehat{\mathbf{v}}_{i-1}) \circ \\ &\qquad\qquad \circ \exp(t_{i+1} \widehat{\mathbf{v}}_{i+1}) \circ \cdots \circ \exp(t_n \widehat{\mathbf{v}}_n)(0,s)\Big] \\ &= \widehat{\mathbf{v}}_i\big|_{\Phi(t,s)}, \end{aligned}$$

the latter equality following from equation (1.6). Therefore,

$$d\Phi\left(\frac{\partial}{\partial t_i}\right) = \widehat{\mathbf{v}}_i, \qquad i = 1, \ldots, n. \qquad (14.5)$$

On the other hand, $\Phi(0,s) = (0,s)$ by definition, so

$$d\Phi\left(\left.\frac{\partial}{\partial s_j}\right|_{(0,s)}\right) = \left.\frac{\partial}{\partial s_j}\right|_{(0,s)}, \qquad j = 1, \ldots, m-n. \qquad (14.6)$$

In particular, at the origin, $\widehat{\mathbf{v}}_i = \partial/\partial y^i$, and hence equations (14.5), (14.6) imply that the differential $d\Phi(0,0)$ is the identity matrix, which proves the first claim. To prove the second claim, we note that (14.5) implies that the tangent space to the submanifold is spanned by the vector fields $\widehat{\mathbf{v}}_i$, so $TN_s = d\Phi(TH_s) = \mathcal{V}$. This suffices to prove that N_s is an integral submanifold (of maximal dimension). *Q.E.D.*

Theorem 14.1 and its proof demonstrate the existence of local integral submanifolds for an involutive vector field system. Moreover, just as the integral curves of a single vector field can be extended to maximal integral curves, so can we extend the integral submanifolds to be maximal. In the sequel, the term "integral submanifold" will always refer to the maximal, connected integral submanifolds of the given differential system. The integral submanifolds of a constant rank vector field system $\mathcal{V}$ provide a foliation of the manifold M by n-dimensional submanifolds — see [**175**; §2.11], [**221**] for the details.

Example 14.2. *Transformation group actions*: If the structure functions in the involutivity conditions (13.6) are constants, then they satisfy the basic conditions (2.14) for the structure constants of a Lie algebra $\mathfrak{g}$. Indeed, Theorem 2.63 states that there is a local action of the associated connected Lie group G on M whose infinitesimal generators are the given vector fields; a proof of this fact can be found in [**195**]. The integral submanifolds provided by Frobenius' Theorem 14.1 coincide with the orbits of the group action. The local coordinates (t, s) constructed in the proof of Theorem 14.1 provide the rectifying coordinates of Theorem 2.23. Moreover, since the orbits of G are just the level sets $\{s = \text{const.}\}$, the coordinate functions $s^1, \ldots, s^{m-n}$ provide a complete set of functionally independent local invariants of the transformation group action. Therefore, our basic Theorem 2.34 on the existence and characterization of local invariants of semi-regular group actions appears as an immediate corollary of the general Frobenius Theorem 14.1.

Example 14.3. Consider the three vector fields $\mathbf{v}_1 = \partial_x$, $\mathbf{v}_2 = \partial_y$, $\mathbf{v}_3 = x\partial_x + zy\partial_y$, which act on $M = \mathbb{R}^3$. Since $[\mathbf{v}_1, \mathbf{v}_2] = 0$, $[\mathbf{v}_1, \mathbf{v}_3] = \mathbf{v}_1$, $[\mathbf{v}_2, \mathbf{v}_3] = z\mathbf{v}_2$, the system spanned by $\mathbf{v}_1, \mathbf{v}_2, \mathbf{v}_3$ is involutive. Indeed, the integral submanifolds are just the planes $N_c = \{z = c\}$ for c constant. Restricted to the integral submanifolds, the vector fields generate nonisomorphic three-parameter planar group actions, $(x, y) \mapsto (\lambda x + \delta, \lambda^z y + \varepsilon)$, corresponding to Case 1.7 with $k = 1$, $z = \alpha$ in our tables. Therefore, $\mathbf{v}_1, \mathbf{v}_2, \mathbf{v}_3$ do not generate a three-parameter group of transformations on $\mathbb{R}^3$. Moreover, one cannot include these vector fields in a finite-dimensional Lie algebra, since $[\mathbf{v}_2, \mathbf{v}_3] = \mathbf{v}_4 = z\partial_y$, $[\mathbf{v}_4, \mathbf{v}_3] = \mathbf{v}_5 = z^2\partial_y$, and so on, hence the successive commutators span an infinite-dimensional Lie algebra of vector fields.

Exercise 14.4. Suppose $\mathcal{V}$ is a commutative vector field system, so $[\mathbf{v}, \mathbf{w}] = 0$ for all $\mathbf{v}, \mathbf{w} \in \mathcal{V}$. Prove that if $\mathcal{V}$ has constant rank n, then there exist local coordinates $(t_1, \ldots, t_n, s_1, \ldots, s_{m-n})$ such that every vector field $\mathbf{v} \in \mathcal{V}$ has the form $\mathbf{v} = \sum_{i=1}^{n} \eta^i(s)\, \partial_{t_i}$.

Exercise 14.5. Let $M = \mathbb{R}^2$ and let $\mathcal{V}$ denote the vector field system spanned by the horizontal vector field ∂_x and all vertical vector fields of the form $f(x)\partial_y$ where f is any smooth scalar function such that it and all of its derivatives vanish at $x = 0$, so $f^{(n)}(0) = 0$, $n = 0, 1, 2, \ldots$. Prove that $\mathcal{V}$ has rank one on the y-axis and rank two elsewhere. Prove that any point on the y-axis is contained in a (nonunique) integral curve C transverse to the y-axis. Therefore the rank of $\mathcal{V}$ at any other point in C is strictly greater than one. This example shows that, for infinitely generated vector field systems of variable rank, one must allow integral submanifolds of nonmaximal dimension. See [**174**] for more details.

Example 14.6. *Relative invariants*: Theorem 3.36 stated a general result on the existence and characterization of relative invariants for multiplier representations of Lie groups. This theorem is not quite an immediate consequence of Frobenius' Theorem 14.1, but, rather, the following extension. Recall first that the group G was extended to act on the Cartesian product $E = M \times U$ of an m-dimensional manifold M with a vector space $U \simeq \mathbb{R}^q$; the group transformations have the form $(x, u) \mapsto (\chi_g(x), \mu(g, x)u)$, where μ is the (matrix) multiplier. In local coordinates, the infinitesimal generators split into a sum $\mathbf{w} = \mathbf{v} + \mathbf{u}$ of a vector field $\mathbf{v}$ on the base M and a linear vector field $\mathbf{u}$ in the fiber coordinates, so that

$$\mathbf{v} = \sum_{i=1}^{m} \xi^i(x) \frac{\partial}{\partial x^i}, \qquad \mathbf{u} = \sum_{\alpha,\beta=1}^{q} h^\alpha_\beta(x) u^\beta \frac{\partial}{\partial u^\alpha}. \tag{14.7}$$

Relative invariants of the dual action on $E^* = X \times U^*$ correspond to *linear* invariants $J(x, u) = \sum_{\alpha=1}^{q} R_\alpha(x) u^\alpha$ of the extended action, and the problem is to determine their precise number. Thus, we need to apply Frobenius' Theorem 14.1 to the vector field system $\mathcal{W}$ on E generated by the vector fields $\mathbf{w}$. The fact that the infinitesimal generators span a Lie algebra implies that $\mathcal{W}$ is involutive, which, in view of the form of the generators, implies that its projection $\mathcal{V}$, which is the vector field system on M generated by the vector fields $\mathbf{v}$, is also involutive. Moreover, we shall assume that G acts regularly on an open subset of M and on a corresponding open subset of E, and that $\mathcal{V}$ and $\mathcal{W}$ have constant rank (orbit dimension) n. According to Frobenius' Theorem 14.1, there exist $m-n$ functionally independent invariants $I_1(x), \ldots, I_{m-n}(x)$ of G on M, and an additional q independent invariants $J_1(x,u), \ldots, J_q(x,u)$ depending on the fiber coordinates, with the property that any other invariant is a function thereof. If we can prove that the latter invariants can be taken to be *linear* in u, so $J_\nu(x,u) = R_\nu(x) \cdot u$, then we will have completed the proof of Theorem 3.36. Indeed, any other relative invariant $R(x)$ determines a linear invariant of the extended action $J(x,u) = R(x) \cdot u$, and can be locally written as a function of the invariants I_κ, J_ν, linear in the J_ν's, so that $R(x) = \sum_{\nu=1}^{q} F_\nu(I_1, \ldots, I_{m-n}) R_\nu$, where each coefficient F_ν is an absolute invariant of G, as desired.

The linearity of the invariants J_ν does not follow automatically from the statement of Frobenius' Theorem, but must be inferred from a more detailed analysis of the proof, keeping careful track of linearity. To begin, we use a linear transformation near a regular point (x_0, u_0) to introduce local coordinates $(y,z,u) = (y^1, \ldots, y^n, z^1, \ldots, z^{m-n}, u^1, \ldots, u^q)$ so that $\mathcal{W}$ is spanned by vector fields of the form

$$\widehat{\mathbf{v}}_i = \frac{\partial}{\partial y^i} + \sum_{l=1}^{m-n} \xi_i^l(y,z) \frac{\partial}{\partial z^l} + \sum_{\beta=1}^{q} \rho_i^\beta(y,z,u) \frac{\partial}{\partial u^\beta}, \qquad i = 1, \ldots, n, \tag{14.8}$$

whose coefficients $\rho_i^\beta(y,z,u)$ are linear functions of the fiber variables u. (Here, in order to preserve linearity, we do not translate (x_0, u_0) to the origin.) Exercise 1.26 implies that the corresponding flow is linear in the fiber variables too. Thus we obtain the rectifying coordinates $(t,s,v) = (t^1, \ldots, t^n, s^1, \ldots, s^{m-n}, v^1, \ldots, v^q)$ by exponentiating the $\widehat{\mathbf{v}}_i$, so that $\mathcal{W}$ is spanned by the coordinate vector fields $\partial/\partial t^1, \ldots, \partial/\partial t^s$, and such that the rectifying fiber coordinates v are *linear* functions of the original fiber coordinates u. The corresponding invariants are the $m+q-n$ coordinates s, v. Clearly, in the rectifying coordinates, the most general linear invariant has the form $J(t,s,v) = \sum_{\alpha=1}^{q} F_\alpha(s)v^\alpha$. In the original coordinates, the $s^i = s^i(x)$ become the functionally independent absolute invariants on M, while each $v^\alpha = \sum_{\beta=1}^{q} H_\beta^\alpha(x)u^\beta$ forms a linear invariant of the extended action, and hence its coefficients provide one of the fundamental relative invariants $R_\nu(x)$ of the multiplier representation. This completes the proof of Theorem 3.36.

Differential Systems

According to Proposition 13.21, the integral submanifolds of a constant rank vector field system and of the dual simply generated exterior ideal coincide. Therefore, an immediate corollary of the vector field version of Frobenius' Theorem 14.1 is the following existence theorem for integral submanifolds of a Pfaffian system, defined by the vanishing of a collection of one-forms, satisfying the involutivity (closure) condition.

Theorem 14.7. *Let $\mathcal{I}$ be a simply generated ideal of constant rank $m-n$. Then $\mathcal{I}$ is n-integrable if and only if $\mathcal{I}$ is closed.*

Therefore, in the regular case when the differential ideal has rank r, the integral submanifolds have as their maximal dimension $m-r$, the corank of the differential system. Thus, the fewer the one-forms in the system, the larger the integral submanifolds. For example, if $\mathcal{I}$ is generated by a single nonvanishing one-form ω, then $\mathcal{I}$ is closed if and only if $d\omega$ is a multiple of ω, or, equivalently, $\omega \wedge d\omega = 0$, so ω has Darboux rank one or two. (By Exercise 1.43 this is the same as ω admitting an integrating factor.)

In general, only at regular points where the rank of the ideal is constant will we be able to construct an integral submanifold associated with a given integral element. The reason for having to impose such a condition is easy to see if we consider an elementary example. Suppose that $\mathcal{I}$ has rank $m-1 = \dim M - 1$ everywhere except for a single point x_0 at which all the one-forms in $\mathcal{I}$ vanish, and hence $\mathcal{I}$ has rank zero at x_0. At each point $x \neq x_0$, then, there is a unique one-dimensional integral element $L|_x \subset TM|_x$. Choosing any (locally defined) vector field $\mathbf{v}$ contained in $L|_x$ at each x, we can integrate $\mathbf{v}$, and the associated integral curves will be the required one-dimensional integral submanifolds of $\mathcal{I}$. However, at the singular point x_0 any tangent vector $\mathbf{v}_0 \in TM|_{x_0}$ will be annihilated by the ideal, and hence define an integral element. But there is no reason to suppose that we can construct an integral curve passing through x_0 in the direction of $\mathbf{v}_0$. For example, let $M = \mathbb{R}^2$ and consider a differential system generated by a single one-form ω with a singularity at the origin. In the case $\omega = y\,dx - x\,dy$, the integral curves are the rays emanating from the origin, so any integral element at 0 generates an integral curve. For $\omega = x\,dx + y\,dy$, the integral curves are the circles centered at the origin, so no tangent vector at 0 generates an integral curve. Finally, for $\omega = (x^2 + y^2)\,dy$, the integral curves are the straight lines parallel to the x-axis, so only the particular tangent vector ∂_x determines an integral curve passing through 0. This situation stands in contrast to the vector field version of Frobenius' Theorem, where the rank can vary from point to point without affecting the existence of integral submanifolds of the appropriate dimension.

Characteristics and Normal Forms

The classical notion of a characteristic direction for a system of partial differential equations — see the discussion following (5.35) — has a natural extension to differential systems defined by the vanishing of a collection of differential forms. The resulting vector field system and its dual Pfaffian system form additional invariant objects associated with the original differential system, which are of importance in the understanding equivalence of differential systems under coordinate changes, cf. [**28**], [**79**].

Definition 14.8. Let $\mathcal{I}$ be a differential ideal. A vector field $\mathbf{v}$ is called a *characteristic direction* for $\mathcal{I}$ if

$$\mathbf{v} \lrcorner \Omega \in \mathcal{I}, \qquad \text{for all} \qquad \Omega \in \mathcal{I}. \tag{14.9}$$

We let $\chi(\mathcal{I})$ denote the vector field system consisting of all characteristic directions for $\mathcal{I}$.

Exercise 14.9. Prove that the characteristic system $\chi(\mathcal{I})$ is involutive. In the constant rank case, its integral submanifolds are called the *Cauchy characteristics* of the system.

Example 14.10. Consider a general first order partial differential equation in one dependent variable:

$$Q(x, u^{(1)}) = 0, \qquad \text{where} \qquad x \in \mathbb{R}^p, \quad u \in \mathbb{R}. \tag{14.10}$$

The differential system whose integral submanifolds describe solutions $u = f(x)$ to (14.10) is obtained by restricting the contact form

$$\theta = du - \sum_{i=1}^{p} u_i\,dx^i \tag{14.11}$$

to the variety $\mathcal{S}_Q \subset \mathrm{J}^1$ described by (14.10). Assuming regularity of equation (14.10), so that $\mathcal{S}_Q$ is a submanifold of J^1, the resulting differential ideal, then, is generated by (14.11) and the additional forms

$$\omega = \sum_{i=1}^{p} \left[\left(\frac{\partial Q}{\partial x^i} + u_i \frac{\partial Q}{\partial u} \right) dx^i + \frac{\partial Q}{\partial u_i} du_i \right], \qquad d\theta = -\sum_{i=1}^{p} du_i \wedge dx^i, \tag{14.12}$$

where $\omega \equiv dQ$ modulo the contact form θ. A vector field

$$\mathbf{v} = \sum_{i=1}^{p} \xi^i(x, u^{(1)}) \frac{\partial}{\partial x^i} + \varphi(x, u^{(1)}) \frac{\partial}{\partial u} + \sum_{i=1}^{p} \chi_i(x, u^{(1)}) \frac{\partial}{\partial u_i} \tag{14.13}$$

on $\mathcal{S}_Q$ will therefore define a characteristic direction if and only if

$$\varphi - \sum_{i=1}^{p} u_i \xi^i = 0, \qquad \sum_{i=1}^{p} \left[\left(\frac{\partial Q}{\partial x^i} + u_i \frac{\partial Q}{\partial u} \right) \xi^i + \frac{\partial Q}{\partial u_i} \chi_i \right] = 0,$$
$$\sum_{i=1}^{p} \left[\xi^i \, du_i - \chi_i \, dx^i \right] = 0. \tag{14.14}$$

The general solution of the system (14.14) is given by

$$\xi^i = -\lambda \frac{\partial Q}{\partial u_i}, \quad \varphi = -\lambda \sum_{i=1}^{p} u_i \frac{\partial Q}{\partial u_i}, \quad \chi_i = \lambda \left\{ \frac{\partial Q}{\partial x^i} + u_i \frac{\partial Q}{\partial u} \right\}, \tag{14.15}$$

where $\lambda(x, u^{(1)})$ is arbitrary. Thus, up to scalar multiple, there is a single characteristic direction. Note that (14.15) is only required to hold on the equation variety $\mathcal{S}_Q$; in particular, choosing $\lambda = 1$, and comparing with (4.60), we find the remarkable fact that, on the differential equation (14.10), *the characteristic direction of a first order partial differential equation* (14.10) *coincides with the contact vector field whose characteristic function is Q itself*! This observation provides added depth to our various choices of the term "characteristic". The characteristic curves (known as *characteristic strips* in the classical literature, cf. [**56**]), are the integral curves traced out by the flow, or one-parameter group of contact transformations, $\exp(t\mathbf{v})(x, u^{(1)})$. According to Exercise 6.17, this flow leaves the equation variety invariant, and hence a characteristic curve starting in $\mathcal{S}_Q$ remains therein. In particular, a quasi-linear first order equation, in which Q is an affine function of the derivative coordinates u_i, has characteristic curves described by prolonged point transformations, which can be unambiguously projected back down to the usual characteristic curves in the base space $E \simeq \mathbb{R}^p \times \mathbb{R}$. This construction elucidates the deep connection between the solution of first order partial differential equations by the method of characteristics and the geometry of contact transformations; see [**56**] for further development of this classical theory.

Definition 14.11. The *Cartan system* $\mathrm{C}(\mathcal{I})$ associated with a differential ideal $\mathcal{I}$ is the Pfaffian system dual to the characteristic system. Specifically,

$$\mathrm{C}(\mathcal{I}) = \{\omega \mid \langle \omega\,; \mathbf{v}\rangle = 0 \quad \text{for all} \quad \mathbf{v} \in \chi(\mathcal{I})\}\,. \tag{14.16}$$

Exercise 14.12. Show that if the differential ideal $\mathcal{I}$ is simply generated, then $\mathcal{I} \subset \mathrm{C}(\mathcal{I})$.

In the constant rank case, the characteristic system, and hence the Cartan system, are involutive, and hence, by Frobenius' Theorem, admit integral submanifolds. We can then eliminate the characteristic directions by restriction to the integral submanifolds; see [**28**; Theorem II.2.2] for a proof of the following result.

Theorem 14.13. *Let M be an m-dimensional manifold. Let $\mathcal{I}$ be a simply generated differential ideal whose Cartan system has constant rank r. Then there exist local coordinates $(y^1, \ldots, y^r, z^1, \ldots, z^{m-r})$ on M such that* a) *the Cauchy characteristics are the slices where z is constant, and* b) *$\mathcal{I}$ is generated by differential forms involving only the coordinates $(z^1, \ldots, z^{m-r})$ and their differentials.*

The differential systems generated by contact forms play a particularly important role in applications. An intrinsic characterization of the contact system, generalizing the Darboux characterization of a single contact form in Theorem 1.42, is known in two particular cases. The first result dates back to Goursat and deals with the contact system on a jet space having one independent and one dependent variable; see [**28**; p. 54] for a proof.

Theorem 14.14. *Let M be an m-dimensional manifold. Let $\mathcal{I}$ be a Pfaffian system generated by one-forms $\omega^1, \ldots, \omega^{m-2}$ which satisfy*

$$\begin{aligned} d\omega^i &\equiv -\omega^{i+1} \wedge \theta \quad \bmod \omega^1, \ldots, \omega^i, \qquad i = 1, \ldots, m-3, \\ d\omega^{m-2} &\not\equiv 0 \quad \bmod \mathcal{I}. \end{aligned} \tag{14.17}$$

Then there exist local coordinates $x, u_0, u_1, \ldots u_{m-2}$*, such that* $\omega^i = du_i - u_{i+1}\, dx$*, and thus* $\mathcal{I}$ *is locally equivalent to the contact ideal* $\mathcal{I}_{m-2}$ *on the jet bundle* $\mathrm{J}^{m-2}E$*, where* $E \simeq \mathbb{R} \times \mathbb{R}$*.*

The second characterization, which deals with the first order contact system in higher dimensions, is much more recent and due to Bryant, [**26**]; see also [**28**; p. 48]. We first define the *wedge length* p of a simply generated differential ideal $\mathcal{I}$ to be the maximal rank of the closed two-forms therein; in other words, p is the smallest integer such that the $(p+1)$-fold wedge product $d\omega \wedge d\omega \wedge \cdots \wedge d\omega \equiv 0$ for all $\omega \in \mathcal{I}^{(1)}$; see (1.29). Note that $p = \left[\frac{s}{2}\right]$, where s is the maximal Darboux rank of the one-forms $\omega \in \mathcal{I}^{(1)}$.

Theorem 14.15. *Let* M *be an* m*-dimensional manifold. Consider a Pfaffian system defined by a differential ideal* $\mathcal{I}$ *having constant rank* $q \neq 2$*, constant wedge length* p*, and derived system* $\delta\mathcal{I} = \{0\}$*. Then its Cartan system satisfies* $\dim \mathrm{C}(\mathcal{I}) = pq + p + q$ *if and only if there exist local coordinates* $x^1, \ldots, x^p$*,* $u^1, \ldots, u^q$*, and* u_i^α*,* $\alpha = 1, \ldots, q$*,* $i = 1, \ldots, p$*, such that* $\mathcal{I}$ *is generated by the one-forms* $\theta^\alpha = du^\alpha - \sum_i u_i^\alpha\, dx^i$*,* $\alpha = 1, \ldots, q$*. Thus,* $\mathcal{I}$ *is locally equivalent to the contact ideal* $\mathcal{I}_1$ *on the first jet bundle* J^1E *where* $E \simeq \mathbb{R}^p \times \mathbb{R}^q$*.*

It is a remarkable fact that in the rank $q = 2$ case, the contact system does *not* provide a unique normal form for such a differential system. Indeed, there are additional invariants associated with such rank two Pfaffian systems.

The Technique of the Graph

We now apply the differential form version of Frobenius' Theorem 14.7 to the construction of equivalence maps between coframes. An important device for proving the existence of smooth maps $\Phi\colon M \to \overline{M}$, such as equivalences, was introduced by Cartan and is now known as the *technique of the graph.* The key idea is that, instead of looking for the required map $\bar{x} = \Phi(x)$ directly, we try to construct its graph $\Gamma_\Phi \equiv \{(x, \Phi(x))\}$, which is an m-dimensional submanifold of the $2m$-dimensional Cartesian product manifold $M \times \overline{M}$. This will be accomplished by realizing Γ_Φ as an integral submanifold of a suitable differential system, and relying on Frobenius' Theorem 14.7 to guarantee its existence. Let us begin by recalling which submanifolds of the Cartesian

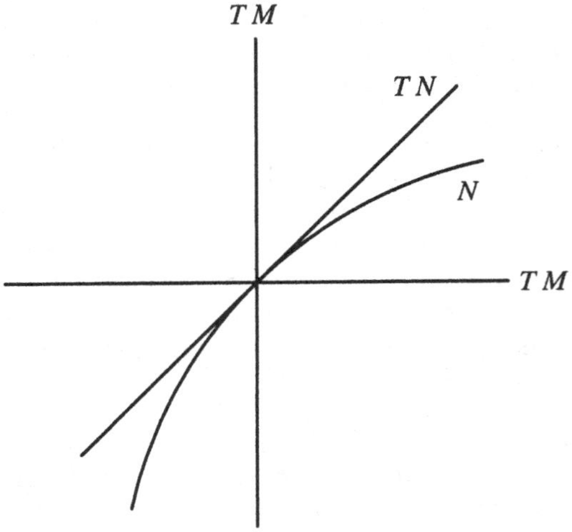

Figure 4. Fully Transverse Submanifold

product space are the graphs of (local) diffeomorphisms.

Definition 14.16. Let M and $\overline{M}$ be m-dimensional manifolds. An m-dimensional submanifold $N \subset M \times \overline{M}$ will be called *fully transverse* at a point $z = (x, \bar{x}) \in N$ if its tangent space $TN|_z$ contains no nonzero horizontal or vertical tangent vectors, so $TN \cap TM = \{0\} = TN \cap T\overline{M}$. (See Figure 4.)

The following key result is a simple consequence of the Implicit Function Theorem; see also Proposition 4.1.

Proposition 14.17. *Let M and $\overline{M}$ be m-dimensional manifolds. If $\Phi: M \to \overline{M}$ is a smooth diffeomorphism, then its graph Γ_Φ is a fully transverse submanifold of $M \times \overline{M}$. Conversely, if $N \subset M \times \overline{M}$ is an m-dimensional submanifold which is fully transverse at a point $z_0 = (x_0, \bar{x}_0) \in N$, then, in a neighborhood of z_0, N coincides with the graph Γ_Φ of a local diffeomorphism $\Phi: M \to \overline{M}$ mapping x_0 to $\bar{x}_0 = \Phi(x_0)$.*

Therefore, Cartan's technique of the graph reduces any existence problem for local diffeomorphisms from M to $\overline{M}$ to the construction of fully transverse m-dimensional submanifolds of $M \times \overline{M}$. As a first illustration of use of this technique, we consider the equivalence problem for a coframe of rank zero, meaning one in which all the structure functions are constant. Let $\boldsymbol{\theta} = \{\theta^1, \ldots, \theta^m\}$ be a coframe defined on an m-dimensional manifold M, whose structure equations take the form

$$d\theta^i = \sum_{\substack{j,k=1 \\ j<k}}^{m} C^i_{jk}\, \theta^j \wedge \theta^k, \qquad i = 1, \ldots, m, \tag{14.18}$$

where the structure functions C^i_{jk} are constants, and hence define the structure constants of a Lie algebra $\mathfrak{g}$ — see the discussion following (8.21). Let $\overline{\boldsymbol{\theta}} = \{\overline{\theta}^1, \ldots, \overline{\theta}^m\}$ be a second coframe, defined on another m-dimensional manifold $\overline{M}$, and having identical structure equations

$$d\overline{\theta}^i = \sum_{\substack{j,k=1 \\ j<k}}^{m} C^i_{jk}\, \overline{\theta}^j \wedge \overline{\theta}^k, \qquad i = 1, \ldots, m, \tag{14.19}$$

with the same constant structure functions. For example, we can take $\overline{\boldsymbol{\theta}}$ to be the Maurer–Cartan coframe on any Lie group G whose Lie algebra $\mathfrak{g}$ has the given structure constants. The general Equivalence Theorem 8.15 in this case states that any two such coframes are always locally equivalent. As a corollary, we deduce that any coframe of rank zero is locally equivalent to the Maurer–Cartan coframe on a suitable Lie group, whose structure constants coincide with the (constant) structure functions of the coframe.

Theorem 14.18. *Let $\boldsymbol{\theta}$ and $\overline{\boldsymbol{\theta}}$ be two coframes of rank zero defined on m-dimensional manifolds M, $\overline{M}$, having the same constant structure functions. Then, for any points $x_0 \in M$ and $\bar{x}_0 \in \overline{M}$, there exists a unique local diffeomorphism $\Phi: M \to \overline{M}$ taking x_0 to $\bar{x}_0 = \Phi(x_0)$, which maps the coframes to each other:*

$$\Phi^*\, \overline{\theta}^i = \theta^i, \qquad i = 1, \ldots, m. \tag{14.20}$$

Proof: Let $\mathcal{I}$ denote the Pfaffian system on $M \times \overline{M}$ generated by the one-forms $\vartheta^i = \overline{\theta}^i - \theta^i$. (Technically speaking, we should write $\vartheta^i = \pi_2^* \overline{\theta}^i - \pi_1^* \theta^i$, where $\pi_1: M \times \overline{M} \to M$ and $\pi_2: M \times \overline{M} \to \overline{M}$ are the obvious projections. Omitting the π's will make the notation simpler, and should not cause any confusion.) It is easy to see that the equivalence conditions (14.20) are satisfied if and only if the forms ϑ^i vanish on the graph Γ_Φ of Φ; in other words, the graph Γ_Φ is an m-dimensional integral submanifold of the exterior ideal $\mathcal{I}$.

Now, in order to apply Frobenius' Theorem 14.7 to deduce the existence of an m-dimensional integral submanifold of the differential system $\mathcal{I}$, we need to show that the ideal $\mathcal{I}$ is closed, and of rank m, since then the integral submanifolds will then have the correct dimension: $\dim(M \times \overline{M}) - \operatorname{rank} \mathcal{I} = 2m - m = m$. The rank condition is immediate: the one-forms ϑ^i are clearly linearly independent since the $\overline{\theta}^i$'s (and the θ^i's) form a coframe. To check closure, we need only show that each differential $d\vartheta^i$ is in $\mathcal{I}$, i.e., can be expressed in terms of the ϑ's. This follows immediately from the two sets of structure equations (14.18), (14.19), which imply

$$\begin{aligned} d\vartheta^i = d\overline{\theta}^i - d\theta^i &= \sum_{j<k} \left\{ C^i_{jk}\, \overline{\theta}^j \wedge \overline{\theta}^k - C^i_{jk}\, \theta^j \wedge \theta^k \right\} \\ &= \sum_{j<k} C^i_{jk} \left\{ (\overline{\theta}^j - \theta^j) \wedge \overline{\theta}^k + \theta^j \wedge (\overline{\theta}^k - \theta^k) \right\} \\ &= \sum_{j<k} C^i_{jk} \left\{ \vartheta^j \wedge \overline{\theta}^k + \theta^j \wedge \vartheta^k \right\}. \end{aligned} \tag{14.21}$$

Therefore, $d\vartheta^i$ belongs to $\mathcal{I}$ and we conclude that the ideal $\mathcal{I}$ is closed. Theorem 14.7 now guarantees the existence of a unique m-dimensional integral submanifold $N \subset M \times \overline{M}$ passing through each point $(x_0, \bar{x}_0)$.

To complete the proof of Theorem 14.18, we need only check that our integral m-dimensional submanifold N really is (locally) the graph of a function $\bar{x} = \Phi(x)$. According to Proposition 14.17, we need only verify that the tangent space TN contains no horizontal tangent vectors, $\mathbf{v} = \sum c_i \partial/\partial\theta^i$, or vertical tangent vectors, $\overline{\mathbf{v}} = \sum \bar{c}_i \partial/\partial\overline{\theta}^i$, which, for convenience, we write in terms of the appropriate dual frame vector fields. Now, if a horizontal tangent vector $\mathbf{v}$ is tangent to N, it must be annihilated by the one-forms in our ideal. But

$$\langle \vartheta^i\,; \mathbf{v} \rangle = \langle \overline{\theta}^i - \theta^i\,; \mathbf{v} \rangle = -\langle \theta^i\,; \mathbf{v} \rangle = c_i, \qquad i = 1, \ldots, m,$$

which will vanish if and only if all of the c_i's are zero, i.e., if and only if $\mathbf{v} = 0$. A similar computation also rules out vertical tangent vectors. This demonstrates the transversality of any integral submanifold, and Proposition 14.17 completes the proof of the theorem. *Q.E.D.*

The symmetry group of a coframe is just the (local) transformation group of self-equivalences. The uniqueness of the integral submanifolds given by Frobenius' Theorem demonstrates that the symmetry group of a rank zero coframe can be identified with the Lie group G whose Maurer–Cartan coframe is locally equivalent to the given coframe. The group G will, of course, act by right multiplication.

Proposition 14.19. *Let G be an m-dimensional Lie group and $\{\theta^1, \ldots, \theta^m\}$ a basis for the Maurer–Cartan one-forms on G. Suppose $\Phi: G \to G$ is a smooth (local) diffeomorphism such that $\Phi^*\theta^i = \theta^i$, $i = 1, \ldots, m$. Then there exists a group element $h \in G$ such that $\Phi(g) = g \cdot h$ for all g in the domain of Φ.*

Proof: Theorem 14.18 and Theorem 14.7 show that through each point $(g, \bar{g}) \in G \times G$ there passes a unique m-dimensional integral submanifold N of the differential system $\mathcal{I}$. On the other hand, since the Maurer–Cartan forms are right-invariant, the graph of right multiplication map R_h with $h = g^{-1} \cdot \bar{g}$ is an integral submanifold which passes through this point since $R_h(g) = g \cdot h = \bar{g}$. Thus, by uniqueness, N is (an open subset of) the graph of this right multiplication map. *Q.E.D.*

Corollary 14.20. *The symmetry group of a rank zero coframe on an m-dimensional manifold is an m-dimensional Lie group whose structure constants are the same as the (constant) structure functions of the coframe.*

The proof of Theorem 8.19 governing the equivalence of general coframes proceeds in the same fashion, and is only complicated by the inclusion of the nonconstant structure functions into the differential system that the graph of the equivalence must satisfy. Recall that the classifying manifold $\mathcal{C}^{(s)}(\boldsymbol{\theta})$ of order $s \geq 0$ of a coframe $\boldsymbol{\theta}$ is the subset of the Euclidean classifying space $\mathbb{K}^{(s)}$ parametrized by the structure map $\mathbf{T}^{(s)}: M \to \mathbb{K}^{(s)}$; see Definition 8.9 and the subsequent discussion for the relevant notation. The rank of the structure map $\mathbf{T}^{(s)}$, which equals the number of functionally independent structure invariants of order at most s, is denoted by ρ_s.

Definition 14.21. A coframe $\boldsymbol{\theta}$ on M is *regular* of *rank* r if, for some $s \geq 0$, the ranks $\rho_s = \rho_{s+1} = r$ are constant and equal on M. The smallest integer s for which this condition holds is called the *order* of the coframe.

Note that this definition of regularity is slightly more general than that used in Chapter 8. Here it is not necessarily true that the order s is bounded by m, the dimension of M. For example, suppose $M = \mathbb{R}$ and we have just one invariant $I(x)$ with covariant derivative $I'(x)$. If $I(x) = x^n$, then $I'(x) = nx^{n-1}$, $I''(x) = n(n-1)x^{n-2}$, etc. In this case the coframe has rank one everywhere, but, at $x = 0$, $\rho_s = 0$ for $s < n$, and n can, of course, be arbitrarily large. On the other hand, in the smooth category, our definition of regularity is slightly stronger than the requirement that the complete system of structure invariants $\mathcal{F} = \{T_\sigma \mid \operatorname{order} \sigma \geq 0\}$ have constant rank r, although, in view of Theorem 1.10, the latter condition does imply that the coframe is locally regular, i.e., on sufficiently small open subsets of M.

Let us first demonstrate that, once the ranks stabilize, they remain fixed.

Proposition 14.22. *Let $\boldsymbol{\theta}$ be a coframe whose s^{th} and $(s+1)^{\text{st}}$ ranks are constant and equal: $\rho_s = \rho_{s+1} = r$. Then $\rho_t = r$ for all $t \geq s$.*

Proof: Locally, we can choose $r = \rho_s$ functionally independent s^{th} order structure invariants $I_1, \ldots, I_r$, which generate the entire family $\mathcal{F}^{(s)}$ of structure invariants of order s. Moreover, since $\mathcal{F}^{(s+1)} \supset \mathcal{F}^{(s)}$ has the same rank, the I_ν's also generate $\mathcal{F}^{(s+1)}$ (possibly on a smaller open subset of the manifold). Let $J \in \mathcal{F}^{(s+1)}$ be any of these invariants, so $J = H(I_1, \ldots, I_r)$ for some classifying function $H(z^1, \ldots, z^r)$. According to the chain rule formula (8.37) for coframe derivatives,

$$\frac{\partial J}{\partial \theta^j} = \sum_{\nu=1}^{r} \frac{\partial H}{\partial z^\nu}(I_1, \ldots, I_r)\, \frac{\partial I_\nu}{\partial \theta^j}\,. \tag{14.22}$$

But, since each I_ν has order at most s, its coframe derivatives $I_{\nu,j} = \partial I_\nu / \partial \theta^j$ have order at most $s+1$, and hence belong to $\mathcal{F}^{(s+1)}$. Consequently, we already know the classifying functions $I_{\nu,j} = H_{\nu,j}(I_1, \ldots, I_r)$ for these structure invariants. Therefore each derived invariant $\partial J / \partial \theta^j$ can also be expressed as a function of the I_ν. By choosing $J = T_\sigma$ to be any of the structure invariants with $\operatorname{order} \sigma = s+1$, we thereby obtain the classifying functions for all the structure invariants $T_{\sigma,j}$ of order $s+2$. Continuing the process, we eventually determine the classifying functions for all the higher order structure invariants. *Q.E.D.*

As a consequence of the proof of Proposition 14.22, we observe that, once we have fixed a fundamental set of structure invariants $I_1, \ldots, I_r$, we need only determine the classifying functions associated with a) any other zeroth order structure functions T^i_{jk}, and b) the first order coframe derivatives $\partial I_\nu / \partial \theta^k$ of the fundamental invariants. The chain rule formula (14.22) implies that all other classifying functions for the associated derived invariants are automatically determined.

Corollary 14.23. *Let $\boldsymbol{\theta}$ and $\overline{\boldsymbol{\theta}}$ be regular coframes of order $s = \bar{s}$. If their $(s+1)^{st}$ order classifying manifolds are the same, $\mathcal{C}^{(s+1)}(\boldsymbol{\theta}) = \mathcal{C}^{(s+1)}(\overline{\boldsymbol{\theta}})$ (or, more generally, are overlapping), then so are all higher order classifying manifolds: $\mathcal{C}^{(s+k)}(\boldsymbol{\theta}) = \mathcal{C}^{(s+k)}(\overline{\boldsymbol{\theta}})$, $k \geq 1$.*

We now state and prove the basic theorem concerning the equivalence of (regular) coframes.

Theorem 14.24. *Suppose M and $\overline{M}$ are smooth m-dimensional manifolds, and let $\boldsymbol{\theta} = \{\theta^1, \ldots, \theta^m\}$, and $\overline{\boldsymbol{\theta}} = \{\overline{\theta}^1, \ldots, \overline{\theta}^m\}$ be regular coframes on them. Then there exists a local diffeomorphism $\Phi\colon M \to \overline{M}$, which maps the coframes to each other, $\Phi^* \overline{\theta}^i = \theta^i$, $i = 1, \ldots, m$, if and only if* a) *the coframes have the same order, s, and* b) *their $(s+1)^{st}$ order classifying manifolds $\mathcal{C}^{(s+1)}(\boldsymbol{\theta})$ and $\mathcal{C}^{(s+1)}(\overline{\boldsymbol{\theta}})$ overlap. Moreover, if $x_0 \in M$ and $\bar{x}_0 \in \overline{M}$ are any points mapping to the same point*

$$z_0 = \mathbf{T}^{(s+1)}(x_0) = \overline{\mathbf{T}}^{(s+1)}(\bar{x}_0) \in \mathcal{C}^{(s+1)}(\boldsymbol{\theta}) \cap \mathcal{C}^{(s+1)}(\overline{\boldsymbol{\theta}}) \qquad (14.23)$$

on the overlap of the two classifying manifolds, then there is a unique local equivalence map Φ mapping x_0 to $\bar{x}_0 = \Phi(x_0)$.

Proof: The proof relies on the graphical technique used to demonstrate the rank zero case in Theorem 14.24. Thus, we begin by introducing the differential system $\mathcal{I}$ on the Cartesian product manifold $M \times \overline{M}$ which is generated by the one-forms $\vartheta^i = \overline{\theta}^i - \theta^i$, $i = 1, \ldots, m$. As it stands, $\mathcal{I}$ is not, in general, involutive, since we have not yet taken the structure invariants into account. However, any equivalence map $\bar{x} = \Phi(x)$ must necessarily satisfy the system of invariance equations

$$F_\sigma(x, \bar{x}) \equiv \overline{T}_\sigma(\bar{x}) - T_\sigma(x) = 0, \qquad \text{order}\, \sigma \leq s+1, \qquad (14.24)$$

which is the same as requiring that x and $\bar{x}$ map to the same point of the $(s+1)^{\text{st}}$ order classifying manifold, cf. (14.23). Therefore, the graph of Φ will be a subset of the variety

$$\mathcal{S} = \left\{ (x, \bar{x}) \;\middle|\; \overline{\mathbf{T}}^{(s+1)}(\bar{x}) = \mathbf{T}^{(s+1)}(x) \right\} \subset M \times \overline{M},$$

defined by the vanishing of the functions $F_\sigma(x, \bar{x})$ for order $\sigma \leq s + 1$. Thus, we should first restrict the differential system $\mathcal{I}$ to the subset $\mathcal{S}$, and *then* prove involutivity.

The first order of business is to prove that $\mathcal{S}$ is a submanifold of $M \times \overline{M}$. This is slightly tricky, since the functions $F_\sigma(x, \bar{x})$ almost never form a regular family and so the Implicit Function Theorem is not immediately applicable! To see why, suppose $\bar{x}$ and x are scalar variables. Then the two functions $\bar{x} - x$ and $\bar{x}^2 - x^2$, arising from the functionally dependent functions x, x^2 and $\bar{x}$, $\bar{x}^2$, do *not* form a regular system, since their Jacobian matrix has rank one on the associated variety $\mathcal{S} = \{\bar{x} = x\}$, but has rank two elsewhere. However, the first function $\bar{x} - x$ by itself is regular, and hence its variety, which is also $\mathcal{S}$, is indeed a submanifold. In general, we proceed in a similar, slightly more cautious fashion. Let $x_0 \in M$ and $\bar{x}_0 \in \overline{M}$ be points mapping to a common point z_0 on the overlap of the two classifying manifolds, cf. (14.23). Since $\mathcal{C}^{(s)}$ is an r-dimensional submanifold of the Euclidean space $\mathbb{K}^{(s)}$, we can choose r of the local coordinate functions, $z_{\sigma_1}, \ldots, z_{\sigma_r}$, order $\sigma_\kappa \leq s$, such that $dz_{\sigma_1} \wedge \cdots \wedge dz_{\sigma_r} | \mathcal{C}^{(s)} \neq 0$ in a neighborhood of z_0. This implies that the z_{σ_κ}'s provide local coordinates on $\mathcal{C}^{(s)}$, which is equivalent to the statement that the invariants $T_{\sigma_1}(x), \ldots, T_{\sigma_r}(x)$ are functionally independent in a neighborhood of $x_0 \in M$. Moreover, since $\mathcal{C}^{(s+1)}$ is also r-dimensional, and projects to $\mathcal{C}^{(s)}$, the *same* coordinate functions remain independent when restricted to $\mathcal{C}^{(s+1)}$. Furthermore, the corresponding barred invariants $\overline{T}_{\sigma_1}, \ldots, \overline{T}_{\sigma_r}$ must also be functionally independent in a neighborhood of $\bar{x}_0 \in \overline{M}$ since they parametrize the self-same classifying manifold. This clearly implies that the differences $F_{\sigma_\kappa}(x, \bar{x})$, $\kappa = 1, \ldots, r$ are functionally independent in a neighborhood of $(x_0, \bar{x}_0)$, and hence form a regular family there. Therefore, we can now use the Implicit Function Theorem to conclude that the variety $\widetilde{\mathcal{S}} = \{(x, \bar{x}) \,|\, F_{\sigma_\kappa}(x, \bar{x}) = 0, \kappa = 1, \ldots, r\}$ forms a $(2m - r)$-dimensional submanifold of $M \times \overline{M}$ near $(x_0, \bar{x}_0)$. However, $\widetilde{\mathcal{S}}$ agrees with $\mathcal{S}$ in this neighborhood, hence we have shown that $\mathcal{S}$ is also a $(2m - r)$-dimensional submanifold of $M \times \overline{M}$.

Now, we need to check that the restriction $\mathcal{I} \,|\, \mathcal{S}$ of our differential system satisfies the hypotheses of Frobenius' Theorem 14.7. First we compute the rank of the restricted system, and verify that it is constant. On $\mathcal{S}$, the differentials dF_{σ_κ} of the local coordinate functions will vanish, as above, which must be taken into account when computing the rank. For any $\sigma = \sigma_\kappa$, we find

$$dF_\sigma = d\overline{T}_\sigma - dT_\sigma = \sum_{j=1}^{m} \left\{ \frac{\partial \overline{T}_\sigma}{\partial \overline{\theta}^j}\,\overline{\theta}^j - \frac{\partial T_\sigma}{\partial \theta^j}\,\theta^j \right\}$$
$$= \sum_{j=1}^{m} \left[\{\overline{T}_{\sigma,j} - T_{\sigma,j}\}\,\overline{\theta}^j + T_{\sigma,j}\,\{\overline{\theta}^j - \theta^j\} \right].$$

Since order $\sigma \leq s$, the difference $\overline{T}_{\sigma,j} - T_{\sigma,j} = F_{\sigma,j}$ vanishes as a consequence of the invariance equations (14.24). Therefore, each differential $dF_\sigma = \sum_j T_{\sigma,j}\,\vartheta^j$ is a linear combination of the one-forms defining our differential system $\mathcal{I}$. Linear independence of the dF_{σ_κ} therefore implies that, at each point of $\mathcal{S}$, precisely $m - r$ of the restricted one-forms $\vartheta^j|\mathcal{S}$ remain linearly independent, and hence the rank of $\mathcal{I}\,|\,\mathcal{S}$ equals $m - r$ at every point of $\mathcal{S}$.

Second, we need to prove that $\mathcal{I}\,|\,\mathcal{S}$ is involutive, and this requires computing the differentials of the generating one-forms, as in (14.21):

$$d\vartheta^i = d\overline{\theta}^i - d\theta^i = \sum_{j<k} \left\{ \overline{T}^i_{jk}\,\overline{\theta}^j \wedge \overline{\theta}^k - T^i_{jk}\,\theta^j \wedge \theta^k \right\}$$
$$= \sum_{j<k} \left[\{\overline{T}^i_{jk} - T^i_{jk}\}\,\overline{\theta}^j \wedge \overline{\theta}^k + T^i_{jk}\,\{\vartheta^j \wedge \overline{\theta}^k + \theta^j \wedge \vartheta^k\} \right].$$

On the submanifold $\mathcal{S}$, the first set of summands vanishes since the T^i_{jk}'s appear among our complete collection of structure invariants T_σ. Therefore $d\vartheta^i|\mathcal{S} \in \mathcal{I}\,|\,\mathcal{S}$, which proves involutivity. Frobenius' Theorem 14.7 guarantees the existence of a unique m-dimensional integral submanifold $N \subset \mathcal{S} \subset M \times \overline{M}$ passing through each point $(x, \overline{x}) \in \mathcal{S}$. The transversality of N follows easily, in the same manner as in the proof of Theorem 14.18. *Q.E.D.*

Exercise 14.25. Prove that Theorem 14.24 holds without change in either statement or proof if we also include additional invariant functions, beyond the structure functions, as discussed at the end of Chapter 8; see also Example 9.13. These and their covariant derivatives are to be included in the collection of basic invariants $\{T_\sigma, J_\tau\}$.

Theorem 14.26. *The symmetry group G of a regular coframe $\boldsymbol{\theta}$ of constant rank r on an m-dimensional manifold M is a local Lie group of transformations of dimension $m - r$.*

Proof: A symmetry is just a self-equivalence, and so we set $\overline{M} = M$, $\overline{\boldsymbol{\theta}} = \boldsymbol{\theta}$. Clearly, symmetries form a local transformation group, since the composition of two symmetries is, where defined, again a symmetry. According to Theorem 14.24, through each point $(x, \bar{x}) \in M \times \overline{M}$ which satisfies the invariance equations (14.24), there passes a *unique* m-dimensional integral submanifold N of the differential system $\mathcal{I}$ determining the graph of a symmetry. Using the Implicit Function Theorem, for each fixed x, there is an $(m-r)$-dimensional submanifold of $\bar{x}$'s which satisfy the restrictions $\{\bar{x} \mid \overline{T}_\sigma(\bar{x}) = T_\sigma(x)\}$, so that the collection of such symmetries forms an $(m - r)$-parameter group. (In particular, the only symmetry of a regular coframe which has a fixed point is the identity map!) Smoothness of the group action follows from the smoothness of the foliation provided by the Frobenius Theorem. *Q.E.D.*

Remark: The level sets $N_z = \{x \in M \mid \mathbf{T}^{(s+1)}(x) = z\} \subset M$ of the fundamental structure invariants of a coframe of order s must be invariant subsets under the action of the associated symmetry group, since any symmetry must leave the invariants fixed. In fact, the proof of Theorem 14.26 implies the group acts transitively and freely on N_z, and hence the coframe reduces to a coframe on N_z satisfying the Maurer–Cartan structure equations for the symmetry group G. What is remarkable is that the reductions to each invariant submanifold N_z are isomorphic coframes, corresponding to one and the same Lie group G. Thus, the different level sets N_z must all be transforming in an "identical" manner under any self-equivalence. See Example 14.3 for a family of group actions for which this is not the case; Theorem 14.26 says that such an example can never arise as the symmetry group of a coframe. By way of contrast, this fact does not necessarily hold for a coframe with an intransitive infinite dimensional symmetry group — the full pseudo-group can restrict to define nonisomorphic actions on the different invariant submanifolds N_z; see [**37**].

Global Equivalence

Global results are quite a bit harder to pin down, in part because covering spaces have identical local structures, so that local equivalence can, at best, only give global equivalence up to a covering map. The prototypical example in this regard is the correspondence between a Lie group and its Lie algebra: Two connected Lie groups have the same Lie algebra if and only if they are locally (in a neighborhood of the identity) isomorphic. Moreover, Theorem 2.56 implies that the two global groups have a common covering Lie group. Of course, the same statements hold if we replace "Lie algebra" by "Maurer–Cartan coframe". The question is to what extent this global correspondence applies to more general locally equivalent coframes.

We begin by recalling the definition of the term "covering map", which means a local diffeomorphism $\Psi\colon \widehat{M} \to M$ from one manifold onto another of the same dimension; see Chapter 1. The covering manifold $\widehat{M}$ is not required to cover M uniformly, although we will require it to be connected so as not to allow too much pathology. In particular, if M is simply connected, then the only connected covering manifold is M itself. Suppose $\boldsymbol{\theta}$ is a coframe on M. A coframe $\widehat{\boldsymbol{\theta}}$ on a connected covering manifold $\widehat{M}$ is called a *covering coframe* for $\boldsymbol{\theta}$ if and only if the covering map $\Psi\colon \widehat{M} \to M$ defines an equivalence, so $\Psi^*\widehat{\boldsymbol{\theta}} = \boldsymbol{\theta}$.

Proposition 14.27. *Let $\widehat{\boldsymbol{\theta}}$ be a covering coframe for $\boldsymbol{\theta}$. Then their classifying manifolds are identical: $\mathcal{C}^{(s)}(\widehat{\boldsymbol{\theta}}) = \mathcal{C}^{(s)}(\overline{\boldsymbol{\theta}})$, $s \geq 0$.*

The proof is immediate. Thus, any pair of coframes having a common covering coframe, or, alternatively, covering a common coframe, will have identical classifying manifolds. An interesting problem is to establish whether or not such coverings exhaust the possible coframes having the same classifying manifolds. We will call two coframes, $\boldsymbol{\theta}$ on M, and $\overline{\boldsymbol{\theta}}$ on $\overline{M}$, *globally equivalent up to covering* if they have a common covering coframe $\widehat{\boldsymbol{\theta}}$, defined on a connected manifold $\widehat{M}$ which covers both M and $\overline{M}$.

The first of our global results applies to the rank zero case, when the coframe is locally equivalent to a Maurer–Cartan coframe on a Lie group.

Theorem 14.28. *Let $\boldsymbol{\theta}$ be a coframe of rank zero on a connected manifold M. Then $\boldsymbol{\theta}$ is globally equivalent, up to covering, to a Maurer–Cartan coframe $\boldsymbol{\alpha}$ on an open subset of a Lie group G.*

Proof: Let $N \subset M \times G$ denote a maximal integral submanifold corresponding to the differential system $\mathcal{I}$ generated by $\vartheta^i = \alpha^i - \theta^i$, $i = 1, \ldots, m$, as described in the proof of Theorem 14.26. We know

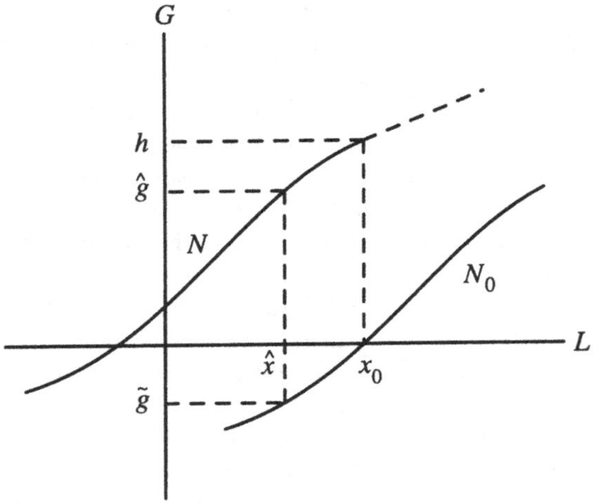

Figure 5. Global Equivalence

that, locally, N is the graph of an equivalence map. Indeed, the two projections $\pi_M: M \times G \to M$, and $\pi_G: M \times G \to G$, determine local diffeomorphisms when restricted to N; moreover, equivalence implies that the given coframe on M and also the Maurer–Cartan coframe $\boldsymbol{\alpha}$ on G both pull back to the same coframe, $\pi_M^* \boldsymbol{\theta} = \pi_G^* \boldsymbol{\alpha}$ on N. I claim that the restricted projection $\pi_M: N \to M$ is a covering map, which requires us to prove that $\pi_M(N) = M$. Assuming this, the theorem follows directly, since N will also cover an open subset of G under π_G, thereby fulfilling the covering conditions. To prove the claim, we first remark that any right translate $R_h[N] = \{(x, g \cdot h) \mid (x, g) \in N\}$ of an integral submanifold N of $\mathcal{I}$, by a group element $h \in G$, is also an integral submanifold; this follows immediately from the right-invariance of the Maurer–Cartan forms. Suppose $\pi_M(N) \neq M$, and let $x_0 \in M \setminus \pi_M(N)$ be a point in the closure of $\pi_M(N)$. According to Theorem 14.18, we can find a local equivalence $\Phi_0: U \to G$, where U is a neighborhood of x_0 in M, mapping x_0 to the identity element $e = \Phi_0(x_0)$. (See Figure 5.) Let N_0 denote the graph of Φ_0, so that N_0 is also an integral submanifold of our differential system passing through (x_0, e). Choose any point $\widehat{x} \in U \cap \pi_M(N)$, so $\widehat{x} = \pi_M(\widehat{x}, \widehat{g})$ for some point $(\widehat{x}, \widehat{g}) \in N$ in the original integral submanifold. Let $\widetilde{g} = \Phi_0(\widehat{x})$, and define $h = \widetilde{g}^{-1} \cdot \widehat{g}$.

Then, by the previous remark, $R_h[N_0]$ is an integral submanifold of the differential system; moreover, the point $(\widehat{x}, \widehat{g}) = (\widehat{x}, \widetilde{g} \cdot h)$ is contained both in $R_h[N_0]$ and in our original integral submanifold N. Therefore, by uniqueness and maximality of N, the integral submanifold $R_h[N_0]$ must be an open submanifold of N. But the point $(x_0, h) = R_h(x_0, e)$ lies in $R_h[N_0]$, and hence in N. This contradicts our original assumption that x_0 was not in the projection $\pi_M(N)$, and so proves the claim. *Q.E.D.*

Example 14.29. To see that, even if M itself is simply connected, it may not be realizable as an open subset of a Lie group, consider the coframe

$$\theta^1 = \cos\varphi\, dr - r\sin\varphi\, d\varphi, \qquad \theta^2 = \sin\varphi\, dr + r\cos\varphi\, d\varphi, \tag{14.25}$$

defined on the half plane $M = \{(r, \varphi) \mid r > 0\}$. Clearly, in terms of polar coordinates, this coframe is locally diffeomorphic to the Maurer–Cartan coframe dx, dy for the two-dimensional abelian Lie group $\mathbb{R}^2$. Indeed, M can be identified with the simply connected covering space for the punctured plane $\mathbb{R}^2 \setminus \{(-1,0)\}$, so that the forms (14.25) are the pull-backs of the coordinate forms under the covering map

$$\pi(r, \varphi) = (r\cos\varphi - 1, r\sin\varphi) \in \mathbb{R}^2 \setminus \{(-1,0)\}. \tag{14.26}$$

(We choose to puncture the plane at $(-1,0)$ so as to retain the origin — the group identity element.) However, it is not difficult to see that the coframe (14.25) is not globally equivalent to the Maurer–Cartan coframe on any open subset of the Lie group $\mathbb{R}^2$.

This particular example is extremely interesting, as it describes the simplest case of a new class of local Lie groups. Locally, M looks like a two-parameter abelian Lie group, and satisfies all the group axioms. Globally, however, M fails to satisfy the associativity axiom! To see this, we identify the point $(1,0) \in M$ as the identity element and define the group product $w = u \cdot v$ of two points $u, v \in M$ so that it projects to the ordinary vector sum $\pi(w) = \pi(u) + \pi(v)$ of their projections, cf. (14.26). This definition can be made in a smooth manner provided one of the summands lies in the same sheet of the covering space as the identity $(1,0)$. A three-fold product $u \cdot v \cdot w$ will be unambiguously defined only if the triangle with vertices $\pi(u)$, $\pi(u) + \pi(v)$, and $\pi(u) + \pi(v) + \pi(w)$, does not contain the singular point $(-1,0)$ in its interior; otherwise the two three-fold products $u \cdot (v \cdot w)$ and $(u \cdot v) \cdot w$ will end up on different sheets

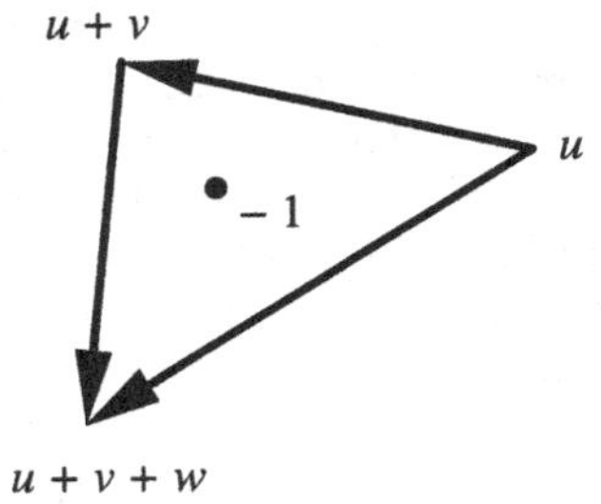

Figure 6. Non–Associativity

of the covering surface and hence are not equal $u \cdot (v \cdot w) \neq (u \cdot v) \cdot w$; see Figure 6. It is possible, by restricting the domain of definition of the group multiplication, to ensure that the associativity axiom holds, but nevertheless there exist four-fold (or "higher-fold") products which are not well defined, e.g., we may find $u \cdot (v \cdot (w \cdot z)) \neq ((u \cdot v) \cdot w) \cdot z$ even though both products are defined. The full details of this construction can be found in the paper [**188**].

In general, a *local Lie group* is given by a manifold, together with a smooth multiplication operation, which is defined provided one or the other multiplicands is sufficiently near the identity, as well as a smooth inversion map, again possibly only defined for group elements sufficiently near the identity. The local group is *regular* provided the left and right multiplication maps are local diffeomorphisms where defined. The prototypical example of a local Lie group is any neighborhood of the identity of a global Lie group. Lie's "Third Fundamental Theorem", cf. [**195**], says that any Lie algebra generates a local Lie group. Cartan, [**35**], proved the global counterpart to Lie's Theorem, appearing here as Theorem 2.56, which implies that, locally, any local Lie group is globalizable, in the sense that some neighborhood of the identity of the local Lie group is isomorphic to a neighborhood of the identity of some global Lie group (having the same Lie algebra). However, the local group of Example 14.29 provides a new, explicit example of a non-globalizable local Lie group. Indeed, Theorem 14.28, can be reinterpreted to state that any regular, local Lie group is a cover of some open subset of a global Lie group.

Mal'cev, [**164**], has pointed out the importance of associativity in understanding the globalizability problem. He proved that a local topological group is contained in a global group if and only if it satisfies the following generalized associativity property: When defined, an n-fold product of group elements does not depend on the order of performing the multiplications. In our context, it can be shown that a local Lie group is globalizable if and only if it also satisfies Mal'cev's global associativity axiom.

Remark: Using a version of the Hopf–Rinow Theorem, Gardner, [**78**; p. 72], shows that a rank zero coframe $\boldsymbol{\theta} = \{\theta^1, \ldots, \theta^m\}$ on a simply connected manifold M which is metrically complete with respect to the Riemannian metric $\sum_i (\theta^i)^2$ induced by the coframe is globally equivalent to a Maurer–Cartan coframe on a global Lie group. Thus, metric completeness is, in some subtle way, related to associativity.

At the other extreme, if a regular coframe $\boldsymbol{\theta}$ has maximal rank $m = \dim M$ then its classifying manifold is itself an m-dimensional submanifold of the classifying space. In fact, this allows one to unambiguously define an equivalent coframe on the classifying manifold itself, in such a way that $\boldsymbol{\theta}$ is a covering coframe of the "classifying coframe". This will imply the following global equivalence theorem for coframes of maximal rank.

Theorem 14.30. *Let $\boldsymbol{\theta}$ on M and $\overline{\boldsymbol{\theta}}$ on $\overline{M}$ be coframes, both defined on m-dimensional manifolds. Assume both coframes are regular, of the same order s, and both have maximal rank m. Then $\boldsymbol{\theta}$ and $\overline{\boldsymbol{\theta}}$ have identical $(s+1)^{\text{st}}$ order classifying manifolds, $\mathcal{C}^{(s+1)}(\overline{\boldsymbol{\theta}}) = \mathcal{C}^{(s+1)}(\boldsymbol{\theta})$, if and only if they both cover a common coframe $\boldsymbol{\vartheta}$.*

Proof: In fact, $\boldsymbol{\vartheta}$ will be a coframe on the common classifying manifold $N = \mathcal{C}^{(s+1)}(\overline{\boldsymbol{\theta}}) = \mathcal{C}^{(s+1)}(\boldsymbol{\theta})$. To construct $\boldsymbol{\vartheta}$, let $\mathbf{T} = \mathbf{T}^{(s+1)} \colon M \to N$ denote the $(s+1)^{\text{st}}$ order structure map. Since $\boldsymbol{\theta}$ has maximal rank, $\mathbf{T}$ is a local diffeomorphism, and hence a covering map from M onto N. Suppose $U \subset M$ is an open subset such that $\mathbf{T}_U \equiv \mathbf{T} \,|\, U$ defines a one-to-one map from U onto some open subset $V = \mathbf{T}(U) \subset N$. Define the coframe $\boldsymbol{\vartheta}_U = (\mathbf{T}_U^{-1})^* \boldsymbol{\theta}$ on V to be the inverse pull-back of our original coframe, restricted to U, so that $\mathbf{T}_U$ defines an equivalence between the two coframes $\boldsymbol{\theta} \,|\, U$ and $\boldsymbol{\vartheta}_U$. I claim that $\boldsymbol{\vartheta}_U$ is just the restriction of a globally defined coframe $\boldsymbol{\vartheta}$ on $\widetilde{M}$, so $\boldsymbol{\vartheta}_U = \boldsymbol{\vartheta} \,|\, V$. To see this, suppose $\widetilde{U} \subset M$ is another open set of the same type, with one-to-one image

$\widetilde{V} = \mathbf{T}(\widetilde{U})$. We need to show that, on the intersection $V \cap \widetilde{V}$, the two coframes $\boldsymbol{\vartheta}_U$ and $\boldsymbol{\vartheta}_{\widetilde{U}}$ agree. By shrinking the two domains U and $\widetilde{U}$ if necessary, we can assume $V = \widetilde{V}$. Theorem 14.24 implies that the composite map $\mathbf{T}_{\widetilde{U}}^{-1} \circ \mathbf{T}_U : U \to \widetilde{U}$ defines an equivalence (i.e., a discrete symmetry) between the restrictions of $\boldsymbol{\theta}$ to U and V, so that

$$\boldsymbol{\theta} \mid U = \left[\mathbf{T}_{\widetilde{U}}^{-1} \circ \mathbf{T}_U\right]^* (\boldsymbol{\theta} \mid \widetilde{U}) = \left[\mathbf{T}_U\right]^* \circ \left[\mathbf{T}_{\widetilde{U}}^{-1}\right]^* (\boldsymbol{\theta} \mid \widetilde{U}).$$

This implies

$$\boldsymbol{\vartheta}_U = \left[\mathbf{T}_U^{-1}\right]^* (\boldsymbol{\theta} \mid U) = \left[\mathbf{T}_{\widetilde{U}}^{-1}\right]^* (\boldsymbol{\theta} \mid \widetilde{U}) = \boldsymbol{\vartheta}_{\widetilde{U}}$$

define the *same* coframe on the common domain $V = \widetilde{V}$, which proves the claim. We conclude the $\boldsymbol{\theta}$ is a covering coframe for $\boldsymbol{\vartheta}$. The same construction clearly works for $\overline{\boldsymbol{\theta}}$, and leads to the same coframe $\boldsymbol{\vartheta}$ on the common classifying manifold N. Therefore, $\boldsymbol{\vartheta}$ serves as the common coframe covered by both $\boldsymbol{\theta}$ and $\overline{\boldsymbol{\theta}}$. *Q.E.D.*

Remark: Given two equivalent coframes of maximal rank, the variety $\mathcal{S} \subset M \times \overline{M}$ defined in the proof of Theorem 14.24 is, in fact, a covering manifold for both M and $\overline{M}$ under the projections π_M and $\pi_{\overline{M}}$. Moreover, the pull-backs of the two coframes to $\mathcal{S}$ are easily seen to agree, $\pi_M^* \boldsymbol{\theta} \mid \mathcal{S} = \pi_{\overline{M}}^* \overline{\boldsymbol{\theta}} \mid \mathcal{S}$, since $\mathcal{S}$ is, itself, locally the same as the graph of an equivalence between the two coframes. However, $\mathcal{S}$ is not necessarily connected, and so can be quite pathological.

The intermediate cases, when the rank of the coframe is neither zero nor maximal, are more delicate. Locally, a coframe of intermediate rank $0 < r < m$ looks like the Cartesian product of its classifying manifold and its symmetry group, $\mathcal{C}^{(s)} \times G$, or, rather, $U \times V$, where $U \subset \mathcal{C}^{(s)}$ and $V \subset G$ are open subsets. Globally, this should remain true up to some form of covering, but I have been unable to successfully formulate a completely general theorem.

Chapter 15

The Cartan–Kähler Existence Theorem

In the situations when Frobenius' Theorem no longer applies, we must resort to more powerful tools. The most general existence result for solutions to systems of differential equations prescribed by the vanishing of a collection of differential forms is known as the Cartan–Kähler Theorem. As before, we will consider a *differential system* defined by a (closed) differential ideal $\mathcal{I}$. Frobenius' Theorem 14.7 deals with the case when $\mathcal{I}$ is simply generated by one-forms, and so prescribes an involutive Pfaffian system. The method of proof is, in essence, to reduce the problem to the integration of a system of ordinary differential equations. When forms of higher degree need to be included among the generators, the system can no longer be attacked by ordinary differential equation methods, and we must appeal to existence theorems for partial differential equations, of which the most basic is the celebrated Cauchy–Kovalevskaya Theorem. The Cartan–Kähler Theorem is a natural (from the view-point of systems of partial differential equations defined by differential forms) extension of the Cauchy–Kovalevskaya Theorem to overdetermined systems of partial differential equations. Of course, the Cauchy–Kovalevskaya Theorem only applies to *analytic* systems of partial differential equations, and hence the Cartan–Kähler Theorem will itself only be applicable to analytic differential systems.

The Cauchy–Kovalevskaya Existence Theorem

Although the Cauchy–Kovalevskaya Theorem has a straightforward generalization to higher order systems of partial differential equations, in our applications we shall only be dealing with first order systems. Therefore, we shall restrict our treatment of this fundamental existence theorem to such cases, and refer the reader to [**56**] for the general case.

Definition 15.1. A system of first order partial differential equations for functions $u(x) = (u^1(x), \ldots, u^q(x))$, $x = (x^1, \ldots, x^p)$ is said to be in *Kovalevskaya form* if it has been solved for the derivatives of the u's with respect to one of the x's, say x^p, so that

$$\frac{\partial u^\alpha}{\partial x^p} = \Delta^\alpha(x, \widetilde{u^{(1)}}), \qquad \alpha = 1, \ldots, q, \tag{15.1}$$

where the right hand side depends on $x^1, \ldots, x^p$, $u^1, \ldots, u^q$, and the first order partial derivatives $u_j^\beta = \partial u^\beta / \partial x^j$ for $\beta = 1, \ldots, q$, $j = 1, \ldots, p-1$. (The fact that the derivatives $\partial u^\alpha / \partial x^p$ appearing on the left hand side do not appear on the right hand side is indicated by the tilde over $u^{(1)}$.)

The Cauchy–Kovalevskaya Theorem concerns the existence and uniqueness of solutions to the Cauchy problem for a system in Kovalevskaya form (15.1), with initial *Cauchy data*

$$u^\alpha(x^1, \ldots, x^{p-1}, x_0^p) = h^\alpha(x^1, \ldots, x^{p-1}), \qquad i = 1, \ldots, q, \tag{15.2}$$

prescribed on an open subset of the initial hypersurface $\{x \mid x^p = x_0^p\}$. The fact that the system can be solved for all the x^p-derivatives means that this hypersurface is *noncharacteristic*. The *Cauchy problem* for the system (15.1) is to find a (local) solution $u(x)$ to the system satisfying the given Cauchy data.

Theorem 15.2. *Let* (15.1) *be an analytic system of partial differential equations in Kovalevskaya form, and let* $h^\alpha(x^1, \ldots, x^{p-1})$ *be analytic functions defined in a neighborhood of* $(x_0^1, \ldots, x_0^{p-1})$. *Then, in a neighborhood of* $x_0 = (x_0^1, \ldots, x_0^p)$, *there exists a unique analytic solution to the Cauchy problem consisting of* (15.1) *and initial data* (15.2).

The analyticity hypothesis in the Cauchy–Kovalevskaya Existence Theorem 15.2 is essential, as an example of H. Lewy, [**149**], makes clear. The proof relies on the construction of a power series solution, and the majorant method is used to prove convergence. We refer the reader to [**56**; §I.7] for the details.

The underdetermined case, in which there are fewer differential equations than unknown functions, will also be of interest to us. A system of first order partial differential equations will be said to be in *underdetermined Kovalevskaya form* if we can only solve for some of the partial derivatives of the u's with respect to one of the coordinates, with the resulting system given by

$$\frac{\partial u^\alpha}{\partial x^p} = \Delta^\alpha(x, \widetilde{u^{(1)}}), \qquad \alpha = 1, \ldots, r, \tag{15.3}$$

for some $r < q$. The right hand side Δ^α now depends on $x^1, \ldots, x^p$, $u^1, \ldots, u^q$, the first order partial derivatives u_j^β for $\beta = 1, \ldots, q$, $j = 1, \ldots, p-1$, as well as the derivatives u_p^γ for $\gamma = r+1, \ldots, q$. In this case, for given Cauchy data (15.2), we can clearly prescribe the extra functions $u^{r+1}(x), \ldots, u^q(x)$ to be *any* analytic functions whatsoever, subject only to the last $q-r$ initial conditions. Once these $q-r$ functions have been specified, the system (15.3) is of the usual Kovalevskaya type, and so the Cauchy–Kovalevskaya Theorem 15.2 will again guarantee the existence of a unique solution.

Theorem 15.3. *Consider an analytic system of partial differential equations in underdetermined Kovalevskaya form* (15.3), *and let* $h^\alpha(x^1, \ldots, x^{p-1})$ *be analytic functions defined in a neighborhood of* $(x_0^1, \ldots, x_0^{p-1})$. *Then, in a neighborhood of* $x_0 = (x_0^1, \ldots, x_0^p)$, *there exists a solution to the Cauchy problem consisting of* (15.3) *and initial data* (15.2). *Moreover, the general solution to the underdetermined Cauchy problem* (15.3), (15.2) *depends on* $q - r$ *arbitrary analytic functions which satisfy the initial conditions.*

Necessary Conditions

Consider a general system of partial differential equations defined by the vanishing of a collection of differential forms. The forms determine a differential ideal, and the goal is to find integral submanifolds $N \subset M$ of a suitable dimension n, which requires that all the differential forms in $\mathcal{I}$ vanish thereon. In the equivalence problem applications, the differential ideal $\mathcal{I}$ defines a Pfaffian system, generated by a collection of one-forms $\omega^1, \ldots, \omega^r$, and their differentials, $d\omega^1, \ldots, d\omega^r$. In the case covered by Frobenius' Theorem, the ideal is closed, so that the two-forms $d\omega^\mu$ belong to the exterior ideal generated by the one-forms. The Cartan–Kähler Theorem will only be required if this does not hold: The ideal is algebraically generated by one-forms ω^μ and one or more of the two forms $d\omega^\mu$. However, it is not much more difficult to treat the completely general case when the differential ideal is (algebraically) generated by a collection of differential forms of various degrees, which will therefore be the focus of this chapter.

As before, let $\mathcal{I}^{(k)}$, $1 \le k \le m$, denote the homogeneous components consisting of all k-forms in the differential ideal $\mathcal{I}$. As remarked earlier in Chapter 14, in the regular case there is no need to include functions (0-forms) in our ideal, so we assume $\mathcal{I}^{(0)} = \{0\}$. Let us assume that, for simplicity, the ideal is finitely generated, and let the one-forms $\omega^1, \ldots, \omega^r$ form a generating set for $\mathcal{I}^{(1)}$, the two-forms $\Omega^1, \ldots, \Omega^s$ form a generating set for $\mathcal{I}^{(2)}$, and, in general, the k-forms $\Omega^1_{(k)}, \ldots, \Omega^{t_k}_{(k)}$ a generating set for $\mathcal{I}^{(k)}$. (The case when $\mathcal{I}$ is not finitely generated is not substantially more difficult, but the notation gets more cumbersome.) Note that, since we are assuming that the ideal is closed under the exterior derivative, the differentials $d\omega^\mu$ of the one-forms in $\mathcal{I}^{(1)}$ must appear among (or as linear combinations of) the generators Ω^ν of $\mathcal{I}^{(2)}$, so that, in fact, we could take the first r of the Ω^ν's to be the differentials of the ω^μ's. Similarly, the first $s-r$ of the generators of $\mathcal{I}^{(3)}$ could be taken as the differentials of the remaining (nonclosed) Ω^ν's, and so on. We do not necessarily assume that the generating forms for the ideal are linearly independent at each point, although this will usually be the case.

The preliminary step is to determine necessary conditions for a submanifold N to be an integral submanifold of the differential ideal $\mathcal{I}$. These will be much more sophisticated versions of our naïve rank condition used in the simply generated case treated in Chapter 13. We begin by analyzing the problem at a single point, thereby deriving a collection of algebraic conditions that must be satisfied before we have any chance of finding an integral submanifold. Given $x \in N$, let $\mathbf{v}_1, \ldots, \mathbf{v}_n$ be a basis for the tangent space $TN|_x$ to N at x. Then the vanishing of the forms in $\mathcal{I}$ on N requires that the basis tangent vectors must satisfy a large system of multi-linear equations:

$$\begin{array}{rll}
\langle \omega^\nu ; \mathbf{v}_i \rangle = 0, & 1 \le i \le n & \nu = 1, \ldots, r = t_1, \\
\langle \Omega^\nu ; \mathbf{v}_{i_1}, \mathbf{v}_{i_2} \rangle = 0, & 1 \le i_1 < i_2 \le n, & \nu = 1, \ldots, s = t_2, \\
\langle \Omega^\nu_{(3)} ; \mathbf{v}_{i_1}, \mathbf{v}_{i_2}, \mathbf{v}_{i_3} \rangle = 0, & 1 \le i_1 < i_2 < i_3 \le n, & \nu = 1, \ldots, t_3, \\
\vdots & \vdots & \vdots \\
\langle \Omega^\nu_{(n)} ; \mathbf{v}_1, \mathbf{v}_2, \ldots \mathbf{v}_n \rangle = 0, & & \nu = 1, \ldots, t_n,
\end{array} \tag{15.4}$$

at the point x. We note that, if any $\Omega^\nu_{(k)}$ can be algebraically re-expressed terms of forms in the ideal of lower degree, then the equation in the system (15.4) corresponding to that form is automatically satisfied as a consequence of lower order equations, and hence we do not need to include it in the system. For instance, if $\Omega^\nu = \sum \eta^\nu_\mu \wedge \omega^\mu$ (the one-forms η^ν_μ are not necessarily in $\mathcal{I}$), then the equation involving Ω^ν in (15.4) is automatically zero as a consequence of the equations involving the ω^μ's. Thus, in the case of a simply generated differential ideal, only the first set of equations in (15.4) is required. In the cases relevant to the Cartan equivalence method, we only need deal with the first two sets of equations in (15.4), since only the generating one-forms and their differentials will be needed.

In order to understand the system of multi-linear equations (15.4) better, let us begin by considering the simplest case of a one-dimensional integral submanifold (an integral curve) C of the differential system. Its tangent vector $\mathbf{v}_1 \neq 0$ at a point $x \in C$ must be annihilated by all the one-forms in our ideal, and so satisfy the system of linear equations

$$\langle \omega^\nu \,; \mathbf{v}_1 \rangle = 0, \qquad \nu = 1, \ldots, r, \tag{15.5}$$

at the point x. In order to express this system in local coordinates, we introduce a local coframe $\theta^1, \ldots, \theta^m$ for M near the point x. (For example, we can use the coordinate coframe $dx^1, \ldots, dx^m$ corresponding to local coordinates $(x^1, \ldots, x^m)$ near x, although in many cases a different local coframe will be more appropriate for the problem under consideration.) We can then express the ω's directly in terms of the coframe:

$$\omega^\nu = \sum_{i=1}^{m} A^\nu_i(x)\, \theta^i, \qquad \nu = 1, \ldots, r. \tag{15.6}$$

Similarly, we express the tangent vector $\mathbf{v}_1$ using the dual frame vector fields $\partial/\partial\theta^i$, so

$$\mathbf{v}_1 = \sum_{i=1}^{m} \xi^i_1 \frac{\partial}{\partial \theta^i}, \qquad \text{where} \qquad \xi^i_1 = \langle \theta^i \,; \mathbf{v}_1 \rangle. \tag{15.7}$$

This allows us to identify the tangent vector $\mathbf{v}_1$ with the column vector $(\xi^1_1, \ldots, \xi^m_1)^T \in \mathbb{R}^m$. Then the conditions (15.5) that the tangent vector to any integral curve must satisfy are explicitly given as a system of r linear equations

$$A\, \mathbf{v}_1 = 0, \tag{15.8}$$

where $A = A(x) = \left(A_i^\nu(x)\right)$ denotes the $r \times m$ coefficient matrix which specifies the one-forms (15.6) relative to the chosen coframe. Now, $\mathbf{v}_1 \neq 0$ is a nonzero vector, so the linear system (15.8) must have at least one nonzero solution, which implies that the rank of the $r \times m$ matrix A is at most $m - 1$. We let

$$\widetilde{s}_0 = \widetilde{s}_0(x) = \operatorname{rank} A(x) \tag{15.9}$$

denote the rank of the matrix A at the point x. Note that, if $\mathcal{I}$ is simply generated, then $\widetilde{s}_0(x)$ is exactly the same as what we called the rank of the ideal at the point x in Chapter 13. We have shown that the inequality

$$\widetilde{s}_0(x) \leq m - 1 \tag{15.10}$$

is an elementary necessary condition for an integral curve to pass through the point x. If the condition (15.10) holds in a neighborhood of x, then there are solutions $\mathbf{v}_1$ to (15.8) depending smoothly on x, and the integral curve to any such vector field determines a one-dimensional integral submanifold to the differential system. (In this case, we do not need to require analyticity, which only plays a role in the existence of higher dimensional integral submanifolds.) A very simple necessary condition for the existence of an n-dimensional integral submanifold N passing through x, then, is that $\widetilde{s}_0(x) \leq m - n$, as this will allow at least n linearly independent solutions to the system (15.8) which could serve as basis tangent vectors to an integral submanifold; see Proposition 13.10. For a simply generated differential ideal, Frobenius' Theorem 14.7 demonstrates that these conditions are also sufficient. However, the presence of higher degree forms, which are not algebraic consequences of the one-forms, necessitates the consideration of a more sophisticated collection of conditions.

Consider a two-dimensional integral submanifold N (an integral surface). Let $\mathbf{v}_1, \mathbf{v}_2$ denote a basis for its tangent space, which we coordinatize as in the previous paragraph, cf. (15.7). Note that any curve lying in N is an integral curve for the system, hence the integral curve inequality (15.10) must already hold. Now we proceed to analyze the further necessary conditions by treating the basis tangent vectors in order. The first tangent vector $\mathbf{v}_1$ must satisfy (15.8). Fixing $\mathbf{v}_1$ for the moment, the second tangent vector must be a solution to the enlarged linear system

$$A\,\mathbf{v}_2 = 0, \qquad B(\mathbf{v}_1)\,\mathbf{v}_2 = 0, \tag{15.11}$$

where the coefficient matrix $B(\mathbf{v}_1) = B(x, \mathbf{v}_1)$ comes from the second set of conditions in (15.4) above. Explicitly, if

$$\Omega^\nu = \sum_{\substack{i,j=1 \\ i<j}}^{m} \widehat{B}^\nu_{ij}(x)\, \theta^i \wedge \theta^j, \qquad \nu = 1, \ldots, s, \tag{15.12}$$

are the coordinate expressions of the generators of the two-forms in $\mathcal{I}^{(2)}$, written in terms of our prescribed coframe, then

$$\langle \Omega^\nu ; \mathbf{v}_1, \mathbf{v}_2 \rangle = \sum_{\substack{i,j=1 \\ i<j}}^{m} \widehat{B}^\nu_{ij}(x) \big(\xi^i_1 \xi^j_2 - \xi^j_1 \xi^i_2 \big), \tag{15.13}$$

where ξ^i_1, ξ^i_2 denote the corresponding coefficients of $\mathbf{v}_1, \mathbf{v}_2$, cf. (15.7). Therefore $B(x, \mathbf{v}_1)$ is the $s \times m$ matrix whose $(\nu, j)^{\text{th}}$ entry is

$$B^\nu_j(x, \mathbf{v}_1) = \sum_{i<j} \widehat{B}^\nu_{ij}(x) \xi^i_1 - \sum_{i>j} \widehat{B}^\nu_{ji}(x) \xi^j_1. \tag{15.14}$$

Now, since N is an integral surface, we know that $\mathbf{v}_2$ is a nonzero solution to the linear system (15.11), which requires that the rank of the corresponding $(r+s) \times m$ coefficient matrix be at most $m-1$. However, we really need more, since the tangent vectors $\mathbf{v}_1$ and $\mathbf{v}_2$ were assumed to be a basis for the tangent space to N, and hence must be linearly independent. Moreover, we already know one nonzero solution to the system (15.11), namely $\mathbf{v}_1$ itself, which, in view of (15.13), trivially satisfies the second set of equations. Therefore, a second, linearly independent solution $\mathbf{v}_2$ will exist if and only if the rank of the coefficient matrix is at most $m-2$. Following Cartan, we keep track of the rank by defining

$$\widetilde{s}_0 + \widetilde{s}_1 = \widetilde{s}_0(x) + \widetilde{s}_1(x, \mathbf{v}_1) = \operatorname{rank} \begin{pmatrix} A(x) \\ B(x, \mathbf{v}_1) \end{pmatrix}. \tag{15.15}$$

The associated necessary condition for ensuring the existence of an integral surface passing through the point x with the prescribed tangent vectors is

$$\widetilde{s}_0 + \widetilde{s}_1 \le m - 2. \tag{15.16}$$

Continuing one step further, after which the general pattern will become manifest, suppose a three-dimensional integral submanifold N passes through the point x. Let $\mathbf{v}_1, \mathbf{v}_2, \mathbf{v}_3$ be a basis for its tangent space at x. The first two tangent vectors $\mathbf{v}_1, \mathbf{v}_2$ must satisfy the linear systems (15.8) and (15.11), and the third tangent vector must satisfy an even larger linear system

$$A\mathbf{v}_3 = 0, \qquad B(\mathbf{v}_1)\mathbf{v}_3 = 0, \qquad B(\mathbf{v}_2)\mathbf{v}_3 = 0, \qquad C(\mathbf{v}_1, \mathbf{v}_2)\mathbf{v}_3 = 0. \tag{15.17}$$

Here A and B are as above, and $C(\mathbf{v}_1, \mathbf{v}_2)$ is the $t_3 \times m$ matrix which stems from any generating three-forms $\Omega^\nu_{(3)}$ in the ideal. (If $\mathcal{I}$ is the closure of a simply generated exterior ideal, as it will be in our applications, then there are no three-forms needed in the generating set, and we do not need to include the matrix C.) We let $\widetilde{s}_0 + \widetilde{s}_1 + \widetilde{s}_2 = \widetilde{s}_0(x) + \widetilde{s}_1(x, \mathbf{v}_1) + \widetilde{s}_2(x, \mathbf{v}_1, \mathbf{v}_2)$ denote the rank of the full $(r+2s+t_3) \times m$ coefficient matrix for the system of linear equations (15.17). As before, both $\mathbf{v}_1$ and $\mathbf{v}_2$ are automatically solutions to this system, hence in order that there exist a third linearly independent solution $\mathbf{v}_3$, the rank must satisfy the inequality $\widetilde{s}_0 + \widetilde{s}_1 + \widetilde{s}_2 \le m - 3$.

In general, in order to have an n-dimensional integral submanifold to our differential system passing through the point x, a basis $\mathbf{v}_1, \ldots, \mathbf{v}_n$ for its tangent space $TN|_x$ will span an n-dimensional integral element for the differential system. The tangent vectors must satisfy a hierarchy of linear systems: First, $\mathbf{v}_1$ must satisfy the system (15.8), with coefficient matrix having rank $s_0 \le m - 1$; second, $\mathbf{v}_2$ must satisfy the system (15.11), with coefficient matrix having rank $\widetilde{s}_0 + \widetilde{s}_1 \le m - 2$; and so on, until the final vector $\mathbf{v}_n$ must satisfy a gigantic linear system whose coefficient matrix has rank

$$\widetilde{s}_0 + \widetilde{s}_1 + \cdots + \widetilde{s}_{n-1} \le m - n, \tag{15.18}$$

where each integer $\widetilde{s}_j$ depends on the point x and the preceding j tangent vectors $\mathbf{v}_1, \ldots, \mathbf{v}_j$. The above system of rank inequalities constitutes evident necessary conditions for the existence of an integral submanifold of the given dimension. Note, however, that these inequalities depend not only on the point x and the tangent space $TN|_x$ to the integral submanifold at x, but also on the choice of basis of the tangent space, and even the choice of ordering of the basis. Different bases, and different orderings, may very well give different sets of necessary conditions(!).

Sufficient Conditions

We now investigate under what additional hypotheses the above necessary conditions for the existence of integral submanifolds can become sufficient conditions, and thereby solve the integration problem: to determine whether an n-dimensional integral element at a point x has an associated integral submanifold. For this to happen, we need to be more restrictive, and impose some genericity assumptions. As before, "infinitesimal" singularities will occur at "points" where the ranks $\widetilde{s}_j$ defined above suddenly change; only at "regular points" where the ranks are constant will we be able to construct an integral submanifold associated with a given integral element. (The reason for having to impose some such condition was already apparent in our treatment of the differential form version of Frobenius' Theorem 14.7, and the same considerations are, a fortiori, valid here.) Since the rank of a variable matrix can decrease at a point (but never suddenly increase), in order to avoid such singularities, we need to restrict our attention to those points and tangent vectors at which the preceding ranks are all maximal.

This motivates the definition of the *Cartan characters* of a differential system. First, $s_0 = \max\{\widetilde{s}_0(x) \mid x \in M\}$ is the maximal rank of the matrix $A(x)$ in (15.8). A point $x \in M$ is called *regular* if $\operatorname{rank} A(x) = s_0$ is maximal. Next, $s_1 = \max\{\widetilde{s}_1(x, \mathbf{v}_1) \mid x \in M,\ \mathbf{v}_1 \in \mathbb{R}^m\}$ is defined so that $s_0 + s_1$ is the maximal rank of the coefficient matrix for the linear system (15.11). A one-dimensional integral element $E \subset TM|_x$ spanned by the nonzero tangent vector $\mathbf{v}_1$ is called *regular* if x is a regular point and its associated coefficient matrix (15.11) has maximal rank $s_0 + s_1$. Continuing, $s_2 = \max\{\widetilde{s}_2(x, \mathbf{v}_1, \mathbf{v}_2) \mid x \in M,\ \mathbf{v}_1, \mathbf{v}_2 \in TM|_x\}$ is defined so that $s_0 + s_1 + s_2$ is the maximal rank of the coefficient matrix for the linear system (15.17). A two-dimensional integral element $E \subset TM|_x$ is called *regular* if x is a regular point and there exists a basis $\mathbf{v}_1, \mathbf{v}_2$ such that $\mathbf{v}_1$ determines a regular one-dimensional integral element, and the rank of the associated coefficient matrix for (15.17) is maximal, i.e., equal to $s_0 + s_1 + s_2$, and so on.

Definition 15.4. The *Cartan characters* $s_0, s_1, \ldots s_{m-1}$ of a differential system $\mathcal{I}$ on an m-dimensional manifold M are defined as the maximum values of the matrix ranks $\widetilde{s}_k$, so

$$s_k = \max\{\, \widetilde{s}_k(x, \mathbf{v}_1, \ldots, \mathbf{v}_k) \mid \mathbf{v}_1, \ldots, \mathbf{v}_k \in TM|_x,\ x \in M\}. \qquad (15.19)$$

Definition 15.5. An integral element $E \subset TM|_x$ with prescribed basis $\mathbf{v}_1, \ldots, \mathbf{v}_n$ is called *regular* if the matrix ranks associated with E are all maximal, equal to the Cartan characters: $\widetilde{s}_k(x, \mathbf{v}_1, \ldots, \mathbf{v}_k) = s_k$, for $k = 0, \ldots, n-1$.

Lemma 15.6. *A differential system has regular n-dimensional integral elements if and only if its Cartan characters $s_0, s_1, \ldots, s_{n-1}$ satisfy the rank inequality*

$$s_0 + s_1 + \cdots + s_{n-1} \leq m - n. \tag{15.20}$$

In this case, we define the nth pseudo-character *$\widehat{s}_n$ by the equation*

$$s_0 + s_1 + \cdots s_{n-1} + \widehat{s}_n = m - n. \tag{15.21}$$

The Cartan–Kähler Theorem states that any regular integral element can be "integrated". Therefore, the rank inequalities (15.20) constitute the basic necessary and sufficient conditions for the existence of (regular) n-dimensional integral submanifolds of a differential system.

Theorem 15.7. *Let $\mathcal{I}$ be an analytic differential system on M. If $E \subset TM|_x$ is a regular n-dimensional integral element for $\mathcal{I}$, then there exists an n-dimensional integral submanifold $N \subset M$ such that $x \in N$ and $TN|_x = E$.*† *Moreover, the general integral submanifold having the prescribed integral element depends on $\widehat{s}_n$ arbitrary (up to specification of the initial conditions at x) analytic functions of n variables. In particular, the integral submanifold is unique if and only if $\widehat{s}_n = 0$.*

The proof is done by induction on the dimension n. As mentioned above, the case $n = 1$ reduces to the existence of integral curves to vector fields. To get from $n-1$ to n, one takes the $(n-1)$-dimensional integral submanifold corresponding to a regular $(n-1)$-dimensional integral subelement of the given integral element as Cauchy data for the system of partial differential equations determining the integral submanifolds of $\mathcal{I}$, and uses the Cauchy–Kovalevskaya Theorem to prove existence of a solution. As an illustration of the general technique of proof, we indicate how to effect the induction step from $n = 1$ to $n = 2$. Thus, suppose that

† Actually, one need only assume that the n-dimensional integral element contains a regular $(n-1)$-dimensional integral element for the theorem to be true.

we are given a point x_0 and a regular two-dimensional integral element $E \subset TM|_{x_0}$, so that E is a two-dimensional subspace of the tangent space, and, moreover, all the two-forms in the differential ideal $\mathcal{I}$ annihilate E. Using our inductive hypothesis, we assume that we are given an integral curve C of the differential system passing through x_0, whose tangent line is contained in the given integral element: $TC|_{x_0} \subset E$. The goal then is to find an integral surface S for $\mathcal{I}$ which contains the given curve C and has tangent plane at x_0 equal to the integral element $E = TS|_{x_0}$.

In order to simplify our computations, we introduce the adapted local coordinates $(x, y, u) = (x, y, u^1, \ldots, u^q)$, where $q + 2 = m$, with the following properties: *a*) the integral curve C has been straightened out in the x direction, $C = \{y = u = 0\}$, and *b*) the integral element E is at the origin, and spanned by the tangent vectors $\partial_x|_0, \partial_y|_0$. In this case, by the Implicit Function Theorem, any surface N whose tangent space at the origin agrees with the integral element E can, in a neighborhood of the origin, be described as the graph of $q = m - 2$ functions

$$N = \Big\{ \big(x, y, u^1(x, y), \ldots, u^q(x, y)\big) \Big\}, \tag{15.22}$$

for (x, y) in a neighborhood of $(0, 0)$. The goal is to determine these functions so that N will be an integral submanifold containing the curve C. If (15.22) is to be an integral submanifold of the differential system described by the ideal $\mathcal{I}$, its tangent space must annihilate all the forms in both $\mathcal{I}^{(1)}$ and $\mathcal{I}^{(2)}$. For N given by (15.22), its tangent space TN will be spanned by the two vector fields

$$\mathbf{v}_1 = \frac{\partial}{\partial x} + \sum_{\alpha=1}^{k} \frac{\partial u^\alpha}{\partial x} \frac{\partial}{\partial u^\alpha}, \qquad \mathbf{v}_2 = \frac{\partial}{\partial y} + \sum_{\alpha=1}^{k} \frac{\partial u^\alpha}{\partial y} \frac{\partial}{\partial u^\alpha}. \tag{15.23}$$

If these are to span an integral element at each point, then, taking them in the given order, $\mathbf{v}_1$ must satisfy the system of linear equations (15.8), where $A = A(x, y, u)$ is found from the generators of $\mathcal{I}^{(1)}$, whereas $\mathbf{v}_2$ must satisfy the expanded system (15.11). We concentrate on the latter system to begin with. Now, by assumption, at the origin the expanded system (15.11) has maximal rank $s_0 + s_1$, hence, by continuity the matrix $B(x, y, u, \mathbf{v}_1)$ will still have maximal rank in a neighborhood of the origin. Since $\mathbf{v}_1, \mathbf{v}_2$ reduce to the basis tangent vectors $\partial_x|_0, \partial_y|_0$ at the origin, by continuity we can solve (15.11) for s_0+s_1 of the derivatives $\partial u^\alpha/\partial y$ in terms of the remaining $q-s_0-s_1$ others; let us assume that we have ordered the coordinates such that these are the first s_0+s_1 of them. Therefore, the solution to (15.11) for the given tangent vectors (15.23) has the form

$$\frac{\partial u^\alpha}{\partial y} = \Phi^\alpha\left(x, y, u^1, \ldots, u^q, \frac{\partial u^1}{\partial x}, \ldots, \frac{\partial u^q}{\partial x}, \frac{\partial u^{s_0+s_1+1}}{\partial y}, \ldots, \frac{\partial u^q}{\partial y}\right), \tag{15.24}$$

for $\alpha = 1, \ldots, s_0+s_1$. This system of partial differential equations is in the underdetermined Kovalevskaya form (15.3). The Cauchy data require that the integral surface contain the integral curve $y = u = 0$, which requires that the functions defining N must satisfy the initial conditions

$$u^1(x,0) = \cdots = u^q(x,0) = 0. \tag{15.25}$$

The functions $u^{s_0+s_1+1}, \ldots, u^q$ can be *arbitrary* analytic functions subject only to the restriction $u^{s_0+s_1+1}(x,0) = \cdots = u^q(x,0) = 0$, and so, by the Cauchy–Kovalevskaya Theorem 15.2, there exists a solution to the Cauchy problem (15.24), (15.25) in a neighborhood of the origin, depending on $q - s_0 - s_1 = m - s_0 - s_1 - 2 = \widehat{s}_2$ analytic functions of two variables.

The claim is that the resulting surface N, given by (15.22), is an integral submanifold. Clearly it contains the given integral curve C, since it is required to satisfy the Cauchy data (15.25). Moreover, the solution satisfies the expanded system (15.11) at all points since that is how we constructed the system of partial differential equations (15.24). Therefore, the only sticky point is whether the system (15.8) for the one-dimensional integral element spanned by the first tangent vector $\mathbf{v}_1$ holds at each point of N. To prove this, we let $\omega = \xi\,dx + \eta\,dy + \sum \zeta_\alpha\,du^\alpha$ be any one-form in $\mathcal{I}^{(1)}$. Restricting to the submanifold,

$$\begin{aligned}\omega \mid N &= \left(\xi + \sum_{\alpha=1}^{q} \zeta_\alpha \frac{\partial u^\alpha}{\partial x}\right) dx + \left(\eta + \sum_{\alpha=1}^{q} \zeta_\alpha \frac{\partial u^\alpha}{\partial y}\right) dy \\ &\equiv \sigma(x,y)\,dx + \tau(x,y)\,dy,\end{aligned}$$

where $\sigma(x,y) = \langle \omega\,; \mathbf{v}_1 \rangle$, $\tau(x,y) = \langle \omega\,; \mathbf{v}_2 \rangle$. Since $\mathbf{v}_2$ satisfies (15.11), the coefficient $\tau = 0$ vanishes on N, so $\omega|N = \sigma\,dx$, and we need to prove that $\sigma = 0$ as well. Now the two-form $d\omega$ belongs to $\mathcal{I}^{(2)}$, hence

$\langle d\omega\,;\mathbf{v}_1,\mathbf{v}_2\rangle = 0$ on N, since this equation appears, by construction, as one of the equations in the second system in (15.11). On the other hand,

$$\frac{\partial \sigma}{\partial y}\, dx \wedge dy = d\omega \mid N = \langle d\omega\,;\mathbf{v}_1,\mathbf{v}_2\rangle\, dx \wedge dy,$$

hence $\partial\sigma/\partial y = 0$. Moreover, since $C \subset N$ is an integral curve, σ vanishes on C, and hence satisfies the initial conditions $\sigma(x,0) = 0$. Uniqueness implies that $\sigma(x,y) = 0$ for all $x, y \in N$, which completes the proof.

The proof for higher dimensional integral submanifolds proceeds in a similar manner. In order to construct an n-dimensional integral submanifold containing a given $(n-1)$-dimensional integral submanifold, coordinates $(x,y,u) = (x^1,\ldots,x^{n-1},y,u^1,\ldots,u^{m-n})$ are introduced, and one realizes N as a graph $N = \{(x,y,u(x,y))\}$. Solving the appropriate expanded system for the last tangent vector leads to a system of first order partial differential equations in Kovalevskaya form for the first $s_0 + \cdots + s_{n-1}$ of the y-derivatives $\partial u^\alpha/\partial y$ in terms of the x's, y, u's, x-derivatives of the u's and the remaining y-derivatives. The Cauchy data are supplied by the initial $(n-1)$-dimensional integral submanifold. Specifying the remaining u's reduces the system to one in Kovalevskaya form, and the Cauchy–Kovalevskaya Theorem produces the desired integral submanifold. It then only remains to show that, on N, the lower order linear systems governing the lower dimensional integral subelements remain valid. This is verified by a process similar to, but a bit more computationally involved than, the preceding case, since the relevant coefficients satisfy a system of partial differential equations, for which the uniqueness part of the Cauchy–Kovalevskaya Theorem is required. See Cartan, [**36**], for an explicit implementation of the step from $n = 2$ to $n = 3$, and [**28**] for a modern proof using more sophisticated machinery. This completes our discussion of the proof of the Cartan–Kähler Theorem 15.7.

Our final comments deal with the degree of generality of the n-dimensional (regular) integral submanifolds of a given differential ideal $\mathcal{I}$. According to the proof of Theorem 15.7, the most general integral curve of $\mathcal{I}$ depends on $\widehat{s}_1 = m - s_0 - 1$ arbitrary analytic functions of a single variable. (If $s_0 = m - 1$, then the integral curves are uniquely determined by their initial points.) For each prescribed integral curve, the most general integral surface depends on $\widehat{s}_2 = m - s_0 - s_1 - 2$ arbitrary functions of two variables, which were required to pin down a solution to the system of partial differential equations (15.24). The functions of two variables will be subject to initial conditions so that the associated integral surface contains the given integral curve. Therefore, as long as $\widehat{s}_2 > 0$, the arbitrary analytic functions of a single variable required to prescribe the general integral curve can be subsumed into the arbitrary functions of two variables that the resulting integral surfaces depend on, so we can only infer that the general integral surface will depend on $\widehat{s}_2$ arbitrary analytic functions of two variables. On the other hand, if $\widehat{s}_2 = 0$, which means $s_0 + s_1 = m - 2$, then the two-dimensional integral submanifolds are uniquely specified by their initial integral curves, which depend on $\widehat{s}_1 = m - 1 - s_0$ arbitrary functions of a single variable. This means that, in this case, the general integral surface will depend on $m - 2 - s_0 = s_1$ arbitrary functions of a single variable; the formula comes from the fact that the set of integral curves belonging to a given integral surface already depends on an arbitrary function of a single variable, which accounts for the loss of one arbitrary function in the overall count. Therefore, proceeding by induction, we deduce the following characterization of the degree of generality of the regular integral submanifolds of a differential system.

Theorem 15.8. *If $\mathcal{I}$ is a differential ideal whose n^{th} pseudo-character $\widehat{s}_n > 0$ is positive, then the general n-dimensional integral submanifold of $\mathcal{I}$ depends on $\widehat{s}_n$ arbitrary analytic functions of n variables. If, on the other hand, $\widehat{s}_n = 0$, then the general n-dimensional integral submanifold of $\mathcal{I}$ depends on s_k arbitrary analytic functions of k variables, where s_k denotes the last nonzero integer in the character sequence $s_1, s_2, \ldots, s_{n-1}$.*

Applications to Equivalence Problems

We now apply the Cartan–Kähler Theorem to the solution of an equivalence problem in the case when the reduction algorithm doesn't end with an invariant coframe and we have the structure equations corresponding to an infinite-dimensional symmetry group. Thus, assume that we are given an equivalence problem based on the lifted coframe $\boldsymbol{\theta} = \{\theta^1, \ldots, \theta^m\}$, which will be one-forms on M whose coefficients depend on the group parameters of the (reduced) structure group, which we denote by G for simplicity. The lifted coframe, then, should really be regarded as a collection of one-forms on the product $M \times G$. Similarly, the second lifted coframe $\overline{\boldsymbol{\theta}} = \{\overline{\theta}^1, \ldots, \overline{\theta}^m\}$ will be a collection of one-forms on $\overline{M} \times \overline{G}$, since they also depend on group parameters, denoted as usual by $\bar{g} \in \overline{G}$, even though $\overline{G} = G$ are one and the same group. The question is: Are the two coframes equivalent, meaning that some specification of the group parameters $g = \Upsilon(x), \bar{g} = \overline{\Upsilon}(\bar{x})$ will make the coframes map to each other? The answer depends on the solvability of an associated differential system.

As in our application of Frobenius' Theorem to the equivalence problem for ordinary coframes, we again try to find the appropriate map $\Phi: M \to \overline{M}$ by Cartan's technique of the graph. As before, the graph Γ_Φ will be an m-dimensional submanifold (satisfying certain transversality properties) of the $2m$-dimensional Cartesian product space $M \times \overline{M}$. However, since we also need to prescribe the values of the group parameters, we should really try to construct a map $\Psi: M \to G \times \overline{M} \times \overline{G}$, which incorporates all three maps $g = \Upsilon(x)$, $\bar{x} = \Phi(x)$, and $\bar{g} = \widetilde{\Upsilon}(x) = \overline{\Upsilon}(\bar{x})$, where $\widetilde{\Upsilon} = \overline{\Upsilon} \circ \Phi$. Thus Ψ determines not only the map Φ that provides the change of variables mapping one coframe to the other, but also the particular expressions for the group parameters g and $\bar{g}$ in each coframe which enable the map Φ to successfully yield an equivalence. We will realize the graph $\Gamma_\Psi \subset M \times G \times \overline{M} \times \overline{G}$ of this enlarged map as an m-dimensional integral submanifold of an appropriate differential system of the $(2m + 2r)$-dimensional product space $M \times G \times \overline{M} \times \overline{G}$.

The required differential system will, as in the Frobenius case, be the differential ideal generated by the differences between the lifted coframe elements $\vartheta^i = \bar{\theta}^i - \theta^i$, $i = 1, \ldots, m$, which are viewed as differential forms on $M \times G \times \overline{M} \times \overline{G}$. (As before, in the formulas for ϑ^i, etc., we omit the pull-backs of the various projections so as to keep the notation uncluttered.) Moreover, they must be restricted to the variety defined by any associated invariant functions (including all their coframe derivatives).

In order to simplify the discussion, let us first make an assumption (which holds in many instances) that the essential torsion coefficients are all constants — the "transitive" case; the more general intransitive case will be discussed later on. We are looking for an (appropriately transversal) m-dimensional submanifold on which all the ϑ^i's vanish. As in the application of Frobenius' Theorem, this necessitates the vanishing of their differentials, which we proceed to compute. Since $d\vartheta^i = d\bar{\theta}^i - d\theta^i$, we need to know the form of the structure equations for the two coframes, which must match up by the necessary conditions imposed by equivalence. After absorption, the structure equations for the first coframe have the form

$$d\theta^i = \sum_{j=1}^{m}\sum_{\kappa=1}^{r} A^i_{j\kappa}\,\pi^\kappa \wedge \theta^j + \sum_{j,k=1}^{m} U^i_{jk}\,\theta^j \wedge \theta^k, \qquad i = 1,\ldots,m, \quad (15.26)$$

where the one-forms $\{\pi^1,\ldots,\pi^r\}$ are equivalent, modulo the θ's to a basis $\{\alpha^1,\ldots,\alpha^r\}$ for the right-invariant Maurer–Cartan one-forms on the group G. More explicitly, $\pi^\kappa = \alpha^\kappa - \sum_j z^\kappa_j(g,x)\,\theta^j$ for certain coefficients z^κ_j — see (10.12). In the transitive (rank 0) case, we are assuming that the remaining essential torsion coefficients U^i_{jk} are all constant, so there is no more torsion left to normalize. Similarly,

$$d\overline{\theta}^i = \sum_{j=1}^{m}\sum_{\kappa=1}^{r} A^i_{j\kappa}\,\overline{\pi}^\kappa \wedge \overline{\theta}^j + \sum_{j,k=1}^{m} \overline{U}^i_{jk}\,\overline{\theta}^j \wedge \overline{\theta}^k, \qquad i = 1,\ldots,m, \quad (15.27)$$

where the $A^i_{j\kappa}$ are necessarily the same constants as in (15.26), since the group parameters were assumed from the outset to enter the two lifted coframes in an identical manner. Therefore, substituting (15.26) and (15.27) we find

$$d\vartheta^i = \sum_{j=1}^{m}\sum_{\kappa=1}^{r} A^i_{j\kappa}\left\{\overline{\pi}^\kappa \wedge \overline{\theta}^j - \pi^\kappa \wedge \theta^j\right\} + \sum_{j,k=1}^{m} \left\{\overline{U}^i_{jk}\,\overline{\theta}^j \wedge \overline{\theta}^k - U^i_{jk}\,\theta^j \wedge \theta^k\right\}. \quad (15.28)$$

In the transitive case, both sets of torsion coefficients must have identical constant values: $U^i_{jk} = \overline{U}^i_{jk}$. Therefore, according to the calculation in (14.21), the second summation in (15.28) is already in the ideal generated by the ϑ^i's. As for the first summation, we replace $\overline{\theta}^j$ by $\vartheta^j + \theta^j$, so that the terms with ϑ's in them are again automatically in the ideal. Therefore, we need only require the vanishing of the two-forms

$$\Theta^i = \sum_{j=1}^{m}\sum_{\kappa=1}^{r} A^i_{j\kappa}\left\{\overline{\pi}^\kappa - \pi^\kappa\right\} \wedge \theta^j, \qquad i = 1,\ldots,m, \quad (15.29)$$

and our differential ideal $\mathcal{I}$ will, in the case under consideration, be generated by *a*) the one-forms ϑ^i spanning $\mathcal{I}^{(1)}$, and *b*) the two-forms Θ^i which, together with the wedge products of the ϑ's with arbitrary one-forms, span $\mathcal{I}^{(2)}$. It is here that the crucial difference between the present set-up and that of Frobenius' Theorem becomes apparent. In the Frobenius case, there were no additional conditions resulting from two-forms in the differential ideal, since they were all expressible in terms of the generating functions and one-forms. Here this is no longer the case, and the inclusion of algebraically independent two-forms in the ideal necessitates the use of the Cartan–Kähler Theorem.

In order to proceed, we need to determine the Cartan characters of our differential system. Clearly $s_0 = m$ since there are m linearly independent one-forms $\vartheta^1, \dots, \vartheta^m$. Now, to compute the remaining characters, let us use the following system of coordinates for tangent vectors $\mathbf{V} \in T[M \times G \times \overline{M} \times \overline{G}] \simeq TM \times TG \times T\overline{M} \times T\overline{G}$:

$$\mathbf{V} = \sum_{i=1}^{m} v^i \frac{\partial}{\partial \theta^i} + \sum_{\kappa=1}^{m} w^\kappa \frac{\partial}{\partial \pi^\kappa} + \sum_{i=1}^{m} \bar{v}^i \frac{\partial}{\partial \bar{\theta}^i} + \sum_{\kappa=1}^{m} \overline{w}^\kappa \frac{\partial}{\partial \overline{\pi}^\kappa}, \tag{15.30}$$

where $v^i = \langle \mathbf{v}\,; \theta^i \rangle$, $w^\kappa = \langle \mathbf{w}\,; \pi^\kappa \rangle$, $\bar{v}^i = \langle \overline{\mathbf{v}}\,; \bar{\theta}^i \rangle$, $\overline{w}^\kappa = \langle \overline{\mathbf{w}}\,; \overline{\pi}^\kappa \rangle$ are the coordinates induced by the respective dual bases for the cotangent spaces. Using these coordinates, we can identify $\mathbf{V}$ with the 4-tuple of vectors $(\mathbf{v}, \mathbf{w}, \overline{\mathbf{v}}, \overline{\mathbf{w}}) \in \mathbb{R}^{2m+2r}$. The condition that the (nonzero) tangent vector $\mathbf{V}$ define a one-dimensional integral element is that it be annihilated by the one-forms $\vartheta^i = \overline{\theta}^i - \theta^i$ which span $\mathcal{I}^{(1)}$; this is the same as requiring that the coefficients $v^i = \bar{v}^i$ be the same. In terms of the chosen coframe, the $m \times (2m + 2r)$ matrix in (15.8) is simply

$$A = (\,\mathbb{1}, 0, -\mathbb{1}, 0\,),$$

where $\mathbb{1}$ denotes the $m \times m$ identity matrix, which is obviously of rank m. Next consider a two-dimensional integral element spanned by the vectors $\mathbf{V}_1$, $\mathbf{V}_2$. Assuming that $\mathbf{V}_1$ is already a one-dimensional integral element, the relevant equations (15.11) for $\mathbf{V}_2$ have the form

$$A\,\mathbf{V}_2 = 0, \qquad B[\,\mathbf{V}_1]\,\mathbf{V}_2 = 0.$$

Here $B[\mathbf{V}_1]$ is the following $m \times (2m + 2r)$ matrix constructed from the two-forms (15.29), as was done in (15.13), (15.14):

$$B[\,\mathbf{V}_1] = \big(K(\overline{\mathbf{w}}_1 - \mathbf{w}_1), L(\mathbf{v}_1), 0, -L(\mathbf{v}_1)\big), \tag{15.31}$$

where $K(\mathbf{w})$ is the $m \times m$ matrix with entries $K^i_j = \sum_\kappa A^i_{j\kappa} w^\kappa$ and $L(\mathbf{v})$ is the $m \times r$ matrix with entries $L^i_\kappa = \sum_j A^i_{j\kappa} v^j$. It helps to introduce a minor simplification at this point: Instead of representing tangent vectors $\mathbf{V}$ as in (15.30), we write

$$\mathbf{V} = \sum_{i=1}^{m} v^i \frac{\partial}{\partial \theta^i} + \sum_{\kappa=1}^{m} w^\kappa \left\{ \frac{\partial}{\partial \pi^\kappa} + \frac{\partial}{\partial \overline{\pi}^\kappa} \right\} + \sum_{i=1}^{m} \overline{v}^i \frac{\partial}{\partial \overline{\theta}^i} + \sum_{\kappa=1}^{m} z^\kappa \frac{\partial}{\partial \overline{\pi}^\kappa}, \tag{15.32}$$

where $z^\kappa = \overline{w}^\kappa - w^\kappa$. We identify $\mathbf{V}$ with the "skewed" 4-tuple of vectors $(\mathbf{v}, \mathbf{w}, \overline{\mathbf{v}}, \mathbf{z}) = (\mathbf{v}, \mathbf{w}, \overline{\mathbf{v}}, \overline{\mathbf{w}} - \mathbf{w})$. In terms of the skewed coordinates, then, the matrix (15.31) has the simpler form

$$B[\mathbf{V}_1] = \big(K(\mathbf{z}_1), 0, 0, L(\mathbf{v}_1)\big), \tag{15.33}$$

relative to the modified basis of $T[M \times G \times \overline{M} \times \overline{G}\,]$. We now compute the Cartan characters $s_1, \ldots, s_{m-1}$. Since the differential system is generated just by one-forms and two-forms, the expanded linear system for a k-dimensional integral element spanned by $\mathbf{V}_1, \ldots, \mathbf{V}_k$ containing an integral element spanned by the first $k-1$ basis vectors $\mathbf{V}_1, \ldots, \mathbf{V}_{k-1}$ is

$$A\,\mathbf{V}_k = 0, \qquad B[\mathbf{V}_1]\,\mathbf{V}_k = 0, \qquad \ldots \qquad B[\mathbf{V}_{k-1}]\,\mathbf{V}_k = 0. \tag{15.34}$$

Therefore, the Cartan characters are defined by the maximal ranks of the coefficient matrices to these successive linear systems,

$$s_0 + s_1 + \cdots + s_k = \max \left\{ \operatorname{rank} \begin{pmatrix} \mathbb{1} & 0 & -\mathbb{1} & 0 \\ K(\mathbf{z}_1) & 0 & 0 & L(\mathbf{v}_1) \\ \vdots & \vdots & \vdots & \vdots \\ K(\mathbf{z}_k) & 0 & 0 & L(\mathbf{v}_k) \end{pmatrix} \right\},$$

for arbitrary vectors $\mathbf{z}_1, \ldots, \mathbf{z}_k \in \mathbb{R}^r$, $\mathbf{v}_1, \ldots, \mathbf{v}_k \in \mathbb{R}^m$, or, since $s_0 = \operatorname{rank} A = m$, in a slightly simpler manner,

$$s_1 + s_2 + \cdots + s_k = \max \left\{ \operatorname{rank} \begin{pmatrix} K(\mathbf{z}_1) & L(\mathbf{v}_1) \\ \vdots & \vdots \\ K(\mathbf{z}_k) & L(\mathbf{v}_k) \end{pmatrix} \right\}. \tag{15.35}$$

According to (15.20), provided $s_0 + s_1 + \cdots + s_{m-1} \le 2m + 2r - m = m + 2r$, we will be able to find a regular m-dimensional integral element to which the Cartan–Kähler Theorem is applicable, the resulting integral submanifold being a candidate for the graph of the desired equivalence map Ψ. Also, since $s_0 = m$, we need just

$$s_1 + s_2 + \cdots + s_{m-1} \leq 2r, \tag{15.36}$$

in order to have integral elements of the desired dimension. Note, in particular, that since the matrices in (15.35) have size $(km) \times (m+r)$, the left hand side of (15.36) is necessarily bounded by $m + r$, and hence (15.36) is automatically satisfied if $m \leq r$. Thus, provided the Cartan characters satisfy the crucial inequality (15.36) guaranteeing the existence of regular m-dimensional integral elements, the Cartan–Kähler Theorem will provide a corresponding m-dimensional integral submanifold. Apparently we are done.

But wait a minute, not so fast! If we have truly demonstrated the existence of integral submanifolds of dimension m, then we have shown that all such coframes are equivalent, and, in fact, equivalent under an infinite-dimensional system of maps depending on arbitrary functions (the functions coming in at each stage of the Cartan–Kähler integration process). Indeed, this argument would have the effect of proving that any equivalence problem with no group-dependent essential torsion would have an infinite dimensional symmetry group depending on these arbitrary functions, but we know that this is definitely not the case — some cases are involutive, whereas others are not and must be prolonged! Something is clearly amiss, and we need to look closer at what we have actually accomplished.

Involutivity and Transversality

The one condition we have not completely dealt with so far is the transversality of the integral submanifold that the Cartan–Kähler Theorem provides us with, which is crucial for it to be the graph of an equivalence between the two coframes. In the Frobenius case, transversality of the integral submanifold followed more or less automatically. Here we need to be quite a bit more careful, since, although the $\bar{x}$'s will also automatically be functions of the x's, the same is not necessarily true of the group parameters. Worse yet, it may very well be the case that even if there are suitably transverse integral elements, they may all fail to satisfy the requisite regularity conditions, and we may be unable to establish the existence of even a single integral submanifold for our system! Indeed, this seemingly innocuous detail turns out to have major consequences and is, in fact, the crux of the whole problem under consideration. The crucial question then is whether (generic) transverse integral elements are regular in the Cartan–Kähler sense. As we shall see, this is a nontrivial condition, and leads directly to the basic distinction between those equivalence problems which have infinite-dimensional symmetry groups (for which the answer is yes) and those which require prolongation (for which the answer is no).

Let us look in detail at the issues. Given a point $y = (x, g, \bar{x}, \bar{g}) \in M \times G \times \overline{M} \times \overline{G}$, consider an m-dimensional integral element $E \subset T[M \times G \times \overline{M} \times \overline{G}]|_y$. The condition that E be *fully transverse*, and hence that it be the tangent space to the graph of a local diffeomorphism $\Psi: M \to G \times \overline{M} \times \overline{G}$, is that it contain no horizontal tangent vectors, $E \cap TM|_y = \{0\}$, or vertical tangent vectors, $E \cap T\overline{M}|_y = \{0\}$. The form of the differential system will show that one of these transversality conditions implies the other, but the initial demonstration is much more problematic than in the earlier Frobenius case. To see what this entails, let us look in detail at the system (15.34) which a k-dimensional integral element with basis $\mathbf{V}_1, \ldots, \mathbf{V}_k$ must satisfy. In order that the integral element be tangent to a transverse integral submanifold, not only must the $\mathbf{V}_i$'s be linearly independent, but also their projections to TM, namely (using the notation in (15.32)), the $\mathbf{v}_i$'s must be linearly independent. Therefore, the linear system (15.34) is not allowed to impose any linear relations on the $\mathbf{v}_i$'s alone. This will be the case if and only if the submatrix consisting of the last r columns of the full coefficient matrix has the same rank as does the full matrix.

These considerations lead us to introduce the following sequence of integers, known as the *reduced characters* of the differential system, which we have already encountered in our discussion of Cartan's test for involutivity (11.13). The first of these is $\widetilde{s}_1' = \widetilde{s}_1'(\mathbf{v}_1, y) = \operatorname{rank} L[\mathbf{v}_1, y]$, where L was defined after (15.31). Similarly, the higher reduced characters are defined in terms of the ranks of the successive submatrices of the full matrices (15.35) used to define the ordinary Cartan characters

$$\widetilde{s}_1' + \widetilde{s}_2' + \cdots + \widetilde{s}_k' = \operatorname{rank} \begin{pmatrix} L[\mathbf{v}_1, y] \\ L[\mathbf{v}_2, y] \\ \vdots \\ L[\mathbf{v}_k, y] \end{pmatrix}, \qquad k = 1, \ldots, m-1, \quad (15.37)$$

so that $\widetilde{s}'_k = \widetilde{s}'_k(\mathbf{v}_1, \ldots, \mathbf{v}_{k-1}, y)$ depends on the previous $k-1$ projected tangent vectors, and on the point $y = (x, g, \bar{x}, \bar{g})$. Note that the matrix whose rank defines the reduced character is a submatrix of that whose rank defines the full character, cf. (15.35), so the reduced characters cannot exceed the ordinary Cartan characters:

$$\widetilde{s}'_k \leq s_k, \qquad k = 0, \ldots, m-1. \tag{15.38}$$

The key result is that if the reduced characters are all equal to the ordinary Cartan characters, then the transverse integral element is regular.

Lemma 15.9. *A transverse integral element is regular if and only if its reduced Cartan characters* $\widetilde{s}'_0, \ldots, \widetilde{s}'_{m-1}$ *satisfy*

$$\widetilde{s}'_k = s_k, \qquad k = 0, \ldots, m-1. \tag{15.39}$$

Proof: Clearly, if (15.39) holds for all k, then the integral element E is regular, since a regular element is one all of whose matrices have maximal rank, equal to s_k for $k < \dim E$. Conversely, suppose the equalities $\widetilde{s}'_j = s_j$ hold for $j = 0, \ldots, l-1$, but $\widetilde{s}'_k < s_k$ for some $k < m$. We then invoke the following simple linear algebra lemma.

Lemma 15.10. *Let C be a $k \times m$ matrix and D a $k \times n$ matrix. Let (C, D) denote the combined $k \times (m+n)$ matrix. Then the homogeneous linear system $C\mathbf{v} + D\mathbf{w} = 0$ has a solution $\mathbf{w} \in \mathbb{R}^n$ for every $\mathbf{v} \in \mathbb{R}^m$ if and only if* $\operatorname{rank} D = \operatorname{rank}(C, D)$.

In other words, the condition $\operatorname{rank} D = \operatorname{rank}(C, D)$ holds if and only if

$$V = \{\mathbf{v} \in \mathbb{R}^m \mid C\mathbf{v} + D\mathbf{w} = 0 \text{ for some } \mathbf{w} \in \mathbb{R}^n\} = \mathbb{R}^m.$$

In our case, if $\widetilde{s}'_k < s_k$, then the rank of the matrix in (15.37) (which plays the role of D in the lemma) is, by assumption, strictly less than the rank of the larger matrix in (15.35). Thus Lemma 15.10 implies that the set $W = \{\mathbf{v}_k \mid \text{there exists a solution } \mathbf{V}_k \text{ to (15.34)}\}$ is not equal to the entire tangent space $TM|_x$. But every tangent vector $\mathbf{V}_j \in E$ must satisfy the system (15.34), so the corresponding projected tangent vectors $\mathbf{v}_j$ all lie in the subspace W, and hence cannot be linearly independent. This contradicts the transversality of the integral element E, and completes the proof. *Q.E.D.*

The generic transverse integral elements will be those for which the associated reduced characters are maximal, and it is those that stand the best chance of being regular, in the Cartan–Kähler sense. We therefore define the *reduced characters* of the differential system:

$$s'_1+s'_2+\cdots+s'_k = \max\left\{ \operatorname{rank}\begin{pmatrix} L[\mathbf{v}_1,y] \\ L[\mathbf{v}_2,y] \\ \vdots \\ L[\mathbf{v}_k,y] \end{pmatrix} \;\middle|\; \begin{matrix} \mathbf{v}_1,\ldots,\mathbf{v}_k \in \mathbb{R}^m \\ y \in M\times G\times \overline{M}\times\overline{G} \end{matrix} \right\}. \tag{15.40}$$

Note that $s'_1 + s'_2 + \cdots + s'_k \le r$ since the matrices have size $(km) \times r$. According to Lemma 15.9, the generic transverse integral element will be regular if and only if its reduced characters agree with the ordinary characters: $s'_k = s_k$, $k = 1, \ldots, m-1$. If this holds, the inequality (15.36) is satisfied (indeed, the left hand side is bounded by r), and hence the Cartan–Kähler Theorem will produce the required *transverse* integral submanifolds to the system. In this case, the differential system is said to be *involutive*.

Remark: The preceding construction is a special case of the general notion of a differential system with independence condition. The *independence condition* is provided by a set of one-forms $\omega^1, \ldots, \omega^n$, and one is required to find an n-dimensional integral submanifold of the differential system on which $\omega^1 \wedge \cdots \wedge \omega^n \neq 0$ does not vanish. In our case, the independence condition is provided by the one-forms $\theta^1, \ldots, \theta^m$, or, equivalently, the coordinate coframe $dx^1, \ldots, dx^m$. A differential system with independence condition is said to be *involutive* if every generic integral element satisfying the independence condition is regular in the Cartan–Kähler sense. The reduced characters are defined by taking the ranks of submatrices obtained from the full coefficient matrices by omitting those columns corresponding to the one-forms ω^μ prescribing the independence condition. The analogue of Lemma 15.9 holds, and so a differential system with independence condition is involutive if and only if its reduced characters are equal to the ordinary Cartan characters.

Fortunately, we can verify the involutivity condition without having to compute both sets of characters, only the computation of the (simpler) reduced characters being required. Consider a generic transverse integral element E with basis $\mathbf{V}_1, \ldots, \mathbf{V}_m$. Transversality and the fact that E

is an integral element implies that we can introduce coordinates so that the basis tangent vectors have the particular form

$$\mathbf{V}_i = \frac{\partial}{\partial \theta^i} + \frac{\partial}{\partial \overline{\theta}^i} + \sum_{\kappa=1}^{r} w_i^\kappa \left\{ \frac{\partial}{\partial \pi^\kappa} + \frac{\partial}{\partial \overline{\pi}^\kappa} \right\} + z_i^\kappa \, \frac{\partial}{\partial \overline{\pi}^\kappa}, \qquad i = 1, \ldots, m. \tag{15.41}$$

The coefficients of $\partial/\partial\theta^i$ and $\partial/\partial\overline{\theta}^i$ agree since each $\mathbf{V}_i$ must span a one-dimensional integral element. The condition that the vector fields (15.41) span an integral element for the differential ideal is that the two-forms Θ^i, defined in (15.29), vanish on any pair. Evaluating $\langle \Theta^i \,; \mathbf{V}_j, \mathbf{V}_k \rangle$, we find the following familiar homogeneous linear system of equations

$$\sum_{\kappa=1}^{r} \left(A^i_{j\kappa} z^\kappa_k - A^i_{k\kappa} z^\kappa_j \right) = 0, \qquad i, j, k = 1, \ldots, m, \quad j < k, \tag{15.42}$$

which constitute the necessary and sufficient conditions for the $\mathbf{V}_i$ to span a transverse integral element of dimension m. Of course, equations (15.42) are nothing but the left hand side of our famous absorption equations (10.14), which have occupied us so much in our pursuit of equivalence! As before, we let $r^{(1)}$ denote the dimension of the space of solutions to this system of homogeneous linear equations, i.e., the degree of indeterminancy of the structure equations; see Definition 11.2. Thus, the generic transverse integral element depends on $r^{(1)}$ independent parameters.

On the other hand, the system (15.42) is just another way of writing the collection of successive linear systems (15.34) for $k = 1, \ldots, m-1$, namely

$$\begin{array}{llll} B[\mathbf{V}_1]\,\mathbf{V}_2 = 0, & & & \\ B[\mathbf{V}_1]\,\mathbf{V}_3 = 0, & B[\mathbf{V}_2]\,\mathbf{V}_3 = 0, & & \\ \quad\vdots & \quad\vdots & \ddots & \\ B[\mathbf{V}_1]\,\mathbf{V}_m = 0, & B[\mathbf{V}_2]\,\mathbf{V}_m = 0, & \ldots & B[\mathbf{V}_{m-1}]\,\mathbf{V}_m = 0. \end{array} \tag{15.43}$$

(We've already satisfied the first set of equations $A\mathbf{V}_k = 0$.) The rank of the linear system (15.43) is the combined ranks of the coefficient matrices to the successive subsystems, which, provided the integral element spanned by the $\mathbf{V}_i$'s is regular, are given in terms of the ordinary characters: s_1, $s_1 + s_2$, $\ldots$, $s_1 + s_2 + \cdots + s_{m-1}$. Therefore, the system (15.43), or equivalently (15.42), has rank $(m-1)s_1+(m-2)s_2+\cdots+s_{m-1}$, and so the space of solutions has dimension equal to the number of unknowns minus the total rank. Therefore, by the well-known duality between rank and dimension of the solution space of a homogeneous linear system, we find that, assuming regularity,

$$r^{(1)} = mr - [(m-1)s_1 + (m-2)s_2 + \cdots + s_{m-1}].$$

But the transverse integral element is regular if and only if the reduced characters and ordinary characters are equal, hence in this case

$$r^{(1)} = mr - (m-1)s'_1 - (m-2)s'_2 - \cdots - s'_{m-1}. \tag{15.44}$$

Conversely, if the integral element is not regular, then at least one of the reduced characters is strictly less than its ordinary counterpart, which, in view of (15.38), implies that

$$r^{(1)} < mr - (m-1)s'_1 - (m-2)s'_2 - \cdots - s'_{m-1}. \tag{15.45}$$

Therefore, the equality (15.44) is both necessary and sufficient for a generic transverse integral element to be regular, and the differential system to be involutive. We have finally deduced Cartan's powerful involutivity test, which we formulate in a slightly different but trivially equivalent way.

Theorem 15.11. *Consider an equivalence problem with degree of indeterminancy $r^{(1)}$ and reduced characters $s'_1, \ldots, s'_{m-1}$ defined as in (15.40) above. Define the final reduced character s'_m by*

$$s'_1 + \cdots + s'_{m-1} + s'_m = r. \tag{15.46}$$

The associated differential system is involutive, which means that every generic transverse integral element is regular, if and only if

$$s'_1 + 2s'_2 + \cdots + ms'_m = r^{(1)}. \tag{15.47}$$

If the system is involutive, then the number of arbitrary analytic functions entering into the general solution is, as a consequence of the general Cartan–Kähler Theorem 15.8, governed by the last nonzero character, which, because of involutivity, equals the last nonzero reduced character. If condition (15.47) is not satisfied, then the system is not involutive, and we cannot apply the Cartan–Kähler Theorem to find integral submanifolds corresponding to *any* of the transverse integral elements. In this case, as we have seen, we must prolong the system before reaching the solution to the equivalence problem.

The intransitive case, when there are nonconstant essential torsion coefficients (not depending on any remaining group parameters) works in a similar fashion as the higher rank Frobenius case. One restricts to the submanifold defined by the difference of the essential torsion coefficients and their coframe derivatives: $\overline{U}_\sigma(\bar{x}) = U_\sigma(x)$. Note that we can assume that none of the differentiated essential torsion coefficients depends explicitly on the group parameters, as otherwise we could use them to further reduce the structure group. We define the classifying manifolds $\mathcal{C}^{(s)} = \mathcal{C}^{(s)}(\boldsymbol{\theta})$ using the essential torsion coefficients as before; the coframe has order s if $\dim \mathcal{C}^{(s)} = \dim \mathcal{C}^{(s+1)}$. The involutivity condition of Theorem 15.11 is the same.

Theorem 15.12. *Let $\boldsymbol{\theta}$ be an analytic lifted coframe on the manifold $M \times G$ with structure equations (15.26), where the group structure coefficients $A^i_{j\kappa}$ are constant, and the essential torsion coefficients U^i_{jk} are independent of the group parameters, as are all their coframe derivatives. Assume $\boldsymbol{\theta}$ is involutive, meaning that the Cartan test (15.47) is satisfied, and of order s. Then $\boldsymbol{\theta}$ is locally equivalent to a second involutive, s^{th} order, analytic coframe $\overline{\boldsymbol{\theta}}$ if and only if their $(s+1)^{\text{st}}$ order classifying manifolds overlap. The set of equivalences depends on s'_k arbitrary analytic functions of k variables, where s'_k is the last nonzero reduced character.*

More generally, in the equivalence problems resulting from normalizations of nonconstant type, the coefficients $A^i_{j\kappa}$, along with their coframe derivatives, might provide yet further structure invariants which must be included in the definition of the classifying manifold. Assuming these are also independent of the remaining group parameters, the statement and proof of Theorem 15.12 goes through unchanged.

This concludes our presentation of the Cartan–Kähler Theorem and its applications to equivalence problems. We refer the reader to [**36**], [**28**], [**230**] for further details, along with a wide range of applications to problems arising in differential geometry and partial differential equations.

Tables

Table 1

Transitive, Imprimitive Lie Algebras of Vector Fields in $\mathbb{C}^2$

	Generators	Dim	Structure
1.1.	$\partial_x, x\partial_x - u\partial_u, x^2\partial_x - 2xu\partial_u$	3	$\mathfrak{sl}(2)$
1.2.	$\partial_x, x\partial_x - u\partial_u, x^2\partial_x - (2xu+1)\partial_u$	3	$\mathfrak{sl}(2)$
1.3.	$\partial_x, x\partial_x, u\partial_u, x^2\partial_x - xu\partial_u$	4	$\mathfrak{gl}(2)$
1.4.	$\partial_x, x\partial_x, x^2\partial_x, \partial_u, u\partial_u, u^2\partial_u$	6	$\mathfrak{sl}(2) \oplus \mathfrak{sl}(2)$
1.5.	$\partial_x, \eta_1(x)\partial_u, \ldots, \eta_k(x)\partial_u$	$k+1$	$\mathbb{C} \ltimes \mathbb{C}^k$
1.6.	$\partial_x, u\partial_u, \eta_1(x)\partial_u, \ldots, \eta_k(x)\partial_u$	$k+2$	$\mathbb{C}^2 \ltimes \mathbb{C}^k$
1.7.	$\partial_x, x\partial_x + \alpha u\partial_u, \partial_u, x\partial_u, \ldots, x^{k-1}\partial_u$	$k+2$	$\mathfrak{a}(1) \ltimes \mathbb{C}^k$
1.8.	$\partial_x, x\partial_x + (ku + x^k)\partial_u, \partial_u, x\partial_u, \ldots, x^{k-1}\partial_u$	$k+2$	$\mathbb{C} \ltimes (\mathbb{C} \ltimes \mathbb{C}^k)$
1.9.	$\partial_x, x\partial_x, u\partial_u, \partial_u, x\partial_u, x^2\partial_u, \ldots, x^{k-1}\partial_u$	$k+3$	$\mathfrak{c}(1) \ltimes \mathbb{C}^k$
1.10.	$\partial_x, 2x\partial_x + (k-1)u\partial_u, x^2\partial_x + (k-1)xu\partial_u,$ $\partial_u, x\partial_u, x^2\partial_u, \ldots, x^{k-1}\partial_u$	$k+3$	$\mathfrak{sl}(2) \ltimes \mathbb{C}^k$
1.11.	$\partial_x, x\partial_x, x^2\partial_x + (k-1)xu\partial_u, u\partial_u,$ $\partial_u, x\partial_u, x^2\partial_u, \ldots, x^{k-1}\partial_u$	$k+4$	$\mathfrak{gl}(2) \ltimes \mathbb{C}^k$

Here $\mathfrak{c}(1) = \mathfrak{a}(1) \oplus \mathbb{C}$.

In Cases 1.5 and 1.6, the functions $\eta_1(x), \ldots, \eta_k(x)$ satisfy a kth order constant coefficient homogeneous linear ordinary differential equation $\mathcal{D}[u] = 0$.

In Cases 1.5 – 1.11 we require $k \geq 1$. Note, though, that if we set $k = 0$ in Case 1.10, and replace u by u^2, we obtain Case 1.1. Similarly, if we set $k = 0$ in Case 1.11, we obtain Case 1.3. Cases 1.7 and 1.8 for $k = 0$ are equivalent to the Lie algebra $\{\partial_x, e^x\partial_u\}$ of type 1.5. Case 1.9 for $k = 0$ is equivalent to the Lie algebra $\{\partial_x, \partial_u, u\partial_u\}$ of type 1.6.

Table 2

Primitive Lie Algebras of Vector Fields in $\mathbb{C}^2$

	Generators	Dim	Structure
2.1.	$\partial_x, \partial_u, x\partial_x - u\partial_u, u\partial_x, x\partial_u$	5	$\mathfrak{sa}(2)$
2.2.	$\partial_x, \partial_u, x\partial_x, u\partial_x, x\partial_u, u\partial_u$	6	$\mathfrak{a}(2)$
2.3.	$\partial_x, \partial_u, x\partial_x, u\partial_x, x\partial_u, u\partial_u,$ $x^2\partial_x + xu\partial_u, xu\partial_x + u^2\partial_u$	8	$\mathfrak{sl}(3)$

Table 3

Intransitive Lie Algebras of Vector Fields in $\mathbb{C}^2$

	Generators	Dim	Structure
3.1.	$\eta_1(x)\partial_u, \ldots, \eta_k(x)\partial_u$	k	$\mathbb{C}^k$
3.2.	$\eta_1(x)\partial_u, \ldots, \eta_k(x)\partial_u, u\partial_u$	$k+1$	$\mathbb{C} \ltimes \mathbb{C}^k$
3.3.	$\partial_u, u\partial_u, u^2\partial_u$	3	$\mathfrak{sl}(2)$

Table 4

Lie Algebras of Contact Transformations in $\mathbb{C}^2$

	Generators	Dim	Structure
4.1.	$1, x, x^2, u_x, xu_x, u_x^2$	6	$\mathfrak{sa}(2) \ltimes \mathbb{C}$
4.2.	$1, x, x^2, u, u_x, xu_x, u_x^2$	7	$\mathfrak{a}(2) \ltimes \mathbb{C}$
4.3.	$1, x, x^2, u, u_x, xu_x, x^2u_x - 2xu,$ $u_x^2, 2uu_x - xu_x^2, 4xuu_x - 4u^2 - x^2u_x^2$	10	$\mathfrak{so}(5) \simeq \mathfrak{sp}(4)$

Table 5

Differential Invariants of Transformation Groups in $\mathbb{C}^2$

	Fundamental differential invariant(s)	Invariant one-form	Lie determinant
1.1.	$u^{-4}(2uu_2 - 3u_1^2)$	$u\,dx$	u^2
1.2.	$(u_1 - u^2)^{-3/2}(u_2 - 6uu_1 + 4u^3)$	$\sqrt{u_1 - u^2}\,dx$	$u_1 - u^2$
1.3.	$Q_2^{-3/2}S_3$	$u^{-1}\sqrt{Q_2}\,dx$	$u\,Q_2$
1.4.	$Q_3^{-3}U_5$	$u_1^{-1}\sqrt{Q_3}\,dx$	$u_1Q_3^2$
1.5.	$W(x)^{-1}\mathcal{D}[u]$	dx	$W(x)$
1.6.	$D_x \log \mathcal{D}[u]$	dx	$W(x)\mathcal{D}[u]$
1.7a.	$u_k^{(\alpha-k)^{-1}-1}u_{k+1}$	$u_k^{-(\alpha-k)^{-1}}\,dx$	u_k
1.7b.	$u_k, \quad u_{k+1}^{-2}u_{k+2}$	$u_{k+1}\,dx$	u_{k+1}
1.8.	$u_{k+1}e^{u_k/k!}$	$e^{-u_k/k!}\,dx$	1
1.9.	$u_{k+1}^{-2}u_ku_{k+2}$	$u_k^{-1}u_{k+1}\,dx$	u_ku_{k+1}
1.10.	$u_k^{-2(k+3)/(k+1)}Q_{k+2}$	$u_k^{2/(k+1)}\,dx$	u_k^2
1.11.	$Q_{k+2}^{-3/2}S_{k+3}$	$u_k^{-1}\sqrt{Q_{k+2}}\,dx$	u_kQ_{k+2}
2.1.	$u_2^{-8/3}R_4$	$u_2^{1/3}\,dx$	u_2^3
2.2.	$R_4^{-3/2}S_5$	$u_2^{-1}\sqrt{R_4}\,dx$	$u_2^2R_4$
2.3.	$S_5^{-8/3}V_7$	$u_2^{-1}S_5^{1/3}\,dx$	$u_2S_5^2$
3.1.	$x, \quad \mathcal{D}[u]$	dx	$W(x)$
3.2.	$x, \quad D_x \log \mathcal{D}[u]$	dx	$W(x)\mathcal{D}[u]$
3.3.	$x, \quad u_1^{-2}Q_3$	dx	u_1^3
4.1	$u_3^{-8/3}\widetilde{R}_5$	$u_3^{1/3}\,dx$	u_3^3
4.2	$\widetilde{R}_5^{-3/2}\widetilde{S}_6$	$u_3^{-1}\sqrt{\widetilde{R}_5}\,dx$	$u_3^2\widetilde{R}_5$
4.3	$T_7^{-5/2}Z_9$	$u_3^{-1}T_7^{1/4}\,dx$	$u_3T_7^2$

In Table 5, for given functions $\eta_1(x), \ldots, \eta_k(x)$, we let $\mathcal{D}$ be a kth order linear ordinary differential operator whose kernel is spanned by $\eta_1(x), \ldots, \eta_k(x)$, and let $W(x)$ denote their Wronskian determinant. Also, we let

$$Q_{k+2} = (k+1)u_k u_{k+2} - (k+2)u_{k+1}^2, \qquad R_4 = 3u_2u_4 - 5u_3^2,$$
$$S_{k+3} = (k+1)^2 u_k^2 u_{k+3} - 3(k+1)(k+3)u_k u_{k+1} u_{k+2} + \\ + 2(k+2)(k+3)u_{k+1}^3,$$
$$\widetilde{R}_5 = 3u_3u_5 - 5u_4^2, \qquad \widetilde{S}_6 = 9u_3^2u_6 - 45u_3u_4u_5 + 40u_4^3,$$
$$T_7 = 10u_3^3u_7 - 70u_3^2u_4u_6 - 49u_3^2u_5^2 + 280u_3u_4^2u_5 - 175u_4^4,$$
$$U_5 = u_1^2\left[Q_3D_x^2Q_3 - \tfrac{5}{4}(D_xQ_3)^2\right] + u_1u_2Q_3D_xQ_3 - (2u_1u_3 - u_2^2)Q_3^2,$$
$$V_7 = u_2^2\left[S_5D_x^2S_5 - -\tfrac{7}{6}(D_xS_5)^2\right] + u_2u_3S_5D_xS_5 - \tfrac{1}{2}(9u_2u_4 - 7u_3^2)S_5^2,$$
$$Z_9 = u_3^2\left[T_7D_x^2T_7 - \tfrac{9}{8}(D_xT_7)^2\right] + u_3u_4T_7D_xT_7 - \tfrac{4}{5}(7u_3u_5 - 5u_4^2)T_7^2.$$

Table 6

Primitive Lie Algebras of Vector Fields in $\mathbb{R}^2$

	Generators	Dim	Structure	$\mathbb{C}$ Type
6.1.	$\partial_x, \partial_u, \alpha(x\partial_x + u\partial_u) + u\partial_x - x\partial_u$	3	$\mathbb{R} \ltimes \mathbb{R}^2$	1.7
6.2.	$\partial_x, x\partial_x + u\partial_u, (x^2 - u^2)\partial_x + 2xu\partial_u$	3	$\mathfrak{sl}(2)$	1.1
6.3.	$u\partial_x - x\partial_u, (1 + x^2 - u^2)\partial_x + 2xu\partial_u,$ $2xu\partial_x + (1 - x^2 + u^2)\partial_u$	3	$\mathfrak{so}(3)$	1.1
6.4.	$\partial_x, \partial_u, x\partial_x + u\partial_u, u\partial_x - x\partial_u$	4	$\mathbb{R}^2 \ltimes \mathbb{R}^2$	1.9
6.5.	$\partial_x, \partial_u, x\partial_x - u\partial_u, u\partial_x, x\partial_u$	5	$\mathfrak{sa}(2)$	2.1
6.6.	$\partial_x, \partial_u, x\partial_x, u\partial_x, x\partial_u, u\partial_u$	6	$\mathfrak{a}(2)$	2.2
6.7.	$\partial_x, \partial_u, x\partial_x + u\partial_u, u\partial_x - x\partial_u,$ $(x^2 - u^2)\partial_x + 2xu\partial_u,$ $2xu\partial_x + (u^2 - x^2)\partial_u$	6	$\mathfrak{so}(3,1)$	1.4
6.8.	$\partial_x, \partial_u, x\partial_x, u\partial_x, x\partial_u, u\partial_u,$ $x^2\partial_x + xu\partial_u, xu\partial_x + u^2\partial_u$	8	$\mathfrak{sl}(3)$	2.3

Table 7

Symmetry Classification of Second Order Ordinary Differential Equations

Symmetry Group	Dim	Type	Invariant Equation
∂_u	1	3.1	$u_{xx} = F(x, u_x)$
∂_x, ∂_u	2	1.5	$u_{xx} = F(u_x)$
$\partial_x, e^x\partial_u$	2	1.5	$u_{xx} - u_x = F(u_x - u)$
$\partial_x, x\partial_x - u\partial_u,$ $x^2\partial_x - 2xu\partial_u$	3	1.1	$u_{xx} = \dfrac{3u_x^2}{2u} + cu^3$
$\partial_x, x\partial_x - u\partial_u,$ $x^2\partial_x - (2xu+1)\partial_u$	3	1.2	$u_{xx} = 6uu_x - 4u^3 +$ $+ c(u_x - u^2)^{3/2}$
$\partial_x, \partial_u, x\partial_x + \alpha u\partial_u$ $\alpha \neq 0, \frac{1}{2}, 1, 2$	3	1.7	$u_{xx} = cu_x^{\frac{\alpha-2}{\alpha-1}}$
$\partial_x, \partial_u, x\partial_x + (x+u)\partial_u$	3	1.8	$u_{xx} = ce^{-u_x}$
$\partial_x, \partial_u, x\partial_x, u\partial_x, x\partial_u, u\partial_u,$ $x^2\partial_x + xu\partial_u, xu\partial_x + u^2\partial_u$	8	2.3	$u_{xx} = 0$

References

[1] Abraham–Shrauner, B., and Guo, A., Hidden symmetries associated with the projective group of nonlinear first order ordinary differential equations, *J. Phys. A* **25** (1992), 5597–5608.

[2] Ackerman, M., and Hermann, R., *Sophus Lie's 1880 Transformation Group Paper*, Math Sci Press, Brookline, Mass., 1975.

[3] Ackerman, M., and Hermann, R., *Sophus Lie's 1884 Differential Invariant Paper*, Math Sci Press, Brookline, Mass., 1976.

[4] Adler, M., On a trace functional for formal pseudo-differential operators and the symplectic structure of the Korteweg-deVries type equations, *Invent. Math.* **50** (1979), 219–248.

[5] Anderson, I.M., Aspects of the inverse problem of the calculus of variations, *Archivum Mathematicum (Brno)* **24** (1988), 181–202.

[6] Anderson, I.M., Introduction to the variational bicomplex, *Contemp. Math.* **132** (1992), 51–73.

[7] Anderson, I.M., *The Variational Bicomplex*, Academic Press, to appear.

[8] Anderson, I.M., and Kamran, N., The variational bicomplex for second order scalar partial differential equations in the plane, *Duke Math. J.*, to appear.

[9] Anderson, I.M., Kamran, N., and Olver, P.J., Internal, external and generalized symmetries, *Adv. in Math.* **100** (1993), 53–100.

[10] Anderson, I.M., and Thompson, G., *The Inverse Problem of the Calculus of Variations for Ordinary Differential Equations*, Memoirs Amer. Math. Soc., Vol. 98, No. 473, Providence, R.I., 1992.

[11] Anderson, R.L., and Ibragimov, N.H., *Lie–Bäcklund Transformations in Applications*, SIAM, Philadelphia, 1979.

[12] Arnol'd, V.I., *Mathematical Methods of Classical Mechanics*, Graduate Texts in Mathematics, Vol. 60, Springer–Verlag, New York, 1978.

[13] Arnol'd, V.I., *Geometrical Methods in the Theory of Ordinary Differential Equations*, Springer–Verlag, New York, 1983.

[14] Arnol'd, V.I., Gusein–Zade, S.M., and Varchenko, A.N., *Singularities of Differentiable Maps*, Birkhäuser, Boston, 1985, 1988.

[15] Bäcklund, A.V., Ueber Flachentransformationen, *Math. Ann.* **9** (1876), 297–320.

[16] Ball, J.M., and Mizel, V.J., One-dimensional variational problem whose minimizers do not satisfy the Euler-Lagrange equation, *Arch. Rat. Mech. Anal.* **90** (1985), 325–388.

[17] Bargmann, V., Irreducible unitary representations of the Lorentz group, *Ann. Math.* **48** (1947), 568–640.

[18] Bianchi, L., *Lezioni sulla Teoria dei Gruppi Continui Finiti di Transformazioni*, Enrico Spoerri, Pisa, 1918.

[**19**] Bleecker, D., *Gauge Theory and Variational Principles*, Addison–Wesley Publ. Co., Reading, Mass., 1981.

[**20**] Bluman, G.W., and Kumei, S., Symmetry-based algorithms to relate partial differential equations: I. Local symmetries, *Euro. J. Appl. Math.* **1** (1990), 189–216.

[**21**] Bluman, G.W., and Kumei, S., Symmetry-based algorithms to relate partial differential equations: II. Linearization by nonlocal symmetries, *Euro. J. Appl. Math.* **1** (1990), 217–223.

[**22**] Bluman, G.W., and Kumei, S., *Symmetries and Differential Equations*, Springer–Verlag, New York, 1989.

[**23**] Bocharov, A.V., Sokolov, V.V., and Svinolupov, S.I., On some equivalence problems for differential equations, preprint, Erwin Schrödinger Institute, Vienna, 1993.

[**24**] Bott, R., and Tu, L.W., *Differential Forms in Algebraic Topology*, Springer–Verlag, New York, 1982.

[**25**] Boyer, C.P., and Plebański, J.F., General relativity and G-structures I. General theory and algebraically degenerate spaces, *Rep. Math. Phys.* **14** (1978), 111–145.

[**26**] Bryant, R.L., *Some Aspects of the Local and Global Theory of Pfaffian Systems*, Thesis, University of North Carolina, Chapel Hill, NC, 1979.

[**27**] Bryant, R.L., On notions of equivalence of variational problems with one independent variable, *Contemp. Math.* **68** (1987), 65–76.

[**28**] Bryant, R.L., Chern, S.-S., Gardner, R.B., Goldschmidt, H.L., and Griffiths, P.A., *Exterior Differential Systems*, Math. Sci. Res. Inst. Publ., Vol. 18, Springer–Verlag, New York, 1991.

[**29**] Bryant, R.L., and Griffiths, P.A., Characteristic cohomology of differential systems I, II, *J. Amer. Math. Soc.* **8** (1995), 507–596, *Duke Math. J.* **78** (1995), 531–676.

[**30**] Bryant, R.L., Griffiths, P.A., and Hsu, L., Hyperbolic exterior differential systems and their conservation laws, Part I, preprint, Selecta Math.; 1 (1995) 21–112.

[**31**] Buchberger, B., Applications of Gröbner bases in non-linear computational geometry, *in: Mathematical Aspects of Scientific Software*, J.R. Rice, ed., IMA Volumes in Mathematics and its Applications, Vol. 14, Springer–Verlag, New York, 1988, pp. 59–87.

[**32**] Carathéodory, C., Über die Variationsrechnung bei mehrfachen Integralen, *Acta Sci. Mat. (Szeged)* **4** (1929), 193–216.

[**33**] Cartan, É., *Leçons sur les Invariants Integraux*, Hermann, Paris, 1922.

[**34**] Cartan, É., *La Méthode du Repére Mobile, la Théorie des Groupes Continus, et les Espaces Généralisés*, Exposés de Géométrie No. 5, Hermann, Paris, 1935.

[**35**] Cartan, É., *La Topologie des Groupes de Lie*, Exposés de Géométrie No. 8, Hermann, Paris, 1936.

[**36**] Cartan, É., *Les Systèmes Différentiels Extérieurs et Leurs Applications Géométriques*, Exposés de Géométrie No. 14, Hermann, Paris, 1945.

[37] Cartan, É., Les sous-groupes des groupes continus de transformations, *in: Oeuvres Complètes*, Part. II, Vol. 2, Gauthier–Villars, Paris, 1953, pp. 719–856.

[38] Cartan, É., Les systèmes de Pfaff à cinq variables aux dérivées partielles du second order, *in: Oeuvres Complètes*, Part. II, Vol. 2, Gauthier–Villars, Paris, 1953, pp. 927–1010.

[39] Cartan, É., Les problèmes d'équivalence, *in: Oeuvres Complètes*, Part. II, Vol. 2, Gauthier–Villars, Paris, 1953, pp. 1311–1334.

[40] Cartan, É., Sur les variétés à connexion projective, *in: Oeuvres Complètes*, Part. III, Vol. 1, Gauthier–Villars, Paris, 1955, pp. 825–861.

[41] Cartan, É., Notice historique sur la notion de parallélisme absolu, *in: Oeuvres Complètes*, Part. III, Vol. 2, Gauthier–Villars, Paris, 1955, pp. 1121–1129.

[42] Cartan, É., La géométrie de l'intégrale $\int F(x, y, y', y'')\,dx$, *in: Oeuvres Complètes*, Part. III, Vol. 2, Gauthier–Villars, Paris, 1955, pp. 1341-1368.

[43] Chakravarty, S., Ablowitz, M.J., and Clarkson, P.A., Reductions of self-dual Yang–Mills fields and classical systems, *Phys. Rev. Lett.* **65** (1990), 1085–1087.

[44] Champagne, B., Hereman, W., and Winternitz, P., The computer calculation of Lie point symmetries of large systems of differential equations, *Comp. Phys. Comm.* **66** (1991), 319–340.

[45] Chazy, J., Sur les équations différentielles du troisième ordre et d'ordre supérieur dont l'intégrale générale a ses points critiques fixes, *Acta Math.* **34** (1911), 317–385.

[46] Chern, S.–S., Sur la géometrie d'un système d'équations différentielles du second ordre, *Bull. Sci. Math.* **63** (1939), 206–212.

[47] Chern, S.–S., The geometry of the differential equation $y''' = F(x, y, y', y'')$, *Sci. Rep. Nat. Tsing Hua Univ.* **4** (1940), 97–111.

[48] Chern, S.–S., On the Euclidean connections in a Finsler space, *Proc. Nat. Acad. Sci.* **29** (1943), 33–37.

[49] Chern, S.–S., Local equivalence and Euclidean connections in Finsler spaces, *Sci. Rep. Nat. Tsing Hua Univ.* **5** (1948), 95–121.

[50] Chern, S.–S., Historical remarks on Gauss-Bonnet, *in: Analysis, Et. Cetera*, P.H. Rabinowitz and E. Zehnder, eds., Academic Press, Boston, 1990, pp. 209–217.

[51] Chrastina, J., On the equivalence of variational problems, I, *J. Diff. Eq.* **98** (1992), 76–90.

[52] Christoffel, E.B., Ueber die Transformation der homogenen Differentialausdrücke zweiten Grades, *J. für die Reine und Angew. Math.* **70** (1869), 46–70.

[53] Ciarlet, P.G., and Rabier, P., *Les Equations de von Kármán*, Lecture Notes in Math., #826, Springer–Verlag, New York, 1980.

[54] Clarkson, P.A., and Olver, P.J., Symmetry and the Chazy equation, *J. Diff. Eq.*, to appear.

[**55**] Cohen, H., Sums involving the values at negative integers of L-functions of quadratic characters, *Math. Ann.* **217** (1975), 271–285.

[**56**] Courant, R., and Hilbert, D., *Methods of Mathematical Physics*, Vol. II, Interscience Publ. Inc., New York, 1953.

[**57**] Crampin, M., and Pirani, F.A.E., *Applicable Differential Geometry*, London Math. Soc. Lecture Note Series, Vol. 59, Cambridge Univ. Press, Cambridge, 1986.

[**58**] Darboux, G., Sur le problème de Pfaff, *Bull. Sci. Math.* (2) **6** (1882), 14–36, 49–68.

[**59**] David, D., and Holm, D.D., Multiple Lie–Poisson structures, reductions, and geometric phases for the Maxwell–Bloch traveling-wave equations, *J. Nonlin. Sci.* **2** (1992), 241–262.

[**60**] Debever, R., Quelques problèmes d'équivalence de formes différentielles alternées, *Bull. Acad. Roy. de Belgique, Classe des Sciences* **31** (1946), 262–277.

[**61**] Dickson, L.E., Differential equations from the group standpoint, *Ann. Math.* **25** (1924), 287–378.

[**62**] Dixmier, J., Quelques aspects de la théorie des invariants, *Gazette des Math.* **43** (1990), 39–64.

[**63**] Doyle, P.W., *Differential Geometric Poisson Bivectors and Quasilinear Systems in One Space Variable*, Ph.D. Thesis, University of Minnesota, 1992.

[**64**] Doyle, P.W., Symmetry classes of quasilinear systems in one space variable, *Nonlinear Math. Phys.* **1** (1994), 225–266.

[**65**] Drinfel'd, V.G., Quantum groups, *in*: *Proc. Int. Congress Math. Berkeley*, A.M. Gleason, ed., Vol. 1., Amer. Math. Soc., Providence, R.I., 1987, pp. 798–820.

[**66**] Ehresmann, C., Introduction a la théorie des structures infinitésimales et des pseudo-groupes de Lie, *in*: *Geometrie Differentielle*, Colloq. Inter. du Centre Nat. de la Recherche Scientifique, Strasbourg, 1953, pp. 97–110.

[**67**] Eisenhart, L.P., *Riemannian Geometry*, Princeton Univ. Press, Princeton, 1926.

[**68**] Elphick, C., Tirapegui, E., Brachet, M.E., Coullet, P., and Ioss, G., A simple global characterization for normal forms of singular vector fields, *Physica* **29D** (1987), 95–127.

[**69**] Fels, M., *Some Applications of Cartan's Method of Equivalence to the Geometric Study of Ordinary and Partial Differential Equations*, Ph.D. Thesis, McGill University, 1993.

[**70**] Fels, M., The equivalence problem for systems of second-order ordinary differential equations, *Proc. London Math. Soc.* **71** (1995), 221–240.

[**71**] Forsyth, A.R., Invariants, covariants, and quotient derivatives associated with linear differential equations, *Phil. Trans. Roy. Soc. London* **179** (1888), 377–489.

[**72**] Forsyth, A.R., The differential invariants of a surface, and their geometric significance, *Phil. Trans. Roy. Soc. London A* **201** (1903), 329–402.

[73] Frobenius, G., Über das Pfaffsche Probleme, *J. für Reine und Angew. Math.* **82** (1877), 230–315.
[74] Fulling, S.A., King, R.C., Wybourne, B.G., and Cummins, C.J., Normal forms for tensor polynomials: I. The Riemann tensor, *Class. Quantum Grav.* **9** (1992), 1151–1197.
[75] Fushchich, W.I., and Yegorchenko, I.A., Second-order differential invariants of the rotation group $O(n)$ and of its extensions: $E(n), P(1,n), G(1,n)$, *Acta Appl. Math.* **28** (1992), 69–92.
[76] Gage, M., and Hamilton, R.S., The heat equation shrinking convex plane curves, *J. Diff. Geom.* **23** (1986), 69–96.
[77] Gardner, R.B., Differential geometric methods interfacing control theory, *in: Differential Geometric Control Theory*, R.W. Brockett et. al., eds., Birkhäuser, Boston, 1983, pp. 117–180.
[78] Gardner, R.B., *The Method of Equivalence and Its Applications*, SIAM, Philadelphia, 1989.
[79] Gardner, R.B., and Kamran, N., Characteristics and the geometry of hyperbolic equations in the plane, *J. Diff. Eq.* **104** (1993), 60–116.
[80] Gardner, R.B., and Shadwick, W.F., Equivalence of one dimensional Lagrangian field theories in the plane I, *in: Global Differential Geometry and Global Analysis*, D. Ferus et. al., eds., Lecture Notes in Math, Vol. 1156, Springer–Verlag, New York, 1985, pp. 154–179.
[81] Gardner, R.B., and Shadwick, W.F., An equivalence problem for a two-form and a vector field on $\mathbb{R}^3$, *in: Differential Geometry, Global Analysis, and Topology*, A. Nicas and W.F. Shadwick, eds., Canadian Math. Soc. Conference Proceedings, Vol. 12, Amer. Math. Soc., Providence, R.I., 1991, pp. 41–50.
[82] Gat, O., Symmetries of third order differential equations, *J. Math. Phys.* **33** (1992), 2966–2971.
[83] Gel'fand, I.M., and Dikii, L.A., Asymptotic behavior of the resolvent of Sturm-Liouville equations and the algebra of the Korteweg-deVries equation, *Russ. Math. Surveys* **30**:5 (1975), 77–113.
[84] Gel'fand, I.M., and Dikii, L.A., A family of Hamiltonian structures connected with integrable non-linear differential equations, preprint, Inst. Appl. Math. Acad. Sci. USSR, No. 136, 1978 (in Russian).
[85] Gel'fand, I.M., and Fomin, S.V., *Calculus of Variations*, Prentice Hall, Englewood Cliffs, N.J., 1963.
[86] Goldschmidt, H., and Sternberg, S., The Hamilton-Cartan formalism in the calculus of variations, *Ann. Inst. Fourier* **23** (1973), 203-269.
[87] Golubitsky, M., Primitive actions and maximal subgroups of Lie groups, *J. Diff. Geom.* **7** (1972), 175–191.
[88] Golubitsky, M., and Guillemin, V., *Stable Mappings and their Singularities*, Springer–Verlag, New York, 1973.
[89] González–Gascón, F., and González–López, A., Symmetries of differential equations. IV, *J. Math. Phys.* **24** (1983), 2006–2021.
[90] González–López, A., *Symmetries of Systems of Ordinary Differential Equations: Direct and Inverse Problems*, Ph.D. Thesis, Universidad Complutense de Madrid, Spain, 1984 (in Spanish).

[**91**] González–López, A., Symmetries of linear systems of second order differential equations, *J. Math. Phys.* **29** (1988), 1097–1105.

[**92**] González–López, A., On the linearization of second order ordinary differential equations, *Lett. Math. Phys.* **17** (1989), 341–349.

[**93**] González–López, A., Symmetry bounds of variational problems, *J. Phys. A* **27** (1994), 1205–1232.

[**94**] González–López, A., Kamran, N., and Olver, P.J., Lie algebras of vector fields in the real plane, *Proc. London Math. Soc.* **64** (1992), 339–368.

[**95**] González–López, A., Kamran, N., and Olver, P.J., Lie algebras of differential operators in two complex variables, *American J. Math.* **114** (1992), 1163–1185.

[**96**] González–López, A., Kamran, N., and Olver, P.J., Quasi–exactly solvable Lie algebras of first order differential operators in two complex variables, *J. Phys. A* **24** (1991), 3995–4008.

[**97**] González–López, A., Kamran, N., and Olver, P.J., Quasi–exact solvability, *Contemp. Math.* **160** (1993), 113–140.

[**98**] Gotay, M., An exterior differential systems approach to the Cartan form, *in*: *Géométrie Symplectique et Physique Mathématique*, P. Donato et. al., eds., Birkhäuser, Boston, 1991, pp. 160–188.

[**99**] Grace, J.H., and Young, A., *The Algebra of Invariants*, Cambridge Univ. Press, Cambridge, 1903.

[**100**] Graustein, W.C., *Differential Geometry*, MacMillan, New York, 1935.

[**101**] Grayson, M., The heat equation shrinks embedded plane curves to round points, *J. Diff. Geom.* **26** (1987), 285–314.

[**102**] Green, M.L., The moving frame, differential invariants and rigidity theorems for curves in homogeneous spaces, *Duke Math. J.* **45** (1978), 735–779.

[**103**] Greene, C., and Kleitman, D.J., Proof techniques in the theory of finite sets, *in*: *Studies in Combinatorics*, G.–C. Rota, ed., Studies in Math., Vol. 17, Math. Assoc. Amer., Washington, D.C., 1978, pp. 22–79.

[**104**] Grissom, C., Thompson, G., and Wilkens, G., Linearization of second order ordinary differential equations via Cartan's equivalence method, *J. Diff. Eq.* **77** (1989), 1–15.

[**105**] Guckenheimer, J., and Holmes, P., *Nonlinear Oscillations, Dynamical Systems, and Bifurcations of Vector Fields*, Appl. Math. Sci., Vol. 42, Springer–Verlag, New York, 1983.

[**106**] Guggenheimer, H.W., *Differential Geometry*, McGraw–Hill, New York, 1963.

[**107**] Gurevich, G.B., *Foundations of the Theory of Algebraic Invariants*, P. Noordhoff Ltd., Groningen, Holland, 1964.

[**108**] Halphen, G.–H., Sur les invariant différentiels, *in*: *Oeuvres*, Vol. 2, Gauthiers–Villars, Paris, 1913, pp. 197–253.

[**109**] Halphen, G.–H., Sur les invariants des équations différentielles linéaires du quatrième ordre, *in*: *Oeuvres*, Vol. 3, Gauthiers–Villars, Paris, 1921, pp. 463–514.

[**110**] Hamermesh, M., *Group Theory and its Application to Physical Problems*, Addison–Wesley Publ. Co., Reading, Mass., 1962.

[**111**] Hawkins, T., Jacobi and the birth of Lie's theory of groups, *Arch. Hist. Exact Sci.* **42** (1991), 187–278.

[**112**] Hereman, W., Review of symbolic software for the calculation of Lie symmetries of differential equations, *Euromath Bull.* **1** (1994), 45–82.

[**113**] Hermann, R., The differential geometry of foliations II, *J. Math. Mech.* **11**(1962), 303–315.

[**114**] Hilbert, D., Über die vollen Invariantensysteme, *in*: *Gesammelte Abhandlungen*, Vol. 2, Springer–Verlag, Berlin, 1933, pp. 287–344.

[**115**] Hilbert, D., *Theory of Algebraic Invariants*, Cambridge Univ. Press, New York, 1993.

[**116**] Hille, E., *Ordinary Differential Equations in the Complex Domain*, John Wiley & Sons, New York, 1976.

[**117**] Hsu, L., and Kamran, N., Classification of second-order ordinary differential equations admitting Lie groups of fiber-preserving symmetries, *Proc. London Math. Soc.* **58** (1989), 387–416.

[**118**] Hsu, L., Kamran, N., and Olver, P.J., Equivalence of higher order Lagrangians II. The Cartan form for particle Lagrangians, *J. Math. Phys.* **30** (1989), 902–906.

[**119**] Humphreys, J.E., *Introduction to Lie Groups and Lie Algebras*, Graduate Texts in Mathematics, Vol. 9, Springer–Verlag, New York, 1976.

[**120**] Ibragimov, N.H., *Essays in the Group Analysis of Ordinary Differential Equations*, Matematika–Kibernetika, Znanie Publ. Moscow, 1991 (in Russian).

[**121**] Ibragimov, N.H., Group analysis of ordinary differential equations and the invariance principle in mathematical physics (for the 150th anniversary of Sophus Lie), *Russian Math. Surveys* **47**:4 (1992), 89–156.

[**122**] Ibragimov, N.H., ed., *CRC Handbook of Lie Group Analysis of Differential Equations*, Vol. 1, CRC Press, Boca Raton, Fl., 1994.

[**123**] Ince, E.L., *Ordinary Differential Equations*, Dover, New York, 1956.

[**124**] Jacobson, N., *Lie Algebras*, Interscience Publ. Inc., New York, 1962.

[**125**] Kalnins, E.G., and Miller, W., Related evolution equations and Lie symmetries, *SIAM J. Math. Anal.* **16** (1985), 221–232.

[**126**] Kalnins, E.G., and Miller, W., Equivalence classes of related evolution equations and Lie symmetries, *J. Phys. A* **20** (1987), 5434–5446.

[**127**] Kamran, N., Contributions to the study of the equivalence problem of Elie Cartan and its applications to partial and ordinary differential equations, *Mem. Cl. Sci. Acad. Roy. Belg.* **45** (1989), Fac. 7.

[**128**] Kamran, N., Lamb, K.G., and Shadwick, W.F.,The local equivalence problem for $d^2y/dx^2 = F(x, y, dx/dy)$ and the Painlevé transcendents, *J. Diff. Geom.* **22** (1985), 139–150.

[**129**] Kamran, N., and Olver, P.J., Equivalence problems for first order Lagrangians on the line, *J. Diff. Eq.* **80** (1989), 32–78.

[**130**] Kamran, N., and Olver, P.J., Equivalence of differential operators, *SIAM J. Math. Anal.* **20** (1989), 1172–1185.

[**131**] Kamran, N., and Olver, P.J., Lie algebras of differential operators and Lie-algebraic potentials, *J. Math. Anal. Appl.* **145** (1990), 342–356.

[**132**] Kamran, N., and Olver, P.J., Equivalence of higher order Lagrangians I. Formulation and reduction, *J. Math. Pures et Appliquées* **70** (1991), 369–391.

[**133**] Kamran, N., and Olver, P.J., Equivalence of higher order Lagrangians III. New invariant differential equations, *Nonlinearity* **5** (1992), 601–621.

[**134**] Kamran, N., and Shadwick, W.F., Équivalence locale des équations aux dérivées partielles quasi linéaires du deuxième ordre et pseudo-groupes infinis, *Comptes Rendus Acad. Sci. (Paris) Série I* **303** (1986), 555–558.

[**135**] Kamran, N., and Shadwick, W.F., A differential geometric characterization of the first Painlevé transcendent, *Math. Ann.* **279** (1987), 117–123.

[**136**] Karlhede, A., A review of the geometric equivalence of metrics in general relativity, *Gen. Rel. Grav.* **12** (1980), 693–707.

[**137**] Karlhede, A., and MacCallum, M.A.H., On determining the isometry group of a Riemannian space, *Gen. Rel. Grav.* **14** (1982), 673–682.

[**138**] Kobayashi, S., *Transformation Groups in Differential Geometry*, Springer-Verlag, New York, 1972.

[**139**] Komrakov, B., Primitive actions and the Sophus Lie problem, preprint, University of Oslo, No. 16, 1993.

[**140**] Komrakov, B., Churyumov, A., and Doubrov, B., Two-dimensional homogeneous spaces, preprint, University of Oslo, No. 17, 1993.

[**141**] Kramer, D., Stephani, H., MacCallum, M., and Herlt, E., *Exact Solutions of Einstein's Field Equations*, V.E.B. Deutscher Verlag Wissen., Berlin, 1980.

[**142**] Krause, J., and Michel, L., Classification of the symmetries of ordinary differential equations, *in: Group Theoretical Methods in Physics*, V.V. Dodonov and V.I. Man'ko, eds., Lecture Notes in Physics, Vol. 382, Springer–Verlag, New York, 1991, pp. 251–262.

[**143**] Kreysig, E., On surface theory in E^3 and generalizations, *Exp. Math.* **12** (1994), 97–123.

[**144**] Kumei, S., and Bluman, G.W., When nonlinear differential equations are equivalent to linear differential equations, *SIAM J. Appl. Math.* **5** (1982), 1157–1173.

[**145**] Kuranishi, M., On E. Cartan's prolongation theorem of exterior differential systems, *Amer. J. Math.* **79** (1957), 1–47.

[**146**] Kuranishi, M., On the local theory of continuous infinite pseudo groups I, *Nagoya Math. J.* **15** (1959), 225–260.

[**147**] Kuranishi, M., On the local theory of continuous infinite pseudo groups II, *Nagoya Math. J.* **19** (1961), 55–91.

[**148**] Laguerre, E., Sur quelques invariants des équations différentielles linéaires, *in*: *Oeuvres*, Vol. 1, Gauthiers–Villars, Paris, 1898, pp. 424–427.

[**149**] Lewy, H., An example of a smooth linear partial differential equation without solution, *Ann. Math.* **64** (1956), 514–522.

[**150**] Lie, S., Klassifikation und Integration von gewöhnlichen Differentialgleichungen zwischen x, y, die eine Gruppe von Transformationen gestatten I–IV, *in*: *Gesammelte Abhandlungen*, Vol. 5, B.G. Teubner, Leipzig, 1924, pp. 240–310, 362–427, 432–448.

[**151**] Lie, S., Gruppenregister, *in*: *Gesammelte Abhandlungen*, Vol. 5, B.G. Teubner, Leipzig, 1924, pp. 767–773.

[**152**] Lie, S., Theorie der Transformationsgruppen, *in*: *Gesammelte Abhandlungen*, Vol. 6, B.G. Teubner, Leipzig, 1927, pp. 1–94; see [**2**] for an English translation.

[**153**] Lie, S., Über Differentialinvarianten, *in*: *Gesammelte Abhandlungen*, Vol. 6, B.G. Teubner, Leipzig, 1927, pp. 95–138; see [**3**] for an English translation.

[**154**] Lie, S., Über Integralinvarianten und ihre Verwertung für die Theorie der Differentialgleichungen, *in*: *Gesammelte Abhandlungen*, Vol. 6, B.G. Teubner, Leipzig, 1927, pp. 664–701.

[**155**] Lie, S., and Engel, F., *Theorie der Transformationsgruppen*, B.G. Teubner, Leipzig, 1888, 1890, 1893.

[**156**] Lie, S., and Engel, F., *Vorlesungen über Differentialgleichungen mit Bekannten Infinitesimalen Transformationen*, B.G. Teubner, Leipzig, 1891.

[**157**] Lie, S., and Scheffers, G., *Geometrie der Berührungstransformationen*, B.G. Teubner, Leipzig, 1896.

[**158**] Littlejohn, L.L., On the classification of differential equations having orthogonal polynomial solutions, *Ann. di Mat.* **138** (1984), 35–53.

[**159**] Macaulay, F.S., Some properties of enumeration in the theory of modular systems, *Proc. London Math. Soc.* **26** (2) (1927), 531–555.

[**160**] Mackey, G.W., *The Theory of Unitary Group Representations*, Univ. of Chicago Press, Chicago, 1976.

[**161**] Magadeev, B.A., On the group classification of nonlinear evolution equations, *St. Petersburg Math. J.* **5** (1994), 345–359.

[**162**] Magri, F., A simple model of the integrable Hamiltonian equation, *J. Math. Phys.* **19** (1978), 1156–1162.

[**163**] Mahomed, F.M., and Leach, P.G.L., The linear symmetries of a nonlinear equation, *Quaes. Math.* **8** (1985), 241–274.

[**164**] Malcev, A.I., Sur les groupes topologiques locaux et complets, *Comptes Rendus Acad. Sci. URSS* **32** (1941), 606–608.

[**165**] Marsden, J.E., and Weinstein, A., Reduction of symplectic manifolds with symmetry, *Rep. Math. Phys.* **5** (1974), 121–130.

[**166**] Martinet, J., Sur les singularitiés des forms différentielles, *Ann. Inst. Fourier* **20** (1970), 95–178.

[**167**] Maschke, H., A new method for determining the differential parameters and invariants of quadratic differential quantics, *Trans. Amer. Math. Soc.* **1** (1900), 197–204.

[**168**] Mikhailov, A.V., Shabat, A.B., and Sokolov, V.V., The symmetry approach to classification of integrable equations, *in*: *What is Integrability?*, V.E. Zakharov, ed., Springer–Verlag, New York, 1990, pp. 115–184.

[**169**] Miller, W., Jr., *Symmetry Groups and their Applications*, Academic Press, New York, 1972.

[**170**] Morikawa, H., Some analytic and geometric applications of the invariant theoretic method, *Nagoya Math. J.* **80** (1980), 1–47.

[**171**] Morrey, C.B., Jr., *Multiple Integrals in the Calculus of Variations*, Springer-Verlag, New York, 1966.

[**172**] Mostow, G.D., The extensibility of local Lie groups of transformations and groups on surfaces, *Ann. Math.* **52** (1950), 606–636.

[**173**] Murillo, R., *Cartan's Equivalence Method and an Application to Second Order Evolution Equations*, Ph.D. Thesis, University of Minnesota, 1994.

[**174**] Nagano, T., Linear differential systems with singularities and an application to transitive Lie algebras, *J. Math. Soc. Japan* **18** (1966), 398–404.

[**175**] Narasimhan, R., *Analysis on Real and Complex Manifolds*, North Holland Publ. Co., Amsterdam, 1968.

[**176**] Neuman, F., *Global Properties of Linear Ordinary Differential Equations*, Kluwer Academic Publ., Boston, 1991.

[**177**] Newell, A., *Solitons in Mathematics and Physics*, Society for Industrial and Applied Mathematics, Philadelphia, 1985.

[**178**] Noether, E., Invariante Variationsprobleme, *Nachr. Konig. Gesell. Wissen. Gottingen, Math.-Phys. Kl.* (1918), 235–257. (See *Transport Theory and Stat. Phys.* **1** (1971), 186–207 for an English translation.)

[**179**] Olver, P.J., Conservation laws and null divergences, *Math. Proc. Camb. Phil. Soc.* **94** (1983), 529–540.

[**180**] Olver, P.J., Invariant theory and differential equations, *in*: *Invariant Theory*, S.S. Koh, ed., Lecture Notes in Mathematics, Vol. 1278, Springer–Verlag, New York, 1987, pp. 62–80.

[**181**] Olver, P.J., The equivalence problem and canonical forms for quadratic Lagrangians, *Adv. Appl. Math.* **9** (1988), 226–257.

[**182**] Olver, P.J., Canonical elastic moduli, *J. Elasticity* **19** (1988), 189–212.

[**183**] Olver, P.J., Classical invariant theory and the equivalence problem for particle Lagrangians. I. Binary Forms, *Adv. in Math.* **80** (1990), 39–77.

[**184**] Olver, P.J., Canonical forms and integrability of biHamiltonian systems, *Phys. Lett. A* **148** (1990), 177–187.

[**185**] Olver, P.J., Canonical anisotropic elastic moduli, *in*: *Modern Theory of Anisotropic Elasticity and Applications*, J.J. Wu, T.C.T. Ting and D.M. Barnett, eds., SIAM, Philadelphia, 1991, pp. 325–339.

[**186**] Olver, P.J., *Applications of Lie Groups to Differential Equations*, Second Edition, Graduate Texts in Mathematics, Vol. 107, Springer–Verlag, New York, 1993.

[**187**] Olver, P.J., Equivalence and the Cartan form, *Acta Appl. Math.* **31** (1993), 99–136.

[**188**] Olver, P.J., Non-associative local Lie groups, *J. Lie Theory* **6** (1996), 23–51.

[189] Olver, P.J., Sapiro, G., and Tannenbaum, A., Differential invariant signatures and flows in computer vision: a symmetry group approach, *in*: *Geometry–Driven Diffusion in Computer Vision*, B. M. Ter Haar Romeny, ed., Kluwer Acad. Publ., Dordrecht, the Netherlands, 1994.

[190] Olver, P.J., Sapiro, G., and Tannenbaum, A., Invariant geometric evolutions of surfaces and volumetric smoothing, *SIAM J. Appl. Math.*, to appear.

[191] Ovsiannikov, L.V., *Group Analysis of Differential Equations*, Academic Press, New York, 1982.

[192] Ovsienko, O.D., and Ovsienko, V.Y., Lie derivatives of order n on the line. Tensor meaning of the Gelfand-Dikii bracket, *Adv. Soviet Math.* **2** (1991), 221–231.

[193] Palais, R.S., *A Global Formulation of the Lie Theory of Transformation Groups*, Memoirs of the Amer. Math. Soc. No. 22, Providence, R.I., 1957.

[194] Pfaff, J.F., Methodus generalis, aequationes differentiarum partialium nec non aequationes differentiales vulgates, ultrasque primi ordinis, inter quotcunque variables, complete integrandi, *Abh. Konig. Akad. Wissen. Berlin* (1814–15), 76–136. (See *Ostwald's Klassiker der Exacten Wissenschaften*, Vol. 129, Verlag von Wilhelm Engelman, Leipzig, 1902, for a German translation.)

[195] Pontrjagin, L., *Topological Groups*, Princeton Univ. Press, Princeton, N.J., 1946.

[196] Rankin, R.A., The construction of automorphic forms from the derivatives of a given form, *J. Indian Math. Soc.* **20** (1956), 103–116.

[197] Riemann, B., Commentatio mathematica, qua respondere tentatur quaestioni ab Illma Academia Parisiensi propositae, *in*: *Gesammelte Mathematische Werke und Wissenschaftlicher Nachlass*, B.G. Teubner, Leipzig, 1892, pp. 391–423.

[198] Rosinger, E.E., and Rudolph, M., Group invariance of global generalized solutions of nonlinear PDEs: A Dedekind order completion method, *Lie Groups and their Appl.* **1** (1994), 203–215.

[199] Rosinger, E.E., and Walus, Y.E., Group invariance of generalized solutions obtained through the algebraic method, *Nonlinearity* **7** (1994), 837–859.

[200] Roždestvenskiĭ, B.L., and Janenko, N.N., *Systems of quasilinear equations and their applications to gas dynamics*, Transl. Math. Monographs, Vol. 55, Amer. Math. Soc., Providence, R.I., 1983.

[201] Sapiro, G., and Tannenbaum, A., On affine plane curve evolution, *J. Func. Anal.* **119** (1994), 79–120.

[202] Se-ashi, Y., A geometric construction of Laguerre–Forsyth's canonical forms of linear ordinary differential equations, *Adv. Studies Pure Math.* **22** (1993), 265–297.

[203] Shakiban, C., A resolution of the Euler operator II, *Math. Proc. Camb. Phil. Soc.* **89** (1981), 501–510.

[204] Singer, I.M., and Sternberg, S., The infinite groups of Lie and Cartan. Part I (the transitive groups), *J. Analyse Math.* **15** (1965), 1–114.

[**205**] Sokolov, V.V., On the symmetries of evolution equations, *Russ. Math. Surveys* **43**:5 (1988), 165–204.

[**206**] Spivak, M., *A Comprehensive Introduction to Differential Geometry*, Vol. 1, Second Ed., Publish or Perish, Inc., Wilmington, Delaware, 1979.

[**207**] Sternberg, S., *Lectures on Differential Geometry*, Prentice-Hall, Englewood Cliffs, N.J., 1964.

[**208**] Strang, G., *Linear Algebra and its Applications*, Third ed., Academic Press, 1986.

[**209**] Stroh, E., Ueber eine fundamentale Eigenschaft des Ueberschiebungsprocesses und deren Verwerthung in der Theorie der binären Formen, *Math. Ann.* **33** (1889), 61–107.

[**210**] Struik, D.J., *Lectures on Classical Differential Geometry*, Addison–Wesley Publ. Co., Reading, Mass., 1950.

[**211**] Sturmfels, B., *Algorithms in Invariant Theory*, Springer–Verlag, New York, 1993.

[**212**] Svinolupov, S.I., and Sokolov, V.V., Representations of contragredient Lie algebras in contact vector fields, *Func. Anal. Appl.* **25** (1991), 146–147.

[**213**] Takhtajan, L.A., A simple example of modular-forms as tau-functions for integrable equations, *Theor. Math. Phys.* **93** (1993), 1308–1317.

[**214**] Tresse, A., Sur les invariants différentiels des groupes continus de transformations, *Acta Math.* **18** (1894), 1–88.

[**215**] Tresse, M.A., *Détermination des Invariants Ponctuels de l'Équation Différentielle Ordinaire du Second Ordre* $y'' = \omega(x, y, y')$, S. Hirzel, Leipzig, 1896.

[**216**] Turnbull, H.W., and Aitken, A.C., *An Introduction to the Theory of Canonical Matrices*, Blackie and Sons, London, 1932.

[**217**] Ushveridze, A.G., *Quasi-exactly Solvable Models in Quantum Mechanics*, Inst. of Physics Publ., Bristol, England, 1994.

[**218**] Vassiliou, P., Coupled systems of nonlinear wave equations and finite-dimensional Lie algebras I, II, *Acta Appl. Math.* **8** (1987), 107–163.

[**219**] Vessiot, E., Sur une théorie nouvelle des problèmes généraux d'intégration, *Bull. Soc. Math. France* **52** (1924), 336–395.

[**220**] Vilenkin, N.J., and Klimyk, A.U., *Representation of Lie Groups and Special Functions*, Kluwer Academic Publishers, Dordrecht, 1991.

[**221**] Warner, F.W., *Foundations of Differentiable Manifolds and Lie Groups*, Scott, Foresman and Co., Glenview, Ill., 1971.

[**222**] Weyl, H., Geodesic fields in the calculus of variations for multiple integrals, *Ann. Math.* **36** (1935), 607–629.

[**223**] Weyl, H., *Classical Groups*, Princeton Univ. Press, Princeton, N.J., 1946.

[**224**] Weyl, H., *The Theory of Groups and Quantum Mechanics*, Dover, New York, 1950.

[**225**] Weyl, H., *Space–Time–Matter*, Dover, New York, 1952.

[**226**] Whitham, G.B., *Linear and Nonlinear Waves*, John Wiley & Sons, New York, 1974.

[**227**] Whittaker, E.T., *A Treatise on the Analytical Dynamics of Particles and Rigid Bodies*, Cambridge Univ. Press, Cambridge, 1937.
[**228**] Wilczynski, E.J., *Projective Differential Geometry of Curves and Ruled Surfaces*, B.G. Teubner, Leipzig, 1906.
[**229**] Wright, J.E., *Invariants of Quadratic Differential Forms*, Hafner Publ. Co., New York, 1908.
[**230**] Yang, K., *Exterior Differential Systems and Equivalence Problems*, Kluwer Academic Publ., Boston, 1992.
[**231**] Zaitsev, V.F., and Polyanin, A.D., *Handbook of Nonlinear Differential Equations*, Fizmatlit Publ., Moscow, 1993 (in Russian).
[**232**] Zhitomirskii, M.Y., Degeneracies of differential 1–forms and Pfaffian structures, *Russ. Math. Surveys* **46**:5 (1991), 53–90.
[**233**] Zhitomirskii, M.Y., *Typical Singularities of Differential 1–forms and Pfaffian Equations*, Transl. Math. Monographs, Vol. 113, Amer. Math. Soc., Providence, R.I., 1992.

Symbol Index

$\mathrm{A}(n)$	affine group, 37
$\mathfrak{a}(n)$	affine Lie algebra, 51
$A^i_{j\kappa}$	group coefficients in structure equations, 306
$B^i_{jk}[\mathbf{z}]$	linear absorption function, 307
C	center of group, 54
$\mathbb{C}$	complex numbers, 8
$\mathrm{C}(\mathcal{I})$	Cartan system, 430
C^i_{jk}	structure constant, 51, 72
$\mathbb{C}^n$	complex n-dimensional space, 8
$\mathbb{CP}^n$	complex projective space, 10
$\mathcal{C}^{(s)}$	classifying set or manifold, 263
C^∞	smooth, 7
d	differential; exterior derivative, 12, 22, 27, 254
D	total differential, 124, 232
$\widehat{\mathrm{D}}$	truncated total differential, 125
$\mathcal{D}$	differential operator, 86, 206, 211, 311
$\mathcal{D}$	invariant differential operator, 147, 154
det	determinant, 25, 76
D_i	total derivative with respect to x^i, 115, 124, 232
$\widehat{D}_i$	truncated total derivative, 125, 130
diag	diagonal matrix, 43
dim	dimension, 11
div	divergence, 70
Div	total divergence, 225
$D(I_1, \ldots, I_p)$	total Jacobian determinant, 166
D_J	higher order total derivative, 116
$(-D)_J$	signed higher order total derivative, 223
$\mathcal{D}_j$	invariant differential operator, 165
d_n	n^{th} order differential, 125
$\widehat{d}_n$	modified n^{th} order differential, 159
D_x	total derivative with respect to x, 115
$\widehat{D}_x$ or D^*_x	adapted total derivative, 316, 320, 324, 340

$\mathcal{D}_{\mathbf{v}}$	infinitesimal generator of multiplier representation, 86
dx	volume form, 25
$\mathbb{D}_x$	Hirota bilinear operator, 100
dx^i	coordinate one-form, 24, 252
dx^I	coordinate k-form, 25
e	identity element of group, 32
E	total space; vector bundle, 10, 84, 106, 112
E	integral element, 412
E	Euler operator, 223, 233
$\widetilde{\mathrm{E}}$	truncated Euler operator, 286, 323
$\mathbf{E}$	Euler-Lagrange matrix, 249
e^A	matrix exponential, 53
$\mathrm{E}(n)$	Euclidean group, 37
exp	exponential map, 53
$\exp(t\mathbf{v})$	flow of vector field; one parameter subgroup, 18, 52
E_α	Euler operator with respect to u^α, 223, 233
f	function, 107, 176
F	differential function, 115
F	Euler operator transformation matrix, 231
$\mathcal{F}$	family of functions, 12, 78
$f^{(n)}$	prolonged function, 112
$\mathcal{F}^{(s)}$	family of sth order structure invariants, 266
g	group element, 32
g	point transformation, 106
G	Lie group, 32, 288
$\mathfrak{g}$	Lie algebra, 49, 58
G^+	identity component of Lie group, 33
G'	derived subgroup, 55
$\mathfrak{g}'$	derived subalgebra, 55
$G^{(1)}$	prolonged structure group, 395
G_H	normalizer subgroup, 55
G/H	quotient group; homogeneous space, 34, 40
g^i_j	structure group parameter, 288
g_{ij}	Riemannian metric components, 387
$\mathfrak{g}_L$	left Lie algebra, 48
$\mathrm{GL}(n)$	general linear group, 33
$\mathfrak{gl}(n)$	general linear algebra, 49
$\mathrm{GL}(n,\mathbb{C})$	complex general linear group, 33

$\mathrm{GL}(n, \mathbb{R})$	real general linear group, 33
G_M	global isotropy group, 38
$g^{(n)}$	prolonged group transformation, 113
$G^{(n)}$	prolonged group, 113
$\mathfrak{g}^{(n)}$	prolonged Lie algebra, 120
$\mathfrak{g}_R$	right Lie algebra, 48
G_x	isotropy group, 38
H	Lie subgroup, 33
H	Hessian, 100
H or H_σ	classifying function, 261, 267
$\mathbf{H}$	Hessian matrix, 345
$\mathfrak{h}$	subalgebra of Lie algebra, 53
^{H}	horizontal component, 123
H^k	cohomology space, 88
H^m	hyperbolic space, 393
h_n	prolonged isotropy group dimension, 139
$\mathbf{H}^{(s)}$	classifying functions, 267
i	quartic invariant, 97
I	invariant; differential invariant, 44, 95, 137
$\mathbf{I}$	vector of differential invariants, 249
$\mathcal{I}$	polynomial ideal; exterior or differential ideal, 168, 410
$\mathcal{I}^{(k)}$	homogeneous component of ideal, 168, 410
i_n	number of independent differential invariants, 139
$\mathcal{I}_n$	contact ideal, 416
j	quartic invariant, 97
J	covariant, 98
J	symmetric multi-index, 112
J	total Jacobian matrix, 230
J^*	fundamental rational covariant, 334
j_n	number of strictly independent differential invariants, 139
j_n	n-jet, 112
J^n	jet space of order n, 112
K^*	fundamental rational covariant, 334
ker	kernel, 56, 207
k_n	number of fundamental differential invariants, 162
k_p	integer sequence, 167
$\mathbb{K}^{(s)}$	classifying space, 263
L	Lagrangian, 222

$\mathcal{L}$	functional, 222
L_g	left multiplication map, 48
lim	limit, 14
m	dimension of manifold, 7
M	manifold, 7, 112
$M^{(1)}$	prolongation manifold, 394
n	order of stabilization, 141
N	submanifold, 13
$\mathbb{N}$	natural numbers, 167
N_c	slice, 42, 275
O	order symbol, 19
$\mathcal{O}$	orbit, 40
$\mathcal{O}^{(n)}$	prolonged group orbit, 140
$\mathrm{O}(n)$	orthogonal group, 34
$\mathrm{O}(p,q)$	pseudo-orthogonal group, 39
$\mathcal{O}_x$	orbit through point x, 40, 295
p	projective coordinate, 9
p	number of independent variables, 106
p	first order derivative variable: u_x, 113
P	symbol polynomial, 168
$\wp$	Weierstrass elliptic function, 198
p_n and $p^{(n)}$	binomial coefficients, 112
$\mathcal{P}^{(n)}$	space of homogeneous polynomials, 79
$\mathrm{PSL}(n,\mathbb{R})$	projective group, 39
q	number of dependent variable, 106
q	second order derivative variable: u_{xx}, 116
Q	homogeneous polynomial or binary form, 79
Q	characteristic, 111
q_n	binomial coefficient, 139
$q^{(n)}$	dimension of jet space fiber, 112
$q_s(m)$	dimension of classifying space, 263
Q^α	component of characteristic, 111
r	dimension of Lie group, 32
r	radius, 42, 137, 158, 172
r	rank of coframe or diferential ideal, 270, 413
r	third order derivative variable: u_{xxx}, 316
R	relative invariant, 91
$\mathbf{R}$	Riemann curvature tensor, 389

$\mathbb{R}$	real numbers, 8
R_g	right multiplication map, 48
$\mathbb{R}^n$	real n-dimensional space, 8, 33
$\mathbb{RP}^n$	real projective space, 9
$r^{(1)}$	degree of indeterminancy, 351
s	orbit dimension; stable orbit dimension, 42, 141
s	order of coframe, 270
s	fourth order derivative variable: u_{xxxx}, 340
S	Schwarzian derivative, 151
$\mathrm{SA}(n)$	special affine group, 37
$\mathrm{SE}(n)$	special Euclidean group, 37
$\mathcal{S}_\mathcal{F}$ or $\mathcal{S}_\mathcal{I}$	variety, 16, 412
sign	sign of real number or permutation, 24, 89
s_k	Cartan character, 455
s'_k	reduced character, 353, 468
$\widetilde{s}_k$	matrix rank, 452–4
$\widetilde{s}'_k$	reduced matrix rank, 466
$\mathrm{SL}(n)$	special linear or unimodular group, 34
$\mathfrak{sl}(n)$	unimodular Lie algebra, 54
$\mathrm{SL}(n, \mathbb{R})$	real special linear group, 34
s_n	prolonged orbit dimension, 139
$\widehat{s}_n$	final pseudo-character, 456
S^n	n-dimensional sphere, 9
$\mathcal{S}^n$	singular variety, 200
$\mathrm{SO}(n)$	special orthogonal group, 34
$\mathfrak{so}(n)$	orthogonal Lie algebra, 54
$\mathrm{Sp}(n)$	symplectic group, 34
$\mathfrak{sp}(n)$	symplectic Lie algebra, 54
$\mathcal{S}_\Delta$	differential equation variety, 176
t	flow parameter, 18, 52
T	Jacobian covariant, 104, 334
^{T}	transpose, 34
TM	tangent bundle of M, 17
T^*M	cotangent bundle of M, 23
T^i_{jk}	structure function; torsion coefficient, 257, 306
T^n	n-dimensional torus, 15
$\mathbf{T}^{(s)}$	structure map, 263
$\widehat{\mathbf{T}}^{(s)}$	extended classifying map, 277

T_σ	structure invariant, 262
tr	trace, 45
u	fiber coordinates; dependent variables, 10, 106
U	space of dependent variables, 106
U	Jacobian covariant, 334
U^i_{jk}	essential torsion coefficient, 309
$u_n = D^n_x u$	higher order derivative, 155
$u^{(n)}$	derivative coordinates up to order n, 116
$U^{(n)}$	space of derivatives of order $\leq n$, 112
$\mathrm{U}(n)$	Heisenberg group, 73
u_x	first derivative of u with respect to x, 113
u^α	dependent variable, 106
u^α_J, $u^\alpha_{J,i}$	derivative coordinates, 112, 116
$\mathcal{V}$	local transformation group domain, 35
$\mathcal{V}$	frame, 67, 253
$\mathcal{V}$	vector field system, 416
$\mathbf{v}$	vector field, 17, 107
$\widehat{\mathbf{v}}$	infinitesimal generator, 55
$\mathbf{v}^{(n)}$	prolonged vector field, 117, 129
V^n	generic subset for prolonged group action, 139
$\mathbf{v}_Q$	evolutionary vector field, 119
$\mathbf{v}^{(n)}_Q$	prolonged evolutionary vector field, 119
$\mathbf{v}(\Omega)$	Lie derivative of differential form Ω, 69
w	rotation invariant, 137, 158
W_α	coordinate chart, 8
x	local coordinates; independent variables, 8, 10, 106
X	space of independent variables, 106
x^i	local coordinate; independent variable, 8, 106
z	point in jet space, 112
$\mathbf{z}$	absorption coefficients, 307
$\mathbf{Z}$	augmented Hessian tensor, 345
$\mathbb{Z}$	integers, 35
$\mathbb{Z}^n$	integer lattice, 35
$z^{(s)}$	point in classifying space, 263
$\widetilde{Z}^{(s)}$	classifying matrix, 278
z^κ_j	absorption coefficient, 307
$\boldsymbol{\alpha}$	Maurer–Cartan coframe, 254

α^i	Maurer–Cartan form, 71, 254
γ	matrix of Maurer–Cartan forms, 72
Γ_f	graph of function f, 15, 107
$\widehat{\Gamma}^l_{pq}$	connection coefficients, 388
$\Gamma_f^{(n)}$	graph of prolonged function $f^{(n)}$, 112
Γ_Φ or Γ_Ψ	graph of equivalence map, 431, 461
Δ	discriminant, 95, 97
Δ	system of differential equations, 176
Δ	Laplacian; Laplace–Beltrami operator, 172, 383
$\Delta = 0$	differential equation, 176
$\Delta/G = 0$	reduced differential equation, 187
$\delta\mathcal{I}$	derived ideal, 415
δ^i_j	Kronecker symbol, 24
$\boldsymbol{\zeta}$	adapted coframe, 328
η	gauge factor, 83, 211, 311
η_L	Cartan form, 244
θ	angular coordinate, 41, 42
θ	contact form, 122, 132
$\boldsymbol{\theta}$	coframe; lifted coframe, 253, 292
θ^i	coframe element, 253
θ^α_J	basic contact form, 123
Θ_C	Carathéodory Cartan form, 344
ι	inversion map, 50
κ	curvature, 137, 250, 272, 380
μ	multiplier, 82
$\mu_{n,k}$	fundamental multiplier, 82
μ_p or μ_p^n	combinatorial function, 167, 168
ξ	base coefficients of vector field, 236
ξ^i	coefficient of vector field, 17, 20, 107
π	projection, 10, 54
π	permutation, 24
$\boldsymbol{\pi}$	modified Maurer–Cartan coframe, 309, 394
π^m_n	jet space projection, 112
π_α	projection, 170
π^κ	modified Maurer–Cartan form, 309, 394
$\boldsymbol{\varpi}$	base coframe; modified Maurer–Cartan coframe, 298, 395
ρ	representation, 76, 77

$\rho_{n,k}$	fundamental representation, 79, 80
ρ_s	rank of classifying map, 263
σ	infinitesimal multiplier, 85
σ	structure invariant multi-index, 262
Υ	quadratic contact form, 233
ϕ	curve, 18
ϕ	rotational angle, 137
Φ	transformation group action, 35
Φ	equivalence map, 288
φ	angular coordinate, 41, 42
φ^α	coefficient of vector field, 107
φ^α_J	coefficient of prolonged vector field, 117, 132
Φ_x	transformation group map, 56
χ	change of variables, 9, 211
$\chi(\mathcal{I})$	characteristic system of differential ideal, 428
χ_α	coordinate map, 8
$\chi_{\alpha\beta}$	overlap map, 8
Ψ	contact transformation, 126
Ψ	prolonged equivalence map, 376, 395, 461
ω	one-form, 23
$\boldsymbol{\omega}$	base coframe, 288
Ω	differential form, 24
Ω	omega process, 99
Ω	integration domain, 222
ω^i	coframe element, 165
$^{-1}$	inverse of function or group element, 8, 12, 32
$\mathbb{1}$	identity matrix, transformation, tensor, 39, 391, 463
∞	point at infinity, 9
$\varnothing$	empty set, 8
∇	gradient, 172
∇	covariant differential, 391
$\widehat{\nabla}_k$	covariant derivative, 389
$\int$	integral, 222
$*$	pull-back; adjoint, 26, 218
#	order of multi-index, 112
$\cap$	intersection, 8
$\cup$	union, 9
$\in$	element of, 8

$\subset$	inclusion or subset, 8
$\setminus$	set theoretic difference, 29
$\circ$	composition, 8
$\cdot$	group multiplication; dot product, 32, 172
$\times$	Cartesian product, 10, 37
$\lrcorner$	interior product, 70
$\ltimes$	semi-direct product, 37, 51
$\oplus$	direct sum, 76
$\otimes$	tensor product, 76
$\bigotimes^k$	tensor power, 390
$\wedge$	wedge product; cross product, 25, 26, 172, 219
$\bigwedge^k$	exterior power, 24, 76
$\vert$	restriction to submanifold, 28
$\vert_x$	evaluation, fiber, 16, 17
$\vert \cdot \vert$	norm, 9
∂	boundary, 222
∂_J	higher order partial derivative, 112
$\dfrac{\partial}{\partial x} = \partial_x$	partial derivative, 11
$\dfrac{\partial}{\partial x^i} = \partial_{x^i}$	coordinate vector field, 17, 252
$\dfrac{\partial(f^1, \ldots, f^k)}{\partial(x^1, \ldots, x^k)}$	Jacobian determinant, 25
$\dfrac{\partial}{\partial \theta^i}$	frame vector field; coframe derivative, 253
$[\cdot, \cdot]$	Lie bracket or commutator, 21, 49, 166
$[\cdot, \cdot, \cdot, \cdot]$	cross-ratio, 46
$\langle \cdot \,; \cdot \rangle$	pairing of vector space and its dual, 23
$\langle \cdot \,; \cdot, \ldots, \cdot \rangle$	evaluation of multi-linear map, 24
$\dbinom{n}{k}$	binomial coefficient, 24, 112
$(\cdot, \cdot)^{(r)}$	transvectant, 99

Author Index

Note: The numbers in brackets refer to the References.

Ablowitz, M.J., 198, [**43**]
Abraham–Shrauner, B., 191, [**1**]
Ackerman, M., [**2**], [**3**]
Adler, M., 102, [**4**]
Ado, I.D., 54
Airy, G.B., 198
Aitken, A.C., 43, [**216**]
Anderson, I.M., 129, 226, 227, 240, 248, 340, 408, [**5–10**]
Anderson, R.L., 119, 129, [**11**]
Arnol'd, V.I., 11, 31, 126, 127, 407, [**12–14**]

Bäcklund, A.V., 119, 128, 129, 136, [**15**]
Ball, J.M., 222, [**16**]
Bargmann, V., 82, [**17**]
Beltrami, E., 383
Bianchi, L., 191, 393, [**18**]
Bleecker, D., 83, [**19**]
Bluman, G.W., 110, 175, 186, 187, 190, 209, 211, [**20–22**], [**144**]
Bocharov, A.V., 407, 408, [**23**]
Bonnet, O.P., 333
Boole, G., 95
Bott, R., 29, [**24**]
Boyer, C.P., 394, [**25**]
Brachet, M.E., 20, [**68**]
Bryant, R.L., 286, 397, 408, 409, 428, 430, 431, 459, 471, [**26–30**]
Buchberger, B., 279, [**31**]
Burgers, J., 184

Carathéodory, C., 345, [**32**]
Cartan, É., xii, xiii, 3–5, 26, 28, 31, 53, 55, 103, 156, 178, 221, 244, 252, 253, 262, 280, 288, 292, 297, 338, 342, 348, 350, 353, 356, 372, 386, 397, 398, 403, 407, 408, 409, 421, 431, 440, 444, 453, 459, 461, 470, 471, [**33–42**]
Cauchy, A.–L., 178, 448
Cayley, A., 60, 99
Chakravarty, S., 198, [**43**]
Champagne, B., 179, [**44**]
Chazy, J., 198, 199, [**45**]
Chern, S.–S., 333, 397, 408, 409, 428, 430, 431, 459, 471, [**28**], [**46–50**]
Chrastina, J., 338, [**51**]
Christoffel, E.B., 4, 386, 392, 408, [**52**]
Churyumov, A., 61, [**140**]
Ciarlet, P.G., 100, [**53**]
Clarkson, P.A., xiii, 61, 198, 199, [**43**], [**54**]
Cohen, H., 102, [**55**]
Cole, J.D., 185
Coullet, P., 20, [**68**]
Courant, R., 429, 430, 447, 448, [**56**]
Cramer, G., 218
Crampin, M., 389, 391, [**57**]
Cummins, C.J., 393, [**74**]

Darboux, G., 30, 367, 414, [**58**]
David, D., 367, [**59**]
Debever, R., 366, 367, 369, [**60**]
deRham, G., 29
deVries, G., 187, 238

Dickson, L.E., 203, [**61**]
Dikii, L.A., 100, 102, [**83**], [**84**]
Dirichlet, P.G.L., 222, 240
Dixmier, J., 103, [**62**]
Doubrov, B., 61, [**140**]
Doyle, P.W., 256, 272, [**63**], [**64**]
Drinfel'd, V.G., 102, [**65**]
Duffing, G., 403

Ehresmann, C., 105, [**66**]
Eisenhart, L.P., 393, [**67**]
Elphick, C., 20, [**68**]
Emden, R., 204
Engel, F., 59, 203, 204, 356, [**155**], [**156**]
Euler, L., 20, 223

Fels, M., 206, 408, [**69**], [**70**]
Finsler, P., 333
Fomin, S.V., 222, 244, 301, 323, [**85**]
Forsyth, A.R., 211, 214, 384, [**71**], [**72**]
Fréchet, R.M., 207
Frobenius, G., xii, 421, [**73**]
Fulling, S.A., 393, [**74**]
Fushchich, W.I., 173, [**75**]

Gage, M., 251, [**76**]
Galileo, G., 184
Galois, E., 4
Gardner, R.B., 252, 267, 296, 333, 346, 359, 367, 369, 371, 397, 409, 428, 430, 431, 445, 459, 471, [**28**], [**77–81**]
Gat, O., 205, [**82**]
Gauss, C.F., 309, 333, 380
Gel'fand, I.M., 100, 102, 222, 244, 301, 323, [**83–85**]
Goldschmidt, H.L., 244, 397, 409, 428, 430, 431, 459, 471, [**28**], [**86**]
Golubitsky, M., 11, 61, [**87**], [**88**]
González–Gascón, F., 203, [**89**]
González–López, A., 61, 88, 90, 203, 206, 242, 407, [**89–97**]
Gotay, M., 244, 246, 345, [**98**]
Goursat, E. J.–B., 430
Grace, J.H., 79, 100, 102, 104, 336, [**99**]
Graustein, W.C., 381, 383, [**100**]
Grayson, M., 251, [**101**]
Green, M.L., 162, 163, 164, 172, [**102**]
Greene, C., 167, [**103**]
Griffiths, P.A., 397, 408, 409, 428, 430, 431, 459, 471, [**28–30**]
Grissom, C., 407, [**104**]
Guckenheimer, J., 20, [**105**]
Guggenheimer, H.W., 37, 155, 156, 157, 174, 380, 381, [**106**]
Guillemin, V., 11, [**88**]
Guo, A., 191, [**1**]
Gurevich, G.B., 97, 100, 102, 104, [**107**]
Gusein–Zade, S.M., 11, [**14**]

Hadamard, J., 346
Halphen, G.–H., 136, 211, [**108**], [**109**]
Hamermesh, M., 75, 82, [**110**]
Hamilton, R.S., 251, [**76**]
Hamilton, W.R., 31, 60
Hausdorff, F., 8
Hawkins, T., 4, 59, 414, [**111**]
Heisenberg, W., 73
Helmholtz, H., 193
Heredero, R., 152
Hereman, W., 179, [**44**], [**112**]
Herlt, E., 394, [**141**]
Hermann, R., 422, [**2**], [**3**], [**113**]
Hesse, L.O., 100
Hilbert, D., 79, 100, 103, 167, 244, 335, 429, 430, 447, 448, [**56**], [**114**], [**115**]
Hille, E., 151, 195, 199, 403, [**116**]
Hirota, R., 100
Holm, D.D., 367, [**59**]
Holmes, P., 20, [**105**]
Hopf, E., 185
Hopf, H., 445

Hsu, L., 246, 401, 402, 403, 407, 408, [**30**], [**117**], [**118**]
Humphreys, J.E., 52, 55, 75, 134, [**119**]

Ibragimov, N.H., 119, 129, 175, 182, 186, 204, 407, [**11**], [**120–122**]
Ince, E.L., 195, 198, [**123**]
Ioss, G., 20, [**68**]

Jacobi, C.G.J., 11, 21
Jacobson, N., 54, 88, [**124**]
Janenko, N.N., 255, [**200**]
Jordan, C., 2, 42, 45

Kähler, E., xii, 178, 348
Kalnins, E.G., 209, [**125**], [**126**]
Kamran, N., xiii, 61, 88, 90, 129, 246, 267, 286, 287, 311, 321, 323, 338, 339, 340, 342, 358, 368, 401, 402, 403, 407, 408, 428, [**8**], [**9**], [**79**], [**94–97**], [**117**], [**118**], [**127–135**]
Karlhede, A., 394, [**136**], [**137**]
Katona, G.O.H., 167
Killing, W. , 55
King, R.C., 393, [**74**]
Klein, F., 37
Kleitman, D.J., 167, [**103**]
Klimyk, A.U., 75, [**220**]
Kobayashi, S., 393, [**138**]
Komrakov, B., 61, [**139**], [**140**]
Korteweg, D.J., 187, 238
Kovalevskaya, S., 178, 448
Kramer, D., 394, [**141**]
Krause, J., 204, [**142**]
Kreysig, E., 387, [**143**]
Kronecker, L., 24, 131, 391
Kruskal, J., 167
Kumei, S., 110, 175, 186, 187, 190, 209, 211, [**20–22**], [**144**]
Kuranishi, M., 356, 397, [**145–147**]

Lagrange, J.-L., 222, 223
Laguerre, E., 211, 214, [**148**]
Lamb, K.G., 403, [**128**]
Lamé, G., 198
Landsberg, G., 333
Laplace, P.S., 115, 318, 383
Leach, P.G.L., 204, [**163**]
Legendre, A.M., 126, 323, 346
Leibniz, G.W., 20, 224
Levi-Civita, T., 389
Lewy, H., 178, 448, [**149**]
Lie, M.S., xiii, 3–5, 21, 53, 59, 61, 119, 126, 127, 131, 134, 136, 146, 152, 187, 200, 203, 204, 205, 221, 238, 242, 318, 327, 356, 401, 407, 408, 444, [**150–157**]
Littlejohn, L.L., 217, [**158**]

Macaulay, F.S., 167, 169, [**159**]
MacCallum, M.A.H., 394, [**137**], [**141**]
Mackey, G.W., 75, 82, [**160**]
Magadeev, B.A., 251, [**161**]
Magri, F., 366, [**162**]
Mahomed, F.M., 204, [**163**]
Mal'cev, A.I., 445, [**164**]
Marsden, J.E., 221, 244, [**165**]
Martinet, J., 31, [**166**]
Maschke, H., 408, [**167**]
Michel, L., 204, [**142**]
Mikhailov, A.V., 186, [**168**]
Miller, W., Jr., 82, 209, [**125**], [**126**], [**169**]
Mills, R.L., 198
Minkowski, H., 393
Mizel, V.J., 222, [**16**]
Möbius, A.F., 10, 36
Morikawa, H., 217, [**170**]
Morrey, C.B., Jr., 346, [**171**]
Mostow, G.D., 41, 61, [**172**]
Murillo, R., 408, [**173**]

Nagano, T., 425, [**174**]
Narasimhan, R., 25, 424, [**175**]
Neuman, F., 211, [**176**]
Newell, A., 100, 186, [**177**]
Newton, I., 225

Olver, P.J., xii, xiii, 22, 31, 32, 61, 88, 90, 100, 103, 106, 117, 119, 129, 155, 168, 175, 178, 180, 184, 186, 187, 190, 191, 198, 199, 207, 221, 226, 227, 232, 236, 238, 244, 246, 247, 249, 251, 286, 287, 311, 321, 323, 327, 333, 335, 336, 338, 340, 342, 345, 346, 358, 366, 444, [**9**], [**54**], [**94–97**], [**118**], [**129–133**], [**179–190**]
Ovsiannikov, L.V., 143, 162, 175, 186, [**191**]
Ovsienko, O.D., 102, [**192**]
Ovsienko, V.Y., 102, [**192**]

Painlevé, P., 198, 403, 407
Palais, R.S., 20, 41, 198, [**193**]
Petrov, A.Z., 394
Pfaff, J.F., 23, 30, 414, [**194**]
Pirani, F.A.E., 389, 391, [**57**]
Plebański, J.F., 394, [**25**]
Pohjanpelto, P.J., 248
Poincaré, H., 29, 286
Poisson, S.D., 102
Polyanin, A.D., 191, [**231**]
Pontrjagin, L., 32, 54, 57, 424, 444, [**195**]
Prandtl, L., 198

Rabier, P., 100, [**53**]
Rankin, R.A., 102, [**196**]
Riccati, J.F., 189
Ricci, G., 393
Riemann, B., 4, 10, 46, 255, 386, 408, [**197**]
Rinow, W., 445
Rosinger, E.E., 107, [**198**], [**199**]
Roždestvenskiĭ, B.L., 255, [**200**]
Rudolph, M., 107, [**198**]

Sapiro, G., xiii, 155, 247, 249, 251, [**189**], [**190**], [**201**]
Scheffers, G., 126, 127, 131, [**157**]
Schrödinger, E., 195, 196, 212, 318
Schwarz, H., 102, 195, 199
Se-ashi, Y., 211, [**202**]
Shabat, A.B., 186, [**168**]
Shadwick, W.F., 346, 367, 368, 369, 371, 403, 408, [**80**], [**81**], [**128**], [**134**], [**135**]
Shakiban, C., 100, [**203**]
Singer, I.M., 356, [**204**]
Sokolov, V.V., 134, 152, 186, 206, 246, 407, 408, [**23**], [**168**], [**205**], [**212**]
Spivak, M., 20, 21, [**206**]
Stephani, H., 394, [**141**]
Sternberg, S., 31, 244, 289, 305, 356, [**86**], [**204**], [**207**]
Strang, G., 309, [**208**]
Stroh, E., 103, [**209**]
Struik, D.J., 381, [**210**]
Sturmfels, B., 102, 103, [**211**]
Svinolupov, S.I., 134, 407, 408, [**23**], [**212**]
Sylvester, J.J., 43

Takhtajan, L.A., 198, [**213**]
Tannenbaum, A., xiii, 155, 247, 249, 251, [**189**], [**190**], [**201**]
Taylor, B., 112, 248
Thompson, G., 226, 227, 407, [**10**], [**104**]
Tirapegui, E., 20, [**68**]
Tresse, M.A., 4, 136, 146, 166, 397, 407, 408, [**214**], [**215**]
Tu, L.W., 29, [**24**]
Turnbull, H.W., 43, [**216**]

Ushveridze, A.G., 88, [**217**]

Varchenko, A.N., 11, [**14**]
Vassiliou, P., 408, [**218**]
Vessiot, E., 408, [**219**]
Vilenkin, N.J., 75, [**220**]
Virasoro, M.A., 102
von Karman, T., 100

Walus, Y.E., 107, [**199**]
Warner, F.W., 12, 28, 32, 33, 383, 424, [**221**]

Weierstrass, K., 198
Weinstein, A., 221, 244, [**165**]
Weyl, H., 31, 75, 80, 82, 83, 345, 393, [**222–225**]
Whitham, G.B., 185, [**226**]
Whittaker, E.T., 31, 127, [**227**]
Wilczynski, E.J., 211, 215, 217, 218, [**228**]
Wilkens, G., 407, [**104**]
Winternitz, P., 179, [**44**]
Wright, J.E., 387, [**229**]
Wronski, H.J.M., 219
Wybourne, B.G., 393, [**74**]

Yang, C.N., 198
Yang, K., 252, 409, 471, [**230**]
Yegorchenko, I.A., 173, [**75**]
Young, A., 79, 100, 102, 104, 336, [**99**]

Zaitsev, V.F., 191, [**231**]

Subject Index

Abelian, 33, 50, 209, 243, 296, 395, 443
Absolute differential invariant, 149
Absolute invariant, 91
Absolute parallelism, 253
Absorption, 309, 374, 394, 469
  equations, 309, 351, 394
Adapted coframe, 328
Adapted total derivative, 316, 324, 340, 400
Adjoint, 218, 219
  representation, 77, 78
Ado's Theorem, 54
Affine, 51
  differential invariant, 220
  function, 108
  geometry, 37, 164
  group, 37, 39, 41, 50, 59, 66, 72, 88, 94, 134, 136, 157, 174, 250, 269, 276, 364
  invariant arc length, 339
  invariant curvature, 251
  transformation, 37
Airy equation, 198
Algebra
  abelian, 51, 54
  computer, 6, 179, 359
  derived, 55
  differential, 100
  Lie, *see* Lie algebra
  symmetry, 88
  Virasoro, 102
Algebraic function, 359
Analytic, 7, 11, 32, 79, 103, 107, 178, 422, 448
  coframe, 271, 355
Angle, 137, 140, 148, 192
Angular coordinate, 41, 42
Angular momentum, 242
Anti-symmetric, 21, 24, 49, 390
Arc, 109, 111
  parabolic, 273
Arc length, 155, 157, 241, 250, 301, 333, 339
  affine-invariant, 339
  group-invariant, 155
Area form, 239
Associativity, 32, 443, 445
Atlas, 8
Augmented Hessian tensor, 345
Automorphic form, 102, 198

Bäcklund's Theorem, 128, 132, 136, 361, 363, 416
Base coordinate, 10
Base transformation, 106
Basic contact form, 122, 123, 130
Basic variable, 351
Basis, 51, 454
  dual, 253
  Gröbner, 279
  Hilbert, 103, 335
Bianchi identity, 393
Bianchi's Theorem, 191
Biform, 296, 346
BiHamiltonian system, 366
Binary form, 79, 150, 229, 302, 333
  symmetry group of, 334
Binomial coefficient, 112, 167
Biquadratic, 346
Boundary
  condition, 222

layer, 198
natural, 198
value, 178, 225
Boussinesq equation, 186, 237
Bracket
Lie, 21, 23, 28, 49, 51, 69, 78, 120, 257, 417,
Poisson, 102
Bundle
cotangent, 23, 91, 253
extended jet, 106
jet, 112, 263
line, 10
principal, 305
tangent, 17, 84, 91
vector, 10, 15, 17, 23, 84, 106
Burgers' equation, 184, 185

Calculus of variations, 100, 222, 227, 285, 321
Canonical form, 2, 9, 42, 44, 91, 296
biforms, 296
curvature tensors, 393
differential operators, 214
group orbits, 42
Jordan, 2, 42, 45
Laguerre–Forsyth, 214
Lie algebras, 59
maps, 11
matrices, 42, 282, 387
one-forms, 30, 281, 414
ordinary differential equations, 19, 204
points, 46
rational, 43
submanifolds, 14
symbols, 296
two-forms, 31, 43, 281
vector fields, 19, 30
vectors, 43
Canonical transformation, 34, 127
Cartan character, 455, 460, 463, 466
Cartan equivalence method, 4, 206, 304, 358
Cartan form, 244, 246, 286, 329, 345, 357, 358
Cartan–Kähler Theorem, 178, 348, 373, 409, 415, 456, 470
Cartan–Kuranishi Theorem, 397
Cartan's Lemma, 26
Cartan's test, 355, 367, 373, 396, 466, 470
Cartan structure equations, 389
Cartan system, 430, 431
Cartesian product, 10, 15, 37, 46, 61, 254, 305, 425, 431
Catastrophe theory, 11
Cauchy characteristic, 428
Cauchy data, 448, 456, 458, 459
Cauchy–Kovalevskaya Theorem, 178, 348, 448, 456, 458
Cauchy problem, 178, 448, 458
Cayley–Hamilton Theorem, 60
Center, 54
Chain rule, 23, 114, 116, 128, 212, 277, 436
Change of coordinates, 8, 18
Character
Cartan, 455, 460, 463, 466
pseudo-, 456, 460
reduced, 353, 355, 373, 466–470
Characteristic, 111, 117, 130, 131, 187, 243
Cauchy, 428
curve, 246, 429
direction, 168, 428
Euler, 20
form, 119, 179
function, 131, 429
method of, 62, 141, 188
polynomial, 60
speed, 255
strip, 429
system, 62, 428, 430
Chart, 8
Chazy equation, 198
Circle, 9, 16, 17, 42, 65, 134, 198, 427
Classical field theory, 244, 345
Classical group, 34, 55

Classical invariant theory, xii, 79, 95, 104, 148–151, 211, 229, 333, 393, 408
Classical mechanics, 31
Classifying
  curve, 272, 318, 327, 332
  function, 262, 267, 371, 436
  manifold, 266, 268, 271, 367, 396, 435, 437, 441, 471
    extended, 277, 302
    reduced, 278, 384
  set, 263, 264
  space, 263, 271, 435
    extended, 277
  surface, 272
Closed
  differential ideal, 414, 419, 427, 434
  form, 29, 31
  subgroup, 33, 40
Closure, 414
Coboundary, 88
Cocyle, 88
Coframe, 253, 288, 297, 390
  adapted, 328
  contact-invariant, 164, 168, 171, 238
  coordinate, 252, 451, 468
  covering, 441
  derivative, 165, 254, 260, 366, 373, 396, 436, 471
  diagonalizing, 282, 378, 385, 387, 391
  eigen-, 255
  equivalence problem for, 4, 256, 271, 292, 296, 367, 433, 437
  G-invariant, 71, 156
  invariant, 71, 155, 157
  lifted, 305, 353, 366, 460
  Maurer–Cartan, 254, 268, 275, 441
  prolonged, 375
  rank of, 367
  regular, 266, 435, 440
  symmetry, 256, 259, 272
  symmetry group of, 274, 347, 367, 396
Cohomology, 29, 88
Commutativity, 32, 422, 425
Commutator, 21, 49, 55, 77, 86, 166
Compact, 77
Compact support, 223
Complete, 445
Complex, 8, 29
Composition, 22, 27, 113
Computer algebra, 6, 179, 359
Computer vision, 221, 246
Conformal
  field theory, 221
  geometry, 164, 385
  group, 136, 173, 385
  symmetry group, 386
Conic section, 205, 220
Conjugate, 34, 38
Conjugation, 45, 76, 77
Connected, 8, 16, 40, 53, 107, 424
  component, 33
  pathwise, 16
  simply, 16, 54, 441, 443
Connection, 4, 263, 387, 389
  projective, 398
Conservation law, 186, 242, 244, 329, 345
Conservative, 242
Constant type, 295
Constraint, 222
Contact, 112, 121
  conditions, 130
  ideal, 416
  symmetry, 180, 251
  system, 430, 431
  vector field, 131, 187, 429
Contact-equivalent, 228, 230, 342
Contact form, 122, 126, 232, 283, 339, 361, 416, 428, 430
  basic, 122, 123, 130
  invariant, 247
  pseudo-, 339
  quadratic, 233
Contact-invariant, 153, 238, 300
  coframe, 164, 168, 171, 238
  form, 235

one-form, 157, 161, 164
Contact transformation, 126, 128, 129, 230, 361, 416, 429
infinitesimal, 129
prolonged, 128, 416
Contact transformation group, 134, 152, 176, 242
Contragredient, 76, 79
Control system, 333
Convex, 29, 167, 323
Coordinate
angular, 41, 42
base, 10
change of, 8, 18
chart, 8
coframe, 252, 451, 468
differential form, 25
fiber, 10
frame, 252
local, 32, 252
polar, 41, 192, 239
projective, 100
rectifying, 19, 42, 63, 187, 243, 423, 426
spherical, 9, 42
Corank, 427
Coset, 34
Cotangent
bundle, 23, 91, 253
space, 23, 71, 252, 253, 390
Covariant, 98, 101, 102, 333
derivative, 323, 339, 389, 391
differential, 391
Jacobian, 100, 104, 149
rational, 334
Covering
coframe, 441
group, 54, 441
manifold, 441
map, 12, 54, 271, 441
space, 443
Cramer's rule, 218
Cross product, 219
Cross-ratio, 46, 97
Cubic
curve, 266
polynomial, 97, 104, 335
Curvature, 137, 140, 155, 192
affine-invariant, 251
Gaussian, 174, 272, 380, 384, 385
group-invariant, 155, 250
invariant, 393
nonconstant, 384
projective, 219
scalar, 393
tensor, 4, 389, 391, 392, 393
Curve, 14, 16, 17, 163, 218, 383, 386
characteristic, 246, 429
classifying, 272, 318, 327, 332
cubic, 266
dual, 219
integral, 18, 411, 418, 424, 427, 429, 451
quadratic, 265
shortening, 251
Cylinder, 10, 17, 63, 384

Darboux rank, 31, 256, 365, 411, 431
Darboux' Theorem, 30, 268, 281, 364, 366, 416
Decomposable, 25
Degenerate, 322
Degree, 24, 26, 80
of generality, 459
of indeterminancy, 351, 355, 373, 394, 470
Dense, 41, 44
Dependent
functionally, 12, 26, 267
linearly, 25, 218
variable, 106, 175, 222
DeRham theory, 29, 226
Derivation, 20, 22
super, 27
Derivative, 112, 139, 175
coframe, 165, 254, 260, 366, 373, 396, 436, 471
covariant, 389, 391
directional, 253

Derivative (*continued*):
Fréchet, 207
Lie, 69, 102, 232
Schwarzian, 151, 195
variational, 223
Derivative covariant, 323, 339
Derived
algebra, 55
ideal, 415
invariant, 260, 262, 437
subgroup, 55
system, 431
Determinant, 33
Jacobian, 25, 27, 100
Lie, 194, 201, 202
total Jacobian, 166
Determinantal
multiplier, 86, 95
representation, 76, 80
variety, 278
Determinate, 373
Determining equations, 179, 181, 183
Diagonalizable, 255
Diagonalizing coframe, 282, 378, 385, 387, 391
Diffeomorphism, 8, 12, 431
Differential, 12, 22, 27, 70, 125, 254, 257, 305, 414
covariant, 391
total, 124, 153, 157, 165, 232
Differential algebra, 100
Differential equation, 115, 175, 186, 210, 247, 251
equivalence problem for, 283, 289, 293, 348, 356, 397, 427
invariant, 199, 201, 202
linear, 60, 211, 218
ordinary, 18, 178, 187, 418
partial, 178, 209, 355, 409, 413, 447
regular, 176
symmetry group of, 136, 179, 185, 191
Differential form, 23, 24, 90, 254, 390, 409
coordinate, 25
equivalence problem for, 415
invariant, 68, 70, 92, 146
Maurer–Cartan, 71, 254, 435
modified Maurer–Cartan, 374, 376, 394, 462
one-form, *see* one-form
Pfaffian, 23
singularity of, 30, 31, 365, 427
two-form, *see* two-form
Differential function, 115, 124, 137, 175
Differential geometry, 377, 421
Differential ideal, 414, 428, 430, 447
closed, 414, 419, 427, 434
rank of, 413, 427, 452
Differential invariant, 137, 141, 146, 165, 168, 192, 238, 384
absolute, 149
affine, 220
Euclidean, 155, 164
functionally independent, 137, 239
fundamental, 147, 153, 157, 161, 162, 171, 193, 202
relative, 149, 154, 216
Differential operator, 86, 102, 116, 206, 209, 211, 219, 311
canonical form for, 214
equivalence problem for, 211, 310
invariant, 147, 153, 157, 165, 171
Jacobian, 150
Lie algebra of, 86, 212
matrix, 91
Differential polynomial, 103, 212
Differential system, 262, 353, 409, 447, 456, 468
equivalence problem for, 415, 428
singularity of, 455
Dihedral triangle, 199
Dimension
orbit, 46, 93, 139
prolonged orbit, 152, 160
stable orbit, 141, 143

Direct equivalence problem, 212, 311
Direct method, 5, 408
Direct sum, 76
Directional derivative, 253
Dirichlet condition, 222
Dirichlet variational problem, 240
Discrete
  subgroup, 35, 54, 102
  symmetry group, 199, 275
Discriminant, 64, 95, 97, 104, 335
Divergence, 70, 90
  equivalence problem, 234, 245, 285, 321, 357
  multiplier, 154
  symmetry, 237, 242
  Theorem, 224, 225
  total, 225, 236
Dual
  basis, 253
  curve, 219
  vector space, 23
Duffing equation, 403
Dynamical space, 247

e-structure, 289
Effective, 38, 86, 145
  locally, 38, 56, 86, 143, 200
Effectively free, 38, 42, 67, 71, 94, 145
Eigen-coframe, 255
Eigen-frame, 255
Eigenvalue, 2, 43, 45, 212, 255, 320
Eigenvector, 255
Elasticity, 100, 255, 346
Electromagnetism, 83
Elementary row operation, 310
Elliptic function, 102, 198
Ellipticity, 346
Emden equation, 204
Energy, 225, 242
Envelope, 187
Equation
  absorption, 309, 351, 394
  Airy, 198
  boundary layer, 198
  Boussinesq, 186, 237
  Burgers', 184, 185
  Chazy, 198
  determining, 179, 181, 183
  differential, *see* differential equation
  diffusion, 186, 210, 247, 251
  Duffing, 403
  Emden, 204
  multiplier, 82
  Euler–Lagrange, *see* Euler–Lagrange equation
  evolution, 206, 247, 249, 251, 255
  heat, 182, 185, 210
  Helmholtz, 193
  homogeneous, 188
  hypergeometric, 199
  invariance, 262, 437
  Korteweg–deVries, 187, 238
  Lamé, 198
  Laplace, 115, 240
  nonlinear diffusion, 186
  ordinary differential, *see* ordinary differential equation
  parabolic, 211
  parametric, 108
  partial differential, *see* partial differential equation
  Riccati, 189, 194, 195, 196
  Schrödinger, 195, 196, 212, 214, 215
  Schwarzian, 195, 214
  structure, 72, 257, 306, 350, 359, 373, 389, 433
  transvectant, 102
  von Karman, 100
  wave, 269, 408
Equi-anharmonic, 97
Equilibrium point, 19
Equivalence map, 431, 442
Equivalence method, 4, 206, 304, 358
Equivalence problem, 1, 9, 43, 280, 288, 304, 432, 460, 470
  binary forms, 333

Equivalence problem (*continued*):
coframes, 4, 256, 271, 292, 296, 367, 433, 437
conformal, 385
contact forms, 361
differential equations, 283, 289, 293, 348, 356, 397, 427
differential forms, 415
differential operators, 211, 310
differential systems, 415, 428
direct, 212, 311
divergence, 234, 245, 285, 321, 357
Euler–Lagrange equations, 234
feedback control, 333
global, 441
homogeneous polynomials, 229
isometric, 377
Lagrangian, 227, 285, 290, 321, 328, 357
isometric, 299, 320
multi-dimensional, 342
local, 257
maps, 11
one-forms, 30, 281, 291, 364
ordinary differential equations, 19, 398, 408
overdetermined, 297, 312
partial differential equations, 408
polynomials, 333
projective, 212, 219
prolonged, 348, 372, 375, 394, 466
Riemannian metrics, 4, 282, 291, 386, 394
second order Lagrangian, 337
standard, 227, 285, 321
submanifolds, 14
two-forms, 31, 281, 291, 367
variational problem, 227, 285
vector fields, 19, 29
Equivalent, 230
gauge, 83, 88, 212, 319
multiplier representation, 87
Essential torsion, 308, 355, 374, 396, 471
Euclidean
differential invariant, 155, 164
division algorithm, 206
geometry, 381
group, 37, 67, 71, 155, 156, 164, 173, 250, 299, 378
metric, 34, 299, 377
plane, 377, 386
space, 7, 29, 33, 37, 106, 164, 277, 393
Euler characteristic, 20
Euler operator, 223, 323
Euler–Lagrange equation, 223, 242, 245, 285, 323, 337, 340
equivalence problem for, 234
matrix, 249
symmetry group of, 236, 238, 240
Evolutionary vector field, 119
Evolution equation, 206, 247, 249, 251, 255
Exact, 29, 31, 187
Exponential, 18, 53
Extended
action, 93
classifying manifold, 277, 302
classifying space, 277
jet bundle, 106
space, 84
structure map, 277
Exterior
derivative, *see* differential
differential system, *see* differential system
ideal, 410, 413, 414
power, 24, 76
Extremal, 222, 230, 244
multi-set, 167
Extrinsic geometry, 380

Family
of functions, 12
of variations, 223, 248
Feedback, 333
Fiber, 10, 106, 112

Fiber-preserving transformation, 106, 176, 227, 321
Field theory, 345
Figure eight, 14
Finitely generated, 410, 416, 422, 450
Finsler metric, 333
First fundamental form, 378
First Fundamental Theorem, 102
First integral, 242, 246
Fixed point, 38, 40, 90, 91
Flow, 18, 36, 52, 55, 69, 129, 256, 422, 429
  heat, 386
  irrational, 41, 44, 45
Fluid mechanics, 18, 198
Foliation, 42, 61, 424, 440
Force field, 225
Form
  area, 239
  automorphic, 102, 198
  basic contact, 122, 123, 130
  binary, 79, 150, 229, 302, 333, 334
  canonical — *see* canonical form
  Cartan, 244, 246, 286, 329, 345, 357, 358
  characteristic, 119, 179
  closed, 29, 31
  contact, 122, 126, 232, 283, 339, 416, 428, 430
  contact-invariant, 235
  differential, *see* differential form
  first fundamental, 378
  Kovalevskaya, 448, 458, 459
  Lagrangian, 222, 230, 233, 285, 342
  Maurer–Cartan, 71, 254, 435
  modified Maurer–Cartan, 374, 376, 394, 462
  modular, 102
  Pfaffian, 23
  row echelon, 310
  symplectic, 34
  volume, 25, 27, 70, 71, 230, 236
Frame, 17, 67, 253, 390
  coordinate, 252
  eigen-, 255
  invariant, 71
  moving, 156
Fréchet derivative, 207
Free, 38, 66, 71, 154, 155
  effectively, 38, 42, 67, 71, 94, 145
  locally, 38, 42, 58, 94
  particle, 206, 326
  variable, 351, 394
Frobenius' Theorem, 42, 178, 348, 422, 427, 434
Fully regular, 178, 266, 268, 271
Fully transverse, 432, 466
Function, 24, 78, 112
  affine, 108
  characteristic, 131, 429
  classifying, 262, 267, 371, 436
  differential, 115, 124, 137, 175
  elliptic, 102, 198
  generating, 127
  Hilbert, 167
  hypergeometric, 102, 199
  invariant, 66, 68, 109
  linear fractional, 156, 195, 205
  potential, 185
  prolonged, 112, 121, 123
  rank of, 12
  scale-invariant, 110, 111
  space, 78
  special, 75
  structure, 257, 261, 347, 433
  symmetric, 45
  transformed, 108
Functional, 222, 235, 285, 337, 342
Functionally dependent, 12, 26, 267
Functionally independent, 12, 26
  differential invariants, 137, 239
  invariants, 62, 110, 267
  strictly, 138, 158, 167
Fundamental
  differential invariant, 147, 153, 157, 161, 162, 171, 193, 202
  invariant, 46
  multiplier, 89, 229

Fundamental (*continued*):
multiplier representation, 95, 120, 148, 215
representation, 80
solution, 184

G–structure, 289, 305
G–valued equivalence problem, 288
Galilean boost, 184
Gas dynamics, 255
Gauge
equivalent, 83, 88, 212, 319
factor, 83, 212
transformation, 107, 219
Gauss–Bonnet Theorem, 333
Gaussian curvature, 174, 272, 380, 384, 385
Gaussian elimination, 309
General linear group, 33, 35, 39, 40, 42, 49, 52, 76, 79, 86, 89, 95, 136, 229
General relativity, 394
Generalized
symmetry, 180, 186
variational symmetry, 244
vector field, 119
Generate, 410
Generating function, 127
Generic integral element, 468
Genuinely nonlinear, 256
Geodesic, 333
Geometric diffusion, 247
Geometric object, 1
Geometry, xi
affine, 37, 164
conformal, 164, 385
differential, 377, 421
Euclidean, 381
extrinsic, 380
hyperbolic, 381
intrinsic, 380
projective, 164, 211
spherical, 381
symplectic, 367
web, 386
Global, 3, 29, 35
equivalence, 441
invariant, 45
isotropy subgroup, 38, 56, 145
Gradient, 222
Graph, 15, 107, 431, 432, 461
prolonged, 112, 121, 123, 126, 176
technique of the, 431, 461
Gravity, 83
Gröbner basis, 279
Group, 32
affine, 37, 39, 41, 50, 59, 66, 72, 88, 94, 134, 136, 157, 174, 250, 269, 276, 364
classical, 34, 55
conformal, 136, 173, 385
covering, 54, 441
Euclidean, 37, 67, 71, 155, 156, 164, 173, 250, 299, 378
general linear, 33, 35, 39, 40, 42, 49, 52, 76, 79, 86, 89, 95, 136, 229
global isotropy, 38, 56, 145
Heisenberg, 73
isometry, 381, 393
isotropy, 38, 40, 58, 139, 140, 163, 295
Lie, 32, 48, 53, 54, 254, 268, 435, 441
local Lie, 53, 444
matrix Lie, 33, 54, 72, 76, 288
one-parameter, 36, 42, 55, 107, 117, 429
orthogonal, 34, 43–45, 54, 55, 73, 282, 291, 302, 387, 389, 393
Poincaré, 136, 173
projective, 39, 57, 59, 155, 181, 241, 250
prolonged, 113
prolonged structure, 395
pseudo-, 4, 355, 466
pseudo-orthogonal, 39, 282, 393
quantum, 102
quotient, 38, 67, 145
reduced structure, 295, 297

rotation, 34, 43, 80, 108, 114, 134, 137, 158, 172, 174, 192, 239, 377
similarity, 141, 157, 250
special affine, 37, 134, 155, 157, 174, 241, 250
special linear, 34, 40, 55, 64, 76, 80, 90, 120, 135, 148, 173, 193, 206, 241, 336
special orthogonal, *see* rotation group
structure, 288, 304, 327, 350, 373, 460
symmetry, *see* symmetry group
symplectic, 34, 39, 54, 55, 134, 282
topological, 445
transformation, *see* transformation group
translation, 59, 88
unimodular, *see* special linear group
Group-invariant
arc length, 155
curvature, 155, 250

Hamiltonian, 31, 102, 256
mechanics, 31, 34, 60, 126, 127, 221
system, 244, 323
multi-, 272, 367
Hausdorff space, 8
Heat equation, 182, 185, 210
Heat flow, 386
Heisenberg group, 73
Helmholtz equation, 193
Hessian, 100, 103, 149, 151, 216, 226, 333
matrix, 173, 345
tensor, 345
Higher order curvature tensor, 392
Hilbert basis, 103, 335
Hilbert function, 167
Hilbert's Basis Theorem, 103
Hilbert's invariant integral, 244
Hirota operator, 100
Hodograph transformation, 210, 231, 331
Hole, 29
Holomorphic, 8
Holonomic, 222
Homogeneous, 187, 410, 450
equation, 188
polynomial, 79, 95, 168, 229, 333
space, 40, 61, 162, 163, 172
Homomorphism, 33, 56, 120
Hopf–Cole transformation, 185
Hopf–Rinow Theorem, 445
Horizontal, 123, 153, 432
component, 123, 124, 156
one-form, 165
Hyperbolic, 255
geometry, 381
partial differential equation, 368
space, 393
system, 259, 272
Hypergeometric function, 102, 199
Hyperplane, 220

Icosahedral, 199
Ideal, 55
contact, 416
derived, 415
differential, 414, 428, 430, 447
exterior, 410, 414
polynomial, 167
Identity, 32, 49
Bianchi, 393
Jacobi, 21, 49, 51, 261, 263, 278, 393
tensor, 391
Image processing, 247, 250, 251
Immersed submanifold, 13
Implicit Function Theorem, 11, 14, 107, 127, 432, 438, 440, 457
Implicit submanifold, 16
Imprimitive, 61
Independence condition, 468
Independent
functionally, 12, 26, 62, 110, 137, 239, 267

Independent (*continued*):
- linearly, 25, 218
- strictly, 138, 158, 167
- variable, 106, 175, 222

Indeterminancy
- degree of, 351, 355, 373, 394, 470

Indeterminate, 394
Inductive method, 328
Inequivalent Lagrangian, 226
Inertia, 43
Infinitesimal, 20, 412
- contact transformation, 129
- generator, 18, 52, 55, 107, 424
- invariance, 65, 92, 110, 141, 179, 236, 237
- multiplier, 85, 87, 92, 236, 247
- normalization, 360
- representation, 77
- symmetry condition, 62, 69, 179, 236, 237

Infinity, 10, 36
Inhomogeneous polynomial, 81, 100, 209, 333
Initial condition, 413
Initial value problem, 178
Integrability condition, 178, 186, 354
Integrable, 226, 411, 413, 417, 422, 427
Integral
- curve, 18, 411, 418, 424, 427, 429, 451
- element, 412, 427, 456, 467, 470
  - regular, 455
  - generic, 468
- invariant, 221, 244
- submanifold, 410, 417, 422, 425, 456, 459, 468
  - maximal, 424
- surface, 411, 452, 458

Integrating factor, 31, 411
Integration by parts, 224, 243
Interior product, 70, 246
Internal symmetry, 129
Intersection, 14
Intransitive, 61, 153, 193, 202, 355, 366, 461, 471
- symmetry group, 204

Intrinsic geometry, 380
Intrinsic method, 358
Invariance equations, 262, 437
Invariant, 2, 44, 46, 62, 91, 109, 110, 258, 424
- absolute, 91
- coframe, 71, 155, 157
- combination, 292
- component, 390
- contact form, 247
- curvature, 393
- derived, 260, 262, 437
- differential, *see* differential invariant
- differential equation, 199, 201, 202
- differential form, 68, 70
- differential operator, 147, 153, 157, 165, 171
- evolution equation, 249
- foliation, 61
- frame, 71
- function, 66, 68, 109
- functionally independent, 62, 110
- fundamental, 46
- global, 45
- integral, 221
- joint, 47
- local, 45
- locally, 64
- one-form, 92, 146
- relative, 91, 92, 93, 95, 109, 247, 249, 425
- solution, 187
- structure, 262, 267, 296, 299, 328, 359, 396, 436, 471
- submanifold, 47, 64
- subset, 40, 44, 47
- subspace, 77
- surface conditions, 110
- system, 64
- translationally, 60, 64
- vector field, 48, 66, 67, 92, 94

volume form, 71
Inverse, 32, 35
Function Theorem, 12, 275, 423
problem, 227
Inversion, 20, 37, 50, 57, 68, 81, 184
Involutive, 355, 367, 417, 419, 422, 468, 470
Irrational flow, 41, 44, 45
Irreducible, 77, 78
representation, 80
Isometric, 386, 392
equivalence, 299, 320, 377
Isometry, 37, 282, 299, 377
group, 381, 393
Isotropy subgroup, 37, 38, 40, 58, 139, 140, 163, 295
global, 38, 56, 145

Jacobian
covariant, 100, 104, 149
determinant, 25, 27, 100
total, 166
differential operator, 150
matrix, 11, 22, 26, 176, 423, 438
top order, 139
total, 166, 230, 235
multiplier representation, 84, 92
total, 165
polynomial, 226
Jacobi identity, 21, 49, 51, 261, 263, 278, 393
Jet, 112, 292
bundle, 112, 263
extended bundle, 106
prolongation, 112, 121, 123, 176
space, 112, 121, 139, 175, 222
Joint invariant, 47
Jordan canonical form, 2, 42, 45

Kernel, 206
Kinetic energy, 225
Korteweg–deVries equation, 187, 238
Kovalevskaya form, 448, 459
Kronecker delta, 24, 131, 343, 391, 392
Kruskal–Katona Theorem, 167

Lagrangian, 222, 230, 240, 285, 342
equivalence problem for, 227, 285, 290, 321, 328, 337, 342, 357
form, 222, 230, 233, 285, 342
inequivalent, 226
nondegenerate, 225, 227, 323
null, 225, 234
quadratic, 241, 346
singularity of, 225, 301, 322, 323
Laguerre–Forsyth normal form, 214
Lamé equation, 198
Landsberg surface, 333
Laplace–Beltrami operator, 383
Laplace equation, 115, 240
Laplacian, 172, 193
radial, 318
Lattice, 35
Left
invariant, 48, 50, 72
Lie algebra, 48, 56
multiplication, 36
Legendre–Hadamard condition, 346
Legendre transformation, 126, 127, 323
Leibniz rule, 20, 224
Lemma
Cartan's, 26
Poincaré, 29, 286
Level set, 16, 44, 174, 440
Lie algebra, 49, 54, 58, 71, 78, 254, 268, 444
canonical form for, 59
cohomology, 88
of differential operators, 86, 212
left, 48, 56
representation, 77
right, 48, 56
of vector fields, 56, 120, 424
Lie–Bäcklund vector field, 119
Lie bracket, 21, 23, 28, 49, 51, 69, 78, 120, 257, 417

Lie derivative, 69, 232
  higher order, 102
Lie determinant, 194, 201, 202
Lie group, 32, 48, 53, 54, 254, 268, 435, 441
  local, 53, 444
  matrix, 33, 54, 72, 76, 288
Lie series, 21
Lie subalgebra, 53
Lie subgroup, 33, 53
Lifted coframe, 305, 353, 366, 460
Line
  bundle, 10
  projective, 20, 36, 81
  segment, 64
Linear, 426
  differential equation, 60, 211, 218
  elasticity, 346
  polynomial, 79
  quasi-, 255, 411, 429
  system, 209
  transformation, 33
Linear fractional
  function, 156, 195, 205
  transformation, 36, 46, 81, 114, 148, 181
Linearization, 186, 209, 400, 406
Linearly dependent, 25
Linearly independent, 218
Local, 3, 10, 35
  coordinate, 32, 252
  diffeomorphism, 12
  equivalence, 257
  invariant, 45
  Lie group, 53, 444
  transformation group, 35
Locally, 10, 35
  effective, 38, 56, 86, 143, 200
  free, 38, 42, 58, 94
  invariant, 64
  regular, 436
  solvable, 177, 209, 413
  transitive, 67
Lowering operator, 80

Macaulay's Theorem, 167, 169
Majorant method, 448
Manifold, 8, 29, 32, 35, 253
  classifying, 266, 268, 271, 367, 396, 435, 437, 441, 471
  covering, 441
  pseudo-Riemannian, 282, 387
  Riemannian, 282, 377, 386, 389
  space-time, 394
Map, 10, 11, 22
  canonical form for, 11
  covering, 12, 54, 271, 441
  equivalence problem for, 11
  exponential, 53
  overlap, 8
  regular, 11
  singularity of, 11
  structure, 263, 270, 277, 367
MAPLE, 309, 359
MATHEMATICA, xv, 309, 359
Matrix, 33, 42, 49
  canonical form for, 42, 282, 387
  differential operator, 91
  Euler–Lagrange, 249
  exponential, 53
  Hessian, 173, 345
  Jacobian, 11, 22, 26, 139, 176, 423, 438
  Lie group, 33, 54, 72, 76, 288
  multiplier, 83
  multiplier representation, 90, 93
  rank of, 455, 467, 469
  skew-symmetric, 34, 39, 43, 54, 73, 76
  symmetric, 43, 44, 282, 387
  total Jacobian, 166, 230, 235
Maurer–Cartan coframe, 254, 268, 275, 441
Maurer–Cartan form, 71, 254, 435
  modified, 374, 376, 394, 462
Maximal
  integral submanifold, 424
  rank, 11, 16, 176, 445
Maximum, 222
Mechanics, 345
  classical, 31

fluid, 18, 198
Hamiltonian, 31, 34, 126, 127, 221
quantum, 31, 75, 78, 211, 212
Method of characteristics, 430
Metric, 377, 385
complete, 445
Euclidean, 34, 299, 377
Finsler, 333
Minkowski, 393
pseudo-Riemannian, 282, 387
Riemannian, 272, 282, 291, 333, 386, 394, 445
tensor, 387
Minimum, 222
Minkowski metric, 393
Möbius band, 10
Möbius transformation, 36
Modular form, 102
Momentum, 31, 242
angular, 242
Monomial, 168, 334
Moving frame, 156
Multi-Hamiltonian system, 272, 367
Multi-index
strictly increasing, 25, 254
symmetric, 112, 223
Multiple integral, 345
Multiple root, 98
Multiplication, 32, 36
operator, 86
Multiplier, 82, 91, 109, 425
determinantal, 86, 95
divergence, 154
equation, 82
fundamental, 89, 229
infinitesimal, 85, 87, 92, 236, 247
matrix, 83
total Jacobian, 165
trivial, 83
Multiplier representation, 82, 93, 106, 425
equivalent, 87
fundamental, 95, 120, 148, 215
Jacobian, 84, 92
matrix, 90, 93
trivial multiplier, 88
Multiply valued, 262, 267

n-integrable, 413
Natural boundary, 198
Newtonian variational problem, 225
Noether's Theorem, 221, 244, 246
Noncharacteristic, 448
Nonclosed two-form, 31, 367
Nonconstant
curvature, 384
type, 294, 306, 350, 471
Nondegenerate, 338, 346
Lagrangian, 225, 227, 323
Nonlocal, 185
Nonsingular, 369
Nontangential, 187
Norm, 34, 37, 382
Normal, 34, 38, 178, 203
Normalizable, 79
Normalization, 291, 295, 299, 305, 309, 359, 395
infinitesimal, 360
Normalizer subgroup, 55, 190
n^{th} order contact, 112
Null Lagrangian, 225, 234
Number theory, 198

Octahedral, 199
Omega process, 99
One-form, 23, 29, 85, 153, 287, 410
canonical form for, 30, 281, 414
contact-invariant, 157, 161, 164
equivalence problem for, 30, 281, 291, 364
horizontal, 165
invariant, 92, 146
rank of, 30
regular, 30
One-parameter
group, 36, 42, 55, 107, 117, 429
subgroup, 52, 53
Open, 8

Operator
  differential, *see* differential operator
  Euler, 223, 323
  Hirota, 100
  Laplace–Beltrami, 172, 193, 383
  lowering, 80
  multiplication, 86
  radial Laplace, 318
  raising, 80
  Schrödinger, 88, 213, 318, 321
Optics, 301, 367
Orbit, 40, 44, 46, 187, 295, 424
  canonical form for, 42
  dimension of, 46, 93, 139
  prolonged, 139, 151
    dimension of, 152, 160
  singular, 44, 98
  stable dimension of, 141, 143
Order, 115, 123, 262, 271, 367, 392, 435
Ordinary differential equation, 18, 178, 187, 418
  canonical form for, 19, 204
  equivalence problem for, 19, 398, 408
  linearization, 400, 406
  normal, 203
  symmetry group of, 180, 187–208, 400–403, 406, 407
Orientation, 253
  preserving, 33, 377
  reversing, 331, 377
Orthogonal group, 34, 43–45, 54, 55, 73, 282, 291, 302, 387, 389, 393
  pseudo-, 39, 282, 393
Orthogonal polynomial, 217
Osculating hyperplane, 220
Overdetermined
  equivalence problem, 297, 312
  system, 63
Overlap, 16, 267, 278, 437, 471
  map, 8
  strictly, 16, 271
Painlevé property, 198
Painlevé transcendent, 403, 407
Parabolic arc, 273
Parabolic equation, 211
Parameter, 32
  space, 13, 28
  variation of, 195
Parametric equation, 108
Parametric method, 359
Parametrized submanifold, 13, 279
Partial derivative, 112
Partial differential, 348
Partial differential equation, 178, 209, 355, 409, 413, 447
  equivalence problem for, 408
  hyperbolic, 368
  linearization, 209
  parabolic, 211
  quasi-linear, 255, 411, 429
  symmetry group of, 182–187
Partial transvectant, 103
Particle physics, 198
Pathwise connected, 16
Pencil, 366
Permutation, 24
Petrov classification, 394
Pfaffian form, 23
Pfaffian system, 414, 427, 428, 430
Plane, 9, 36, 61, 79, 443
Plate, 100
Poincaré group, 136, 173
Poincaré Lemma, 29, 286
Point
  canonical form for, 46
  equilibrium, 19
  fixed, 38, 40, 90, 91
  at infinity, 10, 36
  regular, 19, 455
Point transformation, 106, 128, 284
  group, 58, 135, 152
  symmetry, 176, 203
  vector field, 131
Poisson bracket, 102
Polar coordinates, 41, 192, 239
Polynomial, 79, 95, 150, 279
  characteristic, 60

cubic, 97, 104, 335
differential, 103, 212
equivalence problem for, 229, 333
homogeneous, 79, 95, 168, 229, 333
ideal, 167
inhomogeneous, 81, 100, 209, 333
linear, 79
orthogonal, 217
quadratic, 64, 79, 80, 95
quartic, 97, 103, 336
Taylor, 112, 113
total Jacobian, 226
Position, 31
Positive definite, 43, 346
Potential, 213, 237, 318
energy, 225
function, 185
symmetry, 186
Power
exterior, 24, 76
series, 448
symmetric, 80
Primitive, 61
Principal bundle, 305
Projection, 170, 176, 271, 433, 442, 461
jet space, 112, 113
stereographic, 9
Projective, 46, 89, 229
connection, 398
coordinate, 100
curvature, 219
equivalence, 212, 219
geometry, 164, 211
group, 39, 57, 59, 155, 181, 241, 250
line, 20, 36, 81
product, 241
representation, 82
space, 9, 36, 39, 41, 218
Prolongation
equivalence problem, 348, 372, 375, 394, 466
formula, 117, 119, 130, 133
jet space, 112, 121, 123, 176
Prolonged
coframe, 375
contact transformation, 128, 416
function, 112, 121, 123
graph, 112, 121, 123, 126, 176
group, 113
orbit, 139, 151
orbit dimension, 152, 160
space, 375
transformation, 113
vector field, 117
Pseudo-character, 456, 460
Pseudo-contact form, 339
Pseudo-group, 4, 355, 466
Pseudo-orthogonal group, 39, 282, 393
Pseudo-Riemannian manifold, 282, 387
Pseudo-stabilization, 142, 153, 162, 170, 172, 193, 202
Pull-back, 26, 28, 68, 122, 291
Punctured plane, 443

Quadratic
contact form, 233
curve, 265
Lagrangian, 241, 346
polynomial, 64, 79, 80, 95
Quadrature, 191, 193, 194, 239
Quantum
electrodynamics, 83
group, 102
mechanics, 31, 75, 78, 211, 212
Quartic polynomial, 97, 103, 336
Quasi-exactly solvable, 88
Quasi-linear, 255, 411, 429
Quotient
group, 38, 67, 145
space, 34

Radial Laplacian, 318
Radius, 44, 137, 140, 158, 172, 192, 239
Raising operator, 80

Rank
of coframe, 367
Darboux, 30, 31, 256, 365, 411, 431
of differential ideal, 427, 452
of exterior ideal, 413
of family of functions, 12
of map, 11, 22
of matrix, 455, 467, 469
maximal, 11, 16, 176, 445
of one-form, 30
of structure map, 266
of two-form, 31, 43, 281, 431
of vector field system, 417, 421
zero, 268, 435
Rational
canonical form, 43
covariant, 334
parametrization, 279
Ray, 301
Rectifying coordinate, 19, 42, 63, 187, 243, 423, 426
Recursive formula, 119
for prolongation, 133
Reduced
character, 353, 355, 373, 466–470
classifying manifold, 278, 384
structure group, 295, 297
Reducible, 77
Reduction, 384
Reflection, 34, 276, 377, 378
Regular, 2
coframe, 266, 435, 440
family of functions, 12
fully, 178, 266, 268, 271
local group, 444
locally, 436
map, 11
one-form, 30
point, 19, 455
semi-, 41, 42, 46, 58, 139, 424
submanifold, 14
system, 16
system of differential equations, 176
transformation group, 41, 42, 46, 110
variety, 16, 412, 438
Relative differential invariant, 149, 154, 216
Relative invariant, 91, 92, 93, 95, 109, 247, 249, 425
Relativity, 394
Representation, 75, 78
determinantal, 76, 80
fundamental, 80
infinitesimal, 77
irreducible, 80
Lie algebra, 77
multiplier, *see* multiplier representation
projective, 82
tensor, 80
theory, 35
Restriction, 28
Riccati equation, 189, 194, 195, 196
Ricci scalar, 394
Ricci tensor, 393
Riemann curvature tensor, 4, 389, 391, 392, 393
Riemann invariant, 255
Riemann sphere, 10, 36, 46
Riemannian manifold, 282, 377, 386, 389
equivalence problem for, 4, 282, 291, 386, 394
metric, 272, 445
pseudo-, 282, 387
surface, 333
symmetry group of, 392
Right
invariant, 48, 50, 56, 71, 156, 254
Lie algebra, 48, 56
multiplication, 36
Root, 82, 96, 97, 104, 168, 336
multiple, 98
Rotation, 23, 108
group, 34, 43, 80, 108, 114, 134, 137, 158, 172, 174, 192, 239, 377

Row echelon form, 310

Scalar curvature, 393
Scale-invariant function, 110, 111
Scaling, 19, 46, 57, 110, 189, 236
  operator, 172
Schrödinger equation, 195, 196, 212, 214, 215
Schrödinger operator, 88, 213, 318, 321
Schwarzian derivative, 151, 195
Schwarzian equation, 195, 214
Second Fundamental Theorem, 103
Section, 15, 107, 121, 390
Self-adjoint, 218
Self-dual, 220
Self-intersection, 14
Semi-direct product, 37, 51
Semi-regular, 41, 42, 46, 58, 139, 424
Separable, 187
Separatrix, 187
Series
  Lie, 21
  power, 448
  Taylor, 248
Signature, 43, 45, 282, 346
Similarity
  group, 141, 157, 250
  solution, 110
Simple, 55, 241, 255
Simply connected, 16, 54, 441, 443
Simply generated, 414, 430, 452
Singular, 3
  orbit, 44, 98
  variety, 200
Singularity
  differential form, 30, 31, 365, 427
  differential system, 455
  Lagrangian, 225, 301, 322, 323
  map, 11
  submanifold, 14
  vector field, 19
Skew-adjoint, 218
Skew-symmetric matrix, 34, 39, 43, 54, 73, 76
Slice, 42, 47, 275, 423, 430
Smooth, 7, 10, 32, 79, 103, 107, 113, 178, 222, 422
Solitary wave, 186
Soliton, 100, 102, 180, 186, 246
Solution, 39, 107, 176
  fundamental, 184
  invariant, 187
  similarity, 110
Solvable, 191, 193, 241
Space
  classifying, 263, 271, 435
  cotangent, 23, 71, 252, 253, 390
  covering, 443
  dynamical, 247
  Euclidean, 7, 29, 33, 37, 106, 164, 277, 393
  extended, 84
  function, 78
  Hausdorff, 8
  homogeneous, 40, 61, 162, 163, 172
  hyperbolic, 393
  jet, 112, 121, 139, 175, 222
  parameter, 13, 28
  projective, 9, 36, 39, 41, 218
  prolonged, 375
  quotient, 34
  tangent, 17, 20, 22, 58, 253, 390, 412
  tensor, 76
  topological, 7
  total, 106, 112, 222
  translationally invariant, 60
  vector, 23
  vertical, 112
Space-time manifold, 394
Special affine group, 37, 134, 155, 157, 174, 241, 250
Special function, 75
Special linear group, 34, 40, 55, 64, 76, 80, 90, 120, 135, 148, 173, 193, 206, 241, 336
Special orthogonal group, *see* rotation group

Sphere, 9, 16, 17, 23, 40, 41, 42, 44, 393
Spherical coordinates, 9, 42
Spherical geometry, 381
Spiral, 42
Stabilization, 141, 269, 436
  order of, 141, 154, 161, 165, 171, 200, 247
  pseudo-, 142, 153, 162, 170, 172, 193, 202
Stable orbit dimension, 141, 143
Standard equivalence problem, 227, 285, 321
Stereographic projection, 9
Strictly independent, 138, 158, 167
Strictly overlap, 16, 271
String theory, 221
Strong ellipticity, 346
Structure
  constant, 51, 72, 261, 424, 433, 435
  equations, 72, 257, 306, 350, 359, 373, 433
    Cartan, 389
  function, 257, 261, 347, 433
  group, 288, 304, 327, 350, 373, 460
    prolonged, 395
    reduced, 295, 297
  invariant, 262, 267, 296, 299, 328, 359, 396, 436, 471
  map, 263, 266, 270, 367
    extended, 277
Subalgebra, 53
Subgroup, 33, 40
  closed, 33, 40
  derived, 55
  discrete, 35, 54, 102
  global isotropy, 38, 56, 145
  isotropy, 37, 38, 40, 58, 139, 140, 163, 295
  Lie, 33, 53
  normalizer, 55, 190
  one-parameter, 52, 53
Submanifold, 13, 28, 33, 107, 172, 438
  canonical form for, 14
  equivalence problem for, 14
  immersed, 13
  implicit, 16
  integral, 410, 417, 422, 424, 425, 456, 459, 468
  invariant, 47, 64
  parametrized, 13, 279
  regular, 14
  self-intersecting, 14
  singularity of, 14
Submaximal symmetry group, 205, 242
Subspace, 77
Super derivation, 27
Super-symmetric, 26
Surface, 172, 264, 333, 380, 411
  classifying, 272
  integral, 411, 452, 458
  invariant conditions, 110
  Landsberg, 333
  of revolution, 383
Sylvester's Theorem, 43
Symbol, 168, 170, 296, 346
Symbolic method, 100, 408
Symmetric, 112, 168
  function, 45
  matrix, 43, 44, 282, 387
  multi-index, 112, 223
  power, 80
  super-, 26
Symmetry, 38, 177, 178
  algebra, 88
  coframe, 256, 259, 272
  contact, 180, 251
  divergence, 237, 242
  fiber-preserving, 176
  generalized, 180, 186, 244
  internal, 129
  point, 176, 203
  potential, 186
  reduction, 187, 190, 203
  variational, 235, 244, 246, 345
Symmetry group, 1, 39, 65
  of binary form, 334
  of coframe, 274, 347, 367, 396
  conformal, 386

of differential equation, 136, 179, 185, 191
discrete, 199, 275
of Euler–Lagrange equation, 236, 238, 240
infinitesimal, 62, 69, 179, 236, 237
intransitive, 204
of ordinary differential equation, 180, 187–208, 400–403, 406, 407
of partial differential equation, 182–187
of Riemannian metric, 392
submaximal, 205, 242
of system of equations, 65
Symplectic, 31, 127, 176, 291
form, 34
geometry, 367
group, 34, 39, 54, 55, 134, 282
Syzygy, 103, 104, 171, 174, 335

Tangent
bundle, 17, 84, 91
space, 17, 20, 22, 58, 253, 390, 412
vector, 17, 22, 432
Taylor polynomial, 112, 113
Taylor series, 248
Technique of the graph, 431, 461
Tensor
bundle, 390
curvature, 4, 389, 391, 392, 393
field, 390
Hessian, 345
identity, 391
metric, 387
product, 76
representation, 80
Ricci, 393
space, 76
Weyl, 393
Tetrahedral, 199
TEX, xv
Theorem
Ado's, 54
Bäcklund's, 128, 132, 136, 361, 363, 416
Bianchi's, 191
Cartan–Kähler, 178, 348, 373, 409, 415, 456, 470
Cartan–Kuranishi, 397
Darboux', 30, 268, 281, 364, 366, 416
Divergence, 224, 225
Egregium, 380
First Fundamental, 102
Frobenius', 42, 178, 348, 422, 427, 434
Gauss–Bonnet, 333
Hilbert's Basis, 103
Hopf–Rinow, 445
Implicit Function, 11, 14, 107, 127, 432, 438, 440, 457
Inverse Function, 12, 275, 423
Kruskal–Katona, 167
Macaulay's, 167, 169
Noether's, 221, 244, 246
Second Fundamental, 103
Sylvester's, 43
Third Fundamental, 444
Third Fundamental Theorem, 444
Topological group, 445
Topological space, 7
Topology, 8, 14, 29
Torsion, 306, 308, 355, 374, 396, 471
Torus, 15, 41, 45
Total derivative, 115, 124, 223, 232
adapted, 316, 324, 340, 400
higher order, 116
truncated, 130, 338
Total differential, 124, 153, 157, 165, 232
Total divergence, 225, 236
Total Jacobian
determinant, 166
matrix, 166, 230, 235
multiplier, 165
polynomial, 226
Total space, 106, 112, 222
Trace, 45
Transcendent, 403, 407

Transform, 100
Transformation
  affine, 37
  base, 106
  canonical, 34, 127
  contact, 126, 128, 129, 230, 361, 416, 429
  fiber-preserving, 106, 227, 321
  gauge, 107, 219
  hodograph, 210, 231, 331
  Hopf–Cole, 185
  infinitesimal contact, 129
  Legendre, 126, 127, 323
  linear, 33
  linear fractional, 36, 46, 81, 114, 148, 181
  Möbius, 36
  orientation-preserving, 33, 377
  orientation-reversing, 331, 377
  point, 106, 128, 284
  prolonged, 113, 128, 416
  volume-preserving, 37, 70, 176, 250, 281, 336
Transformation group, 5, 35, 44, 55, 136, 424
  contact, 134, 152, 176, 242
  local, 35
  point, 58, 135, 152
  regular, 41, 42, 46, 110
Transformed function, 108
Transitive, 40, 44, 58, 61, 135, 155, 200, 295, 355, 359, 461
  locally, 67
Translation, 19, 37, 57, 88
  group, 59, 88
Translationally invariant, 60, 64
Transvectant, 99, 100, 101, 149, 199, 216, 219
  equation, 102
  partial, 103
Transverse, 67, 107, 109, 123, 126, 461, 467, 470
  fully, 432, 466
Triangle, 199, 443
Trivial
  multiplier, 83
  multiplier representation, 88
  symmetry, 188
Truncated
  total derivative, 130, 338
  total differential, 125
Two-form, 31, 43, 281, 369
  equivalence problem for, 31, 281, 291, 367
  nonclosed, 31, 367
  rank of, 31, 43, 281, 431
Type, 390
  constant, 295
  nonconstant, 294, 306, 350, 471

Underdetermined, 287, 297, 448
  Kovalevskaya form, 448, 458
Unimodular, 34
  group, *see* special linear group

Variable
  basic, 351
  dependent, 106, 175, 222
  free, 351, 394
  independent, 106, 175, 222
Variation, 223, 248
Variational derivative, 223
Variational problem, 222, 230, 235, 285, 321, 337, 342
  Dirichlet, 240
  equivalence problem for, 227, 285
  Newtonian, 225
Variational symmetry, 235, 246, 345
  generalized, 244
Variational symmetry group, 236, 238, 240, 242, 245, 332
Variation of parameters, 195
Variety, 16, 24, 39, 176, 412
  determinantal, 278
  regular, 16, 412, 438
  singular, 200
Vector
  canonical form for, 43
  tangent, 17, 22, 432
  triple product, 172

Vector bundle, 10, 15, 17, 23, 84, 106
Vector field, 17, 20, 23, 36, 55, 69, 84, 107, 117, 253, 369
 canonical form for, 19, 30
 contact, 131, 187, 429
 equivalence problem for, 19, 29
 evolutionary, 119
 generalized, 119
 invariant, 48, 66, 67, 92, 94
 Lie algebra of, 56, 120, 424
 Lie–Bäcklund, 119
 point, 131
 prolonged, 117
 singularity of, 19
Vector field system, 408, 416, 421, 427, 428
 rank of, 417, 421
Vector space, 23
Vertical, 107, 123, 209, 432
 space, 112
Virasoro algebra, 102
Vision, 221, 246
Volume form, 25, 27, 70, 230, 236
 invariant, 71
Volume-preserving, 37, 70, 176, 250, 281, 336
Von Karman equations, 100

Water, 186, 255
Wave equation, 269, 408
Web geometry, 386
Wedge length, 431
Wedge product, 25, 26, 30, 253
Weight, 80, 83, 91, 95, 101, 215
Weyl tensor, 393
Wronskian, 196, 198, 219

Zero rank, 268, 435